Artificial Intelligence for Unmanned Aerial Vehicles

IEEE Press

445 Hoes Lane

Piscataway, NJ 08854

Artificial Intelligence for Unmanned Aerial Vehicles

Sensing, Communication, and Computing

Shuyan Hu
College of Electronics and Information Engineering, Tongji University, Shanghai, China

Xin Yuan
Data61, Commonwealth Scientific and Industrial Research Organization, Marsfield, NSW, Australia

Kai Li
Technische Universität Berlin, Berlin, Germany and Research Centre in Real-Time and Embedded Computing Systems (CISTER), Porto, Portugal

Wei Ni
School of Engineering, Edith Cowan University, Perth, WA, Australia and School of Computer Science and Engineering, University of New South Wales, Sydney, NSW, Australia

Xin Wang
College of Future Information Technology, Fudan University, Shanghai, China

IEEE Press
WILEY

Published by John Wiley & Sons, Inc., Hoboken, New Jersey.

For general information on our other products and services or for technical support, please contact our Customer Care Department within the United States at (800) 762-2974, outside the United States at (317) 572-3993 or fax (317) 572-4002.

Wiley also publishes its books in a variety of electronic formats. Some content that appears in print may not be available in electronic formats. For more information about Wiley products, visit our web site at www.wiley.com.

Library of Congress Cataloging-in-Publication Data has been applied for:

Hardback ISBN: 9781394369454
ePDF ISBN: 9781394369478
ePub ISBN: 9781394369461
oBook ISBN: 9781394369485

Cover Design: Wiley
Cover Image: © wacomka/stock.adobe.com

Contents

List of Figures

List of Tables

About the Authors

Shuyan Hu (Member, IEEE) received the BE degree in electrical engineering from Tongji University, Shanghai, China, in 2014, and the PhD degree in electronic science and technology from Fudan University, Shanghai, China, in 2019. She is currently an Associate Professor with Tongji University. She was a Postdoctoral Research Fellow with Fudan University from 2019 to 2024. Her research interests include machine learning and convex optimization, and their applications to autonomous aerial vehicle (AAV) networks, wireless communication, and intelligent systems.

Xin Yuan (Senior Member, IEEE) received the BE degree from Taiyuan University of Technology, Taiyuan, Shanxi, China, in 2013, and the dual PhD degree from Beijing University of Posts and Telecommunications (BUPT), Beijing, China, and the University of Technology Sydney (UTS), Sydney, NSW, Australia, in 2019 and 2020, respectively. She is currently a Senior Research Scientist with CSIRO, Sydney, NSW, Australia. She is also an Adjunct Senior Lecturer with the University of New South Wales, Sydney, NSW, Australia. Her research interests include machine learning and optimization, and their applications to intelligent systems.

Kai Li (Senior Member, IEEE) received the BE degree from Shandong University, Jinan, China, in 2009, the MS degree from the Hong Kong University of Science and Technology in 2010, and the PhD degree in computer science from the University of New South Wales, Sydney, NSW, Australia, in 2014. He is currently a Visiting Research Scholar with the School of Electrical Engineering and Computer Science at TU Berlin, Germany, and a Senior Research Scientist with the CISTER Research Center, Porto, Portugal. He also serves as a CMU-Portugal Research Fellow, jointly supported by Carnegie Mellon University (CMU),

Pittsburgh, PA, USA, and the Foundation for Science and Technology (FCT), Lisbon, Portugal. From November to December 2025, he was a Visiting Research Scholar with the Department of Electrical and Computer Engineering, CMU, under the CMU-Portugal Visiting Faculty and Researchers Program. From 2023 to 2024, he was a Visiting Research Scientist with the Division of Electrical Engineering, Department of Engineering, University of Cambridge, UK. In 2022, he was a Visiting Research Scholar with the CyLab Security and Privacy Institute, CMU. Prior to this, he was a Postdoctoral Research Fellow with the SUTD-MIT International Design Centre, Singapore University of Technology and Design, from 2014 to 2016. He also held positions as a Visiting Research Assistant at the ICT Centre, CSIRO, Brisbane, QLD, Australia, from 2012 to 2013, and as a Research Assistant at the Mobile Technologies Centre, The Chinese University of Hong Kong, from 2010 to 2011. Dr. Li has served as an Associate Editor for several journals, including *Internet of Things* (Elsevier, since 2024), *Nature Computer Science* (Springer, since 2023), *Computer Communications* (Elsevier, since 2021), *Ad Hoc Networks* (Elsevier, since 2021), and *IEEE ACCESS* (2018–2024).

Wei Ni (Fellow, IEEE) received the BE and PhD degrees in electronic engineering from Fudan University, Shanghai, China, in 2000 and 2005, respectively. He is the Associate Dean (Research) in the School of Engineering, Edith Cowan University, Perth, WA, Australia and a Conjoint Professor at the University of New South Wales, Sydney, NSW, Australia. He also serves as Technical Expert at Standards Australia in the area of artificial intelligence and big data. He was a Deputy Project Manager at the Bell Labs, Alcatel/Alcatel-Lucent from 2005 to 2008; a Senior Research Engineer at Devices R&D, Nokia from 2008 to 2009; and a Senior Principal Research Scientist and Group Leader at the Commonwealth Scientific and Industrial Research Organisation (CSIRO) from 2009 to 2025. He has co-authored 4 books, 11 book chapters, over 550 technical papers, 27 patents, and 10 standard proposals accepted by IEEE. His research interests include machine learning, online learning, stochastic optimization, and their applications to system efficiency and integrity.

Dr. Ni has been an Editor for *IEEE Transactions on Wireless Communications* since 2018, *IEEE Transactions on Vehicular Technology* since 2022, *IEEE Transactions on Information Forensics and Security* and *IEEE Communication Surveys and Tutorials* since 2024, and *IEEE Transactions on Network Science and Engineering* since 2025. He served first as Secretary, then Vice-Chair and Chair of the IEEE VTS NSW Chapter from 2015 to 2022, Track Chair for VTC-Spring 2017, Track Co-chair for IEEE VTC-Spring 2016, Publication Chair for BodyNet 2015, and Student Travel Grant Chair for WPMC 2014.

Xin Wang (Fellow, IEEE) received the BSc and MSc degrees in electrical engineering from Fudan University, Shanghai, China, in 1997 and 2000, respectively, and the PhD degree in electrical engineering from Auburn University, Auburn, AL, USA, in 2004. From September 2004 to August 2006, he was a Postdoctoral Research Associate with the Department of Electrical and Computer Engineering, University of Minnesota, Minneapolis, MN, USA. In August 2006, he joined the Department of Electrical Engineering, Florida Atlantic University, Boca Raton, FL, USA, as an Assistant Professor, and then was promoted to a tenured Associate Professor in 2010. He is currently a Professor with Fudan University. His research interests include stochastic network optimization, energy-efficient communications, cross-layer design, and signal processing for communications. He is a member of the Signal Processing for Communications and Networking Technical Committee of the *IEEE Signal Processing Society* and a Distinguished Speaker of the *IEEE Vehicular Technology Society*. He is a Senior Area Editor of *IEEE Transactions on Signal Processing* and an Editor of *IEEE Transactions on Wireless Communications*. In the past, he served as an Associate Editor of *IEEE Transactions on Signal Processing*, as an Editor of *IEEE Transactions on Vehicular Technology*, and as an Associate Editor of *IEEE Signal Processing Letters*.

Preface

Unmanned aerial vehicles (UAVs) are rapidly evolving into intelligent, autonomous agents revolutionizing applications in surveillance, environmental monitoring, disaster response, and wireless communications. This book provides a comprehensive and systematic treatment of the foundational and emerging technologies that empower UAVs—focusing on their sensing, communication, computing, and artificial intelligence (AI) capabilities. It offers an in-depth exploration of UAV networks in three-dimensional environments, including trajectory design, interference mitigation, energy efficiency, and secure communication. The book uniquely addresses adversarial UAV behavior, analyzing eavesdropping threats and cooperative swarm attacks from an information-theoretic and signal-processing perspective.

A core emphasis is placed on the transformative role of AI—including deep learning, reinforcement learning, and graph neural models—in enabling adaptive path planning, real-time decision-making, edge computing, and autonomous surveillance. Cutting-edge topics such as reconfigurable intelligent surfaces, joint sensing and communication, and UAV-enabled covert monitoring are rigorously analyzed using advanced optimization and learning frameworks, such as deep deterministic policy gradients (DDPG) and twin-delayed DDPG (TD3). Extensive simulation results and performance benchmarks are presented to validate key concepts and assess their practicality in real-world deployments.

By bridging theoretical foundations with applied research, this book not only consolidates current advancements but also lays a roadmap for future innovations in resilient, intelligent, and collaborative UAV systems. It envisions a new era where UAVs seamlessly integrate AI, edge computing, and next-generation wireless networks to operate autonomously in complex and dynamic environments.

As aerial platforms become increasingly central to smart cities, emergency response, environmental sustainability, and sixth-generation (6G) communications, this work equips researchers, engineers, and practitioners with the knowledge to pioneer the next wave of breakthroughs in aerial autonomy and intelligent skyborne infrastructure.

Please note that all the figures in the book are printed in grayscale; however, the color rendition of them is available in the online version of the book on Wiley's website – please visit https://www.wiley.com/en-us/Artificial+Intelligence+for+Unmanned+Aerial+Vehicles%3A+Sensing%2C+Communication%2C+and+Computing-p-9781394369454.

Acronyms

2D	Two-dimensional
3D	Three-dimensional
AI	Artificial intelligence
AWGN	Additive white Gaussian noise
BCD	Block coordinate descent
BPP	Binomial point process
CDF	Cumulative distribution function
CNN	Convolutional neural network
CSCG	Circularly symmetric complex Gaussian
CSI	Channel state information
CSIR	CSI-at-the-Rx
CSIT	CSI-at-the-transmitter
DDPG	Deep deterministic policy gradients
DL	Deep learning
DNN	Deep neural network
DPG	Deterministic policy gradient
DQN	Deep Q-network
DRL	Deep reinforcement learning
EdgeIoT	Edge Internet-of-Things
FP	Flight Plan
FPGA	Field Programmable Gate Array
GEO	Geosynchronous Earth orbit
GN	Ground node
GNN	Graph neural network
GPS	Global positioning system

HITL	Hardware-in-the-loop
HRWS	High-resolution wide-swath
JCAS	Joint communication and radar sensing
KKT	Karush–Kuhn–Tucker
LEO	Low Earth orbit
LHS	Left-hand side
LoS	Line-of-sight
LSTM	Long short-term memory
MDP	Markov decision process
MEC	Mobile edge computing
MHC	Matérn hardcore
MIMO	Multi-input multi-output
ML	Machine learning
MMSE	Minimum mean-square error
MO	Manifold optimization
MPC	Model-predictive control
MRC	Maximal ratio combining
MSE	Mean square error
NR	New radio
OU	Ornstein–Uhlenbeck
PDF	The probability density function
PG	Policy gradient
RD	Random direction
RF	UAVs for radio frequency
RHC	Receding horizon control
RHS	Right-hand side
RIS	Reconfigurable intelligent surface
RL	Reinforcement learning
RMS	Root mean square
RWP	Random way-point
SAD	Straight ascending or descending
SAR	Synthetic aperture radar
SC	Selection combining
SCA	Successive convex approximation
SDP	Semidefinite programming
SINR	Signal-to-interference-plus-noise ratio
SITL	Software-in-the-loop
SLF	Straight-and-level flight
SNR	Signal-to-noise ratio
SPPP	Spatial Poisson point process

SRCM	Semi-random circular movement
ST	Smooth turn
TD	Temporal difference
TMR	Task missing rate
UAV	Unmanned aerial vehicle
URA	Uniform rectangular array
U2G	UAV-to-ground
U2U	UAV-to-UAV
VTOL	Vertical take-off and landing

1

Introduction

1.1 UAV Applications

UAVs, known for their agility, rapid deployment capabilities, and autonomous operational characteristics, have increasingly emerged as critical assets across various civilian and military applications. In recent years, UAV technologies have demonstrated substantial advantages across numerous fields, driven by advancements in navigation systems, control algorithms, and hardware capabilities.

In environmental monitoring, UAVs are extensively deployed for real-time data collection, facilitating continuous surveillance and precise measurements of ecological parameters in regions otherwise difficult or hazardous to access [88]. UAVs equipped with advanced sensors and imaging technologies contribute significantly to biodiversity conservation, atmospheric pollution monitoring, and agricultural assessments, thereby enabling informed decision-making and rapid responses to environmental changes.

Disaster management scenarios have particularly benefited from UAV integration, leveraging their capability to perform swift and comprehensive aerial reconnaissance immediately following natural disasters, such as earthquakes, floods, and hurricanes [91]. UAVs provide critical situational awareness to rescue teams, allowing them to identify survivors, assess damage to infrastructure, and plan efficient emergency response strategies, significantly enhancing the effectiveness of rescue and relief efforts.

In the logistics and delivery sector, UAV technologies have revolutionized transportation and distribution systems, especially in remote or underdeveloped areas lacking reliable road infrastructure [164, 165]. UAV-based delivery services ensure

Artificial Intelligence for Unmanned Aerial Vehicles: Sensing, Communication, and Computing, First Edition. Shuyan Hu et al.

the timely and efficient transport of critical supplies, such as medical goods, emergency equipment, and commercial products. This capability not only increases service accessibility but also reduces delivery costs and operational risks associated with traditional logistics methods.

Military applications have been profoundly transformed by UAV technologies, offering unprecedented capabilities for aerial combat, reconnaissance, surveillance, and strategic monitoring [96]. UAVs are extensively utilized for border surveillance, intelligence gathering, and combat support missions due to their stealth capabilities, endurance, and minimal risk to personnel. Autonomous UAV systems equipped with AI further enhance mission adaptability, allowing real-time decision-making and effective responses in dynamic combat situations.

Moreover, UAVs have found valuable applications in the fields of infrastructure inspection and maintenance, providing accurate, safe, and cost-effective alternatives to traditional methods. UAVs perform inspections of structures such as bridges, power lines, pipelines, and wind turbines, significantly reducing operational hazards and downtime. Additionally, in agriculture, UAVs facilitate precision farming by conducting detailed crop assessments, soil analysis, and targeted spraying of fertilizers and pesticides, greatly improving crop yield and resource efficiency.

These diverse and impactful applications have motivated extensive academic and industrial research aimed at enhancing UAV performance, operational reliability, and regulatory frameworks. As UAV technologies continue to evolve, it is expected that their integration into everyday life will expand, creating broader societal impacts and opening new horizons for innovation and technological advancement.

1.2 Sensing, Communication, and Computing of UAVs

The effectiveness and versatility of UAVs depend significantly on their capabilities in sensing, communication, and computing, which together determine the quality and scope of UAV-enabled applications. Recent advances in these areas have greatly enhanced UAVs' functionality, enabling more sophisticated missions and improving operational efficiency.

In terms of sensing, UAVs are increasingly equipped with various advanced sensor payloads, such as high-resolution cameras, LiDAR, thermal imaging sensors, radar systems, and environmental monitoring instruments. These advanced sensing technologies enable UAVs to conduct highly precise and comprehensive environmental observations, infrastructure inspections, and target detections, even under adverse conditions. Furthermore, the integration of advanced sensing systems allows for real-time data acquisition and processing, supporting immediate decision-making processes during complex operational scenarios such as disaster management, precision agriculture, and surveillance operations.

Building on these sensing capabilities, context awareness has become a crucial aspect of intelligent UAV operation. It refers to the UAV's ability to perceive and interpret information about its environment, mission requirements, and network conditions in real time. By understanding factors such as terrain, weather, signal quality, and surrounding activity, UAVs can adapt their flight paths, sensing behaviors, and communication strategies accordingly. This enables more responsive and efficient operation, particularly in dynamic or uncertain environments such as disaster zones, congested airspace, or remote monitoring sites. Context awareness thus enhances the UAV's ability to make informed decisions and fulfill complex mission objectives effectively.

Communication capabilities are another critical component of UAV systems, playing a pivotal role in mission coordination, data transmission, and real-time control. Robust and reliable communication links between UAVs and ground stations (U2G), as well as among multiple UAVs (U2U), are essential to ensure continuous operation and timely data dissemination. Recent studies have emphasized the need to analyze and optimize UAV communication under challenging conditions, such as interference from ground-based transmitters and imperfect channel state information (CSI). Additionally, current research explores the communication capacity in three-dimensional (3D) environments, considering random UAV trajectories and dynamic network topologies, which significantly differ from traditional two-dimensional (2D) or static UAV deployments.

The computing capabilities onboard UAVs facilitate efficient data processing, autonomous navigation, and complex mission execution. As UAV missions become more sophisticated, onboard edge computing platforms have emerged as crucial components, enabling UAVs to perform data-intensive computations locally. This significantly reduces latency and bandwidth demands for data transmission. Edge computing capabilities are especially beneficial in time-critical applications, including disaster response, surveillance, and environmental monitoring, where immediate and accurate data analysis directly affects mission success. Moreover, advanced computing architectures allow UAVs to adapt dynamically to changing environments and make real-time decisions, enhancing the robustness and autonomy of UAV operations.

Collectively, advances in sensing, communication, and computing technologies significantly enhance UAV performance, enabling UAVs to operate effectively in complex and dynamic environments. Ongoing research continues to address remaining challenges and further improve these critical capabilities, thus broadening UAVs' applicability and operational effectiveness in diverse practical scenarios.

1.3 Mobility Models of UAVs

The mobility characteristics of UAVs play a critical role in determining the efficiency and effectiveness of aerial communication, networking, and sensing applications. UAVs are broadly categorized into fixed-wing and rotary-wing platforms, each governed by distinct aerodynamic principles and operational capabilities, which in turn shape their respective mobility models.

Fixed-wing UAVs are designed for high-speed, long-endurance missions, often used in wide-area surveillance and data collection tasks. However, they are subject to non-holonomic constraints, requiring continuous forward motion to maintain lift. As such, their mobility is often modeled using waypoint-based linear trajectories, where the UAV follows a sequence of predetermined points with fixed or adaptive speeds. Continuous, smooth trajectories that respect Newton's laws and the physical turning limitations of fixed-wing UAVs are constructed, offering flexibility for path optimization in cluttered environments.

In contrast, rotary-wing UAVs, such as quadcopters, possess vertical take-off and landing (VTOL) capabilities and are capable of hovering in place. This flexibility allows for precise maneuvering, making them well-suited for applications like infrastructure inspection, environmental monitoring, and delivery in urban environments. Their mobility is often captured using models, such as the random waypoint model, where UAVs select random destinations and move with randomized speeds and pauses, commonly used in simulation studies. More realistic models like the hover-and-fly pattern depict the typical stop-and-go nature of rotary-wing UAV missions, alternating between stationary observation and dynamic repositioning.

The choice of mobility model depends on the operational context and system design objectives. Accurate modeling is essential for analyzing communication link dynamics, evaluating network performance, and optimizing energy consumption in UAV-enabled systems.

1.4 UAVs Powered with AI

Recent advancements in AI have significantly expanded the operational capabilities and intelligence levels of UAV systems. By integrating AI-driven methods such as machine learning (ML), deep learning (DL), and reinforcement learning (RL), UAVs can autonomously address complex tasks with enhanced efficiency, adaptability, and robustness. This integration represents a transformative shift from conventional manually controlled or preprogrammed UAV operations toward more intelligent, autonomous, and adaptive missions.

One major application area of AI-aided UAVs is trajectory optimization and path planning. Utilizing RL algorithms, UAVs can autonomously identify optimal flight paths and adjust their routes dynamically in response to environmental changes, mission constraints, or unexpected events. Such adaptability is critical for missions in uncertain or hostile environments, enabling UAVs to maintain optimal communication links, avoid obstacles, and minimize energy consumption without continuous human intervention.

Moreover, AI techniques enhance the UAV's perception capabilities, significantly improving object detection, classification, and tracking accuracy. DL algorithms, particularly convolutional neural networks (CNNs), empower UAVs to interpret sensory data swiftly and precisely, which is indispensable for applications such as real-time surveillance, precision agriculture, infrastructure inspection, and disaster response. For instance, AI-powered UAVs are capable of autonomously detecting infrastructure defects or monitoring crop health, dramatically enhancing operational efficiency and reducing human oversight requirements.

AI-driven resource management and optimization further elevate UAV system performance. ML algorithms enable UAVs to effectively manage limited onboard resources, such as battery life and communication bandwidth. Predictive analytics allow UAVs to estimate energy consumption, anticipate potential system faults, and optimize resource allocation in real-time, ensuring mission sustainability and reliability. This ability is particularly important for prolonged missions requiring efficient energy use and high reliability.

In addition, AI plays a crucial role in enabling UAV swarms, a cutting-edge paradigm in UAV technology. AI algorithms facilitate coordinated decision-making, collective behavior, and distributed sensing among U2U, thereby achieving complex mission objectives that individual UAVs cannot efficiently accomplish alone. Swarm intelligence methods enable UAV groups to perform sophisticated surveillance, reconnaissance, environmental monitoring, and communication relay tasks with enhanced resilience to individual UAV failures or external interference.

AI is also a core enabler of adaptive, real-time decision-making across the UAV networks in sensing, communication, and edge computing. In particular, in AI-aided sensing, UAVs can autonomously interpret raw data from onboard sensors (e.g., cameras, LiDAR, and radar) using computer vision and DL techniques. This empowers them to recognize patterns, detect anomalies, or infer high-level environmental context without relying solely on remote processing. Such intelligent perception is crucial for tasks like target tracking, infrastructure inspection, or smart agriculture.

In AI-driven communication, learning-based methods such as RL, deep Q-networks, and graph neural networks (GNNs) enable UAVs to optimize spectrum access, trajectory-aware beamforming, and interference management in real time. This dynamic adaptation is essential in multi-UAV and UAV-to-ground networks, where link quality and topology vary frequently. In AI-enabled computing, onboard or edge-deployed AI models support distributed inference, caching, and decision-making with minimal latency. Federated learning and online learning approaches further allow UAVs to collaboratively improve their intelligence while preserving data privacy.

Together, these AI-aided capabilities enable UAV networks to operate more autonomously, efficiently, and securely, transforming them into resilient platforms for intelligent sensing-communication-computing fusion. This architectural paradigm not only enhances system performance but also expands the feasibility of complex missions in dynamic, adversarial, or resource-constrained environments.

Finally, the integration of AI with emerging technologies, such as reconfigurable intelligent surfaces (RISs) and edge computing, further extends the capabilities of UAV systems. AI methods optimize RIS configurations and UAV positions to significantly enhance communication coverage, link reliability, and signal quality. Similarly, edge computing augmented with AI provides real-time onboard processing capabilities, reducing latency and improving responsiveness, essential for time-sensitive operations like disaster response and tactical military missions.

1.5 Book Contributions

This book provides a systematic and comprehensive integration of current methodologies and theories in sensing, communication, and computing for UAV systems, with a particular emphasis on the application of advanced AI techniques. Its key contributions are reflected in the following aspects:

First, the book offers an in-depth survey and critical evaluation of UAV communication and connectivity in 3D environments. Unlike traditional studies that rely on simplified 2D or deterministic UAV deployment models, this work thoroughly addresses the complexities of 3D networks, including random UAV trajectories, network densification, interference management, and uncertainties in CSI. This structured review establishes a solid theoretical foundation for advancing the field.

Second, it systematically investigates existing methodologies in UAV trajectory optimization, energy efficiency, and beamforming strategies. By comparing various optimization techniques, the book highlights their respective advantages, limitations, and suitability for different operational scenarios. These insights are intended to assist both researchers and practitioners in selecting the most appropriate approaches for specific mission requirements.

Third, the role of AI in enhancing UAV intelligence is explored through a detailed review of state-of-the-art approaches in AI-driven trajectory planning, resource allocation, and autonomous decision-making. Techniques such as RL and DL are examined for their effectiveness in improving UAV autonomy, adaptability, and resource utilization.

Fourth, the book provides a thorough examination of UAV-enabled edge computing, including task offloading, scheduling strategies, and adaptive communication protocols. These discussions underscore the critical role of edge computing in enabling timely and efficient decision-making for time-sensitive UAV missions.

Fifth, emerging technologies, such as RISs, are also addressed. The book reviews AI-based optimization methods for integrating RISs into UAV communication systems, evaluating their ability to improve performance and reliability in dynamic and interference-prone environments.

Finally, the book includes extensive simulation results and performance evaluations derived from the existing literature. These empirical findings are critically analyzed to assess the practicality, efficiency, and limitations of the reviewed methodologies in real-world deployments.

By synthesizing and analyzing a broad spectrum of current research, this book serves as a reference for scholars, engineers, and industry professionals working with UAV systems. It consolidates existing knowledge and outlines promising research directions to guide future developments in this rapidly evolving field.

Several books have explored UAV communications, intelligence, and applications in recent years. Yet, they did not fully cover AI-empowered integrated sensing, communication, and computing for UAV networks, as done in this book. For instance, [197] focuses on the architecture, protocols, and performance aspects of UAV-based wireless networks, without addressing AI-driven optimization or sensing applications in depth. [206] offers a more recent look into computational intelligence methods applied to UAV communication, without considering sensing and in-depth AI computing. [135] covers theoretical modeling and practical UAV applications for different scenarios, without covering AI-driven architectures or communication-sensing-computing fusion. [11] focuses on urban applications of UAVs, without a generalizable or technical treatment of AI-enhanced UAV network design, or coverage of resilience, privacy, and real-time autonomy.

Unlike the aforementioned existing books, this book uniquely focuses on AI-enhanced architectures that tightly integrate sensing, communication, and edge computing within UAV networks. It emphasizes real-time autonomy, adaptive decision-making, context awareness, and resilience against security threats, supported by both theoretical modeling and AI-driven methods. This integrated approach is critical for the next generation of UAV systems operating in complex, adversarial, or resource-constrained environments, and fills a notable gap in current literature.

1.6 Book Outline

This book is structured to systematically review and analyze critical aspects of UAV technologies in the domains of sensing, communication, computing, and AI integration. In particular, each chapter is dedicated to a specific topic, ensuring comprehensive coverage of both theoretical foundations and practical implications.

Chapters 2 to 4 analyze the utility and integrity of UAV networks, and the threat of adversarial UAVs to terrestrial networks. Chapter 2 delves into the communication capacity and connectivity of UAV networks within complex 3D operational environments. It provides an in-depth analysis of U2U and U2G links, emphasizing challenges posed by random UAV trajectories, interference management, and network densification effects.

Chapter 3 investigates the vulnerability of terrestrial wireless communications to aerial eavesdropping from an information-theoretic perspective. The chapter presents a rigorous analysis of the ergodic secrecy rate and the average ϵ-outage secrecy rate for a ground-based transmitter-receiver pair intercepted by an aerial eavesdropper flying a random 3D trajectory with smooth turns within a spherical region.

In Chapter 4, the security risk is studied with cooperative UAV adversaries. This chapter focuses on the model that is extended to analyze the impact of coordinated eavesdropping strategies involving a swarm of UAVs using diversity techniques, such as selection combining (SC) and maximal ratio combining (MRC). The impact of oscillator phase noise and finite memory at the eavesdroppers is also incorporated. Closed-form approximations for ergodic and ϵ-outage secrecy capacities are shown to be asymptotically tight as the number of eavesdroppers grows.

Chapters 5 to 7 delineate the integration of AI techniques to UAV navigation and surveillance. Chapter 5 aims to address the problem of jointly optimizing UAV navigation and task offloading decisions in real time, under the constraint that the UAV lacks prior knowledge of IoT devices' task queues, battery levels, and channel conditions. To tackle the uncertainty in network states, a GNN-assisted advantage actor-critic model is developed to capture the underlying relational structure among devices and guides the training of the UAV's decision-making process.

Chapter 6 addresses the use of UAVs for radio frequency (RF) and video surveillance applications, emphasizing the integration of advanced communication technologies. It reviews approaches for trajectory optimization, power control, and stealth performance. This chapter illustrates how UAV systems have been optimized in previous research to balance performance and operational security effectively.

Chapter 7 discusses the emerging role of UAV-enabled legitimate monitoring systems via joint trajectory design and energy management. The system includes a suspicious transmission link with a terrestrial transmitter and a terrestrial receiver, and a UAV to monitor the suspicious link. Capitalizing on the successive convex approximation method, an efficient iterative approach is put forth to find a feasible solution fulfilling the Karush–Kuhn–Tucker (KKT) conditions. Numerical results demonstrate that with UAV trajectory optimization, the overall energy consumption can be greatly suppressed.

Chapters 8 and 9 describe the role of UAV and AI in emerging technologies (e.g., joint sensing and communications (JSAC) and RIS). In Chapter 8, a novel BS-UAV bistatic synthetic aperture radar (SAR) platform is designed, where the flight path of the UAV is optimized to minimize its propulsion energy and guarantee the required sensing resolutions on a series of landmarks. A new deep RL (DRL)-based online control scheme for visual-based UAV-on-UAV tracking and monitoring is proposed, where a solar-powered, fixed-wing UAV tracks a suspicious UAV target by having the target inside its effective visual range.

Chapter 9 presents a unified DRL framework for RIS-assisted UAV communications, addressing two critical scenarios: Covert radio surveillance and anti-jamming operations. By leveraging the latest twin delayed deep deterministic policy gradient (TD3) algorithm, a dual-mode intelligent control system is designed, which can dynamically optimize UAV trajectory and RIS configuration based solely on observed communication metrics, eliminating the need for explicit CSI.

Finally, Chapter 10 concludes the book by summarizing key findings, providing a synthesized view of current technological advances, and identifying open challenges and promising research directions in UAV sensing, communication, computing, and AI integration. This concluding chapter serves as a comprehensive guide for researchers and practitioners, highlighting critical insights and prospects for continued innovation in UAV technologies.

This book presents a unified exploration of integrated sensing and communication in UAV networks, supported by theoretical modeling and AI techniques. While not every chapter focuses explicitly on AI, the progression of the content builds a coherent path from fundamental modeling to AI-driven solutions. Chapters 2–4 provide foundational models for UAV communication and sensing—such as 3D mobility, link capacity, and interference management—under complex dynamics. These theoretical results are critical for designing scalable AI-driven strategies later in the book. Chapters 5, 8, and 9 explicitly implement AI techniques, including RL and deep neural networks, to optimize UAV trajectories, resource allocation, and collaborative behavior.

Chapters 6 and 7 focus on sensing and beamforming in challenging environments, offering insight into system-level constraints and opportunities for AI-based adaptation. Together, the chapters form a layered framework where fundamental analysis enables the deployment of intelligent decision-making in UAV-assisted networks. The book concludes with a discussion on limitations, real-world applicability, and directions for future research—particularly around security, online learning, and context-awareness.

2

Capacity and Connectivity of UAVs in Three-dimensional Spaces

2.1 Introduction

UAVs can be broadly classified into quadrotors, unmanned miniature helicopters, unmanned airships, and fixed-wing UAVs. Equipped with autonomous navigation capabilities and high maneuverability, UAVs have found extensive applications in environmental and natural resource monitoring, patrolling, disaster response, search and rescue operations, goods delivery, and even military engagements such as air combat [18, 92, 146]. These versatile applications have driven growing demand for reliable communication, both among UAVs themselves and between UAVs and ground stations [21, 22].

UAV swarms, leveraging swarm intelligence and collective behavior, are increasingly being adopted for complex, large-scale missions. Unlike terrestrial networks constrained to 2D space, UAV swarms operate in fully 3D airspace, offering unprecedented flexibility in deployment. However, this added freedom also introduces unique challenges. In particular, swarm communication can be severely affected by interference from terrestrial transmitters, potentially disrupting intra-swarm links and degrading the overall mission performance.

Despite the growing interest in swarm-based UAV systems, the theoretical foundations—especially regarding connectivity, spatial density, interference resilience, and swarm coordination—remain underdeveloped. Most existing research in swarm robotics focuses on 2D connectivity control [48, 78, 249], with limited attention to realistic 3D flight dynamics or dynamic network behavior in aerial environments. Studies such as [5, 6, 86, 277, 307] primarily consider static or disconnected UAV trajectories and rarely address communication resilience within the swarm under interference. Furthermore, trajectory optimization studies [174, 217, 218, 255] tend to overlook intra-swarm connectivity altogether.

Artificial Intelligence for Unmanned Aerial Vehicles: Sensing, Communication, and Computing, First Edition. Shuyan Hu et al.

Current research on communication performance between autonomous UAVs in 3D aerial space remains limited. Many existing studies assume static UAV placements or rely on overly simplified and deterministic flight paths. For example, the ABSOLUTE project [47] utilized static or semi-static UAVs as LTE-A aerial base stations to restore communication in post-disaster scenarios. These UAVs were integrated into a heterogeneous public-safety network architecture [185], and their deployment was shown to enhance both network capacity and coverage [193, 240].

Under stationary UAV configurations, researchers derived outage probability expressions over Nakagami-m fading channels [5], and analyzed bit rate and outage metrics in cooperative UAV networks operating over generalized Gaussian-Finite-Mixture fading channels [6]. Hybrid satellite-terrestrial relay networks were also studied, with ergodic capacity bounds and approximations presented in [17]. However, these efforts mostly focused on high-level network performance indicators, such as throughput and delay scaling laws, while ignoring the effects of UAV flight dynamics and wireless channel conditions. Similarly, the work in [277] analyzed total throughput and average delay for aeronautical networks under various assumptions, but did not incorporate the impact of physical flight parameters or realistic channel conditions, implicitly assuming constant link capacities.

In a different yet relevant context of terrestrial wireless communications, a lot of studies have been conducted on analyzing the ergodic capacity, outage probability, and outage capacity in 2D spaces. For example, Baccarelli and Fasano [24] derived the upper and lower bounds for the capacity of fading channels by exploiting the Jensen's inequality. A recurrence expression for the capacity of a Nakagami-m fading channel with binary phase shift keying (BPSK) modulation was developed in [310], with the assumption that channel state information (CSI) was available at the receiver. Rezki and Alouini [224] investigated the capacity of flat Rayleigh fading channels with ideal CSI at both the transmitter and receiver with asymptotically low signal-to-noise ratio (SNR) and established the scaling law of $\rho \log(1/\rho)$. In [161], the capacity of an $\alpha{-}\eta{-}\kappa{-}\mu$ fading channel was studied by using infinite series theory. In [202] and [95], the impact of MAC protocols on the capacity of vehicular networks was studied with a focus on the backoff timer designs of channel-sense multiple-access with collision avoidance (CSMA/CA). Hong *et al.* [103] studied the statistics of the capacity of wide-band indoor channels, based on experimental measurements. All these works were based on 2D environments, and cannot incorporate 3D trajectories and mobility, which are distinguishing features of UAVs.

This section analyzes the capacity of the U2U links between autonomous UAVs flying random 3D trajectories with practical smooth turns in Rician fading channels. This is distinctively different from the aforementioned existing works under the assumption of either 2D or deterministic trajectories, and particularly interesting to applications, such as surveillance and air combat.

A key contribution of the chapter is that we geometrically derive the probability distributions of the distances between a pair of UAVs that are assigned to serve the same 3D spaces but fly autonomously in an uncoordinated fashion. By exploiting Jensen's inequality, the distributions are translated to the closed-form bounds for the ergodic capacity and outage capacity between the UAVs in 3D spaces. The analysis can also capture U2U links at fixed altitudes and U2G links between UAVs with 3D trajectories and static ground stations.

Another important contribution is that we extrapolate our analysis to dense 3D networks of UAVs, and quantify the impact of network densification on the capacity and coverage of the networks. By exploiting order statistics, the distances between adjacent UAVs are evaluated, and the closed-form lower bound for the multi-hop ergodic capacity is established.

Other contributions of the chapter also include the lower bounds for the ergodic capacity of U2U links in the presence of imperfect CSI at the receiver, and for the signal-to-interference-plus-noise ratio (SINR) in the presence of nonnegligible interference from ground transmitters. Validated by simulations, our analysis reveals that a U2U link with random 2D trajectories is superior in terms of outage capacity due to its short average link distance. It is also shown that a U2G link can incur substantially lower capacity than a U2U link, even in the case that the 3D coverage of the UAVs is the same. This results from the longer average length of U2G links. Analyzing link capacity under these realistic, nondeterministic flight patterns is critical for designing robust communication protocols and for training AI models capable of adapting to nonstationary aerial environments. The derived capacity bounds thus serve as building blocks for intelligent UAV decision-making strategies developed in later chapters.

The rest of the chapter is organized as follows. In Section 2.2, the system model and performance metrics are presented. In Section 2.3, the ergodic capacity and outage capacity of the U2U and U2G links are analyzed by evaluating the distributions of the link lengths. In Section 2.4, the analysis is extrapolated to study the impact of network densification on the capacity. In Section 2.8, the analyses are numerically validated by simulations, followed by the extensions of the analysis to the cases of imperfect CSI and nonnegligible interference. In Section 2.10, conclusions are provided.

2.2 System Model and Performance Metrics

2.2.1 System Model

As illustrated in Figure 2.1, the first scenario that we consider involves two UAVs, each equipped with an effective omni-directional antenna. To eliminate the body blockage resulting from the mobility and gesture changes of UAVs, the omni-directional antenna can be effectively implemented by deploying an array

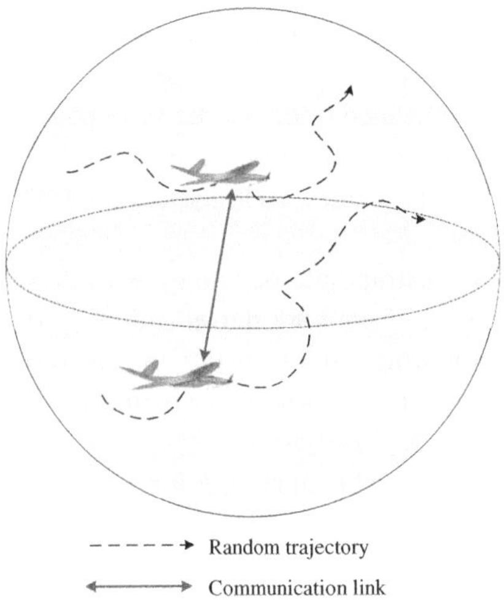

Figure 2.1 Illustration on a pair of autonomous UAVs flying random 3D trajectories with smooth turns in a 3D sphere.

of antenna elements around the perimeter of a UAV, or attaching a pair of patch antennas on the upper and lower surfaces of the UAV. The different antenna elements, or patch antennas, can be aggregated at a radio frequency (RF) combiner/splitter before being connected to a single RF chain (including power amplifier, filter, up- and down-convertor, and analog-to-digital/digital-to-analog convertor) and then a baseband digital signal processor, producing effectively a single omnidirectional antenna free of body blockage.

The UAVs have independent 3D flight trajectories following a Smooth Turn (ST) Mobility Model, which incorporates centripetal and tangential accelerations and preserves the smoothness of aerial trajectories [268, 296]. This assumption is motivated by the need to capture the intrinsic maneuverability constraints of fixed-wing and rotary UAVs, which cannot execute abrupt direction changes due to aerodynamic and inertial limitations. The ST model provides a significant advantage over simplistic mobility models (e.g., random walk or random waypoint) by preserving trajectory continuity and reflecting the smooth turning behavior observed in real UAV operations, such as surveillance, patrolling, and area coverage missions. The dynamics and constraints on the maneuverability of UAVs are captured. Moreover, the model has the important feature of uniform stationary distributions of UAV positions [268]. This feature facilitates tractable performance analysis of communication metrics, such as coverage probability or link capacity, and supports fair spatial sampling over time.

We also assume that there is no collision between the UAVs. The assumption is legitimate, provided the dimension of the UAVs is so small and negligible to the flight region. This is because of the continuous nature of the uniform node distributions, even in the case that the UAVs are uncoordinated and independently distributed, as dictated by the aforementioned property of the ST mobility model. The probability of UAV collisions approaches zero. In practice, collision-avoidance techniques are typically based on sensing and ranging. Broadcasting safety messages can also prevent collisions. Our study on the channel capacity of UAVs can contribute to the prevention of collisions.

When one of the UAVs transmits to the other UAV, the received signal at the latter is given by

$$y_r(t) = \sqrt{P} h_1 x(t) + n(t), \tag{2.1}$$

where P denotes the transmit power of the UAV, $x(t)$ is the transmit symbol[1], and $n(t)$ is the zero-mean additive white Gaussian noise (AWGN) at the UAV with $\mathbb{E}\left[|n(t)|^2\right] = \sigma^2$. h_1 is the complex channel coefficient between the pair of UAVs, and it is assumed to be precisely known to the receiver unless otherwise specified.

The LoS links between UAVs are available in open space. The U2U link is modeled to experience Rician fading, and the Rician factor K accounts for the influence of scattering and reflection from the surrounding environments. Particularly, K is defined to be the ratio between the signal power of the LoS path and the power of other scattered paths. We assume that the perfect CSI is available at the receiver. The probability density function (PDF) of the received SNR, denoted by γ, can be written as [21]

$$f_\gamma(x) = \frac{1+K}{\bar{\gamma}} \exp\left[-K - \frac{(1+K)x}{\bar{\gamma}}\right] I_0\left(2\sqrt{\frac{K(K+1)}{\bar{\gamma}}x}\right), \tag{2.2}$$

where $\bar{\gamma} = \frac{P}{\sigma^2 l^\alpha}$ is the statistically average SNR; α is the large-scale path loss exponent; $I_0(x) = \sum_{n=0}^{\infty} \frac{(x/2)^{2n}}{n!\Gamma(n+1)}$ is the 0-th order modified Bessel function of the first kind; and $\Gamma(\cdot)$ is the gamma function.

We also consider the scenario where a swarm of N UAVs fly autonomously within a 3D spherical space centered at a point of interest. The sphere, denoted by V, has radius r and volume $|V|$. An interferer (or a jammer) is located on the ground, as illustrated in Figure 2.2. The point of interest is h_0 (in meters) above the interferer's antenna. $h_0 > r$. The size of the sphere indicates the coverage of the UAV swarm. All UAVs are assumed to fly independent trajectories following a Smooth Turn (ST) Mobility Model [268], e.g., at the initial or transition stage when the swarm is just formed or deciding to reform. As proved in [268], the positions

1 It is assumed that $x(t)$ follows a circularly symmetric complex Gaussian distribution $\mathcal{CN}(0,1)$, i.e., $\mathbb{E}[|x(t)|^2] = 1$, where $\mathbb{E}[\cdot]$ denotes an expectation operation.

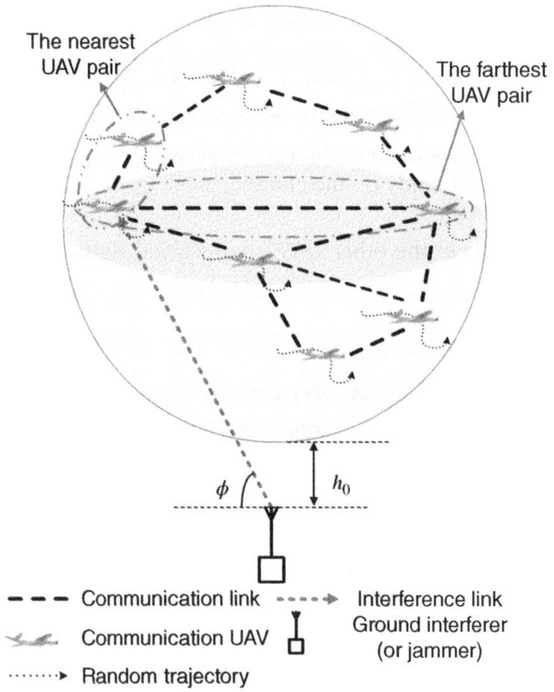

Figure 2.2 A UAV swarm flies autonomously within a 3D sphere in the presence of external interference from the ground, where h_0 is the distance from the ground interferer to the bottom of the 3D sphere, and ϕ is the elevation angle from the ground interferer to the UAV in the swarm.

of the UAVs under the ST mobility model follow a random and uniform stationary distribution within the 3D sphere.

The ST model is adopted for two reasons. On the one hand, the model is specifically designed for UAVs to produce random trajectories with practical smooth turns [268]. The smooth turns rationalize the unique maneuverability of UAVs, such as mechanical and aerodynamic constraints, high mobility, and safe separation distance [268]. On the other hand, the ST mobility model can be considered as the worst-case mobility pattern, as the connectivity between UAVs can be improved by further optimizing the mobility pattern. We assume that the UAVs do not collide, provided the dimension of such a UAV can be very small as compared to the 3D spherical flight space.

Note that the UAVs can also be equipped with collision-avoidance sensors such as a camera, so that a minimum safety distance can be enforced between UAVs to avoid collisions even when the UAVs are not so small. As discussed in Appendix 1, the proposed analysis can still apply.

We can calculate the distance between two UAVs and the distance between a UAV and the ground interferer. Let $L_{i,\{i=1,\ldots,N-1\}}$ denote the distances from a UAV to the other $(N-1)$ UAVs in the swarm. Given that the N UAVs fly randomly within the spherical space with smooth turns, at any moment, the distance of a UAV from its nearest neighbor in the swarm is

$$l_{\min} = \min_{\{i=1,\ldots,N-1\}} \{L_i\};$$

and its expectation can be obtained as follows.

Lemma 2.1 *The expectation of the distance between a UAV and its nearest neighbor in the swarm, i.e., $\mathbb{E}[l_{\min}]$, is given by*

$$\mathbb{E}[l_{\min}] = \Gamma\left(\frac{4}{3}\right) N^{-\frac{1}{3}} r; \tag{2.3}$$

Proof. See Appendix 2.

Let l_I denote the distance from the ground interferer to a designated UAV in the swarm. The expectation of l_I, i.e., $\mathbb{E}[l_I]$, is provided in the following lemma.

Lemma 2.2 *The expectation of the distance from the ground interferer to a UAV in the swarm, i.e., $\mathbb{E}[l_I]$, is given by*

$$\mathbb{E}(l_I) = \frac{(h_0 + 2r)^4(3r - h_0) + h_0^4(5r - 9h_0)}{40r^3(h_0 + r)}. \tag{2.4}$$

If $h_0 = 0$, then $\mathbb{E}(l_I) = \frac{6}{5}r$.

Proof. See Appendix 3.

As reported in [21, 22], the aerial receivers are vulnerable to the interference coming from the ground transmitters. We proceed to discuss the impact of the interference from a fixed ground station on the link performance between a pair of UAVs flying random 3D trajectories with smooth turns.

The received signal at each UAV is rewritten as

$$y(t) = \sqrt{Pl^{-\alpha}}h(t)x(t) + \sqrt{P_I l_I^{-\alpha_I}}h_I(t)x(t) + n(t), \tag{2.5}$$

where P_I is the transmit powers of the ground interferer, respectively; $h_I(t)$ is the channel coefficient for the ground-to-UAV (G2U) interference link; α_I ($\alpha_I \geq \alpha$) is the path loss exponent for the G2U interference link; and l_I is the distance between a UAV and the ground interferer.

The height of a UAV in the swarm can affect the propagation characteristics of the interference link, including the antenna gain and the path loss model. The ground interferer may be equipped with a directional antenna, and it points at the center of the swarm. The antenna gain depends on the elevation angle of the transmitting UAV, i.e., $\phi\left(0 \leq \phi \leq \frac{\pi}{2}\right)$. In such a case, the antenna gain of the ground interferer can be written as $\mathcal{G}(\phi)$, as shown in Figure 2.2.

The path loss of the G2U interference link is also dependent on the elevation angle. Therefore, the path loss exponent for the interference link can be modeled as a function of the elevation angle ϕ [10, 21], as given by

$$\alpha_I(\phi) = a_1 \cdot \mathcal{P}_{\text{LoS}}(\phi) + b_1, \tag{2.6}$$

where $\mathcal{P}_{\text{LoS}}(\phi) = \frac{1}{1 + a_2 \exp(-b_2\phi)}$; and the model parameters a_1, a_2, b_1, and b_2 are dependent on the carrier frequency and the surrounding environment [10, 21]

$$\begin{aligned} a_1 &= \frac{\alpha_{\frac{\pi}{2}} - \alpha_0}{\mathcal{P}_{\text{LoS}}\left(\frac{\pi}{2}\right) - \mathcal{P}_{\text{LoS}}(0)}; \quad b_1 = \alpha_0 - a_1 \mathcal{P}_{\text{LoS}}(0); \\ a_2 &= \sum_{n=0}^{3} \sum_{m=0}^{3-n} c_{mn}^a (\beta_1\beta_2)^m \beta_3^n; \quad b_2 = \sum_{n=0}^{3} \sum_{i=0}^{3-n} c_{mn}^b (\beta_1\beta_2)^m \beta_3^n, \end{aligned} \tag{2.7}$$

where the coefficients c_{mn}^a and c_{mn}^b, $m, n \in \{0, 1, 2, 3\}$, are provided in [10, Tables I and II]; β_1, β_2, and β_3 are environment-dependent variables[2]; we set $\alpha_0 = \alpha_I(0)$ and $\alpha_{\frac{\pi}{2}} = \alpha_I\left(\frac{\pi}{2}\right)$.

The SINR, denoted by ζ, can be represented as

$$\zeta = \frac{Pl^{-\alpha}|h|^2}{P_I {l_I}^{-\alpha_I(\phi)} |h_I|^2 \mathcal{G}(\phi) + \sigma^2} = \frac{\gamma}{\gamma_I + 1}, \tag{2.8}$$

where $\gamma = \frac{Pl^{-\alpha}|h|^2}{\sigma^2}$ and $\gamma_I = \frac{P_I {l_I}^{-\alpha_I(\phi)} |h_I|^2 \mathcal{G}(\phi)}{\sigma^2}$.

Also, assume that the interference link experiences the Rician fading with the Rician factor K_I. Likewise, the PDF of the received interference power γ_I is given by

$$f_{\gamma_I}(x) = \frac{1 + K_I}{\bar{\gamma}_I} \exp\left(-K_I - \frac{(1 + K_I)x}{\bar{\gamma}_I}\right) I_0\left(2\sqrt{\frac{K_I(K_I + 1)}{\bar{\gamma}_I}x}\right),$$

where $\bar{\gamma}_I = \frac{P_I {l_I}^{-\alpha_I(\phi)} \mathcal{G}(\phi)}{\sigma^2}$.

2.2.2 Definitions of Performance Metrics

The performance metrics that we are particularly interested in are defined as follows.

---

2 In [10], β_1, β_2, and β_3 for four selected urban environment are provided, including Urban (0.3, 500, 15), Suburban (0.1, 750, 8), High-rise Urban (0.5, 300, 50), and Dense Urban (0.5, 300, 20), respectively.

2.2.2.1 Ergodic Capacity (in bits/s/Hz)

This is a critical measure of the link between a pair of UAVs with random trajectories, and can be achieved by taking the average of the instantaneous capacity over all fading states. It is defined as [63]

$$C_{\text{erg}}(\gamma) = \mathbb{E}\left[\log_2(1+\gamma)\right] = \int_0^{\infty} \log_2(1+x)\, f_\gamma(x)\; dx, \tag{2.9}$$

which, given the Rician fading channel, can be approximated to [63]

$$\begin{aligned} C_{\text{erg}}(\bar{\gamma}) &\approx \frac{1}{\ln(2)}\left[\ln(1+\bar{\gamma}) - \frac{(2K+1)\,\bar{\gamma}^2}{2(1+K)^2(1+\bar{\gamma})^2}\right], \\ &= \frac{1}{\ln(2)}\left[\ln\left(1+\frac{P}{\sigma^2}l^{-\alpha}\right) - \frac{(2K+1)}{2(1+K)^2\left(1+\frac{\sigma^2}{P}l^{\alpha}\right)^2}\right] \\ &\triangleq C_{\text{erg1}}(l), \end{aligned} \tag{2.10}$$

given the statistically averaged SNR $\bar{\gamma} = \frac{P}{\sigma^2 l^\alpha}$. The accuracy of this approximation was validated in [63, Figure 1], and have been widely accepted and approved in the literature, e.g., [18].

We can prove that $C_{\text{erg1}}(l)$ is convex. This is because $\frac{d^2 C_{\text{erg1}}(l)}{dl^2} \geq 0$ for $\alpha \leq \frac{1}{2}$ or $\alpha \geq \frac{121}{79}$, while the path loss exponent α is no less than 2 in practice. The detailed proof is provided in Appendix 2.

2.2.2.2 Outage Probability

This metric defines the probability that the instantaneous SNR at the receiver is below a threshold γ_{th} required for successful reception [5, 6]. In the Rician fading channel, it can be written as [21]

$$\begin{aligned} P_{\text{out}} &= \mathbb{P}(x \leq \gamma_{th}) \\ &= \int_0^{\gamma_{\text{th}}} f_\gamma(x)\; dx \\ &= \int_0^{\gamma_{\text{th}}} \frac{1+K}{\bar{\gamma}} \exp\left(-K - \frac{1+K}{\bar{\gamma}}x\right) I_0\left(2\sqrt{\frac{K(K+1)}{\bar{\gamma}}x}\right) dx \\ &= 1 - Q\left(\sqrt{2K}, \sqrt{\frac{2\gamma_{\text{th}}(1+K)}{\bar{\gamma}}}\right), \end{aligned} \tag{2.11}$$

where $Q(\sqrt{a}, \sqrt{b}) = \int_b^\infty \frac{1}{2}\exp(-\frac{x+a}{2}) I_0(\sqrt{ax})dx$ is the first-order Marcum Q-function. The SNR threshold γ_{th} is set up in prior, based on quality-of-service (QoS) requirements.

The first-order Marcum Q-function $Q(\sqrt{a}, \sqrt{b})$ can be approximated to [37]

$$\begin{aligned} &Q\left(\sqrt{2K}, \sqrt{\frac{2\gamma_{\text{th}}(1+K)\sigma^2 l^\alpha}{P}}\right) \\ &\approx \exp\left[-e^{\nu(\sqrt{2K})}\left(\frac{2\sigma^2\gamma_{\text{th}}(1+K)l^\alpha}{P}\right)^{\frac{1}{2}\mu(\sqrt{2K})}\right] \triangleq \tilde{Q}(l) \end{aligned} \tag{2.12}$$

where $\nu(\sqrt{2K})$ and $\mu(\sqrt{2K})$ are nonnegative parameters depending on K.

2.2.2.3 Outage Capacity (in bits/s/Hz)

This metric defines the constant data rate which can be achieved with an outage probability less than the SNR threshold γ_{th} [216]. It can be written as

$$\begin{aligned} C_{\text{out}} &= (1 - P_{\text{out}}) \cdot \log_2(1+\gamma_{\text{th}}) \\ &= Q\left(\sqrt{2K}, \sqrt{\frac{2\gamma_{\text{th}}(1+K)}{\bar{\gamma}}}\right) \cdot \log_2(1+\gamma_{\text{th}}) \\ &\approx \tilde{Q}(l)\log_2(1+\gamma_{\text{th}}). \end{aligned} \tag{2.13}$$

Note that both the ergodic capacity and outage capacity provide the theoretical limits of practically achievable data rates of UAV communication systems under realistic channel conditions. The ergodic capacity specifies the capacity of a channel, irrespective of the capability or operations of the transmitter and receiver (such as modulations). The outage capacity captures the capability of the receiver, and accounts for the receiver sensitivity γ_{th}, which further depends on the modulation-and-coding schemes available to the transmitter-receiver pair. Moreover, recent advances in coding theory are continuously closing the gap between the capacity and practically achievable data rate. For example, designs of Low-Density Parity-Check Codes were reported to achieve the channel capacity within 0.0045 dB of the Shannon Limit [56]. Thus, evaluating both capacities is of practical value, which not only aids in understanding the potential of UAV communication links, but also informs practical decisions in protocol and hardware development.

2.3 Analysis of UAVs with Random Trajectories

2.3.1 U2U Links with Random 3D Trajectories

We start with a general 3D scenario where two UAVs fly randomly in a 3D spherical region with radius r_s, as illustrated in Figure 2.3. Given the uniform stationary distribution of each of the UAVs under the ST model, closed-form expressions for the ergodic capacity and outage capacity between the pair of UAVs can be evaluated.

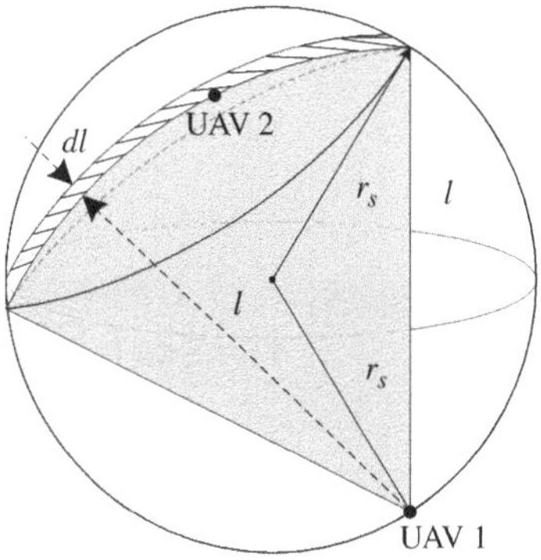

Figure 2.3 Two UAVs in the spherical region (Special case: UAV 1 is located on the surface of the 3D spherical region).

A key step of our analysis is to evaluate the distance between the pair of UAVs by exploiting the Crofton Fixed Point Theorem [60]. First, the Crofton Fixed Point Theorem is recalled in Lemma 2.3.

Lemma 2.3 *Suppose that N points ζ_i, $i = 1, 2, \dots, N$, are randomly and independently distributed in a domain A with the volume $|A|$, and H depends on $\zeta_1, \dots, \zeta_N$. Let $A' \subset A$, and δA is an infinitesimal boundary of A, but not in A'. Then the following relation holds*

$$d\,\Pr\{H\} = N\,(\Pr\{H \mid \zeta_i \in \delta A\} - \Pr\{H\})\,|A|^{-1} d|A|, \tag{2.14}$$

where $\Pr\{H \mid \zeta_i \in \delta A\}$ is the probability that H occurs when one of the random points ζ_i is on the boundary δA of A [60].

By exploiting the Crofton Fixed Point Theorem and Jensen's inequality, we are able to establish the following theorem, namely, Theorem 2.1, on the ergodic capacity and outage capacity between the pair of UAVs flying random 3D trajectories with smooth turns.

Theorem 2.1 *For a pair of UAVs flying random trajectories with practical smooth turns in a 3D spherical region with Rician fading, their ergodic capacity is lower bounded by*

$$C^*_{\text{erg1}} = \frac{1}{\ln(2)}\left[\ln\left(1 + \frac{P}{\sigma^2}\left(\frac{36}{35}r_s\right)^{-\alpha}\right) - \frac{(2K+1)}{2(1+K)^2\left(1 + \frac{\sigma^2}{P}\left(\frac{35}{36}r_s\right)^{\alpha}\right)^2}\right]; \tag{2.15}$$

and the outage capacity satisfies

$$\begin{cases} C_{\text{out1}} \geq C^*_{\text{out1}}, & \mathbb{E}[L_1] < l_{\text{th}} \\ C_{\text{out1}} \leq C^*_{\text{out1}}, & \mathbb{E}[L_1] \geq l_{\text{th}}, \end{cases} \tag{2.16}$$

where $l_{\text{th}} = \sqrt[\tau_3]{\frac{\tau_3 - 1}{\tau_1\tau_2\tau_3}}$, $\tau_1 = e^{\nu(\sqrt{2K})}$, $\tau_2 = \left(\frac{2\sigma^2\rho_{\text{th}}(1+K)}{P}\right)^{\frac{1}{2}\mu(\sqrt{2K})}$, $\tau_3 = \frac{1}{2}\alpha\mu\left(\sqrt{2K}\right)$, *and*

$$C^*_{\text{out1}} = \log_2(1+\gamma_{\text{th}}) \times \exp\left[-e^{\nu(\sqrt{2K})}\left(\frac{2\sigma^2\gamma_{\text{th}}(1+K)\left(\frac{36}{35}r_s\right)^{\alpha}}{P}\right)^{\frac{1}{2}\mu(\sqrt{2K})}\right]. \tag{2.17}$$

Proof. We start by proving (2.15). To do this, we use L_1 to denote the Euclidean distance between the pair of UAVs, and $f_{L_1}(l)$ the PDF of L_1. We also let $\mathcal{P}$ denote the probability of two UAVs separated by distance l, and $\mathcal{P}_1$ denotes this probability in the case that one of the UAVs is located on the surface of the sphere, as shown in Figure 2.3.

According to the Crofton Fixed Point Theorem; see Lemma 2.3, we can have

$$d\mathcal{P} = 2\left(\mathcal{P}_1 - \mathcal{P}\right)|V|^{-1}d|V|, \tag{2.18}$$

where $|V|$ denotes the volume of the 3D sphere, i.e., $|V| = \frac{4}{3}\pi r_s^3$ and $d|V| = 4\pi r_s^2 dr_s$.

To evaluate $\mathcal{P}_1$, we set that UAV 1 is located on the surface of the sphere, then the other UAV, UAV 2, is located on the spherical cap centered at UAV 1 and with radius l in the sphere, as illustrated by the shaded part in Figure 2.3. Let dl denote the thickness of the spherical cap. Therefore, the volume of the spherical cap is $2\pi l^2\left(1 - \frac{l}{2r_s}\right)dl$. $\mathcal{P}_1$ can be obtained as

$$\mathcal{P}_1 = \frac{2\pi l^2\left(1 - \frac{l}{2r_s}\right)dl}{\frac{4}{3}\pi r_s^3} = \frac{3l^2\left(2r_s - l\right)dl}{4r_s^4}. \tag{2.19}$$

Substituting (2.19) into (2.18), (2.18) can be rewritten as

$$d\mathcal{P} = 2\left(\frac{3l^2\,(2r_s - l)\,dl}{4r_s^4} - \mathcal{P}\right)\frac{3dr_s}{r_s}, \tag{2.20}$$

which, by mathematical manipulations, is further rewritten as

$$r_s^6 d\mathcal{P} + 6r_s^5\mathcal{P}dr_s = \left(9l^2r_s^2 - \frac{9}{2}l^3r_s\right)dldr_s. \tag{2.21}$$

Integrating both sides with respect to r_s, we obtain

$$\mathcal{P}r_s^6 = \int\left(9l^2r_s^2 - \frac{9}{2}l^3r_s\right)dldr_s = 9l^2dl\left(\frac{1}{3}r_s^3 - \frac{1}{4}lr_s^2\right) + c_1, \tag{2.22}$$

where c_1 is a constant to be determined.

We note that in the case $r_s = \frac{l}{2}$, both UAVs are uniquely located at the two ends of a diameter, and both on the surface of the sphere. Given the continuous nature of $\mathcal{P}$, the probability of $r_s = \frac{l}{2}$ is 0, i.e., $\mathcal{P} = 0$. By substituting this into (2.22), $c_1 = \frac{3}{16}l^5dl$, and the PDF of l can be given by

$$f_{L_1}(l) = \frac{3l^2}{r_s^3} - \frac{9l^3}{4r_s^4} + \frac{3l^5}{16r_s^6}, \quad 0 \le l \le 2r_s, \tag{2.23}$$

The average distance between the UAVs can be obtained as

$$\mathbb{E}[L_1] = \int_0^{2r_s} l \cdot f_{L_1}(l)\; dl = \frac{36}{35}r_s. \tag{2.24}$$

Based on Jensen's inequality and the aforementioned convexity of $C_{\text{erg1}}(l)$, as discussed in Section 2.2.2.1, the lower bound of the ergodic capacity can be established as

$$\mathbb{E}\left[C_{\text{erg1}}(l)\right] \ge C_{\text{erg1}}(\mathbb{E}[L_1]) = \frac{1}{\ln(2)}\left[\ln\left(1 + \frac{P}{\sigma^2}\left(\frac{36}{35}r_s\right)^{-\alpha}\right) - \frac{(2K+1)}{2(1+K)^2\left(1 + \frac{\sigma^2}{P}\left(\frac{35}{36}r_s\right)^{\alpha}\right)^2}\right] \triangleq C^*_{\text{erg1}}. \tag{2.25}$$

This concludes the proof of (2.15).

We proceed to prove (2.16). To do this, it is important to first evaluate the convexity and concavity of $\tilde{Q}(l)$. According to [37], the values of $\nu(\sqrt{2K})$ and $\mu(\sqrt{2K})$ in (2.12) can be obtained for the Rician factor of interest, i.e., $1 \leq K \leq 10$:

1. In the case of $1 \leq K \leq 10$, $\nu(\sqrt{2K})$ and $\mu(\sqrt{2K})$ are given by [37]

$$\begin{aligned}\mu(\sqrt{2K}) &= 2.174 - 0.592\sqrt{2K} + 0.593(\sqrt{2K})^2 \\ &\quad - 0.092(\sqrt{2K})^3 + 0.005(\sqrt{2K})^4,\end{aligned} \tag{2.26}$$

$$\begin{aligned}\nu(\sqrt{2K}) &= -0.840 + 0.372\sqrt{2K} - 0.74(\sqrt{2K})^2 \\ &\quad + 0.083(\sqrt{2K})^3 - 0.004(\sqrt{2K})^4.\end{aligned} \tag{2.27}$$

2. In the case of $K = 0$, $\mu_0 = 2$ and $\nu_0 = -\ln 2$.

In both cases, we have $\nu(\sqrt{2K}) < -1$ and $\mu(\sqrt{2K}) \geq 2$. The convexity and concavity of $\tilde{Q}(l)$ can be judiciously evaluated. This is critical to the development of the bounds of the outage capacity with the use of Jensen's inequality. To evaluate the convexity and concavity of $\tilde{Q}(l)$, the second-order derivative of $\tilde{Q}(l)$ is given by

$$\frac{d^2\tilde{Q}(l)}{dl^2} = \left[(\tau_1\tau_2\tau_3)^2 l^{2\tau_3-2} - \tau_1\tau_2\tau_3(\tau_3 - 1)l^{\tau_3-2}\right] e^{-\tau_1\tau_2 l^{\tau_3}}, \tag{2.28}$$

where, for notation simplicity, $\tau_1 = e^{\nu(\sqrt{2K})}$, $\tau_2 = \left(\frac{2\sigma^2\gamma_{\mathrm{th}}(1+K)}{P}\right)^{\frac{1}{2}\mu(\sqrt{2K})}$, and $\tau_3 = \frac{1}{2}\alpha\mu\left(\sqrt{2K}\right)$.

Clearly, the sign of $\frac{d^2\tilde{Q}(l)}{dl^2}$ solely depends on $\left[(\tau_1\tau_2\tau_3)^2 l^{2\tau_3-2} - \tau_1\tau_2\tau_3(\tau_3 - 1)l^{\tau_3-2}\right]$. In the case that $l < \sqrt[\tau_3]{\frac{\tau_3 - 1}{\tau_1\tau_2\tau_3}} \triangleq l_{\mathrm{th}}$, $\frac{d^2\tilde{Q}(l)}{dl^2} < 0$ and $\tilde{Q}(l)$ is concave with respect to l. Therefore, if $\mathbb{E}[L_1] < l_{\mathrm{th}}$, the upper bound for the expectation of the outage capacity, denoted by $\mathbb{E}[C_{\mathrm{out1}}]$, can be obtained by taking the expectation of (2.13), as given by

$$\begin{aligned}\mathbb{E}[C_{\mathrm{out1}}] &\approx \mathbb{E}[\tilde{Q}(l)] \cdot \log_2(1+\gamma_{\mathrm{th}}) \\ &\leq \tilde{Q}\left(\mathbb{E}[L_1]\right) \cdot \log_2(1+\gamma_{\mathrm{th}}) \\ &= \log_2(1+\gamma_{\mathrm{th}}) \exp\left[-e^{\nu(\sqrt{2K})}\left(\frac{2\sigma^2\gamma_{\mathrm{th}}(1+K)\left(\frac{36}{35}r_s\right)^\alpha}{P}\right)^{\frac{1}{2}\mu\left(\sqrt{2K}\right)}\right]. \\ &\triangleq C^*_{\mathrm{out1}}, \text{ if } \mathbb{E}[L_1] < l_{\mathrm{th}},\end{aligned} \tag{2.29}$$

where the inequality is due to the exploitation of the Jensen's inequality and the concavity of $\tilde{Q}(l)$, and the last equality is achieved by first plugging (2.12) and then substituting $\mathbb{E}[L_1] = \frac{36}{35}r_s$ from (2.24).

In the case that $l \geq l_{\text{th}}$, $\frac{d^2\tilde{Q}(l)}{dl^2} \geq 0$ and $\tilde{Q}(l)$ is convex with respect to l. Therefore, if $\mathbb{E}[L_1] \geq l_{\text{th}}$, the upper bound for $\mathbb{E}\left[C_{\text{out1}}\right]$ can be obtained as

$$\mathbb{E}\left[C_{\text{out1}}\right] \geq C^*_{\text{out1}}, \text{ if } \mathbb{E}[L_1] \geq l_{\text{th}}. \tag{2.30}$$

This concludes the proof of (2.16).

2.3.2 U2U Links with Random 2D Trajectories at Fixed Altitudes

The above analysis of random 3D trajectories with smooth turns is general, and can be extended to different interesting scenarios. One of the interesting scenarios is that two UAVs fly random 2D trajectories with smooth turns in a 2D sphere (or in other words, a disk) with radius r_s at a fixed altitude, as illustrated in Figure 2.4. This 2D scenario is of interest to surveillance and monitoring applications. By extending Theorem 2.1, the following corollary can be obtained for the 2D scenario.

Corollary 2.1 *In the case that two UAVs fly random trajectories with smooth turns in a 2D spherical region with Rician fading, the lower bound for the ergodic capacity between the UAVs can be established as*

$$C^*_{\text{erg2}} = \frac{1}{\ln(2)}\left[\ln\left(1+\frac{P}{\sigma^2}\left(\frac{128}{45\pi}r_s\right)^{-\alpha}\right) - \frac{(2K+1)}{2(1+K)^2\left(1+\frac{\sigma^2}{P}\left(\frac{45\pi}{128}r_s\right)^{\alpha}\right)^2}\right]. \tag{2.31}$$

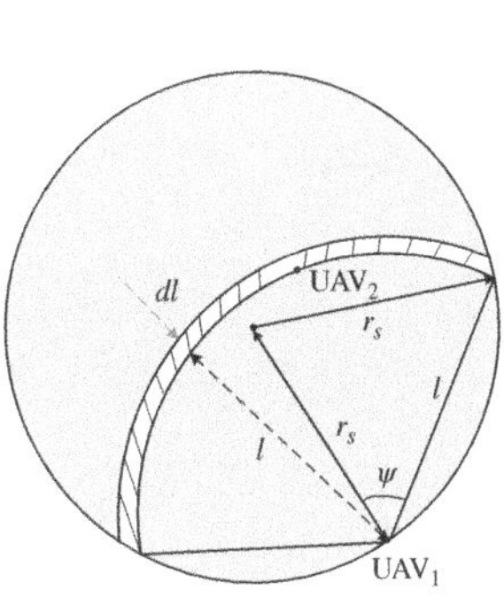

Figure 2.4 Two UAVs in a 2D disk (Special case: UAV 1 is located on the boundary of the disk).

and the outage capacity between the UAVs satisfies

$$\begin{cases} C_{\text{out2}} \geq C^*_{\text{out2}}, & \mathbb{E}[L_2] < l_{\text{th}} \\ C_{\text{out2}} \leq C^*_{\text{out2}}, & \mathbb{E}[L_2] \geq l_{\text{th}}, \end{cases} \tag{2.32}$$

where

$$C^*_{\text{out2}} = \log_2(1+\gamma_{\text{th}}) \exp\left[-e^{\nu(\sqrt{2K})} \left(\frac{2\sigma^2 \gamma_{\text{th}}(1+K)\left(\frac{128}{45\pi} r_s\right)^{\alpha}}{P} \right)^{\frac{1}{2}\mu(\sqrt{2K})} \right]. \tag{2.33}$$

Proof. Let L_2 denote the distance between the pair of UAVs. With reference to (2.18), the PDF of L_2, denoted by $f_{L_2}(l)$, can be obtained by exploiting the Crofton fixed points theorem. Let $\mathcal{P}'$ denote the probability that the two UAVs are separated by the distance l, and $\mathcal{P}_2$ denote the same probability that one of the UAVs is on the boundary of the disk. By referring to (2.18), we can obtain

$$d\mathcal{P}' = 2(\mathcal{P}_2 - \mathcal{P}')|S|^{-1} d|S|, \tag{2.34}$$

where $|S|$ is the area of the disk, i.e., $|S| = \pi {r_s}^2$ and $d|S| = 2\pi r_s dr_s$. UAV 1 is located on the boundary of the disk, and UAV 2 is located on the part of an arch centered at UAV 1, with radius l, and inside the disk, as illustrated by the shaded part in Figure 2.4. We use dl to denote the thickness of the arch. The area of the arch is $2\psi l dl$. $\mathcal{P}_2$ can be obtained as

$$\mathcal{P}_2 = \frac{2l dl \arccos\left(\frac{l}{2r_s}\right)}{\pi {r_s}^2}. \tag{2.35}$$

By referring to the evaluation of $f_{L_1}(l)$, the PDF of L_2 can be given by

$$f_{L_2}(l) = \frac{2l}{r_s^2}\left(\frac{2}{\pi}\cos^{-1}\left(\frac{l}{2r_s}\right) - \frac{l}{\pi r_s}\sqrt{1-\frac{l^2}{4r_s^2}}\right), 0 \leq l \leq 2r_s. \tag{2.36}$$

From (2.36), the expectation of L_3 can be evaluated by

$$\begin{aligned} \mathbb{E}[L_2] &= \int_0^{2r_s} l \cdot f_{L_2}(l) dl \\ &= \int_0^{2r_s} \frac{2l^2}{{r_s}^2}\left(\frac{2}{\pi}\cos^{-1}\left(\frac{l}{2r_s}\right) - \frac{l}{\pi r_s}\sqrt{1-\frac{l^2}{4{r_s}^2}}\right) dl \\ &= \frac{128}{45\pi} r_s. \end{aligned} \tag{2.37}$$

Given $\mathbb{E}[L_2]$, the rest of this proof can follow the proof of Theorem 2.1, and be suppressed here.

2.3.3 U2G Link

Another interesting scenario is that one UAV flies a random 3D trajectory with smooth turns, and communicates with a static ground station. This scenario can be analyzed by extending Theorem 2.1 such that the ground station is fixed on the surface of the aforementioned 3D sphere while the UAV flies within the sphere.

Different from the analysis of a pair of UAVs, the altitude of the UAV can have a strong impact on the propagation characteristics of the U2G link, since the LoS path condition, and the environment between the UAV and the ground station, can alter as the elevation angle ϕ, $0 \leq \phi \leq \frac{\pi}{2}$. In other words, the Rician factor K can vary as a function of ϕ, i.e., $K = K(\phi)$. According to [21, 23], the Rician factor can be modeled as a nonincreasing function of ϕ. A larger value of $\phi \in [0, \frac{\pi}{2}]$ leads to a higher LoS contribution and less multipath scattering at the receiver, resulting in a larger Rician factor K. For the typically range of the Ricain factor K, $1 \leq K \leq 10$, the Rician factor can be modeled as [21]

$$K(\phi) = \kappa_0 \cdot \exp\left[\frac{2}{\pi} \ln\left(\frac{\kappa_{\frac{\pi}{2}}}{\kappa_0}\right)\phi\right], \tag{2.38}$$

where $\kappa_0 = K(0) = 1$ and $\kappa_{\frac{\pi}{2}} = K\left(\frac{\pi}{2}\right) = 10$ are environment- and frequency-dependent parameters.

Given the uniform stationary distribution of the UAV in the 3D sphere, the expectation of K can then be computed as

$$\begin{aligned} \bar{K} &= \mathbb{E}[K(\phi)] \\ &= \int_0^{\frac{\pi}{2}} \phi\kappa_0 \exp\left[\frac{2}{\pi} \ln\left(\frac{\kappa_{\frac{\pi}{2}}}{\kappa_0}\right)\phi\right] d\phi \\ &= \frac{\pi^2\kappa_{\frac{\pi}{2}}}{4\ln\left(\frac{\kappa_{\frac{\pi}{2}}}{\kappa_0}\right)}\left[1 - \frac{1}{\ln\left(\frac{\kappa_{\frac{\pi}{2}}}{\kappa_0}\right)} + \frac{\kappa_0}{\kappa_{\frac{\pi}{2}} \ln\left(\frac{\kappa_{\frac{\pi}{2}}}{\kappa_0}\right)}\right]. \end{aligned} \tag{2.39}$$

By incorporating this elevation angle-dependent Rician fading model into the analysis of Theorem 2.1, the following corollary can be established.

Corollary 2.2 *In the case that a UAV flies a random 3D trajectory with smooth turns in a Rician fading channel, the lower bound of the ergodic capacity between the UAV and a static ground station can be established as*

$$C^*_{\text{erg3}} = \frac{1}{\ln(2)} \left[\ln\left(1 + \frac{P}{\sigma^2}\left(\frac{6}{5}r_s\right)^{-\alpha}\right) - \frac{(2\bar{K}+1)}{2(1+\bar{K})^2\left(1+\frac{\sigma^2}{P}\left(\frac{5}{6}r_s\right)^{\alpha}\right)^2} \right], \quad (2.40)$$

where $\bar{K} = \frac{\pi^2 \kappa_{\frac{\pi}{2}}}{4\ln\left(\frac{\kappa_{\frac{\pi}{2}}}{\kappa_0}\right)} \left[1 - \frac{1}{\ln\left(\frac{\kappa_{\frac{\pi}{2}}}{\kappa_0}\right)} + \frac{\kappa_0}{\kappa_{\frac{\pi}{2}} \ln\left(\frac{\kappa_{\frac{\pi}{2}}}{\kappa_0}\right)} \right]$ *is the expectation of the Rician factor* K, $\kappa_0 = 1$ *and* $\kappa_{\frac{\pi}{2}} = 10$.

The outage capacity of the link satisfies

$$\begin{cases} C_{\text{out3}} \geq C^*_{\text{out3}}, & \mathbb{E}[L_3] < l_{\text{th}} \\ C_{\text{out3}} \leq C^*_{\text{out3}}, & \mathbb{E}[L_3] \geq l_{\text{th}}, \end{cases} \quad (2.41)$$

where

$$C^*_{\text{out3}} = \log_2(1+\gamma_{\text{th}}) \cdot \frac{2}{\pi}\int_0^{\frac{\pi}{2}} \exp\left[-e^{\nu(\sqrt{2K(\phi)})} \left(\frac{2\sigma^2\gamma_{\text{th}}(1+K(\phi))\left(\frac{6}{5}r_s\right)^{\alpha}}{P} \right)^{\frac{1}{2}\mu(\sqrt{2K(\phi)})} \right] d\phi. \quad (2.42)$$

Proof. Let L_3 denote the distance between the UAV and the ground station. Based on (2.19), we can obtain the PDF of the distance L_3, as given by

$$f_{L_3}(l) = \frac{\mathcal{P}_1}{dl} = \frac{3l^2\,(2r_s - l)}{4r_s^4}, \quad 0 \leq l \leq 2r_s. \quad (2.43)$$

The expectation of L_3 is given by

$$\mathbb{E}(L_3) = \int_0^{2r_s} l f_{L_3}(l)dl = \int_0^{2r_s} \left(\frac{3l^3}{2r_s^3} - \frac{3l^4}{4r_s^4}\right) dl = \frac{6}{5}r_s. \quad (2.44)$$

With reference to the proof of Theorem 2.1, the lower bound of C_{erg3} can be obtained as

$$\begin{aligned}
\mathbb{E}_{K,L_3}\left[C_{\text{erg1}}\right] &\geq \mathbb{E}_{L_3}[C_{\text{erg1}}\,[\mathbb{E}(K)]] \\
&\geq C_{\text{erg1}}\,[\mathbb{E}(K),\mathbb{E}(L_3)] \\
&= \frac{1}{\ln(2)}\left[\ln\left(1+\frac{P}{\sigma^2}\left(\frac{6}{5}r_s\right)^{-\alpha}\right) - \frac{(2\bar{K}+1)}{2(1+\bar{K})^2\left(1+\frac{\sigma^2}{P}\left(\frac{5}{6}r_s\right)^{\alpha}\right)^2}\right] = C^*_{\text{erg3}},
\end{aligned} \tag{2.45}$$

where $\mathbb{E}_{K,L_3}[\cdot]$ takes expectation over K and L_3.

In the case that $\mathbb{E}[L_3] < l_{\text{th}}$, the lower bound for the expectation of the outage capacity, denoted by $\mathbb{E}[C_{\text{out3}}]$, can be obtained by referring to (2.29), as given by

$$\begin{aligned}
\mathbb{E}_{K,L_3}\left[C_{\text{out3}}\right] &\leq \mathbb{E}_K[\tilde{Q}_1\,(\mathbb{E}[L_3])]\cdot\log_2(1+\gamma_{\text{th}}) \\
&= \log_2(1+\gamma_{\text{th}})\frac{2}{\pi} \\
&\quad \int_0^{\frac{\pi}{2}} \exp\left[-e^{\nu(\sqrt{2K(\phi)})}\left(\frac{2\sigma^2\gamma_{\text{th}}(1+K(\phi))\left(\frac{6}{5}r_s\right)^{\alpha}}{P}\right)^{\frac{1}{2}\mu\left(\sqrt{2K(\phi)}\right)}\right]d\phi \\
&\triangleq C^*_{\text{out3}}, \text{ if } \mathbb{E}[L_3] < l_{\text{th}}.
\end{aligned} \tag{2.46}$$

In the case that $\mathbb{E}[L_3] \geq l_{\text{th}}$, the upper bound for $\mathbb{E}[C_{\text{out3}}]$ can be obtained as

$$\mathbb{E}\left[C_{\text{out3}}\right] \geq C^*_{\text{out3}}, \text{ if } \mathbb{E}[L_3] \geq l_{\text{th}}. \tag{2.47}$$

This concludes the proof of Corollary 2.2.

2.4 Network Densification and Relay

Another interesting extension of our analysis in Section 2.3 is to understand the impact of network densification on the capacity and coverage of UAVs with random 3D trajectories. Considering N UAVs with random 3D trajectories within the aforementioned 3D sphere, we are particularly interested in the case where the farthest two of the UAVs are the source and the destination, denoted by S and D, respectively; and the other UAVs act as relays, denoted by R_i, $i = 1, \ldots, N-2$.

2.4.1 Direct Link

Let l_{SD} denote the distance between the farthest pair of UAVs, i.e., S and D. l_{SD} is the longest of the distances between any pairs of the N UAVs in the 3D sphere. In the case that N is large, the cumulative distribution function (CDF) of l_{SD} can be given by [184, Theorem 1.1]

$$F_{l_{\mathrm{SD}}}(x) = \Pr\{l_{\mathrm{SD}} < x\} \to \exp\left\{-\frac{3}{4}(2r_s - x)^3 N^2\right\}. \tag{2.48}$$

The expectation of l_{SD} can be calculated as

$$\begin{aligned}
\mathbb{E}\left[l_{\mathrm{SD}}\right] &\to \int_0^{2r_s} x\, dF_{l_{\mathrm{SD}}}(x) \\
&= \int_0^{2r_s} \frac{9}{4} N^2 (2r_s - x)^2 x \exp\left[-\frac{3}{4}(2r_s - x)^3 N^2\right] dx \\
&\overset{(a)}{=} \int_0^{2r_s} \frac{9}{4}(2r_s t^2 - t^3) N^2 \exp\left(-\frac{3}{4}N^2 t^3\right) dt \\
&\overset{(b)}{=} \underbrace{\int_0^{2r_s} \frac{9}{2} N^2 r_s t^2 \exp\left(-\frac{3}{4}N^2 t^3\right) dt}_{S_1} - \underbrace{\int_0^{2r_s} \frac{9}{4} N^2 t^3 \exp\left(-\frac{3}{4}N^2 t^3\right) dt}_{S_2},
\end{aligned} \tag{2.49}$$

where (a) is obtained by setting $t = 2r_s - x$. S_1 and S_2 can be further given in closed form by

$$\begin{aligned}
S_1 &= 2r_s\left[1 - \exp(-6N^2 r_s^3)\right] \\
&\overset{(c)}{=} \int_0^{6N^2 r_s^3} \left(\frac{4}{3}\right)^{\frac{1}{3}} N^{-\frac{2}{3}} v^{\frac{1}{3}} \exp(-v) dv \\
&\overset{(d)}{=} \sqrt[3]{\frac{4}{3}} N^{-\frac{2}{3}} \gamma\left(\frac{4}{3}, 6N^2 r_s^3\right),
\end{aligned} \tag{2.50}$$

where (c) is obtained by setting $v = \frac{3}{4}N^2 t^3$; (d) is due to the identity integration $\int_0^u x^{v-1}\exp(-\varsigma x)dx = \varsigma^{-v}\gamma(v, \varsigma u)$ in [90, EH I 266(22), EH II 133(1)]; and $\gamma(a, b) = \int_0^b e^{-t} t^{a-1} dt$ is the incomplete gamma function.

Finally, we can rewrite (2.49) as

$$\mathbb{E}\left[l_{\mathrm{SD}}\right] = 2r_s\left[1 - \exp(-6N^2 r_s^3)\right] - \sqrt[3]{\frac{4}{3}} N^{-\frac{2}{3}} \gamma\left(\frac{4}{3}, 6N^2 r_s^3\right). \tag{2.51}$$

By exploiting the Jensen's inequality and the convexity of (2.10), the lower bound of the ergodic capacity of the farthest pair of UAVs, denoted by C^*_{erg4}, can be given by substituting (2.51) into (2.10).

2.4.2 Relayed Link

Based on Lemma 2.1, we have the expectation of the distance between a UAV and its nearest neighbor in the swarm, i.e., $\mathbb{E}\left[l_{\min}\right] = \Gamma\left(\frac{4}{3}\right) N^{-\frac{1}{3}} r$.

Based on the Jensen's inequality and the convexity of (2.10), the lower bound of the average multi-hop ergodic capacity between the farthest pair of UAVs, denoted by C^*_{erg5}, can be given by substituting $\mathbb{E}\left[l_{\min}\right]$ into $\frac{1}{N-1} C_{\text{erg1}}(l)$, where $C_{\text{erg1}}(l)$ is given by (2.10) and $\frac{1}{N-1}$ is due to the worst-case propagation through all $(N-2)$ relays; or in other words, $(N-1)$ hops. Decode-and-Forward relay is assumed here, due to the consideration of the fast-changing topology of the UAVs; other relay strategies, such as amplify-and-forward, typically require stable topologies and are less relevant in this case.

2.5 Instantaneous Outage Probability of UAVs Under Ground Interference

In this section, we derive the outage probability of any UAV pair in the presence of ground interference, given the distances between the UAV pair and between the pair and the ground interferer. The outage probability defines the probability that the received SINR is below a threshold γ_{th}, which is necessary for successful reception [5, 6]. Accordingly, the outage probability is given by

$$P_{\text{out}} = \Pr\left(\zeta \le \gamma_{\text{th}}\right) = \Pr\left(\frac{\gamma}{\gamma_I + 1} \le \gamma_{\text{th}}\right) \tag{2.52a}$$

$$= \Pr\left(\gamma - \gamma_I \gamma_{\text{th}} \le \gamma_{\text{th}}\right) \tag{2.52b}$$

$$= \Pr\left(Z \le \gamma_{\text{th}}\right), \tag{2.52c}$$

where (2.52c) is obtained by defining $Z \triangleq \gamma - \gamma_I \gamma_{\text{th}}$.

Theorem 2.2 *Suppose that both the U2U link and the G2U interference link experience the Rician fading with the Rician factor K and K_I, respectively. The outage probability of a UAV pair is given in* (2.53),

$$P_{\text{out}} \approx Q\left(\sqrt{\frac{2K_I(1+K)\bar{\gamma}_I\gamma_{\text{th}}}{(1+K_I)\bar{\gamma} + (1+K)\bar{\gamma}_I\gamma_{\text{th}}}}, \sqrt{\frac{2K(1+K_I)\bar{\gamma}}{(1+K_I)\bar{\gamma} + (1+K)\bar{\gamma}_I\gamma_{\text{th}}}}\right)$$
$$- \frac{(1+K_I)\bar{\gamma}}{(1+K_I)\bar{\gamma} + (1+K)\bar{\gamma}_I\gamma_{\text{th}}} \times \exp\left[-\frac{K_I(1+K)\bar{\gamma}_I\gamma_{\text{th}} + K(1+K_I)\bar{\gamma}}{(1+K_I)\bar{\gamma} + (1+K)\bar{\gamma}_I\gamma_{\text{th}}}\right]$$
$$I_0\left(\frac{2\sqrt{KK_I(1+K)(1+K_I)}}{(1+K_I)\bar{\gamma} + (1+K)\bar{\gamma}_I\gamma_{\text{th}}}\right). \tag{2.53}$$

where $Q(\sqrt{a}, \sqrt{b}) = \int_b^\infty \frac{1}{2}\exp\left(-\frac{x+a}{2}\right) I_0(\sqrt{ax})dx$ *is the first-order Marcum Q-function [21, equation 8].*

Proof. Given that h and h_I follow the Rician distribution with the respective Rician factors K and K_I, the outage probability is given in (2.54),

$$P_{\text{out}} = \Pr\{\gamma - \gamma_{\text{th}}\gamma_I \le \gamma_{\text{th}}\} = 1 - \int_0^\infty f_{\gamma_I}(y)\left[\int_{\gamma_{\text{th}}(y+1)}^\infty f_\gamma(x)dx\right]dy \quad (2.54a)$$

$$= 1 - \frac{1+K_I}{\bar{\gamma}_I}\exp(-K_I)\int_0^\infty \exp\left(-\frac{1+K_I}{\bar{\gamma}_I}y\right) I_0\left(2\sqrt{\frac{K_I(1+K_I)y}{\bar{\gamma}_I}}\right)$$

$$Q\left(\sqrt{2K}, \sqrt{\frac{2\gamma_{\text{th}}(1+K)(y+1)}{\bar{\gamma}}}\right)dy, \quad (2.54b)$$

$$\approx 1 - \frac{1+K_I}{\bar{\gamma}_I}\exp(-K_I)\int_0^\infty \exp\left(-\frac{1+K_I}{\bar{\gamma}_I}y\right) I_0\left(2\sqrt{\frac{K_I(1+K_I)y}{\bar{\gamma}_I}}\right)$$

$$Q\left(\sqrt{2K}, \sqrt{\frac{2\gamma_{\text{th}}(1+K)y}{\bar{\gamma}}}\right)dy, \quad (2.54c)$$

where

$$F_\gamma(\gamma_{\text{th}}(y+1)) = \int_0^{\gamma_{\text{th}}(y+1)} f_\gamma(x)dx$$
$$= 1 - Q\left(\sqrt{2K}, \sqrt{\frac{2\gamma_{\text{th}}(1+K)(y+1)}{\bar{\gamma}}}\right);$$

based on [213, equation 3.5], (2.54c) can be approximated as

$$P_{\text{out}} \approx Q\left(\frac{A_1 u}{\sqrt{1+u^2}}, \frac{B_1}{\sqrt{1+u^2}}\right) - \frac{1}{1+u^2}\exp\left[-\frac{A_1^2u^2 + B_1^2}{2(1+u^2)}\right] I_0\left(\frac{A_1 B_1 u}{1+u^2}\right), \quad (2.55)$$

where $A_1 = \sqrt{2K_I}$, $B_1 = \sqrt{2K}$, and $u = \sqrt{\frac{(1+K)\bar{\gamma}_I\gamma_{\text{th}}}{(1+K_I)\bar{\gamma}}}$. By substituting A_1, B_1, and u into (2.55), (2.53) can be achieved.

Remark 2.1 *Theorem 2.2 provides a closed-form expression for the outage probability in the presence of ground interference, i.e., (2.53), which needs to evaluate the first-order Marcum-Q function, and a product of an exponential function and the zeroth order modified Bessel function of the first kind. This obscures the impact of different parameter settings on the connectivity of the UAV swarm. For this reason, we proceed to develop bounds for the outage probability with much simpler expressions.*

Remark 2.2 *The Marcum Q-function is used to provide a precise and concise expression in (2.54). This is because semi-infinite integrands are required to evaluate the outage probability of a wireless link experiencing Rician fading in (2.54a). The resulting integral turns out to be a Marcum Q-function; see (2.54b). By further utilizing the properties of the Marcum Q-function, such as strict monotonicity, we are able to obtain the upper bound of the outage probability; see (2.23).*

There is unlikely to be a direct path between the UAV and ground interferer since the interference link from the ground interferer to the UAV is susceptible to the blockage from the UAV itself and the surrounding infrastructure. Therefore, the G2U channel model can be assumed to undergo Rayleigh fading, i.e., $K_I = 0$. In this case, the outage probability is presented as follows.

Corollary 2.3 *Suppose that the U2U link experiences the Rician fading with the Rician factor K, and the G2U interference channel experiences the Rayleigh fading, i.e.,* $|h_I|^2 \sim \exp(1)$ *(or in other words,* $K_I = 0$*). The outage probability of a UAV pair is given in* (2.57)*, which is upper bounded by*

$$P_{\text{out}} \leq P'_{\text{out}} + \Omega_m \exp\left[-K + K\Omega_m + \frac{\sigma^2\gamma_{\text{th}}}{\bar{\gamma}_I} - e^{\nu\left(\sqrt{2K}\right)}\left(\frac{2(1+K)\sigma^2\gamma_{\text{th}}}{\bar{\gamma}\Omega_m}\right)^{\frac{1}{2}\mu\left(\sqrt{2K}\right)}\right] \triangleq \Phi\left(l, l_I\right), \tag{2.56}$$

where $P'_{\text{out}} = 1 - Q\left(\sqrt{2K}, \sqrt{\frac{2\gamma_{\text{th}}(1+K)\sigma^2 l^\alpha}{P}}\right)$ *is the outage probability in the absence of ground interference [324, equation 5]; and* $\Omega_m = \frac{(1+K)\bar{\gamma}_I}{\bar{\gamma} + (1+K)\bar{\gamma}_I}$ *for* $0 \leq \Omega_m \leq 1$.

Proof. Given that h follows the Rician distribution with K and $|h_I|^2 \sim \exp(1)$, the PDF of Z can be written as (2.57), where $Y' = \gamma_{\text{th}}\gamma_I$; (2.57b) is obtained by setting $y_1 = y + z$; (2.57c) is obtained by setting $y_2 = Ay_1$,

$$f_Z(z) = \int_0^{\infty} f_\gamma(z+y) f_{Y'}(y) dy \tag{2.57a}$$

$$= \frac{(1+K)\exp(-K)}{\bar{\gamma}\bar{\gamma}_I\gamma_{\text{th}}} \int_z^{\infty} \exp\left(-\frac{1+K}{\bar{\gamma}}y_1\right) I_0\left(2\sqrt{\frac{K(1+K)y_1}{\bar{\gamma}}}\right)$$

$$\exp\left(-\frac{y_1 - z}{\gamma_{\text{th}}\bar{\gamma}_I}\right) dy_1 \tag{2.57b}$$

$$= \frac{B\exp\left(-K+\frac{B}{2}\right)}{2K\bar{\gamma}_I\gamma_{\text{th}}} \exp\left(\frac{z}{\gamma_{\text{th}}\bar{\gamma}_I}\right) \int_{Az}^{\infty} \frac{1}{2}\exp\left(-\frac{y_2+B}{2}\right) I_0\left(\sqrt{By_2}\right) dy_2 \tag{2.57c}$$

$$= \frac{B\exp\left(-K+\frac{B}{2}\right)}{2K\bar{\gamma}_I\gamma_{\text{th}}} \exp\left(\frac{z}{\gamma_{\text{th}}\bar{\gamma}_I}\right) Q\left(\sqrt{B}, \sqrt{Az}\right), \tag{2.57d}$$

$A = \frac{2\left[\bar{\gamma} + (1+K)\gamma_{\text{th}}\bar{\gamma}_I\right]}{\bar{\gamma}\bar{\gamma}_I\gamma_{\text{th}}}$ and $B = \frac{2K(1+K)\gamma_{\text{th}}\bar{\gamma}_I}{\bar{\gamma} + (1+K)\gamma_{\text{th}}\bar{\gamma}_I}$; and based on the definition of the Marcum Q-function, (2.57d) is obtained.

We can evaluate the outage probability. Based on [204, equation 42], the outage probability is given by

$$P_{\text{out}} = \Pr\left(Z \le \gamma_{\text{th}}\right)$$

$$= 1 - \int_{\gamma_{\text{th}}}^{\infty} f_Z(z) dz$$

$$= 1 - \int_{\gamma_{\text{th}}}^{\infty} \frac{B\exp\left(-K+\frac{B}{2}+\frac{z}{\bar{\gamma}_I\gamma_{\text{th}}}\right)}{2K\bar{\gamma}_I\gamma_{\text{th}}} Q\left(\sqrt{B}, \sqrt{Az}\right) dz$$

$$= \underbrace{1 - Q\left(\sqrt{2K}, \sqrt{\frac{2(1+K)\gamma_{\text{th}}}{\bar{\gamma}}}\right)}_{P_{\text{out}}} + \underbrace{\frac{B}{2K}\exp\left(-K+\frac{B}{2}+\frac{1}{\bar{\gamma}_I}\right) Q\left(\sqrt{B}, \sqrt{A\gamma_{\text{th}}}\right)}_{\Delta P}. \tag{2.58}$$

By substituting A and B into (2.58), (2.59) can be achieved as

$$P_{\text{out}} = 1 - Q\left(\sqrt{2K}, \sqrt{\frac{2(1+K)\gamma_{\text{th}}}{\bar{\gamma}}}\right) + \frac{(1+K)\gamma_{\text{th}}\bar{\gamma}_I}{\bar{\gamma} + (1+K)\gamma_{\text{th}}\bar{\gamma}_I} \exp\left(-\frac{K\bar{\gamma}}{\bar{\gamma} + (1+K)\bar{\gamma}_I\gamma_{\text{th}}}\right)$$
$$\times Q\left(\sqrt{\frac{2K(1+K)\bar{\gamma}_I\gamma_{\text{th}}}{\bar{\gamma} + (1+K)\bar{\gamma}_I\gamma_{\text{th}}}}, \sqrt{\frac{2\left[\bar{\gamma} + (1+K)\bar{\gamma}_I\gamma_{\text{th}}\right]}{\bar{\gamma}\bar{\gamma}_I}}\right). \tag{2.59}$$

Let $\Omega_m = \frac{(1+K)\bar{\gamma}_I\gamma_{\text{th}}}{\bar{\gamma} + (1+K)\bar{\gamma}_I\gamma_{\text{th}}}$, with $0 < \Omega_m < 1$. We can rewrite $A = \frac{2(1+K)}{\bar{\gamma}\Omega_m}$ and $B = 2K\Omega_m$. ΔP can be rewritten as

$$\Delta P = \Omega_m \exp\left(-K + K\Omega_m + \frac{1}{\bar{\gamma}_I}\right) \times Q\left(\sqrt{2K\Omega_m}, \sqrt{\frac{2(1+K)\gamma_{\text{th}}}{\bar{\gamma}\Omega_m}}\right). \tag{2.60}$$

Given the fact that the Marcum-Q function is strictly monotonic [256], we can obtain

$$Q\left(\sqrt{2K\Omega_m}, \sqrt{\frac{2(1+K)\gamma_{\text{th}}}{\bar{\gamma}\Omega_m}}\right) \leq Q\left(\sqrt{2K}, \sqrt{\frac{2(1+K)\gamma_{\text{th}}}{\bar{\gamma}\Omega_m}}\right) \tag{2.61a}$$

$$\approx \exp\left[-e^{\nu(\sqrt{2K})}\left(\frac{2(1+K)\gamma_{\text{th}}}{\bar{\gamma}\Omega_m}\right)^{\frac{1}{2}\mu(\sqrt{2K})}\right], \tag{2.61b}$$

where (2.61a) is because, for $b > 0$, $Q(\sqrt{a}, \sqrt{b})$ strictly increases with $a \in [0, \infty)$, i.e., $Q(\sqrt{a_1 + a_2}, \sqrt{b}) > Q(\sqrt{a_1}, \sqrt{b})$ for all $a_1 \geq 0$, $a_2 > 0$ and $b > 0$ [256]; and (2.61b) is obtained by substituting the approximation of Q-function, i.e., $Q\left(\sqrt{a}, \sqrt{b}\right) \approx \exp\left[-e^{\nu(\sqrt{a})} b^{\frac{1}{2}\mu(\sqrt{a})}\right]$ [37].

Since $\bar{\gamma} \geq 0$, then $\Omega_m \leq 1$ and $2K\Omega_m \leq 2K$ always hold for any $\alpha_I \geq \alpha$. By substituting (2.61) into (2.58), the upper bound of P'_{out} can be obtained as

$$P_{\text{out}} \leq P'_{\text{out}} + \Omega_m \exp\left(-K + K\Omega_m + \frac{1}{\bar{\gamma}_I}\right) \exp\left[-e^{\nu(\sqrt{2K})}\left(\frac{2(1+K)\gamma_{\text{th}}}{\bar{\gamma}\Omega_m}\right)^{\frac{1}{2}\mu(\sqrt{2K})}\right]$$
$$\triangleq \Phi\left(l, l_I\right), \tag{2.62}$$

which concludes the proof.

Corollary 2.4 *In the high SNR regime, the outage probability can be approximated by*

$$P_{\text{out}} \approx \frac{(1+K)\bar{\gamma}_I \gamma_{\text{th}}}{\bar{\gamma}+(1+K)\bar{\gamma}_I \gamma_{\text{th}}} \exp\left[-\frac{K\bar{\gamma}}{\bar{\gamma}+(1+K)\bar{\gamma}_I \gamma_{\text{th}}}\right] \triangleq \Phi_0\left(l, l_I\right). \tag{2.63}$$

Proof. This can be proved by substituting $K_I = 0$ into (2.53).

2.6 Ergodic Outage Probability of UAVs Under Ground Interference

In this section, we analyze the average ergodic outage probability of a UAV from its nearest neighbor in the presence of ground interference, where the UAVs fly random trajectories with smooth turns inside the 3D spherical region. We first consider a simple case where the ground interferer has an omni-directional antenna and the path loss for the G2U interference link is isotropic, i.e., $\mathcal{G}(\phi) = 1$ and $\alpha_I(\phi) = \alpha_I$. Then, we extend the results of the simple case to the general cases where both the directional antenna gain $\mathcal{G}(\phi)$ and the path loss $\alpha_I(\phi)$ depend on the elevation angle.

Given the expectations of the distances between the UAVs and between the UAVs and the ground interferer, we can establish the lower bounds for the average outage probability of a UAV from its nearest neighbor at any moment in the presence of the ground interference, as follows.

Theorem 2.3 *Suppose that a swarm of N UAVs fly randomly with smooth turns inside a 3D spherical region. The wireless links experience Rician fading. In the presence of ground interference, the average outage probability of a UAV from its nearest neighbor satisfies*

$$\begin{aligned}\mathbb{E}\left[\Phi_0\left(l_{\min}, l_I\right)\right] &\geq \Phi_0\left(\mathbb{E}\left[l_{\min}\right], \mathbb{E}\left[l_I\right]\right)\\ &= \frac{K' P_I \mathbb{E}[l_I]^{-\alpha_I}}{P\mathbb{E}[l_{\min}]^{-\alpha} + K' P_I \mathbb{E}[l_I]^{-\alpha_I}} \exp\left(-\frac{KP\mathbb{E}[l_{\min}]^{-\alpha}}{P\mathbb{E}[l_{\min}]^{-\alpha} + K' P_I \mathbb{E}[l_I]^{-\alpha_I}}\right),\end{aligned} \tag{2.64}$$

where $K' = (1+K)\gamma_{\text{th}}$.

Given an outage probability requirement/threshold ϵ_0, the number of UAVs in the swarm is lower bounded by

$$N \geq N_{\min} \triangleq \Gamma^3 \left(\frac{4}{3}\right) \left[\frac{PW_0(\epsilon_0)}{K'\left(K - W_0(\epsilon_0)\right) P_I \mathbb{E}[l_I]^{-\alpha_I}}\right]^{-\frac{3}{\alpha}} r^3, \tag{2.65}$$

where $W_0(\epsilon_0) = \mathcal{W}_0\left[\epsilon_0 K \exp(K)\right]$, and $\mathcal{W}_0(\cdot)$ is the principal branch of the Lambert W function. $N_{\min}$ specifies the size of an effective, uninterruptedly functioning swarm.

Proof. Based on (2.63), let $\chi = \frac{\bar{\gamma}}{\bar{\gamma}_I} = \frac{Pl^{-\alpha}}{P_I l_I^{-\alpha_I}}$, and $\Psi(m) \triangleq \frac{(1+K)\gamma_{\text{th}}}{m + (1+K)\gamma_{\text{th}}} \exp\left[-\frac{Km}{m + (1+K)\gamma_{\text{th}}}\right]$. By substituting $\bar{\gamma} = Pl^{-\alpha}/\sigma^2$ and $\bar{\gamma}_I = P_I \gamma_{\text{th}} l_I^{-\alpha_I}/\sigma^2$ into $\Psi(\bar{\gamma}, \bar{\gamma}_I)$, we can write the outage probability (2.63) as a function of l/l_I, denoted by $\Phi_0(l, l_I)$.

We find that

$$\frac{\partial \Psi(\chi)}{\partial \chi} = \left[\frac{-K'}{(\chi + K')^2} - \frac{KK'^2}{(\chi + K')^2}\right] \exp\left(-\frac{K\chi}{\chi + K'}\right) \leq 0;$$

$$\frac{\partial \Psi^2(\chi)}{\partial \chi^2} = \left[\frac{2K'}{(\chi + K')^3} + \frac{6KK'^2}{(\chi + K')^4} + \frac{K^2 K'^3}{(\chi + K')^5}\right] \exp\left(-\frac{K\chi}{\chi + K'}\right) \geq 0.$$

Therefore, $\Psi(\chi)$ is a monotonically decreasing and convex function of χ. By using Jensen's inequality, the expectation of $\Psi(\chi)$ can be lower bounded by $\mathbb{E}\left[\Psi(\chi)\right] \geq \Psi\left(\mathbb{E}\left[\chi\right]\right)$.

Since $\chi = \frac{Pl^{-\alpha}}{P_I l_I^{-\alpha_I}}$ is a function of l and l_I, and $\alpha, \alpha_I > 2$, we have

$$\frac{\partial \chi}{\partial l} = -\alpha \frac{Pl^{-\alpha-1}}{P_I l_I^{-\alpha_I}} \leq 0; \quad \frac{\partial^2 \chi}{\partial l^2} = \alpha(\alpha + 1) \frac{Pl^{-\alpha-2}}{P_I l_I^{-\alpha_I}} \geq 0;$$

$$\frac{\partial \chi}{\partial l_I} = \alpha_I \frac{Pl^{-\alpha-1}}{P_I l_I^{1-\alpha_I}} \geq 0; \quad \frac{\partial^2 \chi}{\partial l_I^2} = \alpha_I(\alpha_I - 1) \frac{Pl^{-\alpha-2}}{P_I l_I^{2-\alpha_I}} \geq 0,$$

where χ is a convex function of both l and l_I. Therefore, we have

$$\begin{aligned} \mathbb{E}\left[\chi\right] &= \mathbb{E}_{\{l, l_I\}}\left[\chi(l, l_I)\right] \\ &\geq \mathbb{E}_{l_I}\left[\chi\left(\mathbb{E}\left[l\right], l_I\right)\right] \geq \chi\left(\mathbb{E}\left[l\right], \mathbb{E}\left[l_I\right]\right), \end{aligned} \tag{2.66}$$

which is obtained by exploiting Jensen's inequality twice in succession.

By substituting $\mathbb{E}[l]$ and $\mathbb{E}[l_I]$ into $\Psi(\mathbb{E}[l], \mathbb{E}[l_I])$, the lower bound of the expectation of $\Psi(m)$ is given by

$$\mathbb{E}[\Psi(\chi)] \geq \Psi(\mathbb{E}[\chi]) \geq \frac{K'P_I\mathbb{E}[l_I]^{-\alpha_I}}{P\mathbb{E}[l]^{-\alpha} + K'P_I\mathbb{E}[l_I]^{-\alpha_I}} \times \exp\left(-\frac{KP\mathbb{E}[l]^{-\alpha}}{P\mathbb{E}[l]^{-\alpha} + K'P_I\mathbb{E}[l_I]^{-\alpha_I}}\right). \tag{2.67}$$

Consider the connectivity of the swarm, the lower bound of the average outage probability of a UAV from its nearest neighbor at every moment can be obtained by replacing $\mathbb{E}[l]$ with $\mathbb{E}[l_{\min}]$ in (2.67), resulting in (2.64). Since the minimum outage probability ϵ_0 is required to ensure the connectivity of a UAV, we have

$$\frac{K'P_I\mathbb{E}[l_I]^{-\alpha_I}}{P\mathbb{E}[l_{\min}]^{-\alpha} + K'P_I\mathbb{E}[l_I]^{-\alpha_I}} \times \exp\left(-\frac{KP\mathbb{E}[l_{\min}]^{-\alpha}}{P\mathbb{E}[l_{\min}]^{-\alpha} + K'P_I\mathbb{E}[l_I]^{-\alpha_I}}\right) \geq \epsilon_0, \tag{2.68}$$

The lower bound of the UAV number in the swarm can be obtained by exploiting the Lambert W function in (2.65).

In Theorem 2.3, the outage probability of the UAV swarm is established in the presence of ground interference, under the simplified channel model, where the ground interferer antenna is omni-directional and the path loss of the interference link is isotropic. We can generalize the theorem to a practical channel model where the ground interferer antenna is directional and the path loss of the interference link relies on the elevation angle ϕ [10, 21].

Corollary 2.5 *In the general case where the path loss of the interference link depends on the elevation angle and the ground interferer antenna is directional, the outage probability of a UAV from its nearest neighbor satisfies*

$$\begin{aligned}\mathbb{E}[\Phi_0(l_{\min}, l_I, \phi)] &\geq \Phi_1(\mathbb{E}[l_{\min}], \mathbb{E}[l_I], \bar{\mu}) \\ &= \frac{K'P_I\mathbb{E}[l_I]^{-\bar{\mu}}}{P\mathbb{E}[l_{\min}]^{-\alpha} + K'P_I\mathbb{E}[l_I]^{-\bar{\mu}}} \exp\left(-\frac{KP\mathbb{E}[l_{\min}]^{-\alpha}}{P\mathbb{E}[l_{\min}]^{-\alpha} + K'P_I\mathbb{E}[l_I]^{-\bar{\mu}}}\right),\end{aligned} \tag{2.69}$$

where $K' = (1+K)\gamma_{\text{th}}$ *and* $\bar{\mu} = \int_0^{\frac{\pi}{2}} \frac{1}{\pi}\left[\alpha_I(\phi) - \frac{\ln(\mathcal{G}(\phi))}{\ln(\mathbb{E}[l_I])}\right] d\phi$.

Proof. When considering the practical channel model, i.e., the elevation angle dependent path loss model and antenna gain, we have $\bar{\gamma}_I = P_I l_I^{-\alpha_I(\phi)} \mathcal{G}(\phi)$. Since l_I is independent of ϕ, following the process of proving Theorem 2.3, we have

$$\mathbb{E}\left[\Phi_0\left(l_{\min}, l_I, \phi\right)\right] \geq \Phi_0\left(\mathbb{E}\left[l_{\min}\right], \mathbb{E}\left[l_I\right], \phi\right) \tag{2.70a}$$

$$= \frac{K' P_I \mathbb{E}[l_I]^{-\alpha_I(\phi)} \mathcal{G}(\phi)}{P\mathbb{E}[l_{\min}]^{-\alpha} + K' P_I \mathbb{E}[l_I]^{-\alpha_I(\phi)} \mathcal{G}(\phi)}$$

$$\exp\left(-\frac{KP\mathbb{E}[l_{\min}]^{-\alpha}}{P\mathbb{E}[l_{\min}]^{-\alpha} + K' P_I \mathbb{E}[l_I]^{-\alpha_I(\phi)} \mathcal{G}(\phi)}\right) \tag{2.70b}$$

$$= \frac{K' P_I \mathbb{E}[l_I]^{-\mu(\phi)}}{P\mathbb{E}[l_{\min}]^{-\alpha} + K' P_I \mathbb{E}[l_I]^{-\mu(\phi)}}$$

$$\times \exp\left(-\frac{KP\mathbb{E}[l_{\min}]^{-\alpha}}{P\mathbb{E}[l_{\min}]^{-\alpha} + K' P_I \mathbb{E}[l_I]^{-\mu(\phi)}}\right) \tag{2.70c}$$

$$\triangleq \Phi_1\left(\mathbb{E}\left[l_{\min}\right], \mathbb{E}\left[l_I\right], \mu\right), \tag{2.70d}$$

where (2.70c) is obtained by letting $\mu(\phi) = \alpha_I(\phi) - \frac{\ln(\mathcal{G}(\phi))}{\ln(\mathbb{E}[l_I])}$, and $\mu(\phi) > 0$ since $\mathcal{G}(\phi) \leq 1$ and $\alpha_I(\phi) > 0$; and (2.70d) defines $\Phi_1\left(\mathbb{E}\left[l_{\min}\right], \mathbb{E}\left[l_I\right], \mu\right)$ by rewriting $\Phi_0\left(\mathbb{E}\left[l_{\min}\right], \mathbb{E}\left[l_I\right], \phi\right)$ as a function of μ.

Therefore, the average outage probability of a UAV from its nearest neighbor is lower bounded by

$$\mathbb{E}_\mu\left[\Phi_1\left(\mathbb{E}\left[l_{\min}\right], \mathbb{E}\left[l_I\right], \mu\right)\right] \geq \Phi_1\left(\mathbb{E}\left[l_{\min}\right], \mathbb{E}\left[l_I\right], \bar{\mu}\right), \tag{2.71}$$

where $\bar{\mu} = \mathbb{E}[\mu(\phi)] = \int_0^{\frac{\pi}{2}} \frac{1}{\pi}\left[\alpha_I(\phi) - \frac{\ln(\mathcal{G}(\phi))}{\ln(\mathbb{E}[l_I])}\right] d\phi$. Here, (2.69) can be proved by first proving that $\Phi_1\left(\mathbb{E}\left[l_{\min}\right], \mathbb{E}\left[l_I\right], \mu\right)$ is a convex function of μ, and then using Jensen's inequality. By substituting $\mathbb{E}\left[l_{\min}\right]$, $\mathbb{E}\left[l_I\right]$ and $\bar{\mu}$ into (2.71), finally (2.69) is achieved.

2.7 Ergodic Outage Probability of UAVs Without Ground Interference

Also consider an interference-free scenario for comparison, where the UAV swarm is far away from the ground (i.e., $h_I = 0$) and there is no external interference from the ground. By exploiting Lemma 2.1 and Jensen's inequality, and based on P'_{out} in Corollary 2.3, the following theorem can be established on the average outage probability of a UAV from its nearest neighbor in the swarm at any moment.

Theorem 2.4 *Suppose that a swarm of N UAVs fly randomly with smooth turns in a 3D sphere in the absence of ground interference, and the wireless channels experience Rician fading. The outage probability of a UAV from its nearest neighbor in the swarm satisfies*

$$\begin{cases} \mathbb{E}[P'_{\text{out}}] \geq P^*_{\text{out}}, \text{ if } \mathbb{E}[l_{\min}] < l_{\text{th}}; \\ \mathbb{E}[P'_{\text{out}}] \leq P^*_{\text{out}}, \text{ if } \mathbb{E}[l_{\min}] \geq l_{\text{th}}, \end{cases} \tag{2.72}$$

where $l_{\text{th}} = \sqrt[\tau_3]{\frac{\tau_3 - 1}{\tau_1 \tau_2 \tau_3}}$, $\tau_1 = e^{\nu(\sqrt{2K})}$, $\tau_2 = \left(\frac{2\gamma_{\text{th}}(1+K)}{P}\right)^{\frac{1}{2}\mu(\sqrt{2K})}$, $\tau_3 = \frac{1}{2}\alpha\mu\left(\sqrt{2K}\right)$, *and*

$$P^*_{\text{out}} = 1 - \exp\left[-e^{\nu(\sqrt{2K})} \left(\frac{2\gamma_{\text{th}}(1+K)\Gamma\left(\frac{4}{3}\right)^{\alpha} N^{-\frac{\alpha}{3}} r^{\alpha}}{P} \right)^{\frac{1}{2}\mu(\sqrt{2K})} \right]. \tag{2.73}$$

The transmit power of the UAVs that guarantees the connectivity of the UAVs is given by

$$\begin{cases} P \geq P^*_1, \text{ if } \mathbb{E}[l_{\min}] < l_{\text{th}}; \\ P \leq P^*_1, \text{ if } \mathbb{E}[l_{\min}] \geq l_{\text{th}}, \end{cases} \tag{2.74}$$

where $P^*_1 = \frac{[-e^{\nu(\sqrt{2K})} \ln(1 - P^*_{\text{out}_1})]^{\frac{2}{\mu(\sqrt{2K})}} N^{\frac{\alpha}{3}}}{2\gamma_{\text{th}}\,(1+K)\,\Gamma^{\alpha}\left(\frac{4}{3}\right) r^{\alpha}}$.

Suppose that the outage probability requirement is ϵ_0, *the number of UAVs in the swarm is*

$$N^* = \Gamma^3\left(\frac{4}{3}\right) r^3 \left[\frac{P}{2(1+K)\gamma_{\text{th}}} \left(-e^{-\nu(2K)} \ln(1 - \epsilon_0)\right)^{-\mu(2K)/2} \right]^{-\frac{3}{\alpha}}. \tag{2.75}$$

Proof. Please refer to the proof of [324, Theorem 1].

2.8 Numerical Evaluation

In this section, simulations are conducted to validate the analytical results. The random trajectories of the UAVs are generated by using the ST mobility model [268]. The transmit power of UAVs, P, is set to be 20 dBm, unless otherwise specified. The noise power, σ^2, is set to be -80 dBm. Without loss of generality, we set the path loss exponent $\alpha = 3$, as assumed in [80, 229] for small UAV systems running data collection and ferrying in civil domains. It is important to note that our analysis is not restricted to a particular value of α, and can take other values for α. Other simulation parameters are listed in Table 2.1 with reference to [80, 191, 230].

Figure 2.5 shows the ergodic capacity of the U2U links in both 2D and 3D spaces, and the U2G link in the 3D spaces, as the radii of the spherical regions, r_s, increase. Along with the analytical results dictated in Theorem 2.1, and Corollaries 2.1 and 2.2, Monte-Carlo simulation results are also plotted. We see that the analytical results coincide with the simulation results, and provide tight lower bounds for the ergodic capacity. This confirms the validity of the proposed theorem and corollaries. As expected, we can see that the ergodic capacity declines with the radius. The decline slows down, as r_s grows. The ergodic capacity of each of the scenarios

Table 2.1 Simulation parameters.

Parameter	Value
Transmit power of UAV P	20 dBm
Noise power σ^2	−80 dBm
Path loss exponent α	2.8, 3 [80, 229]
Rician factor K	0, 2, 5, 10 dB
SNR/SINR threshold	0, 5, 10 dB

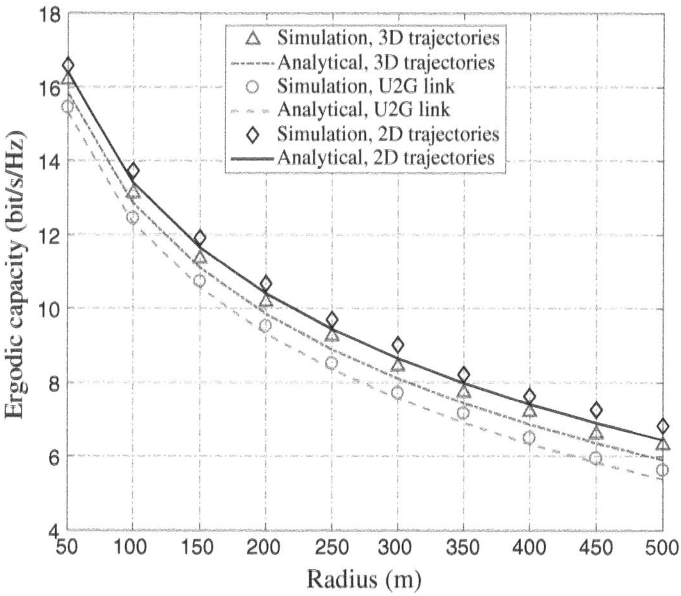

Figure 2.5 The ergodic capacity vs. radius, with the Rician factor $K = 5$ dB, $P = 0.1$ W.

asymptotically converges to 0, as $r_s \to \infty$, as can be revealed in the theorem and corollaries.

In Figure 2.5, we also see that the ergodic capacity of the U2U links is greater than that of the U2G link. One reason for this is that the U2G link suffers more scattering from the ground, resulting in larger fading. Another reason is that the average distance between the UAV and the fixed ground station, $\mathbb{E}[L_3] = \frac{6}{5}r_s$, is far longer than the distances between two UAVs with uncoordinated random trajectories, i.e., $\mathbb{E}[L_1] = \frac{36}{35}r_s$ in 3D spaces and $\mathbb{E}[L_2] = \frac{128}{45\pi}r_s$ in 2D spaces; see Section 2.3. For the same reason, the U2U link with 2D random trajectories has a larger ergodic capacity than that with 3D random trajectories, since $\mathbb{E}[L_2] > \mathbb{E}[L_1]$.

Figure 2.6 plots the outage capacity of the U2U links in both 2D and 3D spaces, and the U2G link in the 3D spaces, as the radii of the spherical regions, r_s, increase. By comparing the analytical results of Theorem 2.1, and Corollaries 2.1 and 2.2 with simulation results, the accuracy of the analysis is validated. We can see that the U2U link with 2D trajectories has the highest outage capacity, followed by the U2U link with 3D trajectories; and the outage capacity decreases, as the radius of the spherical regions grows. These are consistent with the results in Figure 2.5.

In Figure 2.6, we also see that the gaps of the outage capacity increasingly enlarge between the three different settings of links. This is because the outage

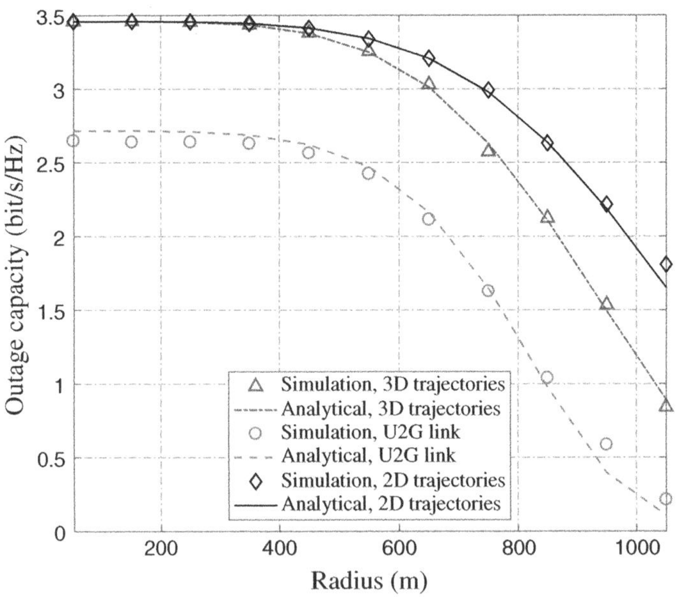

Figure 2.6 The outage capacity vs. radius, with the Rician factor $K = 5$ dB, $P = 0.1$ W, and $\gamma_{th} = 10$ dB.

capacity is defined to be the multiplicative product of the outage probability and ergodic capacity, both of which decrease with the growth of r_s. This multiplicative coupling effect can increasingly enlarge the differences between the link settings.

Figure 2.7 plots the outage capacity of the U2U links in both the 2D and 3D spaces, and the U2G link in the 3D space against the SNR threshold γ_{th}, where the Rician factor $K = 5$ dB, $P = 0.1$ W and the radius of the spherical region $r_s = 500$ m. We can see that the outage capacity of each of the link conditions increases almost linearly with the growth of γ_{th}. The difference between the U2U links with the 2D and 3D trajectories is marginal, especially for the low SNR thresholds. The difference between the U2U link and the U2G link is much larger, and increasingly enlarges with the growth of γ_{th}.

Last but not least, Figure 2.8 evaluates the impact of network densification on the capacity and coverage of the UAVs with 3D trajectories, where the multi-hop ergodic capacity and the direct ergodic capacity of the farthest pair of UAVs in the sphere, C_{erg5} and C_{erg4}, are plotted against the radius (or the coverage) of the sphere r_s. We can see C_{erg4} decreases rapidly with the growth of r_s. In contrast, the multi-hop ergodic capacity decreases slowly with the growth of r_s, and it also decreases with the growth of the number of UAVs N in the sphere. As an effective measure against network densification, the use of relay techniques can start to outperform

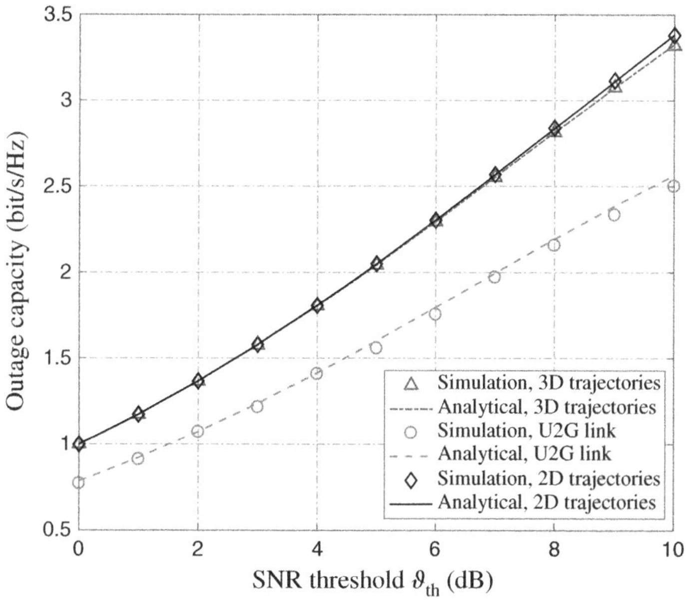

Figure 2.7 The outage capacity vs. SNR threshold, with the Rician factor $K = 5$ dB, $P = 0.1$ W, and $r_s = 500$ m.

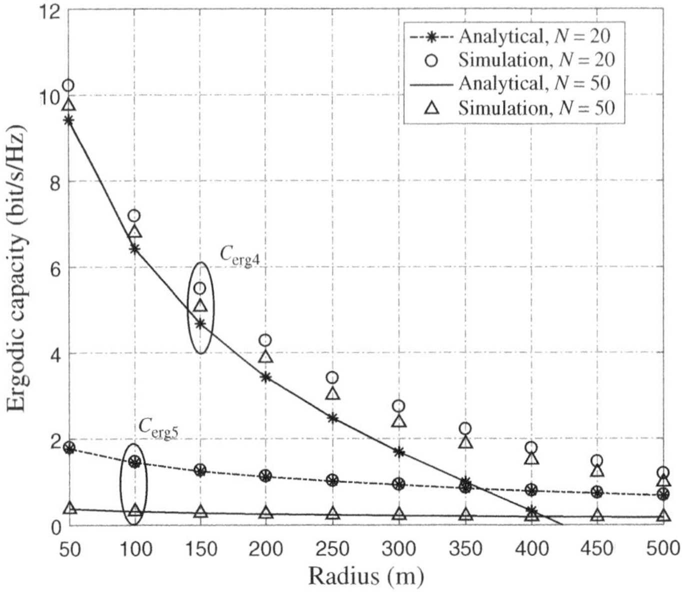

Figure 2.8 The ergodic capacity vs. SNR threshold, with the Rician factor $K = 5$ dB, $P = 0.01$ W, and $\gamma_{\text{th}} = 0$ dB.

the direct link when $r_s = 350$ m in the case of $N = 20$ and $r_s = 400$ m in the case of $N = 50$, as shown in the figure.

We also see that the analytical results of the multi-hop ergodic capacity are fairly accurate, as an extension of Theorem 2.1 whose accuracy is validated in Figure 2.5. However, the analytical results of the direct-link ergodic capacity are not tight. This is because the analysis is based on the assumption of $N \to \infty$, as discussed in Section 2.4.1.

2.9 Impact of Imperfect CSI at UAV Receiver

As discussed in [25, 226], the imperfect linear channel estimation based on a minimum mean-square error (MMSE) criterion can be performed, and channel estimation errors occur. The imperfect CSI can be modeled as [84]: $h_1 = \hat{h}_1 + h_e$, where $\hat{h}_1$ is the estimated CSI at the receiver, and h_e is a zero-mean, complex Gaussian estimation error which is independent of h_1 and has the variance ϵ. The received average SNR becomes

$$\bar{\gamma}' = \frac{PL_1{}^{-\alpha}|\hat{h}_1|^2}{PL_1{}^{-\alpha}\epsilon + \sigma^2} = \frac{P|\hat{h}_1|^2}{P\epsilon + \sigma^2 L_1{}^{\alpha}}. \tag{2.76}$$

The ergodic capacity, C_{erg6}, can be written as

$$\begin{aligned} C_{\text{erg6}} &= C_{\text{erg1}}(\bar{\gamma}') \\ &= \frac{1}{\ln(2)}\left[\ln\left(1+\frac{P}{P\epsilon+\sigma^2 L_1^{\alpha}}\right) - \frac{(2K+1)}{2(1+K)^2\left(1+\epsilon+\frac{\sigma^2}{P}L_1^{\alpha}\right)^2}\right]. \end{aligned} \tag{2.77}$$

Given the convexity of $C_{\text{erg1}}(l)$, the lower bound of the ergodic capacity can be obtained by using Jensen's inequality and (2.24), as given by

$$\begin{aligned} \mathbb{E}[C_{\text{erg6}}(L_1)] &\geq C_{\text{erg6}}(\mathbb{E}[L_1]) \\ &= \frac{1}{\ln(2)}\left[\ln\left(1+\frac{P}{P\epsilon+\sigma^2\left(\frac{36}{35}r_s\right)^{\alpha}}\right) - \frac{(2K+1)}{2(1+K)^2\left(1+\epsilon+\frac{\sigma^2}{P}\left(\frac{36}{35}r_s\right)^{\alpha}\right)^2}\right] \\ &\triangleq C^*_{\text{erg6}}. \end{aligned} \tag{2.78}$$

Extensive simulations are carried out to validate the accuracy of our analysis. Our simulation is based on the MATLAB platform. We assume that there is a swarm of N UAVs ($N = 20$ unless otherwise specified) in a 3D spherical region with radius r (e.g., 500 m), and the sphere is h_0 (e.g., h_0 ranges from 0 to 50 m) above the ground interferer. The 3D ST mobility model is adopted for each UAV. As noted in [268], the ST mobility model can be viewed as a random direction (RD) mobility model equipped with smooth turns. In our simulation, we pre-generate random trajectories, for each of which we first generate waypoints using a 3D random uniform distribution within the 3D spherical region and then connect the waypoints (e.g., by keeping linking the nearest unconnected waypoint to the end of the trajectory, as done in the RD mobility model). Then, we uniquely and geometrically specify the arc between every two successive waypoints along the trajectory on the xy-plane (as illustrated in Figure 2.9), and connect linearly the waypoints along the z-axis. The resultant trajectories have smooth turns accounting for

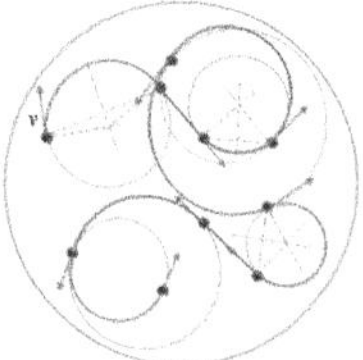

Figure 2.9 A top view of part of a 3D ST trajectory, where there are 10 waypoints marked by black dots, the trajectory is in red solid line, and the arcs between consecutive waypoints (i.e., the turn centers and radii) are uniquely determined geometrically.

aerodynamics, and randomly uniformly distributed waypoints. Figure 2.10 provides an example of the trajectories. Given the trajectories, the statistics of the turning radii and the lengths of the arcs can be geometrically analyzed, e.g., the average speed and the variation of the speed. Some examples of these parameters were provided in [297].

The transmit power of a UAV is $P = 20$ dBm, and the transmit power of the ground interferer is $P_I = 20$ dBm; unless otherwise specified. With reference to [80, 191], other system parameters are provided in Table 2.1.

Figure 2.11 shows the average outage probability of a UAV from its nearest neighbor in the swarm in the presence and absence of the ground interference, where different Rician factors K are considered. As shown in the figure, our analytical results provide lower bounds for the corresponding simulation results, with and without the ground interferences. With the increase of N, the lower bound becomes asymptotically tight. This is because $\tilde{Q}(l)$ exhibits weak curvature. The gap between the lower bound (2.64) and the actual outage probability is small, and increasingly negligible, as N grows. For small N, e.g., $N = 10$, the analytical results of (2.64) are less accurate due to the approximation. The figure shows that the average outage probability decreases with the growing number of UAVs, N, since the expectation of the distance between a UAV and its nearest neighbor in the swarm decreases as N grows. The figure also shows that the average outage probability declines as K increases. This is because the LoS path becomes increasingly dominant and reduces the outage.

Figure 2.12 plots the average outage probability of a UAV from its nearest neighbor in the swarm in both the presence and absence of the ground interference, where the number of UAVs, N, and the SNR/SINR thresholds, γ_{th}, vary. We see that the analytical result of (2.64) provides increasingly tight lower bounds for the simulation results, as N increases. We also find that the average outage probability decreases with the growth of N, since the distance between UAVs declines as N increases, and becomes increasingly tight and accurate, as γ_{th} grows.

In Figure 2.13, the average outage probabilities of the UAV swarm in both the presence and absence of the ground interference are plotted against the increasing

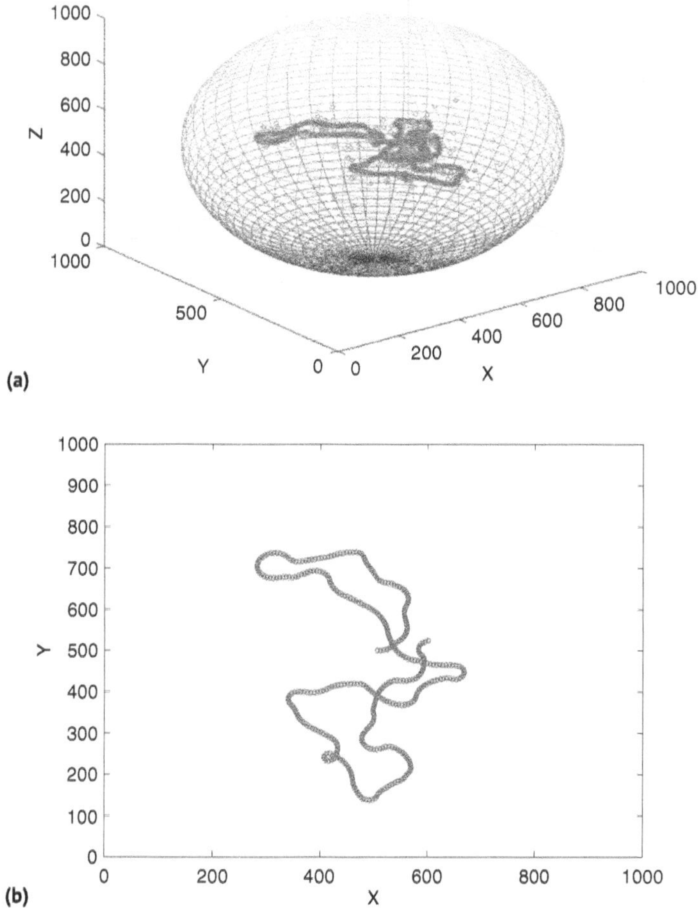

Figure 2.10 An example of a 3D ST trajectory with 3000 waypoints. (a) An example of a segment of a 3D ST trajectory of a UAV within a 3D sphere, where there are 3000 waypoints in the segment to keep the figure clear. (b) The projection of the 3D trajectory on the *xy*-plane.

transmit power of UAVs, P, where different SNR thresholds γ_{th} are considered. We see that the average outage probability decreases with the growth of P. With the growth of γ_{th}, the analytical result of (2.73) becomes increasingly accurate. We also show the transmit powers required to achieve the minimum outage probability of 8×10^{-3} under different values of γ_{th}. The maximum transmit powers required are consistent with the simulation results, as analyzed in Sections 2.6 and 2.7.

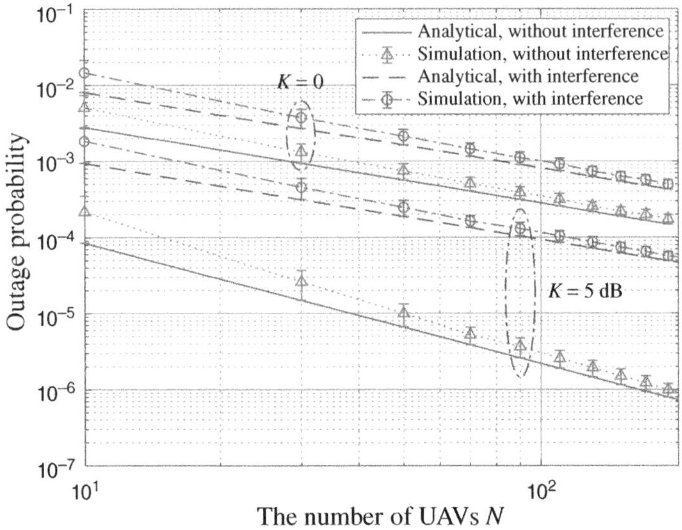

Figure 2.11 The outage probability of an arbitrary UAV vs. the number of UAVs, N, under different Rician factors, where the number of waypoints $N_m = 10,000$, $r = 500$ m, $\gamma_{\text{th}} = 5$ dB, and $\alpha_I = 3.5$.

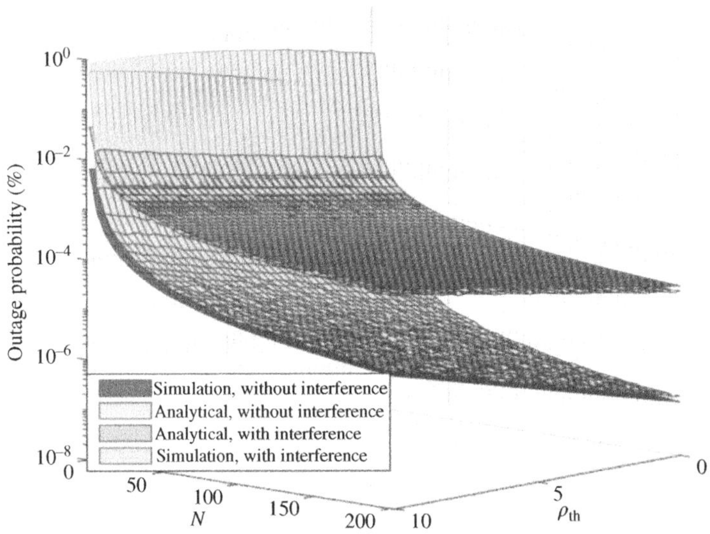

Figure 2.12 The outage probability of an arbitrary UAV vs. the number of UAVs under different SNR/SINR thresholds, where $K = K_I = 0$ and $r = 500$ m.

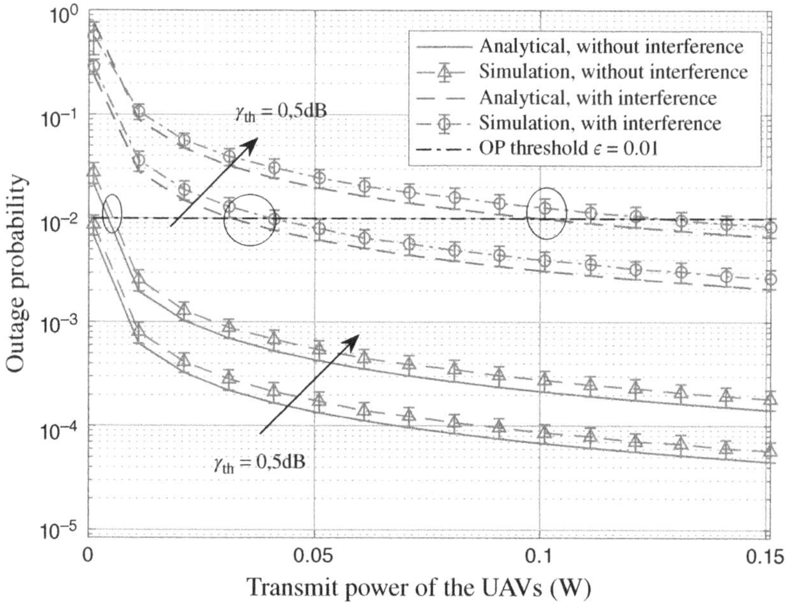

Figure 2.13 The outage probability of an arbitrary UAV vs. the transmit power of UAVs for different values of SNR threshold, where $r = 500$ m, $N = 50$, $\alpha = 2.8$, $\alpha_I = 3.0$, and the Rician factor is $K = 0$.

It is worth pointing out that the y-axis of Figure 2.13 (and Figure 2.14) is displayed in a dB scale. Despite the gap between the simulation and analytical results looking unchanged with the growing transmit power, the actual gap decreases significantly. For example, when there is ground interference and the SINR threshold γ_{th} is 5 dB, the gap decreases from 1% to 0.15% as the transmit power increases from 20 to 50 mW.

In Figure 2.14, the average outage probabilities of a UAV from its nearest neighbor in the swarm are plotted in both the presence and absence of interference from the ground, as the 3D spherical flight zone increases. The figure shows the probability under different numbers of UAVs in the swarm, i.e., N. We can see that our analytical lower bounds (2.64) and (2.73), are increasingly tight with the enlarging size of the 3D sphere (or in other words, the coverage area of the swarm). We also find that the analytical upper bound of the outage probability becomes increasingly accurate as the number of UAVs increases. To this end, the lower bounds (2.64) and (2.73) are useful to measure the connectivity of a UAV swarm over a large 3D space.

Figure 2.15 plots the average outage probability in the presence of the ground interference against the altitude of the swarm (relative to the ground interferer),

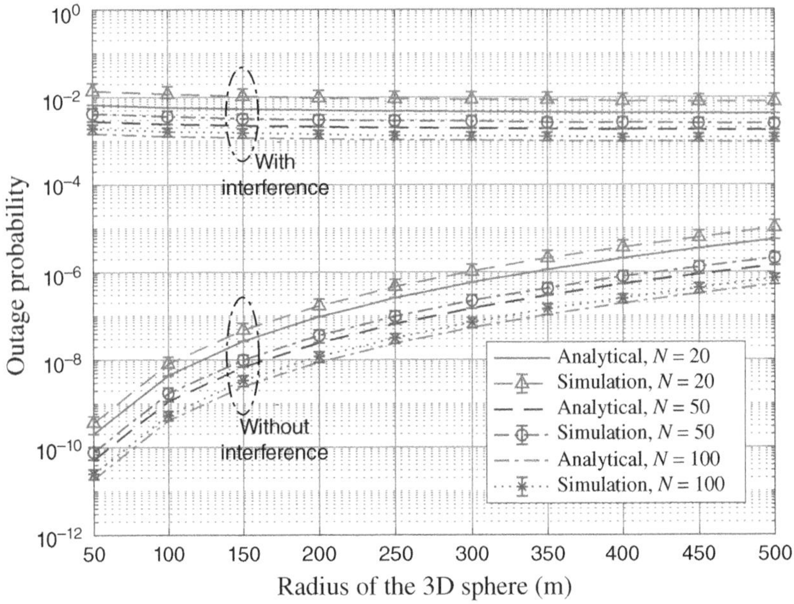

Figure 2.14 The outage probability of an arbitrary UAV vs. the radius of the sphere r, under different total numbers of UAVs, where the number of waypoints $N_m = 10,000$, the Rician factor $K = 5$ dB, $\alpha = 2.8$, $\alpha_I = 3$, $P = 0.1$ W and $\gamma_{\text{th}} = 5$ dB.

under different values of the Rician factor. We find that, with the growth of both the height h_0 of the 3D sphere from the ground interferer and the Rician factor K between the UAVs, the average outage probability decreases. For example, given $h_0 = 25$ m, the average outage probabilities are around 0.0275, 0.016, and 0.007 for $K = 0$, $K = 2$ dB, and $K = 5$ dB, respectively. Given $K = 2$ dB, the probabilities are around 0.016 and 0.015 for $h_0 = 20$ m and $h_0 = 30$ m, respectively. We also find that the average outage probability declines increasingly faster, as the Rician factor K declines. Specifically, the absolute value of the slope of the green solid line for $K = 0$ is about 0.005/45, which is more than twice that of the green solid line for $K = 5$ dB, around 0.002/45.

We note that the simulation plots in general match the analytical results in the above figures. The reason is that the 3D ST mobility model can be viewed as an RD mobility model with smooth turns [268], and the waypoints of the 3D ST model exhibit the same random uniform stationary distributions as those of the RD mobility model (which is a variation of the classical random walk model). Our analysis of the outage probability is based on the random uniform stationary distribution of waypoints. An increasing number of simulated waypoints per UAV does

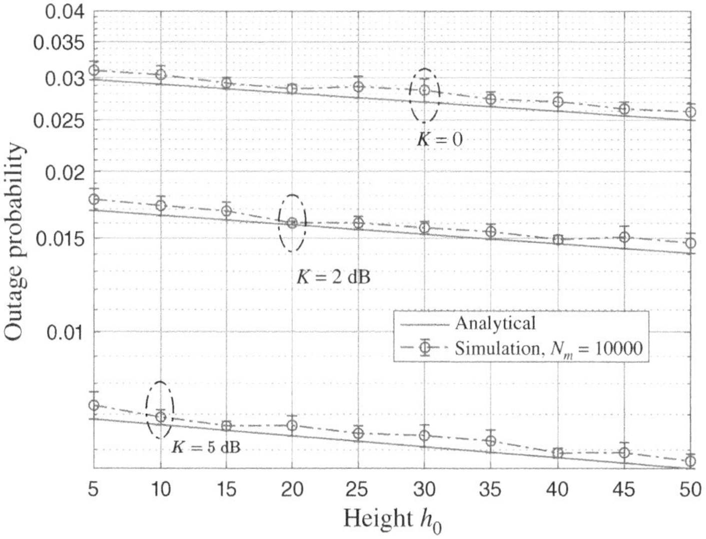

Figure 2.15 The outage probability in the presence of the ground jammer vs. the height h_0 under different Rician factor K, where $r = 500$ m, $N = 20$, $K_I = 0$, $\gamma_{\text{th}} = 5$ dB, $\alpha = 2.8$, and $\alpha_I = 3.0$.

not affect the results of the average outage probability. Nevertheless, the increasing number does help reduce the error bars (or the confidence intervals), as shown in Figures 2.11–2.15.

Figure 2.16 plots the lower bounds for the number of UAVs in the swarm $N_{\min}$ against the swarm size, i.e., the radius of the swarm r, in the presence and absence of the ground interference, under different swarm height h_0 and outage probability threshold ϵ_0. In the absence of the ground interference, $N_{\min}$ grows with both r and ϵ_0. In the presence of the ground interference, $N_{\min}$ decreases with the growth of r and ϵ_0. This is because the interference significantly diminishes the received signal power at the receiving UAV, and the decrease of the received signal power slows down with the growth of r. In consequence, $N_{\min}$ reduces. $N_{\min}$ decreases as h_0 increases. We also see that $N_{\min}$ increases first and then decreases, and finally converges to the same value achieved in the absence of the ground interference. This is because, when r is small, the height h_0 has a strong impact on the average outage probability and subsequently $N_{\min}$. With the increase of r, the impact of h_0 on the average outage probability and $N_{\min}$ declines, and eventually diminishes when r is large enough.

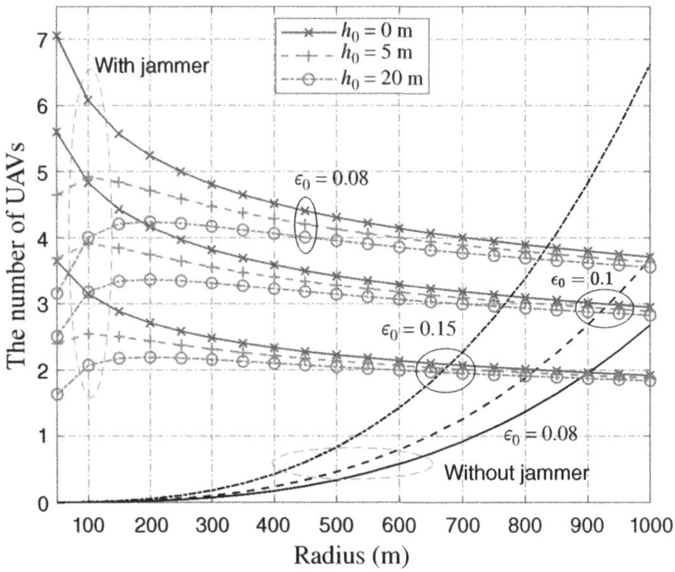

Figure 2.16 The number of UAVs in the swarm both in the presence and absence of the ground jammer vs. radius under outage probability threshold ϵ_0, where $r = 500$ m, $K = 2$ dB, $K_I = 0$, $\gamma_{th} = 5$ dB, $\alpha = 2.8$, and $\alpha_I = 3.0$.

2.10 Conclusion

In this chapter, we presented a detailed analytical study of wireless communication performance in autonomous UAV networks operating in 3D space. We derived closed-form expressions for both the link capacity and average outage probability, focusing on U2U and U2G communications under random trajectory models. The analysis incorporated key practical factors such as network densification, imperfect CSI, and external interference sources. Our findings highlight that U2U links, particularly those constrained to random 2D trajectories, exhibit higher average capacity due to shorter link distances. In contrast, U2G links experience significantly reduced capacity, even when the UAVs provide equivalent 3D coverage, due to their inherently longer link lengths. Additionally, we derived tight approximations for the outage probability of UAV swarms confined within a 3D spherical space and evaluated how interference from terrestrial sources affects their connectivity. Extensive simulations validated the accuracy of the proposed models, demonstrating their effectiveness in predicting performance bounds for UAV swarm density and wireless connectivity. Overall, this chapter underscores the importance of trajectory design, spatial deployment, and interference mitigation in enabling resilient and high-capacity communication for next-generation UAV networks.

3

Secrecy Risks Posed by Individual UAV Adversary

3.1 Introduction

Mini-drones and UAVs have attracted growing interest in civil and military applications, such as environmental monitoring, patrol and surveillance, search and rescue, and goods delivery [50, 92, 146]. UAVs have extensive applications due to their relatively small size, excellent flexibility, maneuverability, and mobility. On the other hand, UAVs are also increasingly posing threats to information security and privacy. UAVs could be used as eavesdroppers to sniff, intercept in an unauthorized manner, jam, and spoof wireless communications between legitimate terrestrial transceivers [93, 241, 247, 286].

It would be of particular interest to understand the robustness of terrestrial wireless communications under exposure to new threats from aerial adversaries, such as aerial eavesdroppers [93]. The attractive advantages or merits of UAVs, such as excellent flexibility, maneuverability, and mobility, as well as elevated positions, can potentially increase security risks that UAVs could adversely pose [247]. Despite a significant amount of research efforts on physical layer security in wireless communications, aerial adversaries have drawn little attention. The secrecy performance, captured by ergodic secrecy capacity, secrecy outage probability, or outage secrecy capacity [127, 132, 267], has yet to be understood in the presence of aerial eavesdroppers with unprecedented mobility and maneuverability in 3D spaces [93].

3.1.1 Traditional Techniques

3.1.1.1 Physical Layer Security in Terrestrial Wireless Networks

Physical layer security has been widely studied in terrestrial wireless communications, particularly over fading channels. The secrecy performance, commonly evaluated through metrics such as ergodic secrecy capacity, secrecy outage

Artificial Intelligence for Unmanned Aerial Vehicles: Sensing, Communication, and Computing, First Edition. Shuyan Hu et al.

probability, and average information leakage rate, is influenced by channel characteristics and the spatial configuration of the transmitter, receiver, and eavesdropper. In [99], three generalized secrecy metrics were proposed for quasi-static fading channels: the generalized secrecy outage probability, an asymptotic lower bound on the eavesdropper's decoding error probability, and the average information leakage rate. These were used to evaluate fixed-rate wiretap codes. Closed-form expressions for secrecy outage probability under various fading conditions were derived in [132, 133], which considered the Fisher–Snedecor $\mathcal{F}$ and cascaded α–μ fading models, respectively, in the presence of active eavesdroppers. The influence of transmission rate and random eavesdropper placement in a 2D ring region centered at the base station was analyzed in [129]. To address eavesdropper location uncertainty, the authors of [87] introduced a secrecy graph model incorporating node and link failures, while the authors of [9] studied V2V secrecy using Wyner's model under uncertain eavesdropper positions. Although these foundational works offer deep insights, they are limited to 2D settings and do not extend naturally to 3D spatial domains or dynamic mobile scenarios.

3.1.1.2 Secure UAV-enabled Wireless Communications

UAVs have emerged as flexible tools for enhancing physical layer security due to their mobility and deployment versatility. In [342], UAV mobility was leveraged for joint trajectory and power control to enhance the secrecy rate in U2G and G2U communications. UAV-based mobile jamming systems were developed in [143], while [273] proposed UAVs as mobile relays. Dual-UAV architectures with separate communication and jamming roles were investigated in [41, 140]. UAV deployment in millimeter wave (mmWave) systems was examined in [361], where a Matérn Hardcore (MHC) spatial point process was used to model UAV locations with safety constraints, and Poisson-distributed ground eavesdroppers were considered. Joint trajectory and transmit power optimization for secure U2G communications was studied in [62]. Aerial eavesdropping was considered in [315], which analyzed the secrecy outage probability and average secrecy rate in a 3D spherical aerial network under a free-space model, where selection combining (SC) was applied at the eavesdroppers. However, these UAV-based studies mainly focus on UAVs as security enablers rather than as potential aerial threats. The secrecy performance of U2U links remains underexplored.

3.1.1.3 3D Wireless Networks Without Security Considerations

Research has also addressed wireless capacity and coverage in 3D networks, though often without considering security. In [280], the link capacity and routing density were analyzed for cognitive wireless networks with uniformly distributed nodes in a 3D cube. Achievable capacity in social networks based on spatial and social parameters was studied in [281]. Optimal UAV placement for maximizing downlink coverage was derived in [193]. UAV coverage using a binomial point process at fixed altitudes was studied in [54], and [239] modeled dynamic UAV altitude control using random waypoint mobility. In our recent work [324], we derived the ergodic and outage capacities for U2U links under both 2D and 3D random mobility models, revealing that 2D motion typically leads to higher link capacity due to shorter average link distances. Despite offering important performance insights, these studies did not address physical layer security.

3.1.1.4 Diversity and Relay Techniques in Secure Terrestrial Networks

Several terrestrial studies have enhanced secrecy through diversity and cooperative relay techniques. In [309], diversity techniques were proposed to improve secrecy in fading environments. The impact of maximal ratio combining (MRC) and SC at relays in two-hop systems was studied in [668], while [28] explored relay selection strategies under Rayleigh fading and multiple eavesdroppers. Secrecy performance under cooperative relay networks with static eavesdroppers was further analyzed in [71, 243] for multi-source and multi-relay scenarios.

These approaches, while effective for static 2D terrestrial networks, often neglect the large-scale fading and line-of-sight (LoS) characteristics that dominate UAV-based communication, limiting their applicability to aerial networks. The existing literature reveals several gaps: most physical layer security studies are constrained to 2D terrestrial models and cannot directly incorporate 3D mobility or dynamic aerial environments. UAV-based security research primarily focuses on enhancing ground communication secrecy, without examining vulnerabilities in UAV-to-UAV links or aerial eavesdropping threats. Moreover, diversity and relay-based approaches lack consideration for LoS-dominant large-scale fading, which is essential in 3D UAV networks. A comprehensive investigation of secure communication among UAVs in 3D environments, considering realistic fading, mobility, and potential aerial eavesdroppers, remains an open and critical research direction.

3.1.2 Contributions

This chapter investigates the vulnerability of terrestrial wireless communications to aerial eavesdropping from an information-theoretic perspective. It presents a rigorous analysis of the ergodic secrecy rate and the average ϵ-outage secrecy rate for a ground-based transmitter-receiver pair intercepted by an aerial eavesdropper flying a random 3D trajectory with smooth turns within a spherical region, with the following key contributions:

- **Secrecy Rate Analysis in 3D Aerial Eavesdropping:** Closed-form asymptotic approximations of the ergodic and ϵ-outage secrecy rates are derived under realistic 3D spatial mobility and fading conditions. The analysis leverages the theory of almost sure (*a.s.*) convergence to decouple the effects of random trajectories from fading-induced variations.

- **Identification of Vulnerability Zones:** The study reveals the existence of critical 3D spherical regions in which the secrecy performance of the ground link collapses entirely. Within these zones, aerial eavesdroppers can overhear all ground transmissions undetected.
- **Analytical Tractability Through Expectation Decoupling:** Despite the complexity due to the coupling of spatial mobility, fading, and diversity strategies, this chapter provides tractable, closed-form approximations by separately analyzing expected link distances and conditioning secrecy performance over these expectations.

These contributions offer theoretical insights and practical guidelines for designing secure terrestrial communication systems in the presence of sophisticated, mobile aerial adversaries. The results highlight the need for new defensive mechanisms, such as UAV-restricted zones and transmitter-receiver synchronization, to safeguard critical infrastructure from high-altitude surveillance threats.

The analysis of this chapter lays a critical foundation for AI-empowered system design by providing fundamental performance bounds and vulnerability insights under realistic adversarial conditions. The study quantifies how mobility patterns and spatial uncertainty impact secure communication. These analytical results serve as benchmarks for ensuing AI-based solutions, such as adaptive beamforming, intelligent jamming, or trajectory planning, that aim to dynamically enhance security in real time. Moreover, understanding the effects of 3D mobility and coordinated attacks enables the development of AI models trained on realistic threat landscapes, improving their robustness and decision-making in complex, adversarial UAV environments.

The rest of this chapter is organized as follows. In Section 3.2, the system model is described. In Section 3.3, we first analyze the ergodic and the average ϵ-outage secrecy rates of a ground transmitter-receiver pair in the presence of an aerial eavesdropper following the ST mobility model, and then extend the analysis to the practical case, where path loss model is non-isotropic and the ground transmitter antenna is not omnidirectional. The concluding remarks are provided in Section 3.5.

3.2 System Model

Assume that an aerial eavesdropper, e.g., a UAV, flies a random 3D trajectory following the ST mobility model [268, 296]. The model decouples the movement of the UAV between the horizontal directions, i.e., along the x- and y-axes, and the vertical direction, i.e., along the z-axis [296]. On the horizontal plane, the UAV randomly chooses a turn center to circle at a constant speed until the next turn center is selected. The duration of the UAV circling a turn center is exponentially distributed. The next turn center is picked up on the line perpendicular to the instantaneous heading direction of the UAV. This ensures the smoothness of the flight trajectories. Furthermore, the ST mobility model assumes that the inverse of the radius of every turn follows a zero-mean Gaussian distribution with a small variance. In the vertical direction, i.e., along the z-axis, the UAV is assumed to maintain a constant acceleration while circling a turning center. The relevant parameters in this model can be estimated from experimental flight test data and measurements. It has been proved in [268, 296] that the ST mobility model has a uniform stationary distribution of the position of the UAV within a 3D finite space.

3.2.1 Mobility Model

The scenario of interest is that the aerial eavesdropper flies with the random ST trajectory within a 3D spherical cap centered at a ground transmitter, as illustrated in Figure 3.1. The 3D spherical cap specifies the coverage region of the ground transmitter. For illustration convenience, the transmitter, the receiver, and the aerial eavesdropper are assumed to be equipped with a single omni-directional antenna; unless otherwise specified.

The PDF of the transmitter-eavesdropper distance l_e, denoted by $f_{l_e}(\tau)$, is given by (see Appendix B.1)

$$f_{l_e}(\tau) = \frac{2(1+\cos\theta)\cos^2\theta + \sin^2\theta}{2(1+\cos\theta)\cos^2\theta + \sin^2\theta\cos^3\theta}\frac{3\tau^2}{r^3}, \quad \text{for } 0 \leq \tau \leq r\cos\theta; \tag{3.1}$$

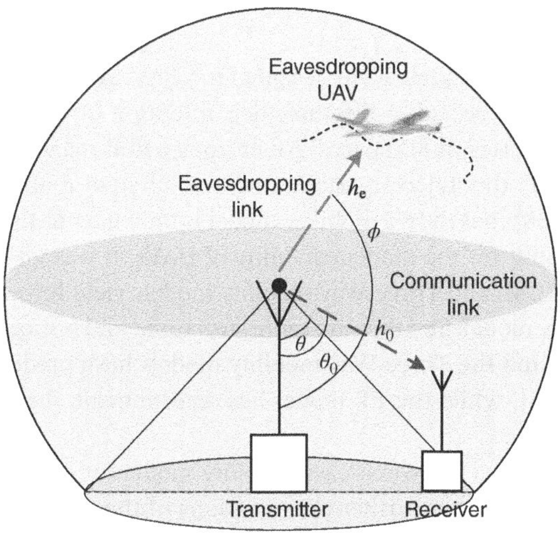

Figure 3.1 The system of interest, where there is a pair of legitimate transmitter and receiver on the ground, and an aerial eavesdropper flying within the transmission range of the transmitter. The eavesdropper follows the ST mobility model.

$$f_{l_e}(\tau) \approx \frac{2(1+\cos\theta)(1-\cos^3\theta)+\sin^2\theta\cos\theta}{2(1+\cos\theta)(1-\cos^3\theta)+\sin^2\theta\cos\theta(1-\cos^3\theta)}\frac{3\tau^2}{r^3}, \quad \text{for } r\cos\theta \le \tau \le r. \tag{3.2}$$

Accordingly, the expectation of the transmitter-eavesdropper distance, denoted by $\bar{l}_e$, is given by (see Appendix B.1)

$$\begin{aligned} \bar{l}_e &\approx \frac{3}{4}\left[1+\frac{\sin^2\theta\cos\theta}{\sin^2\theta\cos\theta+2\,(1+\cos\theta)}\right]r, \text{ if } 0<\theta<\frac{\pi}{2}; \\ \bar{l}_e &= \frac{3}{4}r, \text{ if } \theta=0 \text{ or } \theta=\frac{\pi}{2}, \end{aligned} \tag{3.3}$$

where r is the radius of the spherical cap accounting for the transmission range of the ground transmitter; and $\theta \in [0, \frac{\pi}{2}]$ is the polar angle of the spherical cap.

Apart from the ST mobility model, other existing mobility models include the Random Directional (RD) [228], Three-Way Random [137], Semi-Random Circular Movement (SRCM) [275], and Flight Plan (FP) models [262]. All the models can support a good level of mobility and maneuverability. They are random processes, except for the FP model, which is deterministic. With the consideration of aerodynamics, the SRCM, Three-Way Random, and FP models were also designed for UAVs with constraints on the turn center and radius, and trajectory of the UAV, respectively [137, 262, 268, 275, 296]. The SRCM has a predefined turn center and various radii [275]. The Three-Way Random model has constant heading speed and turn radius [137]. The FP model has predefined trajectory plans [262]. In contrast, the RD model is a generic mobility model. Although extended for UAVs, the RD model undergoes sharp turns and may not comply with aerodynamic and mechanical constraints [228].

None of the existing UAV mobility models were designed for UAV eavesdroppers. The UAV eavesdroppers are expected to disguise their intention of eavesdropping by flying random trajectories in 3D spaces. An entropy-based measure was developed in [268] to quantify the trajectory randomness of different mobility models. It was reported in [268] that the ST mobility model is only next to the RD model, which does not account for the maneuverability of UAVs. It was also reported in [268] that the FP, SRCM, and Three-Way mobility models yield lower randomness than the ST mobility model, and they are more structured and noticeable. This is because the SRCM and the Three-Way mobility models have predefined turn centers and radii [296], while the FP model has deterministic flight patterns [262], as mentioned earlier.

The ST mobility model can serve as the worst-case mobility model for practical aerial eavesdroppers, in the sense that further optimizations of the mobility patterns can improve the eavesdropping performance. Such optimizations would require both the eavesdropping rate and the furtiveness of the eavesdroppers to be taken into account, and have yet to be studied in the literature.

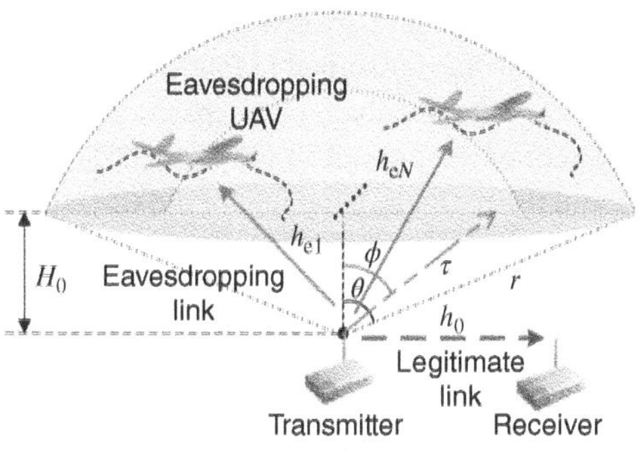

Figure 3.2 Illustration of the system model, where N aerial eavesdroppers fly random ST mobility within a spherical cap with polar angle 2θ, $\theta \in \left[0, \frac{\pi}{2}\right]$.

We also extend the above scenario to the case with a legitimate transmitter-receiver pair on the ground, and N autonomous aerial eavesdroppers flying within a 3D space, as illustrated in Figure 3.2. For illustration convenience, we first set the minimum safety elevation of the eavesdroppers, denoted by H_0, to zero. We consider that the N eavesdroppers fly within a 3D hemispherical space with the radius r and the center at the ground transmitter, since the radio signals of the ground transmitter propagate hemispherically in the 3D space. The aerial eavesdroppers attempt to eavesdrop on the transmissions from the ground transmitter. They fly independent and random trajectories to disguise their intention of eavesdropping.

There is no need for precise real-time synchronization (e.g., by handshaking) among the aerial eavesdroppers (in either time or frequency). Each aerial eavesdropper only needs to synchronize its receive frequency with the transmit frequency of the ground transmitter, e.g., by using a phase-locked loop (PLL) circuit. The received signals of the eavesdroppers can be carried back after the eavesdropping mission to the operator of the eavesdroppers, where the signals are processed offline (e.g., by running SC or MRC). The ground transmitter and receiver, and the aerial eavesdroppers are assumed to be equipped with a single omnidirectional antenna. Spatial diversity (more explicitly, macro diversity) is still captured through the use of SC or MRC eavesdropping. SC and MRC can achieve full diversity of wireless channels [39]. Macro diversity is particularly interesting, as the eavesdropping channels are typically dominated by LoS paths, and the use of multiple antennas per eavesdropper would provide little extra diversity [345].

3.2.2 Channel Model

The channels from the ground transmitter to the receiver and aerial eavesdropper consist of small-scale channel fading and path loss. The small-scale channel coefficient between the ground transmitter and receiver, denoted by h_0, is assumed to follow the Rician fading with the Rician factor $K \geq 0$ (where $K = 0$ for the

Rayleigh fading). The small-scale channel coefficient between the ground transmitter and the aerial eavesdropper, denoted by h_e, is assumed to follow the Rician fading with the Rician factor $K_e > 0$, since LoS prevails in open spaces. L and α are the distance and the path loss exponent of the ground-to-ground (G2G) link, respectively. l_e and α_e are the distance and the path loss exponent of the G2U link, respectively.

As modeled in [10, 21], the path loss from the ground transmitter to the aerial eavesdropper depends on the elevation angle in-between. Let $\phi \in \left[-\frac{\pi}{2}, \frac{\pi}{2}\right]$ denote the elevation angle. For $0 \leq \phi \leq \frac{\pi}{2}$, the path loss exponent, denoted by $\alpha_e(\phi)$, can be modeled as [21]

$$\alpha_e(\phi) = a_1 \cdot \mathcal{P}_{\text{LoS}}(\phi) + b_1, \tag{3.4}$$

where $\mathcal{P}_{\text{LoS}}(\phi) = \frac{1}{1 + a_2 \exp(-b_2\phi)}$; and the model parameters a_1, a_2, b_1, and b_2 depend on the environment and the carrier frequency [21, 10]

$$\begin{aligned} a_1 &= \frac{\alpha_{\frac{\pi}{2}} - \alpha_0}{\mathcal{P}_{\text{LoS}}\left(\frac{\pi}{2}\right) - \mathcal{P}_{\text{LoS}}(0)}; \quad b_1 = \alpha_0 - a_1\mathcal{P}_{\text{LoS}}(0); \\ a_2 &= \sum_{j=0}^{3}\sum_{i=0}^{3-j} c_{ij}^{a}(\beta_1\beta_2)^i\beta_3^j; \qquad b_2 = \sum_{j=0}^{3}\sum_{i=0}^{3-j} c_{ij}^{b}(\beta_1\beta_2)^i\beta_3^j, \end{aligned} \tag{3.5}$$

where coefficients c_{ij}^a and c_{ij}^b, $\forall\, i, j$, are given in [10, Tables I and II]; β_1, β_2, and β_3 are environment-dependent variables, and four selected urban environments, i.e., Suburban (0.1, 750, 8), Urban (0.3, 500, 15), Dense Urban (0.5, 300, 20), and High-rise Urban (0.5, 300, 50) for β_1, β_2, and β_3, respectively, are provided in [10]. We set $\alpha_0 = \alpha_e(0)$ and $\alpha_{\frac{\pi}{2}} = \alpha_e\left(\frac{\pi}{2}\right)$. For $-\frac{\pi}{2} \leq \phi < 0$, we assume $\alpha_e(\phi) = \alpha_e(0)$.

The ground transmitter may also be equipped with a directional antenna. Let $\mathcal{G}(\phi)$ denote the antenna gain of the ground transmitter in the direction of the aerial eavesdropper. It is reasonable to assume that $\mathcal{G}(\phi)$ first increases in $\left[-\frac{\pi}{2}, \theta_0\right)$ and then decreases in $\left(\theta_0, \frac{\pi}{2}\right]$. θ_0 is the pointing direction of the antenna of the ground transmitter, as depicted in Figure 3.1.

Let x_s denote the signals from the ground transmitter to the receiver. The signals at the receiver and the i-th eavesdropper, denoted by y_r and y_{e_i}, respectively, can be written as

$$y_r = \sqrt{PL^{-\alpha}}h_0x_s + n_r; \quad y_{e_i} = \sqrt{Pl_i^{-\alpha_e}}h_{e_i}x_s + n_{e_i},$$

where $i = 1, \dots, N$; x_s follows a circularly symmetric complex Gaussian distribution $\mathcal{CN}(0, 1)$, i.e., $\mathbb{E}\left[|x_s|^2\right] = 1$, $|\cdot|$ stands for norm, and $\mathbb{E}\left[\cdot\right]$ stands for expectation; P is the transmit power of the ground transmitter; n_r and n_{e_i} are the zero-mean additive white Gaussian noises (AWGN) with $\mathbb{E}\left[|n_r|^2\right] = \sigma_r^2$ and $\mathbb{E}\left[\left|n_{e_i}\right|^2\right] = \sigma_{e_i}^2 = \sigma_e^2$, respectively.

The instantaneous SNR at the ground receiver and the i-th aerial eavesdropper ($i = 1, \dots, N$) are

$$\zeta_r = Pl^{-\alpha}|h_0|^2/\sigma_r^2; \quad (3.6)$$

$$\zeta_{e_i} = Pl_i^{-\alpha_e}|h_{e_i}|^2/\sigma_e^2,\ i = 1, 2, \dots, N. \quad (3.7)$$

In a special case, $a_2 = 0$ (in turn, $\alpha_e(\phi) = \alpha_e$) and $\mathcal{G}(\phi) = 1$; in other words, both the path loss and the transmitter antenna gain are isotropic. (3.7) can be rewritten as

$$\zeta_e = Pl_e^{-\alpha_e}|h_e|^2/\sigma_e^2. \quad (3.8)$$

It is assumed that no CSI of the legitimate and eavesdropping channels is available at the transmitter, i.e., no CSI-at-the-transmitter (CSIT). This is because the aerial eavesdropper flies a random trajectory to disguise its intention of eavesdropping. Leave alone returning any CSI of the eavesdropping channel. The legitimate transmitter may not be aware of the existence of the eavesdropper, and keeps transmitting the full power P. It is also assumed that perfect CSI is available at the ground receiver and the aerial eavesdropper. Both the receiver and the eavesdropper can estimate their CSI based on pilot signals, and then detect the information signals by using coherent detection. In the rest of this chapter, the achievable ergodic and outage secrecy rates with no CSIT are analyzed, and shown to asymptotically converge to their counterparts with perfect CSIT in high SNR regimes. The 3D regions are identified within which an aerial eavesdropper with the ST mobility model can have the legitimate ground link completely exposed.

3.3 Achievable Secrecy Rate Under Aerial Eavesdropper

In this section, closed-form expressions for the (spatio-temporally) achievable ergodic and the average ϵ-outage secrecy rates of the G2G link are derived in the presence of an aerial eavesdropper following the ST mobility model. The research approach we take here is to first analyze the temporally ergodic secrecy rate and the instantaneous outage secrecy rate with respect to a specific distance of the eavesdropper to the ground transmitter l_e. Then, we apply the *a.s.* convergence to replace l_e with its expectation and derive closed-form approximations or bounds for the (spatio-temporally) ergodic secrecy rate and the average outage secrecy rate with the mobility of the aerial eavesdropper captured.

For illustration convenience, we start with a simple case where the path loss between the ground transmitter and the aerial eavesdropper is isotropic and the ground transmitter has an omnidirectional antenna (i.e., $\alpha_e(\phi) = \alpha_e$ and $\mathcal{G}(\phi) = 1$). Then, we show that the conclusions drawn in the simple case can be readily extended to the general cases, where the path loss depends on the elevation angle (as modeled in [10, 21]), and the ground transmitter antenna is not omnidirectional.

3.3.1 Ergodic Secrecy Rate

The achievable secrecy rate can be defined as [36]

$$C_{\text{erg}}^{s} = \mathbb{E}\left[C^{s}\right] = \mathbb{E}\left[\left[\log_2\left(1+\zeta_r\right) - \log_2\left(1+\zeta_e\right)\right]^{+}\right], \tag{3.9}$$

where $[x]^{+} = \max\{x, 0\}$. (3.9) measures the secrecy loss of the legitimate link to a seemingly harmless UAV to which the legitimate transmitter and receiver pay little attention.

In high SNR regimes (i.e., $\zeta_r, \zeta_e \gg 1$), the achievable ergodic secrecy rate with no CSIT can asymptotically approach that with perfect CSIT, with the growth of the SNR. This is because the legitimate transmitter and receiver, and the eavesdropper are assumed to be equipped with a single antenna, and the ergodic secrecy rate with perfect CSIT would be achieved by optimizing the transmit power P based on the perfect CSIT of both the legitimate and eavesdropping channels. When $\zeta_r, \zeta_e \gg 1$, however, $\mathbb{E}\left[\max_P \left[\log_2\left(1+\zeta_r\right) - \log_2\left(1+\zeta_e\right)\right]^{+}\right] \rightarrow \mathbb{E}\left[\left(\log_2\left(\frac{\zeta_r}{\zeta_e}\right)\right)^{+}\right] = \mathbb{E}\left[\left(\log_2\left(\frac{|h_0|^2 L^{-\alpha}/\sigma^2}{|h_e|^2 l_e^{-\alpha_e}/\sigma_e^2}\right)\right)^{+}\right]$. The ergodic secrecy rate with perfect CSIT becomes independent of P. The achievable ergodic secrecy rate with no CSIT converges to that with perfect CSIT.

Given the eavesdropper's location, we first evaluate the temporally ergodic secrecy rate with no CSIT. In Rician fading channels and an isotropic UAV path loss model, (3.9) can be given as follows.

Lemma 3.1 Suppose that the channel coefficients h_0 and h_e follow the Rician distributions with parameters K and K_e, respectively. Given L and l_e, the temporally ergodic secrecy rate of the G2G link with no CSIT can be approximated by

$$C_{\text{erg}}^{s} \approx \log_2\left(\frac{P\sigma_e^2 L^{-\alpha} + \sigma_r^2\sigma_e^2}{P\sigma_r^2 l_e^{-\alpha_e} + \sigma_r^2\sigma_e^2}\right) \triangleq C_{\text{erg}}\left(L, l_e\right). \tag{3.10}$$

If $\sigma_r^2 = \sigma_e^2 = \sigma^2$, *then* $C_{\text{erg}}^{s} \approx \log_2\left(\frac{PL^{-\alpha} + \sigma^2}{Pl_e^{-\alpha_e} + \sigma^2}\right)$.

Proof. See Appendix B.2.

In Lemma 3.1, the approximation is increasingly accurate with the *a.s.* convergence, i.e., $C_{\text{erg}}^{s} \xrightarrow{a.s.} C_{\text{erg}}(L, l_e)$, as the numbers of antennas increase at the ground receiver and the aerial eavesdropper where MRC is carried out. The proof of the lemma is based on the *a.s.* convergence, when these antenna numbers are large; see Appendix B.3. The approximation is taken here, since we set the antenna numbers to be one to focus on the impact of the eavesdropper's mobility on the confidentiality of the legitimate ground link. Lemma 3.1 can be asymptotically accurate, e.g., in future mmWave communications with increasingly large numbers of antennas at the legitimate receiver and the eavesdropper.

Given $C_{\text{erg}}(L, l_e)$, we can derive the (spatio-temporally) ergodic secrecy rate $\mathbb{E}_{l_e}[C_{\text{erg}}(L, l_e)]$ with the mobility of the aerial eavesdropper captured, by nontrivial mathematical manipulations.

Theorem 3.1 In the presence of an aerial eavesdropper with a random 3D trajectory following the ST mobility model (i.e., the location of the eavesdropper has a random uniform distribution), the (spatio-temporally) ergodic secrecy rate of the G2G link with no CSIT can be approximated by

$$\mathbb{E}_{l_e}\left[C_{\text{erg}}\left(L, l_e\right)\right] \approx C_{\text{erg}}\left(L, \bar{l}_e\right) = \log_2\left(\frac{P\sigma_e^2 L^{-\alpha} + \sigma_r^2\sigma_e^2}{P\sigma_r^2\left(\bar{l}_e\right)^{-\alpha_e} + \sigma_r^2\sigma_e^2}\right), \tag{3.11}$$

where the approximation error can asymptotically diminish with the increase of the radius of the aerial eavesdropper's flight zone, r, and the antenna numbers of the legitimate receiver and aerial eavesdropper.

Proof. See Appendix B.4.

Suppose that $\sigma_e^2 = \sigma_r^2 = \sigma^2$. The radius of the aerial eavesdropper's flight region, r_c, within which the ergodic secrecy rate is zero, is $r_c \approx \frac{2}{3}\left[\frac{\sin^2\theta\cos\theta + 2(1+\cos\theta)}{\sin^2\theta\cos\theta + \cos\theta + 1}\right] L^{\frac{\alpha}{\alpha_e}}$. If $r > r_c$, then $C_{\text{erg}}\left(L, \bar{l}_e\right) > 0$; otherwise, $C_{\text{erg}}\left(L, \bar{l}_e\right) = 0$. The G2U link between the aerial eavesdropper and the ground transmitter is likely to be an LoS path. In most cases, $\alpha_e \leq \alpha$. Under these circumstances, we have $r_c \geq \frac{2}{3}\left[\frac{\sin^2\theta\cos\theta + 2(1+\cos\theta)}{\sin^2\theta\cos\theta + \cos\theta + 1}\right] L$. As long as flying randomly in the 3D space with the radius $r < \frac{2}{3}\left[\frac{\sin^2\theta\cos\theta + 2(1+\cos\theta)}{\sin^2\theta\cos\theta + \cos\theta + 1}\right] L$, the aerial eavesdropper can overhear all the transmission of the ground transmitter.

In high SNR regimes, i.e., $P \gg \sigma^2$, the achievable ergodic secrecy rate with no CSIT depends on the distances L and $\bar{l}_e$, and path loss exponents α and α_e, and is independent of the transmit power of the ground transmitter P, as can also be shown in (3.11). As mentioned earlier, the achievable ergodic secrecy rate with no CSIT can asymptotically converge to that with perfect CSIT in the high SNR regimes.

3.3.2 Average ϵ-Outage Secrecy Rate

The secrecy outage probability defines the probability that the instantaneous secrecy rate with no CSIT is lower than a secrecy rate threshold $x > 0$, and can be evaluated [99]

$$\begin{aligned}\mathcal{P}^{s}_{\text{out}}(x) &= \Pr\left(C^{s} < x\right)\\ &= \Pr\left\{\left[\log_2\left(1+\zeta_r\right) - \log_2\left(1+\zeta_e\right)\right]^{+} < x\right\}.\end{aligned} \tag{3.12}$$

The ϵ-outage secrecy rate is the secrecy rate threshold x, denoted by C^{s}_{out}, under which the secrecy outage probability $\mathcal{P}^{s}_{\text{out}}\left(C^{s}_{\text{out}}\right) = \epsilon$. By substituting (3.6) and (3.8) into (3.12), the secrecy outage probability can be rewritten as

$$\Pr\left\{\Psi < 2^{C^{s}_{\text{out}}} - 1\right\} = \epsilon, \tag{3.13}$$

where $\Psi = \zeta_r - 2^{C^{s}_{\text{out}}}\zeta_e = \frac{PL^{-\alpha}|h_0|^2}{\sigma_r^2} - \frac{2^{C^{s}_{\text{out}}}Pl_e^{-\alpha_e}|h_e|^2}{\sigma_e^2}$.

The ground transmitter-receiver pair is likely to experience non-LoS (NLoS). The fading is typically assumed to follow the Rayleigh distribution, i.e., $K = 0$ [246]. Given a location of the aerial eavesdropper and the isotropic UAV path loss model, the ϵ-outage secrecy rate of the ground transmitter-receiver pair is given in the following lemma.

Lemma 3.2 Given a location of the eavesdropper and an outage probability ϵ, the ϵ-outage secrecy rate of the G2G link with no CSIT can be approximated by

$$\begin{aligned}C^{s}_{\text{out}} &\approx C_{\text{out}}\left(L, l_e\right)\\ &= \begin{cases}\log_2\left[\dfrac{\left(1+K_e\right)\left(K_e - W_0\right)\sigma_e^2 L^{-\alpha}}{W_0\sigma_r^2 l_e^{-\alpha_e}}\right], & \text{if } K_e > 0;\\[2ex] \log_2\left[\dfrac{\sigma_r^2\sigma_e^2 + \epsilon P\sigma_e^2 L^{-\alpha}}{\sigma_r^2\sigma_e^2 + \left(1-\epsilon\right)P\sigma_r^2 l_e^{-\alpha_e}}\right], & \text{if } K_e = 0,\end{cases}\end{aligned} \tag{3.14}$$

where $W_0 = \mathcal{W}_0\left(\frac{K_e\exp\left(K_e\right)\left(1-\epsilon\right)PL^{-\alpha}}{\sigma_r^2 + PL^{-\alpha}}\right)$, and $\mathcal{W}_0\left(\cdot\right)$ is the principal branch of the Lambert W function.[1]

Proof. See Appendix B.5.

With Lemma 3.2, we can use the similar techniques used to prove Theorem 3.1 here to achieve the average ϵ-outage secrecy rate of the G2G link in which the mobility of the aerial eavesdropper is taken into account.

---

1 There are countable branches of the Lambert W function, denoted by $\mathcal{W}_k\left(\cdot\right)$ for integer k; $\mathcal{W}_0\left(\cdot\right)$ is the principal branch.

Theorem 3.2 In the presence of an aerial eavesdropper following the ST mobility model, the average ϵ-outage secrecy rate of the G2G link with no CSIT can be approximated by

$$\mathbb{E}_{l_e}\left[C_{\mathrm{out}}\left(L, l_e\right)\right] \approx C_{\mathrm{out}}\left(L, \bar{l}_e\right) = \begin{cases} \log_2\left[\dfrac{\left(1+K_e\right)\left(K_e-W_0\right)\sigma_e^2 L^{-\alpha}}{W_0 \sigma_r^2\left(\bar{l}_e\right)^{-\alpha_e}}\right], & \text{if } K_e>0; \\ \log_2\left[\dfrac{\sigma_r^2 \sigma_e^2+\epsilon P \sigma_e^2 L^{-\alpha}}{\sigma_r^2 \sigma_e^2+(1-\epsilon) P \sigma_r^2\left(\bar{l}_e\right)^{-\alpha_e}}\right], & \text{if } K_e=0. \end{cases} \tag{3.15}$$

The approximation error of (3.15) can diminish asymptotically with the increase of the radius of the eavesdropper's flight zone r and the antenna numbers of the legitimate receiver and aerial eavesdropper in high SNR regimes.

Proof. See Appendix B.6.

Suppose that $\sigma_e^2 = \sigma_r^2 = \sigma^2$. We evaluate the radius of the eavesdropper's flight region, r_c', within which the average ϵ-outage secrecy rate is zero.

$$r_c' \approx \begin{cases} c\left[\dfrac{W_0 L^{\alpha}}{\left(1+K_e\right)\left(K_e-W_0\right)}\right]^{\frac{1}{\alpha_e}}, & \text{if } K_e>0; \\ c\left[\dfrac{(1-\epsilon) L^{\alpha}}{\epsilon}\right]^{\frac{1}{\alpha_e}}, & \text{if } K_e=0, \end{cases} \tag{3.16}$$

where $c=\frac{2}{3}\left[\frac{\sin^2\theta \cos\theta+2(1+\cos\theta)}{\sin^2\theta \cos\theta+\cos\theta+1}\right]$. If $r>r_c'$, $C_{\mathrm{out}}\left(L, \bar{l}_e\right)>0$; otherwise, $C_{\mathrm{out}}\left(L, \bar{l}_e\right)=0$.

As discussed in Section 3.3.1, in most cases, $\alpha_e \leq \alpha$ and

$$r_c' \geq r_c^* \triangleq \begin{cases} c\left[\dfrac{W_0}{\left(1+K_e\right)\left(K_e-W_0\right)}\right]^{\frac{1}{\alpha_e}} L, & \text{if } K_e>0; \\ c\left(\dfrac{1-\epsilon}{\epsilon}\right)^{\frac{1}{\alpha_e}} L, & \text{if } K_e=0. \end{cases}$$

In other words, if the aerial eavesdropper flies randomly in a 3D spherical cap with the radius $r < r_c^*$, it can overhear a $(1-\epsilon)$ ratio of the transmissions of the ground transmitter.

In high SNR regimes, i.e., $P \gg \sigma^2$, the achievable average ϵ-outage secrecy rate is independent of the transmit power of the ground transmitter P, as the ergodic secrecy rate is in Section 3.3.1. However, different from the ergodic secrecy rate, the ϵ-outage secrecy rate also depends on the Rician factor of the eavesdropping link K_e and ϵ (in addition to the distances L and $\bar{l}_e$, and the path loss exponents α and α_e on which the ergodic secrecy rate depends).

3.3.3 Ergodic and Outage Secrecy Rates Under Practical UAV Channel Model

We note that our analyses are a nontrivial extension of the typical wiretap channel, typically involving a stationary eavesdropper. In particular, Lemma 3.1 gives the temporally achievable ergodic secrecy rate of the legitimate ground link given a position of the eavesdropper, like the typical wiretap channel. Based on Lemma 3.1, Theorem 3.1 derives an asymptotic expression for the (spatio-temporally) achievable ergodic secrecy rate by translating the mobility of the eavesdropper to a random point process. Different from the typical wiretap channel, the theorem devises the upper and lower bounds for the (spatio-temporally) achievable ergodic secrecy rate and proves their asymptotic convergence with the expansion of the eavesdropper's flight zone. Likewise for Lemma 3.2 and Theorem 3.2.

Theorems 3.1 and 3.2 establish the ergodic and outage secrecy rates of the ground link, under the simplified UAV channel model where the UAV path loss model is isotropic and the ground transmitter antenna is omni-directional. The theorems can be readily extended to the practical model where the path loss depends on the elevation angle of the eavesdropper [10, 21] and the ground transmitter antenna is not omni-directional, as described in Section 3.2.

Given l_e and ϕ, the temporally ergodic secrecy rate in Lemma 3.1 is updated as

$$C^s_{\text{erg}} \approx \log_2\left(\frac{P\sigma_e^2 L^{-\alpha} + \sigma_r^2\sigma_e^2}{P\sigma_r^2 l_e^{-\alpha_e(\phi)}\mathcal{G}(\phi) + \sigma_r^2\sigma_e^2}\right) \triangleq \mathcal{C}_{\text{erg}}\left(l_e, \phi\right), \tag{3.17}$$

which can be proved by substituting $\bar{\zeta}_r = \frac{PL^{-\alpha}}{\sigma_r^2}$ and $\bar{\zeta}_e = \frac{Pl_e^{-\alpha_e(\phi)}\mathcal{G}(\phi)}{\sigma_e^2}$ into (B.6) in the proof of Lemma 3.1, and then proved in the same way as Lemma 3.1. This is due to the fact that, given ϕ, $\alpha_e(\phi)$ and $\mathcal{G}(\phi)$ are fixed and, with l_e given, $l_e^{-\alpha_e(\phi)}\mathcal{G}(\phi)$ is also fixed (as $l_e^{-\alpha_e}$ is in Lemma 3.1). We can replace $l_e^{-\alpha_e}$ with $l_e^{-\alpha_e(\phi)}\mathcal{G}(\phi)$ here, and then evaluate the effect of the Rician fading, as done in Lemma 3.1.

Since l_e is independent of ϕ, given ϕ, $\mathbb{E}_{l_e}\left[\mathcal{C}_{\text{erg}}\left(l_e, \phi\right)\right]$ complies with (3.11), and can be approximated by

$$\begin{aligned}\mathbb{E}_{l_e}\left[\mathcal{C}_{\text{erg}}\left(l_e, \phi\right)\right] &\approx \log_2\left(\frac{P\sigma_e^2 L^{-\alpha} + \sigma_r^2\sigma_e^2}{P\sigma_r^2\left(\bar{l}_e\right)^{-\alpha_e(\phi)}\mathcal{G}(\phi) + \sigma_r^2\sigma_e^2}\right)\\ &= \log_2\left(\frac{P\sigma_e^2 L^{-\alpha} + \sigma_r^2\sigma_e^2}{P\sigma_r^2\left(\bar{l}_e\right)^{-\eta(\phi)} + \sigma_r^2\sigma_e^2}\right),\end{aligned} \tag{3.18}$$

which can be proved in the same way as Theorem 3.1. $\eta(\phi) = \alpha_e(\phi) - \frac{\ln(\mathcal{G}(\phi))}{\ln(\bar{l}_e)}$ defines the effective path loss exponent with the non-isotropic path loss model and the transmitter antenna gain incorporated. $\eta(\phi) > 0$ due to $\mathcal{G}(\phi) \leq 1$ and $\alpha_e(\phi) > 0$.

Therefore, the (spatio-temporally) achievable ergodic secrecy rate can be approximated by

$$\begin{aligned}\mathbb{E}_{l_e,\phi}\left[C_{\text{erg}}(l_e,\phi)\right] &= \mathbb{E}_{\phi}\left\{\mathbb{E}_{l_e}\left[C_{\text{erg}}\left(l_e,\phi\right)\right]\right\} \\ &\approx \log_2\left(\frac{P\sigma_e^2 L^{-\alpha} + \sigma_r^2\sigma_e^2}{P\sigma_r^2\left(\bar{l}_e\right)^{-\bar{\eta}} + \sigma_r^2\sigma_e^2}\right),\end{aligned} \tag{3.19}$$

where $\bar{\eta} = \mathbb{E}[\eta(\phi)] = \int_{-\frac{\pi}{2}}^{\frac{\pi}{2}} \frac{1}{\pi}\left[\alpha_e(\phi) - \frac{\ln(\mathcal{G}(\phi))}{\ln(\bar{l}_e)}\right] d\phi$. (3.19) can be proved by first proving that $C_{\text{erg}}(l_e,\phi)$ is a decreasing and convex function of $\eta(\phi)$ and an increasing and concave function of $1/\eta(\phi)$, and then following the proof of Theorem 3.1.

Likewise, the average ϵ-outage secrecy rate of the G2G link is approximated by

$$\mathbb{E}_{l_e,\phi}\left[C_{\text{out}}(l_e,\phi)\right] \approx \begin{cases} \log_2\left[\dfrac{(1+K_e)(K_e - W_0)\,\sigma_e^2 L^{-\alpha}}{W_0\sigma_r^2\left(\bar{l}_e\right)^{-\bar{\eta}}}\right], & \text{if } K_e > 0; \\ \log_2\left[\dfrac{\sigma_r^2\sigma_e^2 + \epsilon P\sigma_e^2 L^{-\alpha}}{\sigma_r^2\sigma_e^2 + (1-\epsilon)\,P\sigma_r^2\left(\bar{l}_e\right)^{-\bar{\eta}}}\right], & \text{if } K_e = 0. \end{cases} \tag{3.20}$$

3.4 Numerical and Simulation Results

In this section, we conduct extensive simulations to validate our analysis for the secrecy rates of the ground transmission in the presence of an aerial eavesdropper with a 3D trajectory. Without loss of generality, we assume the same receiver noise powers at the ground receiver and the aerial eavesdropper. $\theta = \frac{\pi}{2}$; unless otherwise specified. Other parameters are given in Table 3.1 with reference to [80, 191].

We note that only the expectation of the transmitter-eavesdropper distance is needed to evaluate the ergodic and ϵ-outage secrecy rates in Theorems 3.1 and 3.2; see (3.11) and (3.15). We refer to the scenario captured in the theorems (see Figure 3.2) as Scenario 1. Theorems 3.1 and 3.2 can be readily applied to evaluate another scenario (Scenario 2), where the eavesdropper attempts to eavesdrop on

Table 3.1 Simulation parameters.

Parameter	Value
Transmit power of the ground transmitter P	20 dBm
Noise power $\sigma_r^2 = \sigma_e^2$	−120 dBm
Path loss exponent α	3.0
Path loss exponent α_e	2.8 [80]
Rician factor K_e	0, 10 dB

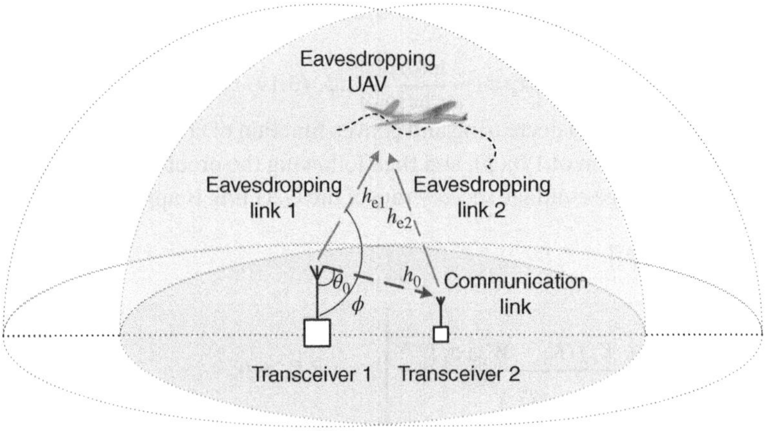

Figure 3.3 The scenario of a bidirectional ground link, where the aerial eavesdropper flies within the overlapping coverage region of both the legitimate ground nodes.

a bidirectional ground link. The eavesdropper flies the ST mobility model within the overlapping coverage region of the two transceivers, as depicted in Figure 3.3. Assume that the two transceivers are identical and their heights are comparatively negligible to their coverage. The expectation of the transmitter-eavesdropper distance can be numerically evaluated by (see Appendix B.2)

$$\bar{l}_e^{(2)} = \frac{1}{2}\left[\int_{\frac{L}{2}}^{r} \frac{48\tau^3 - 24L\tau^2}{16r^3 - 12Lr^2 + L^3} d\tau \right.$$
$$\left. + \int_{\frac{L}{2}}^{r} \int_{0}^{\cos^{-1}\left(\frac{L}{2\tau}\right)} \frac{\sqrt{L^2 + \tau^2 - 2L\tau \cos x}\left(48\tau^2 - 24L\tau\right)}{\cos^{-1}\left(\frac{L}{2\tau}\right)(16r^3 - 12Lr^2 + L^3)} dx d\tau \right], \tag{3.21}$$

which can be substituted into (3.11) and (3.15) to evaluate the secrecy rate of the bidirectional link.

Figure 3.4 plots the ergodic secrecy rate in both Scenarios 1 and 2. Along with the analytical results provided in Theorem 1, Monte-Carlo simulation results (i.e., $\mathbb{E}\left[C_{\text{erg}}^{s}\right]$) are provided. We simulate the ST mobility model of the eavesdropper for the considered volume and evaluate the expected value of the expression in (3.9), i.e., $\mathbb{E}\left[C_{\text{erg}}^{s}\right]$. The number of samples used in the simulations is 5000. We can see that the analytical results coincide with the simulation results, and provide accurate approximations to the ergodic secrecy rate. We also see that the ergodic

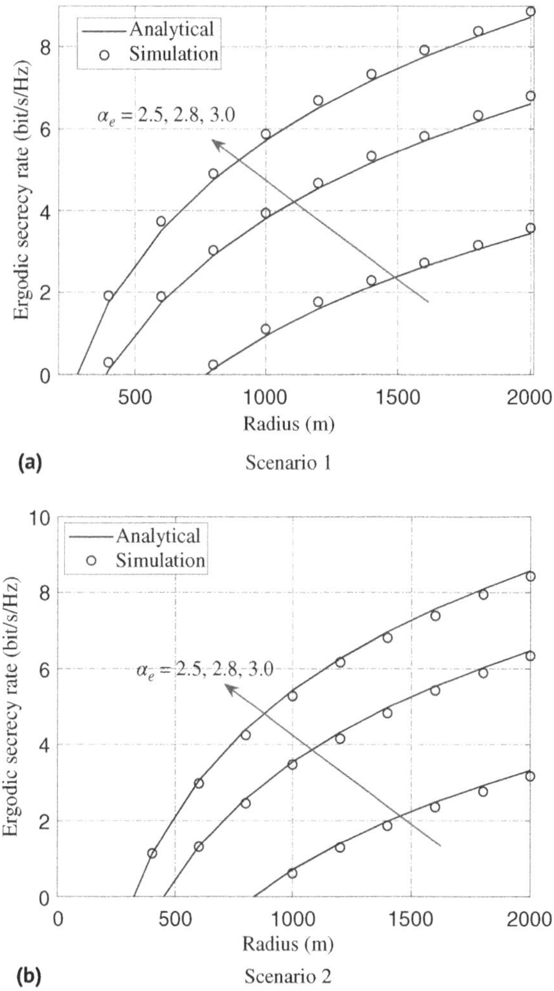

Figure 3.4 The ergodic secrecy rate vs. the radius of the eavesdropper's flight region, r, under different values of path loss, α_e, where $L = 200$ m, and $\alpha = 3$.

secrecy rate increases with r. The ergodic secrecy rate also increases with α_e, while the threshold of r, under which the ergodic secrecy rate approaches zero, decreases with the growth of α_e. In other words, the ground transmission is vulnerable to aerial eavesdropping, especially for the reason that the eavesdropper can implement wiretapping at a distance without being easily noticed or identified.

Figure 3.5 plots the ergodic secrecy rate of the G2G link, as $\frac{\alpha_e}{\alpha}$ increases. We can see that the secrecy rate increases linearly with $\frac{\alpha_e}{\alpha}$, and the increase rate of

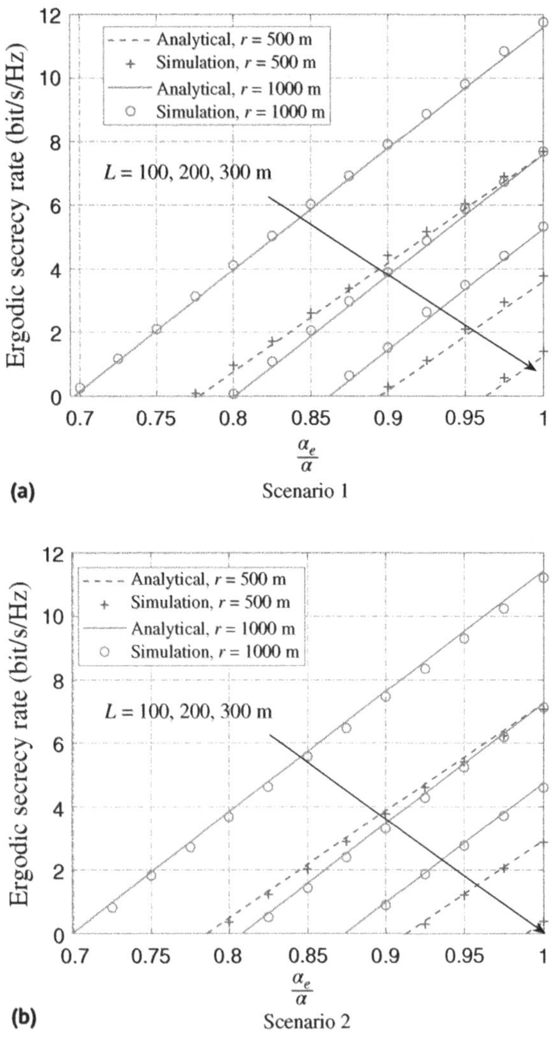

Figure 3.5 The ergodic secrecy rate vs. $\frac{\alpha_e}{\alpha}$, in the presence of an aerial eavesdropper under different values of path loss exponent, α_e, where $\alpha = 4$.

the ergodic secrecy rate remains unchanged under different distances between the ground stations, i.e., $L = 100, 200, 300$ m. Moreover, the growth rate rises slowly with the growth of r. With the growth of the distance between the ground stations, the ergodic secrecy rate decreases. We note that the simulation results in Figures 3.4 and 3.5 are for the exact ergodic secrecy rate, and the approximations developed in Theorem 3.1 provide fine accuracy, even when the legitimate receiver and the aerial eavesdropper are equipped with a single antenna.

Figure 3.6 plots the analytical and simulation results for the average ϵ-outage secrecy rate of the G2G link, as the radius of the 3D space the aerial eavesdropper flies, i.e., r, increases. Different values of the Rician factor and path loss

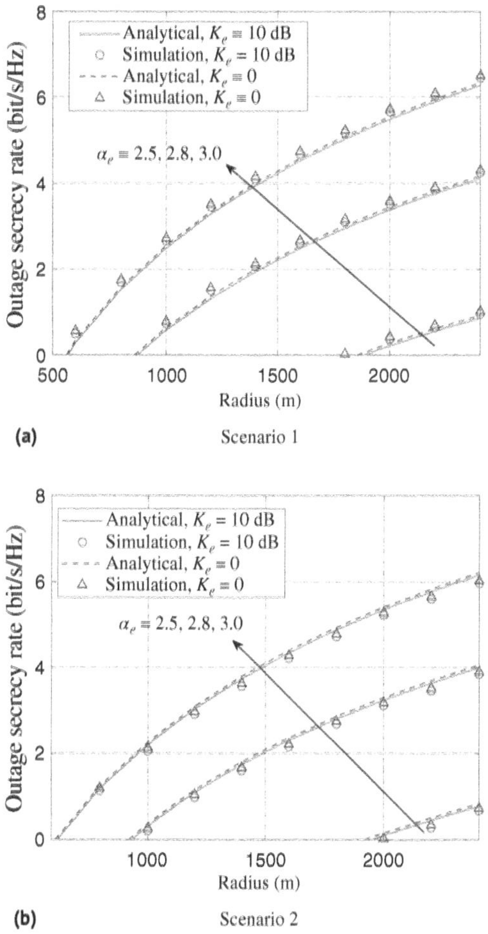

Figure 3.6 The ϵ-outage secrecy rate vs. the radius of the eavesdropper's flight region, r, under different values of both path loss exponent and Rician factors, where $\alpha = 3$, $L = 200$ m, and $\epsilon = 0.1$.

exponent are taken for the G2U eavesdropping link. The Rician factor $K_e = 0$ corresponds to the case where there is no LoS path and the channel undergoes the Rayleigh fading. We also set $K_e = 10$ dB in the case where there is an LoS path and NLoS paths between the ground transmitter and the aerial eavesdropper. The NLoS can result from the body blockage of the UAV and the reflection and scattering stemming from buildings or other surrounding infrastructures. The average ϵ-outage secrecy rate is slightly higher in the case of $K_e = 10$ dB than it is in the case of $K_e = 0$. The reason is that the LoS available between the ground transmitter and the aerial eavesdropper facilitates eavesdropping. By comparing the analytical results of Theorem 3.2, i.e., (3.15), with the simulation results, the accuracy of the analysis is validated.

From Figure 3.6, we also see that the average ϵ-outage secrecy rate increases with r. The average ϵ-outage secrecy rate decreases as the difference between the G2G path loss exponent, α, and the G2U path loss exponent, α_e, increases. The conclusion drawn is that the eavesdropping from the air can be a particularly severe issue, especially in cases when the eavesdropping link has a strong LoS path with a small path loss exponent.

Figure 3.7 plots the average ϵ-outage secrecy rate against the outage probability, ϵ, under different values of the G2U path loss exponent, α_e. We see that the average ϵ-outage secrecy rate increases with ϵ. Particularly, the larger ϵ is, the higher the secrecy outage rate is required to secure reliable communication. For different values of α_e, the average ϵ-outage secrecy rate has a similar growth rate, and increases quickly with α_e. We also find that the average ϵ-outage secrecy rate decreases with the growth of K_e.

Figure 3.8 depicts the average ϵ-outage secrecy rate against $\frac{\alpha_e}{\alpha}$, under various values of the Rician factor, K_e, and the radius of the 3D space in which the aerial eavesdropper flies, r. We can see that the average ϵ-outage secrecy rate increases linearly with the growth of $\frac{\alpha_e}{\alpha}$, and the increase rate of the secrecy rate remains unchanged at different distances between the ground stations, i.e., $L = 100, 150, 200$ m. The growth rate rises slowly with the growth of r. We also see that the average ϵ-outage secrecy rate decreases slightly with the growth of the Rician factor K_e, since the LoS path becomes increasingly dominant with the growth of K_e.

Figure 3.9 plots the ergodic and outage secrecy rates of the ground link under a non-isotropic UAV path loss model described in [10] and a typically assumed squared cosine antenna pattern [321]

$$\mathcal{G}_{\cos}(\phi) = \begin{cases} \cos^2(\phi - \theta_0), & \text{if } |\phi| \leq \frac{\pi}{2}; \\ 0, & \text{otherwise.} \end{cases}$$

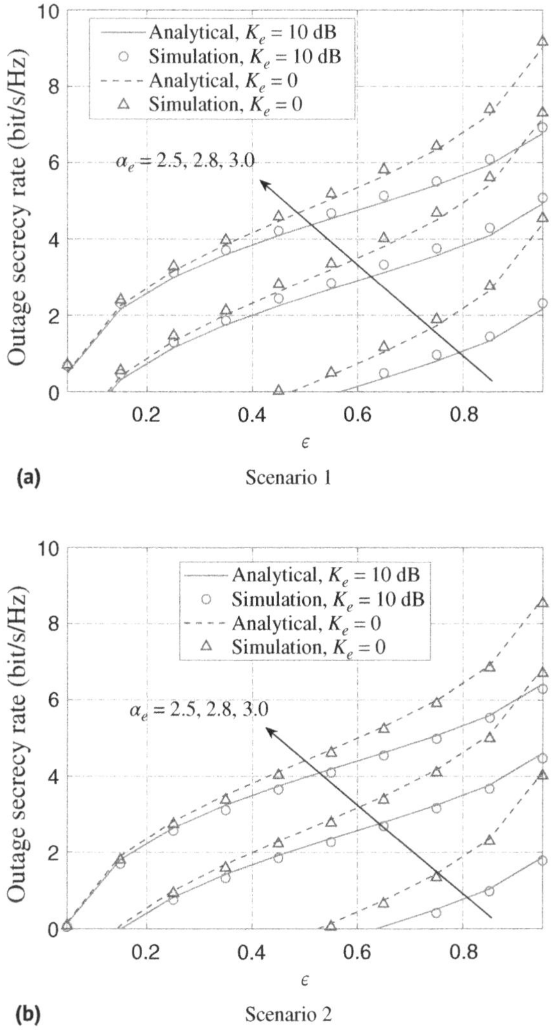

Figure 3.7 The ϵ-outage secrecy rate vs. the outage probability, ϵ, in the presence of an aerial eavesdropper flying in 3D spherical spaces, where $L = 200$ m, $r = 800$ m, and $\alpha = 3$.

Figure 3.9a shows the change of the path loss exponent α_e against the elevation angle ϕ, where the path loss exponent is $\alpha_0 = 3$ when the eavesdropper and the transmitter are at the same height. We take different values for $\alpha_{\frac{\pi}{2}}$, the path loss exponent when the eavesdropper is right above the transmitter. The smaller $\alpha_{\frac{\pi}{2}}$ is, the clearer the space is and the less influential the ground structures are on the aerial eavesdropper's channel. Figure 3.9b and c plot the ergodic and outage

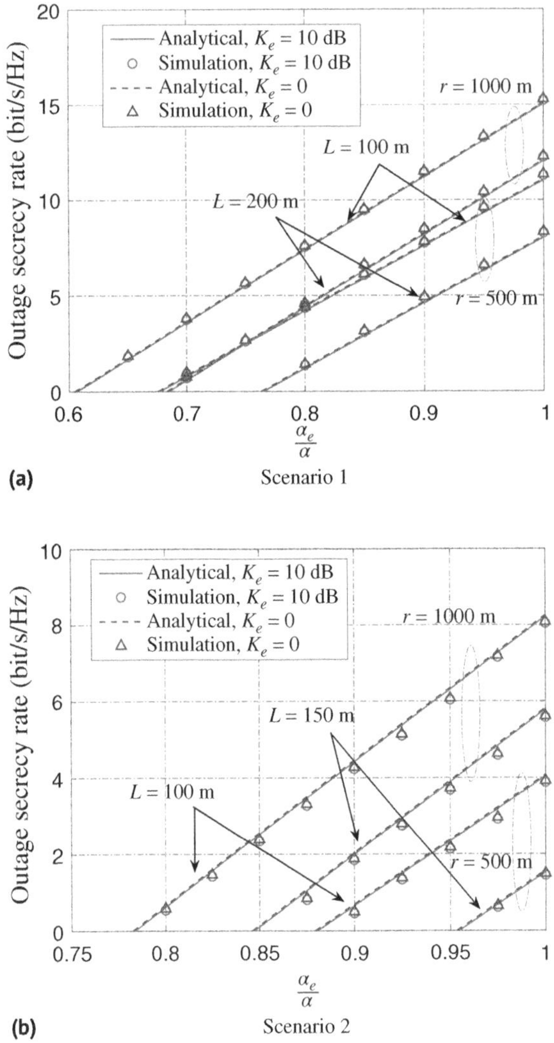

Figure 3.8 The ϵ-outage secrecy rate vs. $\frac{\alpha_e}{\alpha}$, under different values of both target secrecy rate and Rician factor.

secrecy rates under different values of $\alpha_{\frac{\pi}{2}}$. In general, the secrecy rates grow with the increase of $\alpha_{\frac{\pi}{2}}$ due to the increasing attenuation over the eavesdropping link.

Despite being proved to be asymptotically accurate with the growth of r and the antenna numbers at the receiver and eavesdropper, the approximations in Theorems 3.1 and 3.2 are shown to provide fine accuracy under isotropic UAV

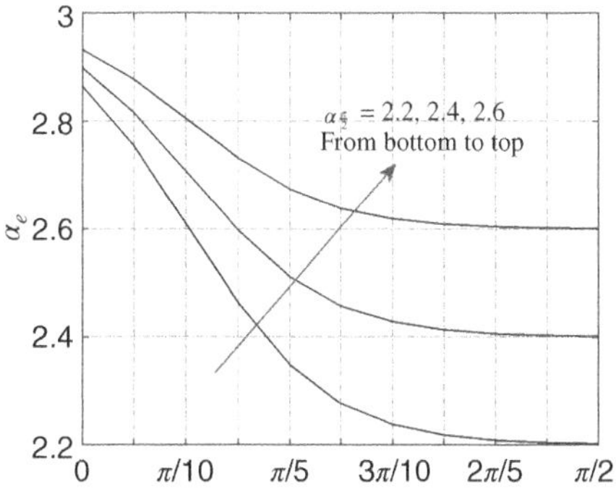

(a) α_e vs. ϕ.

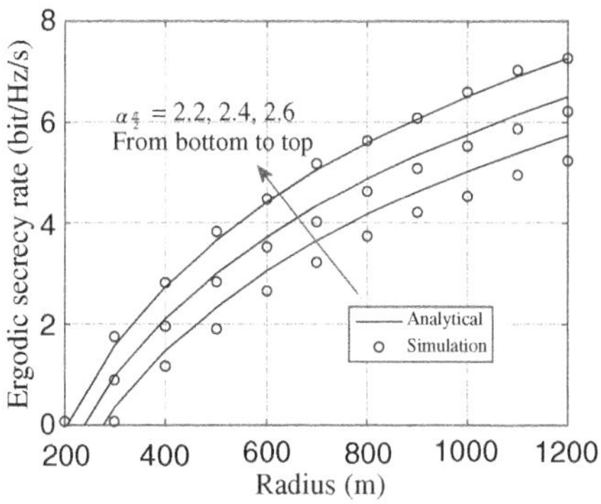

(b) Ergodic secrecy rate.

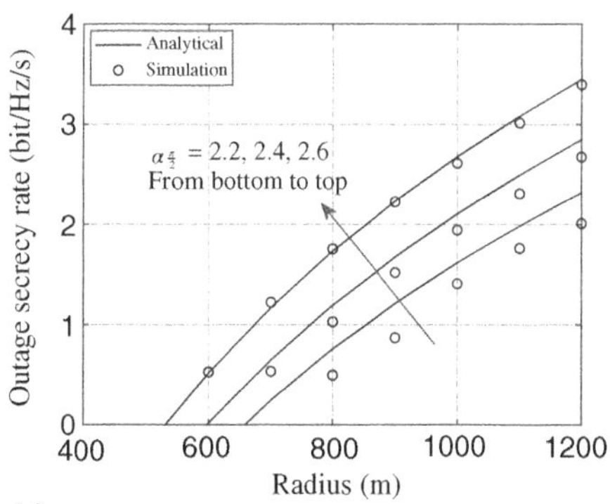

(c) Outage secrecy rate.

Figure 3.9 The secrecy rates (Scenario 1) under different $\alpha_{\frac{\pi}{2}}$ values, where $L = 200$ m, $K = 0$, $K_e = 10$ dB, and $\alpha_0 = 3$.

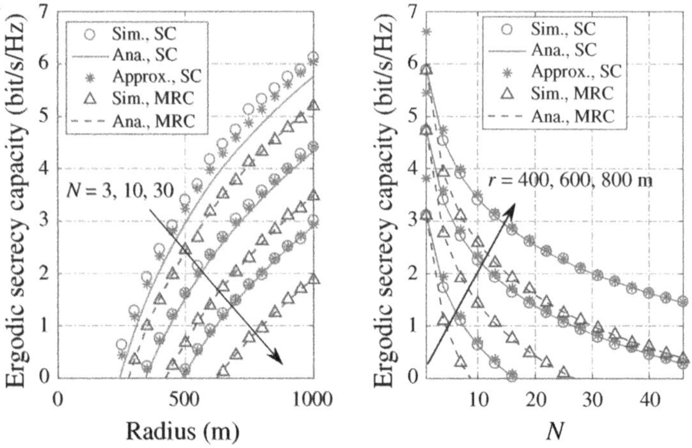

(a) Ergodic secrecy capacity vs. r. **(b)** Ergodic secrecy capacity vs. N.

Figure 3.10 Ergodic secrecy capacity of the eavesdroppers conducting SC and MRC vs. r and N, where $K_e = 0$, $\alpha = 3$, and $\alpha_e = 2.8$.

path loss models even with a single antenna at the receiver and eavesdropper in Figures 3.4–3.8. The approximations are less accurate in Figure 3.9b and c under non-isotropic UAV path loss models, due to the additional step dealing with ϕ in (3.19). Better accuracy can be observed when the difference between $\alpha_{\frac{\pi}{2}}$ and α_0 is reduced in Figure 3.9b and c. Ergodic secrecy capacities using SC or MRC under different values of N or r are shown in Figure 3.10a and b, respectively. Better approximation accuracy can be observed when N or r increases.

3.5 Conclusion

This chapter presented a comprehensive analytical framework for evaluating the secrecy performance of terrestrial wireless communications under the threat of aerial eavesdropping in 3D environments. We analyzed both the ergodic secrecy rate and the average ϵ-outage secrecy rate of a ground transmitter-receiver pair in the presence of UAVs acting as eavesdroppers following random trajectories within a bounded 3D spherical region. Closed-form expressions and asymptotic approximations were derived using advanced mathematical tools, including Jensen's inequality and the Lambert W function, and were validated through simulations. The results demonstrated that aerial eavesdropping poses a significant risk to ground communications, especially due to the ability of UAVs to perform passive eavesdropping from elevated and distant positions without being easily detected. Our analysis revealed critical 3D spatial regions where the secrecy rate of the ground link collapses, effectively rendering the transmission vulnerable.

4

Secrecy Risks Posed by Cooperative UAV Adversaries

4.1 Ergodic and ϵ-Outage Secrecy Capacity Against Cooperative Aerial Eavesdroppers

In cooperative UAV eavesdropping scenarios, multiple malicious UAVs can simultaneously intercept the transmitted signal and jointly process the received information to improve eavesdropping performance. Two commonly considered signal combining strategies at the eavesdroppers are MRC and SC.

MRC involves coherently combining the signals received at each UAV in such a way that the SNR is maximized. Each received signal is weighted proportionally to its SNR and aligned in phase before summation. This technique leverages all available spatial diversity, providing the best possible SNR at the eavesdropper and thereby significantly enhancing its interception capability. In essence, MRC allows the cooperative UAVs to act like a distributed antenna array, yielding strong improvements in eavesdropping efficiency.

In contrast, SC is a simpler yet effective approach where only the signal from the UAV with the best reception quality (i.e., highest SNR) is used for decoding, while the others are discarded. Though SC does not exploit the full spatial diversity as MRC, it requires less coordination and computational overhead, making it practical for loosely coupled or resource-constrained UAV adversaries.

Both MRC and SC amplify the threat posed by cooperative UAV eavesdroppers by increasing their ability to successfully decode intercepted messages, thus underscoring the importance of designing robust secrecy-enhancing transmission schemes. MRC eavesdropping is typically known to be more destructive than SC eavesdropping. But their difference has yet to be quantified in the literature. In this section, we analyze the secrecy capacities of the ground link in the presence of SC and MRC eavesdroppers.

Artificial Intelligence for Unmanned Aerial Vehicles: Sensing, Communication, and Computing, First Edition. Shuyan Hu et al.

4.1.1 Secrecy Capacity Against Aerial SC Eavesdropping

We first consider the SC at the eavesdroppers, where at any moment the eavesdropper with the highest received SNR is chosen to output the data [15]. The received SNR of SC, denoted as ζ_c^{sc}, is $\zeta_c^{sc} = \max_{i=1,\dots,N}\{\zeta_{e_i}\}$. Since the N G2U links are assumed to follow i.i.d. Rician distributions, it is reasonable to assume that the closest eavesdropper to the ground transmitter achieves the highest SNR at any moment. This is because the path loss has a dominating effect on the SNR. Let l_i, $i = 1, \dots, N$, collect the Euclidean distance from the ground transmitter to the i-th aerial eavesdropper. $\mathcal{L} = \min_{i=1,\dots,N}\{l_i\}$ is the one with the shortest distance. The expectation of the shortest distance can be given in the following lemma.

Lemma 4.1 *The expectation of the shortest distance between the ground transmitter and aerial eavesdroppers,* $\mathbb{E}[\mathcal{L}]$, *is given by*

$$\mathbb{E}[\mathcal{L}] = \frac{\Gamma\left(\frac{4}{3}\right)\Gamma(N+1)}{\Gamma\left(N+\frac{4}{3}\right)} r \xrightarrow{N\to\infty} \Gamma\left(\frac{4}{3}\right) N^{-\frac{1}{3}} r, \tag{4.1}$$

where $\Gamma(z) = \int_0^\infty \frac{t^{z-1}}{e^t} dt$ *is the Gamma function.*

Proof. See Appendix C.1; cf. (C.5).

4.1.1.1 Ergodic Secrecy Capacity

The instantaneous secrecy capacity of a channel realization between the ground transmitter and receiver is given by [36]

$$C^s = \left[\log_2(1+\zeta_r) - \log_2(1+\zeta_c)\right]^+,$$

where $[x]^+ = \max\{x, 0\}$. Based on Lemma 4.1, the ergodic secrecy capacity can be established in the presence of N aerial eavesdroppers exploiting SC, as stated in Theorem 4.1.

Theorem 4.1 *In the presence of N aerial SC eavesdroppers with random 3D trajectories, the ergodic secrecy capacity of the G2G link is given by*

$$\begin{aligned} &\mathbb{E}_{\mathcal{L}}\left[C_{\text{erg}}^{\text{sc}}(l,\mathcal{L})\right] \xrightarrow{a.s.} C_{\text{erg}}^{\text{sc}}(l,\mathbb{E}[\mathcal{L}]) \\ &= \log_2\left\{\frac{P\sigma_e^2 l^{-\alpha} + \sigma_r^2\sigma_e^2}{P\sigma_r^2\left[\Gamma\left(\frac{4}{3}\right)N^{-\frac{1}{3}}r\right]^{-\alpha_e} + \sigma_r^2\sigma_e^2}\right\}, \text{ if } N \le N_c^{\text{sc}}, \end{aligned} \tag{4.2}$$

where N_c^{sc} defines the cut-off number of SC eavesdroppers beyond which the ergodic secrecy capacity of the ground link vanishes. The approximation in (4.2) is asymptotically tight. If $\sigma_r^2 = \sigma_e^2 = \sigma^2$, $N_c^{sc} = \left[\Gamma\left(\frac{4}{3}\right) r l^{-\frac{\alpha}{\alpha_e}}\right]^3$. Here, N_c^{sc} increases cubically with the radius of the eavesdroppers' flying zone r (and hence linearly with the size of the zone). As a result, SC eavesdropping has a consistent minimum, cut-off density of the SC eavesdroppers, $\rho^{sc} = \frac{3}{2\pi}\Gamma^3\left(\frac{4}{3}\right) l^{-\frac{3\alpha}{\alpha_e}}$, which allows the SC eavesdroppers to overhear all the traffic of the ground transmitter.

Proof. See Appendix C.2.

4.1.1.2 ϵ-Outage Secrecy Capacity

The secrecy outage probability defines the probability that the instantaneous secrecy capacity is lower than a secrecy rate threshold $\mathcal{R}_{th}$ [133]

$$
\begin{aligned}
\mathcal{P}_{out}^{s}(\mathcal{R}_{th}) &= \Pr\left(\mathcal{C}^{s} < \mathcal{R}_{th}\right) \\
&= \Pr\left\{\log_2\left(1+\zeta_r\right) - \log_2\left(1+\zeta_c\right) < \mathcal{R}_{th}\right\}.
\end{aligned}
\tag{4.3}
$$

The ϵ-outage secrecy capacity is defined as the maximum secrecy rate threshold, denoted by, $\mathcal{C}_{out}^{s}$, against which the outage probability of the channel is lower than a given probability ϵ, that is, $\mathcal{P}_{out}^{s}\left[\mathcal{C}_{out}^{s}\right] = \epsilon$. Based on Lemma 4.1, the ϵ-outage secrecy capacity can be established in the presence of N aerial eavesdroppers exploiting SC, as provided in Theorem 4.2.

Theorem 4.2 *In the presence of N aerial SC eavesdroppers with random 3D trajectories, the ϵ-outage secrecy capacity of the G2G link is given by*

$$
\begin{aligned}
&\mathbb{E}_{\mathcal{L}}\left[\mathcal{C}_{out}^{sc}\left(l,\mathcal{L}\right)\right] \xrightarrow{a.s.} \mathcal{C}_{out}^{sc}\left(l,\mathbb{E}\left[\mathcal{L}\right]\right) \\
&= \begin{cases}
\log_2\left\{\dfrac{(1+K_e)(K_e-W_0)\,\sigma_e^2 l^{-\alpha}}{W_0\sigma_r^2\left[\Gamma\left(\frac{4}{3}\right)N^{-\frac{1}{3}}r\right]^{-\alpha_e}}\right\}, & \text{if } K_e>0 \text{ and } N \le N_c^{sc'}; \\[2ex]
\log_2\left\{\dfrac{\sigma_r^2\sigma_e^2 + \epsilon P\sigma_e^2 l^{-\alpha}}{\sigma_r^2\sigma_e^2 + (1-\epsilon)P\sigma_r^2\left[\Gamma\left(\frac{4}{3}\right)N^{-\frac{1}{3}}r\right]^{-\alpha_e}}\right\}, & \text{if } K_e=0 \text{ and } N \le N_c^{sc'},
\end{cases}
\end{aligned}
\tag{4.4}
$$

where $W_0 = \mathcal{W}_0\left(\frac{K_e \exp(K_e)(1-\epsilon)Pl^{-\alpha}}{(\sigma_r^2 + Pl^{-\alpha})}\right)$. $N_c^{sc'}$ *defines the cut-off number of SC eavesdroppers beyond which the* ϵ*-outage secrecy capacity of the ground link vanishes. If* $\sigma_r^2 = \sigma_e^2 = \sigma^2$, $N_c^{sc'} = \left[\Gamma\left(\frac{4}{3}\right)\eta_{sc}^{-1} r l^{-\frac{\alpha}{\alpha_e}}\right]^3$, *where*

$$\eta_{sc} = \begin{cases} \left(\frac{W_0}{(1+K_e)(K_e - W_0)}\right)^{\frac{1}{\alpha_e}}, & \text{if } K_e > 0; \\ \left(\frac{1-\epsilon}{\epsilon}\right)^{\frac{1}{\alpha_e}}, & \text{if } K_e = 0. \end{cases}$$

Here, $N_c^{sc'}$ *increases linearly with the size of the eavesdroppers' flying zone. The SC eavesdropping has a consistent cut-off density of the eavesdroppers* $\rho^{sc'} = \frac{3}{2\pi}\Gamma^3\left(\frac{4}{3}\right)\eta_{sc}^{-3} l^{-\frac{3\alpha}{\alpha_e}}$.

Proof. See Appendix C.4.

4.1.2 Secrecy Capacity Against Aerial MRC Eavesdropping

We consider that N aerial eavesdroppers carry out MRC to eavesdrop on the G2G link. This is achieved by weighting the eavesdropped signals with their root-mean-square (RMS) SNRs and then taking the sum [271]. The use of MRC allows for the highest output SNR [31]. The SNR at the output of the MRC, denoted by ζ_c^{mrc}, is $\zeta_c^{mrc} = \sum_{i=1}^N \zeta_{e_i}$ [31]. The PDF of ζ_c^{mrc} is the noncentral Chi-square distribution with $2N$ degrees-of-freedom (DoFs) and noncentral parameter $\frac{K_e}{K_e+N}$, as given by [31]

$$\begin{aligned} f_{\zeta_c^{mrc}}(x) = {} & \left(\frac{N+K_e}{N\Omega_c}\right)^{\frac{N+1}{2}}\left(\frac{x}{K_e}\right)^{\frac{N-1}{2}} \exp\left(-K_e - \frac{(N+K_e)x}{N\Omega_c}\right) \\ & \times I_{N-1}\left(\sqrt{\frac{4K_e(N+K_e)}{N\Omega_c}x}\right), \end{aligned} \tag{4.5}$$

where $\Omega_c = \frac{\bar{\zeta}_c^{mrc}}{N} = \frac{1}{N}\sum_{i=1}^N \bar{\zeta}_{e_i}$ and $\bar{\zeta}_{e_i} = \frac{Pl_i^{-\alpha_e}}{\sigma_e^2}$; $I_{N-1}(\cdot)$ is the modified $(N-1)$-th order Bessel function of the first kind, and $I_{N-1}(x) = \sum_{k=0}^{\infty} \frac{(x/2)^{N-1+2k}}{k!(N+k-1)!}$. To analyze $\bar{\zeta}_{e_i}$ and Ω_c, we first give the average distance in Lemma 4.2.

Lemma 4.2 *The expectation of the distance between an aerial eavesdropper within a 3D hemispherical space and the ground transmitter is* $\mathbb{E}\,[l_i] = \frac{3}{4}r,\, i = 1, \dots, N.$

Proof. See Appendix C.1; see (C.2).

4.1.2.1 Ergodic Secrecy Capacity

Based on Lemma 4.2, the ergodic secrecy capacity is established in the presence of N MRC aerial eavesdroppers with random 3D trajectories, as stated in Theorem 4.3.

Theorem 4.3 *In the presence of N aerial MRC eavesdroppers with random 3D trajectories, the ergodic secrecy capacity of the G2G link is given by*

$$\mathbb{E}_{l_i}\left[C_{\text{erg}}^{\text{mrc}}(l, l_i)\right] \xrightarrow{a.s.} \log_2\left[\frac{P\sigma_e^2 l^{-\alpha} + \sigma_r^2\sigma_e^2}{NP\sigma_r^2\left(\frac{3}{4}r\right)^{-\alpha_e} + \sigma_r^2\sigma_e^2}\right], \quad \text{if } N \leq N_c^{\text{mrc}}, \tag{4.6}$$

where N_c^{mrc} *defines the cut-off number of MRC eavesdroppers. If* $\sigma_r^2 = \sigma_e^2 = \sigma^2$, $N_c^{\text{mrc}} = l^{-\alpha}\left(\frac{3}{4}r\right)^{\alpha_e}$. *Here,* N_c^{mrc} *depends on the radius of the eavesdroppers' flying zone* r, *and the propagation condition within the zone* α_e. *As a result, the cut-off density of MRC eavesdroppers is* $\rho^{\text{mrc}} = \frac{3}{2\pi}\left(\frac{3}{4}\right)^{\alpha_e} l^{-\alpha} r^{\alpha_e - 3}$, *which allows the MRC eavesdroppers to intercept all the traffic of the ground transmitter.*

Proof. See Appendix C.5. The use of MRC eavesdropping can help reduce the density of the eavesdroppers, as compared to the SC eavesdropping. Given the fact that the aerial channels are typically dominated by line-of-sight (LoS) with relatively low signal attenuation, the MRC eavesdropping typically requires a smaller number of eavesdroppers than the SC eavesdropping; i.e., $N_c^{\text{mrc}} \leq N_c^{\text{sc}}$; since $\left(\frac{4}{3}\right)^{\alpha_e}\Gamma^3\left(\frac{4}{3}\right) r^{3-\alpha_e} l^{\frac{\alpha(3-\alpha_e)}{\alpha_e}} > 1$ especially for small path loss exponents (e.g., $\alpha_e \leq 3$).

4.1.2.2 ϵ-Outage Secrecy Capacity

Based on Lemma 4.2, the ϵ-outage secrecy capacity is derived in the presence of N aerial eavesdroppers using MRC, as presented in Theorem 4.4.

Theorem 4.4 *In the presence of N aerial MRC eavesdroppers with random 3D trajectories, the ϵ-outage secrecy capacity of the G2G link is given by*

$$\mathbb{E}_{l_i}\left[C_{\text{out}}^{\text{mrc}}(l, l_i)\right] \xrightarrow{a.s.} C_{\text{out}}^{\text{mrc}}(l, \mathbb{E}[l_i])$$
$$= \begin{cases} \log_2\left[\dfrac{(N+K_e)(K_e-W_c)\,\sigma_e^2 l^{-\alpha}}{NW_c\sigma_r^2\left(\frac{3}{4}r\right)^{-\alpha_e}}\right], & \text{if } K_e > 0 \;\&\; N \le N_c^{\text{mrc}'}; \\ \log_2\left[\dfrac{\sigma_r^2\sigma_e^2 - P\sigma_e^2 l^{-\alpha}\ln(1-\epsilon)}{\sigma_r^2\sigma_e^2 + NP\sigma_r^2\left(\frac{3}{4}r\right)^{-\alpha_e}}\right], & \text{if } K_e = 0 \;\&\; N \le N_c^{\text{mrc}'}, \end{cases} \tag{4.7}$$

where $W_c = N\mathcal{W}_0\left[\dfrac{K_e}{N}\left(\dfrac{\exp(K_e)(1-\epsilon)Pl^{-\alpha}}{\sigma_r^2 + Pl^{-\alpha}}\right)^{\frac{1}{N}}\right]$*, and* $N_c^{\text{mrc}'}$ *defines the cut-off number of MRC eavesdroppers beyond which the ϵ-outage secrecy capacity of the ground link vanishes.*

If $\sigma_r^2 = \sigma_e^2 = \sigma^2$ *and* $K_e = 0$, $N_c^{\text{mrc}'}$ *is given by*

$$N_c^{\text{mrc}'} = \eta_{\text{mrc}}^{-\alpha_e} \cdot \frac{\left(\frac{3}{4}r\right)^{\alpha_e}}{l^\alpha}, \text{ where } \eta_{\text{mrc}} = \left[-\frac{1}{\ln(1-\epsilon)}\right]^{\frac{1}{\alpha_e}},$$

where $N_c^{\text{mrc}'}$ *depends on the radius of the eavesdroppers' flying zone r, and the propagation condition within the zone* α_e*. As a result, the cut-off density of MRC eavesdroppers within the zone is* $\rho^{\text{mrc}'} = \frac{3}{2\pi}\left(\frac{3}{4}\right)^{\alpha_e}\eta_{\text{mrc}}^{-\alpha_e} l^{-\alpha} r^{\alpha_e - 3}$.

Proof. Refer to Appendix C.6.

4.1.3 Safety Height and Collision Avoidance

In practice, UAVs usually need to fly above a minimum altitude, denoted by H_0, to avoid obstacles, as illustrated in Figure 3.1. The flight region of the aerial eavesdroppers is the part of the hemisphere above H_0. The expectations of the distances and the shortest of the distances between the ground transmitter and the aerial eavesdroppers, and the shortest of the distances, are given in Lemma 4.3, where $\mathbb{E}[l_i]$ and $\mathbb{E}[\mathcal{L}]$ can be substituted in Theorems 4.1 and 4.2 to evaluate the ergodic and outage secrecy capacities under the safety altitude H_0.

Lemma 4.3 *The expectation of the distance between any eavesdropper i and the ground transmitter l_i is* $\mathbb{E}[l_i] = \frac{3r^4 - 4H_0 r^3 + H_0^4}{2V_1'}$, *where* $V_1' = 2r^3 - 3H_0 r^2 + H_0^3$.

The expectation of the shortest of distances is given by

$$\mathbb{E}[\mathcal{L}] = H_0 + \left(1 - \frac{H_0^3}{V_1'}\right)^N \sum_{k=0}^{N} \binom{N}{k} \frac{V_1' - H_0^3}{3H_0(2k+1)}$$
$$\times \left[r^{2k+1} \left(\frac{3H_0 - 2r}{V_1' - H_0^3}\right)^{k+1} {}_2F_1\left(1, 3k+2; 2(k+1); \frac{2r}{3H_0}\right) \right.$$
$$\left. - \frac{H_0^{3k+2}}{(V_1' - H_0^3)^{k+1}} {}_2F_1\left(1, 3k+2; 2(k+1); \frac{2}{3}\right) \right], \tag{4.8}$$

where ${}_2F_1(a, b; c; z) = \sum_{n=0}^{\infty} \frac{(a)_n (b)_n}{(c)_n} \cdot \frac{z^n}{n!}$ *is the Gaussian hypergeometric function.*

Proof. See Appendix C.7.

Remark 4.1 Our analysis can be used to evaluate the setup with all the eavesdroppers on the ground (traveling random trajectories), as long as the expectations of the eavesdropping link distances are finite and available. This is due to the fact that the secrecy capacities are decoupled from the distances in the analysis using the theory of *a.s.* convergence.

Collision-avoidance can also be taken into account by enforcing a minimum safety distance between UAVs [268, Sec. III]. Existing protocols designed for the RD mobility model, such as the sense-and-stop (S&S) protocol [171], can be extended to the ST mobility model: At any time, if the waypoints of a pair of UAVs are within the safety distance of each other, the UAV to the relative right hovers (not stopping) and the other maintains its original mobility pattern until they are the safety distance away. The S&S protocol has been proved to maintain the stationary uniform distribution of the waypoints of each UAV in the RD mobility model [171, Theorem 3]. Therefore, the waypoints of each UAV are uniformly distributed in the ST mobility model, since the RD and ST mobility models use the same way to deploy the waypoints.

The joint distribution of the UAVs may not be uniform [171], but can be given by

$$f_{\mathbf{X}_1,\ldots,\mathbf{X}_N}(\boldsymbol{x}_1, \ldots, \boldsymbol{x}_N) = f_{\mathbf{X}_N}(\boldsymbol{x}_N) \times \prod_{i=1}^{N-1} f_{\mathbf{X}_i|\mathbf{X}_{i+1},\ldots,\mathbf{X}_N}(\boldsymbol{x}_i|\boldsymbol{x}_{i+1}, \ldots, \boldsymbol{x}_N), \tag{4.9}$$

where $\mathbf{X}_i = (l_i, \theta_i, \phi_i)$ is the polar coordinates of the i-th UAV, l_i is the distance of the i-th UAV to the center of the hemisphere, and θ_i and ϕ_i are the elevation and azimuth angles; $\boldsymbol{x}_i = (\tau_i, \beta_i, \varphi_i)$ is a random realization of $\mathbf{X}_i$; the conditional probability distributions can be evaluated as

$$f_{\mathbf{X}_i|\mathbf{X}_{i+1},\ldots,\mathbf{X}_N}(\boldsymbol{x}_i|\boldsymbol{x}_{i+1},\ldots,\boldsymbol{x}_N) = \frac{\mathcal{S}_{\tau_i,\boldsymbol{x}_{i+1},\ldots,\boldsymbol{x}_N}}{2\pi\tau_i^2}, \tag{4.10}$$

with $\mathcal{S}_{\tau_i,\boldsymbol{x}_{i+1},\ldots,\boldsymbol{x}_N}$ being the surface area of the co-centric 3D hemisphere with the radius τ_i excluding the areas inside the 3D spheres centered at $\boldsymbol{x}_j$ for $j = i+1, \ldots, N$ and with radius of the safety distance (as illustrated in Figure 4.1); and the PDF is $f_{\mathbf{X}_N}(\boldsymbol{x}_N) = \frac{3}{2\pi r^3}$.

Closed-form expressions for (4.10) and, in turn, for (4.9) may exist, but can be very tedious to obtain. Specifically, $\mathcal{S}_{\tau_i,\boldsymbol{x}_{i+1},\ldots,\boldsymbol{x}_N}$ can have different expressions under different values of $\boldsymbol{x}_{i+1}, \ldots, \boldsymbol{x}_N$. The number of expressions grows combinatorially with $(N-i)$. Nevertheless, (4.10) can be readily and numerically studied by evaluating the area of $\mathcal{S}_{\tau_i,\boldsymbol{x}_{i+1},\ldots,\boldsymbol{x}_N}$. The average and the minimum of the distances from the N UAVs to the center of the hemisphere (the ground transmitter) can be obtained by numerically evaluating

$$\mathbb{E}\left\{\frac{\sum_{i=1}^{N} l_i}{N}\right\} = \oiint \frac{1}{N}\sum_{i=1}^{N}\tau_i f_{\mathbf{X}_1,\ldots,\mathbf{X}_N}(\boldsymbol{x}_1,\ldots,\boldsymbol{x}_N)\prod_{i=1}^{N} d\boldsymbol{x}_i; \tag{4.11}$$

$$\mathbb{E}\left\{\min_{i=1,\ldots,N} l_i\right\} = \oiint \min_{i=1,\ldots,N} \tau_i f_{\mathbf{X}_1,\ldots,\mathbf{X}_N}(\boldsymbol{x}_1,\ldots,\boldsymbol{x}_N)\prod_{i=1}^{N} d\boldsymbol{x}_i, \tag{4.12}$$

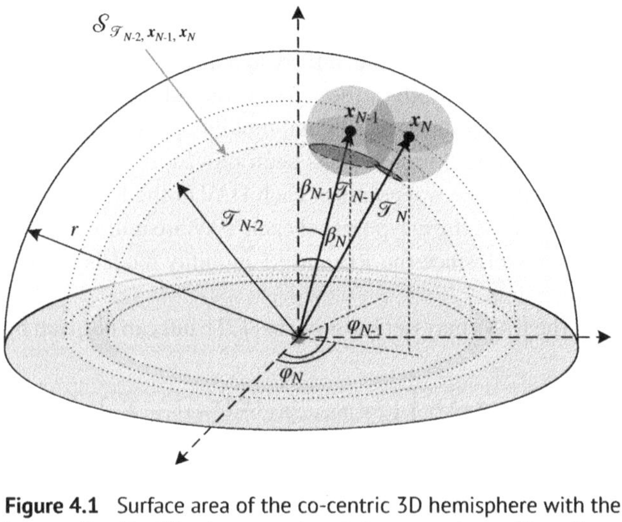

Figure 4.1 Surface area of the co-centric 3D hemisphere with the radius τ_{N-2} intersecting the 3D spheres centered at $\boldsymbol{x}_{N-1}$ and $\boldsymbol{x}_N$ with radius of the safety distance.

which can be substituted in our analysis, i.e., Theorems 4.1–4.4. The analysis would remain valid, but can hardly provide insightful closed-form expressions of the radius of the hemisphere r, and can be increasingly computationally demanding with the growing number of eavesdroppers.

4.2 Secrecy Under Imperfect Oscillator and Finite Memory of Eavesdroppers

4.2.1 Nonnegligible Oscillator Phase Noises

Consider that the aerial eavesdroppers have independent oscillator phase noise processes due to the imperfections in the circuitry of their oscillators. ξ_i, $i = 1, \dots, N$ denotes the oscillator phase noise process at the i-th eavesdropper. Each oscillator phase noise process can be modeled as an independent discrete-time Wiener process [211]: $\xi_i[k] = \xi_i[k-1] + \omega_i^t[k]$, where $\omega_i^t[k] \sim \mathcal{N}(0, \sigma_\xi^2)$ are i.i.d. zero-mean Gaussian increments at time k with variance $\sigma_\xi^2 \triangleq 4\pi^2 f_c^2 c_\xi T_s$. f_c is the carrier frequency. T_s is the symbol interval. c_ξ is a constant that depends on the specific oscillators, and is typically set to 4.7×10^{-18} $(\text{rad} \cdot \text{Hz})^{-1}$. Based on [211, Sec. IV], the instantaneous SNR (more appropriately, the signal-to-interference-plus-noise ratio, SINR, due to the self-interference resulting from the oscillator phase noises) at the i-th aerial eavesdropper is given by

$$\zeta'_{e_i} = \frac{e^{-\sigma_\xi^2} P l_i^{-\alpha_e} |h_{e_i}|^2}{(1 - e^{-\sigma_\xi^2}) P l_i^{-\alpha_e} |h_{e_i}|^2 + \sigma_e^2}, \quad i = 1, \dots, N. \tag{4.13}$$

The eavesdropping SINR at the output of SC and MRC eavesdropping can be given by

$$\zeta_c^{\text{sc}'} = \max_{i=1,2,\dots,N} \{\zeta'_{e_i}\} = \frac{e^{-\sigma_\xi^2} P \mathcal{L}^{-\alpha_e} |h_e|^2}{(1 - e^{-\sigma_\xi^2}) P \mathcal{L}^{-\alpha_e} |h_e|^2 + \sigma_e^2};$$

$$\zeta_c^{\text{mrc}'} = \sum_{i=1}^{N} \zeta'_{e_i} = \sum_{i=1}^{N} \frac{e^{-\sigma_\xi^2} P l_i^{-\alpha_e} |h_e|^2}{(1 - e^{-\sigma_\xi^2}) P l_i^{-\alpha_e} |h_e|^2 + \sigma_e^2}, \quad i = 1, \dots, N.$$

The average SINR of SC and MRC eavesdropping can be approximated by [141]

$$\begin{aligned}\bar{\zeta}_c^{\mathrm{sc}'} &\approx \frac{e^{-\sigma_\xi^2} P\mathcal{L}^{-\alpha_e}\mathbb{E}[|h_e|^2]}{(1-e^{-\sigma_\xi^2})P\mathcal{L}^{-\alpha_e}\mathbb{E}[|h_e|^2]+\sigma_e^2} \\ &= \frac{e^{-\sigma_\xi^2} P\mathcal{L}^{-\alpha_e}}{(1-e^{-\sigma_\xi^2})P\mathcal{L}^{-\alpha_e}+\sigma_e^2};\end{aligned} \tag{4.14}$$

$$\begin{aligned}\bar{\zeta}_c^{\mathrm{mrc}'} &= \sum_{i=1}^{N} \bar{\zeta}'_{e_i} \approx \sum_{i=1}^{N} \frac{e^{-\sigma_\xi^2} P l_i^{-\alpha_e}\mathbb{E}[|h_e|^2]}{(1-e^{-\sigma_\xi^2})P l_i^{-\alpha_e}\mathbb{E}[|h_e|^2]+\sigma_e^2} \\ &= \sum_{i=1}^{N} \frac{e^{-\sigma_\xi^2} P l_i^{-\alpha_e}}{(1-e^{-\sigma_\xi^2})P l_i^{-\alpha_e}+\sigma_e^2},\end{aligned} \tag{4.15}$$

since $\mathbb{E}[|h_0|^2]=1$ and $\mathbb{E}[|h_{e_i}|^2]=1$.

As a result, the ergodic secrecy capacity of the G2G link under the SC and MRC eavesdropping can be obtained as

$$C_{\mathrm{erg}}^{\mathrm{sc}'}(\mathcal{L}) \xrightarrow{a.s.} \log_2\left\{\frac{(Pl^{-\alpha}+\sigma_r^2)\left[(1-e^{-\sigma_\xi^2})P\mathcal{L}^{-\alpha_e}+\sigma_e^2\right]}{P\sigma_r^2\mathcal{L}^{-\alpha_e}+\sigma_r^2\sigma_e^2}\right\};$$

$$C_{\mathrm{erg}}^{\mathrm{mrc}'}(l_i) \xrightarrow{a.s.} \log_2\left[\frac{1+\frac{Pl^{-\alpha}}{\sigma_r^2}}{1+\sum_{i=1}^{N}\frac{e^{-\sigma_\xi^2} P l_i^{-\alpha_e}}{(1-e^{-\sigma_\xi^2})P l_i^{-\alpha_e}+\sigma_e^2}}\right], i=1,\dots,N.$$

The corresponding ergodic secrecy capacities are given by

$$\begin{aligned}\mathbb{E}\left[C_{\mathrm{erg}}^{\mathrm{sc}'}(\mathcal{L})\right] &\xrightarrow{a.s.} C_{\mathrm{erg}}^{\mathrm{sc}'}(\mathbb{E}[\mathcal{L}]) \\ &= \log_2\left\{\frac{(Pl^{-\alpha}+\sigma_r^2)\left[(1-e^{-\sigma_\xi^2})P\left(\Gamma\left(\frac{4}{3}\right)N^{-\frac{1}{3}}r\right)^{-\alpha_e}+\sigma_e^2\right]}{\sigma_r^2\left[P\left(\Gamma\left(\frac{4}{3}\right)N^{-\frac{1}{3}}r\right)^{-\alpha_e}+\sigma_e^2\right]}\right\}, \\ &\qquad \text{if } N \le N_{\mathrm{n}}^{\mathrm{sc}};\end{aligned} \tag{4.16}$$

$$\mathbb{E}\left[\mathcal{C}_{\text{erg}}^{\text{mrc}'}(l_i)\right] \rightarrow \mathcal{C}_{\text{erg}}^{\text{mrc}'}(\mathbb{E}[l_i])$$
$$= \log_2 \left\{ \frac{(Pl^{-\alpha} + \sigma_r^2)\left[\left(1 - e^{-\sigma_\xi^2}\right) P\left(\frac{3}{4}r\right)^{-\alpha_e} + \sigma_e^2\right]}{\sigma_r^2 \left[\left(1 + (N-1)e^{-\sigma_\xi^2}\right) P\left(\frac{3}{4}r\right)^{-\alpha_e} + \sigma_e^2\right]} \right\}, \tag{4.17}$$
$$\text{if } N \le N_{\text{n}}^{\text{mrc}},$$

where N_{n}^{sc} and $N_{\text{n}}^{\text{mrc}}$ are the cut-off number of eavesdroppers. By assuming $\sigma_r^2 = \sigma_e^2 = \sigma^2$, N_{n}^{sc} and $N_{\text{n}}^{\text{mrc}}$ are given by

$$N_{\text{n}}^{\text{sc}} = \Gamma^3\left(\frac{4}{3}\right) r^3 \left[l^\alpha e^{-\sigma_\xi^2} - \frac{P}{\sigma^2}\left(1 - e^{-\sigma_\xi^2}\right)\right]^{-\frac{3}{\alpha_e}};$$

$$N_{\text{n}}^{\text{mrc}} = e^{\sigma_\xi^2}\left[\frac{(1 - e^{-\sigma_\xi^2})Pl^{-\alpha}}{\sigma^2} + \left(\frac{3}{4}r\right)^{\alpha_e} l^{-\alpha}\right].$$

Particularly, (4.16) is achieved by first letting $\mu = \mathcal{L}^{-\alpha_e}$ and $\nu = \mathcal{L}^{\alpha_e}$, and rewriting $\mathcal{C}_{\text{erg}}^{\text{sc}'}(\mathcal{L})$ as

$$\mathcal{H}_3(\mu) = \log_2 \left\{ \frac{(Pl^{-\alpha} + \sigma_r^2)\left[(1 - e^{-\sigma_\xi^2})P\mu + \sigma_e^2\right]}{P\sigma_r^2\mu + \sigma_r^2\sigma_e^2} \right\};$$

$$\mathcal{H}_4(\nu) = \log_2 \left\{ \frac{(Pl^{-\alpha} + \sigma_r^2)\left[(1 - e^{-\sigma_\xi^2})P + \sigma_e^2\nu\right]}{P\sigma_r^2 + \sigma_r^2\sigma_e^2\nu} \right\}.$$

We can find that $\frac{\partial \mathcal{H}_3}{\partial \mu} \le 0$, $\frac{\partial^2 \mathcal{H}_3}{\partial \mu^2} \ge 0$, $\frac{\partial \mathcal{H}_4}{\partial \nu} \ge 0$, and $\frac{\partial^2 \mathcal{H}_4}{\partial \nu^2} \le 0$. Therefore, $\mathcal{H}_3(\mu)$ is a monotonically decreasing and convex function of μ; and $\mathcal{H}_4(\nu)$ is a monotonically increasing and concave function of ν. By following the proof of Theorem 1, (4.16) can be derived. Likewise, (4.17) can be obtained.

4.2.2 Effective Eavesdropping Rate

Based on [338], an eavesdropping performance metric is the effective eavesdropping rate defined as

$$R_{\rm ev} = \begin{cases} R_r, & \text{if } R_e \geq R_r; \\ 0, & \text{otherwise,} \end{cases}$$

where $R_r = \log_2(1+\zeta_r)$ and $R_e = \log_2(1+\zeta_c)$.

We can accordingly define the eavesdropping probability as $\mathcal{P}_{\rm ev} \triangleq \Pr\{R_e - R_r \geq 0\}$, and the effective eavesdropping rate can be written as $\mathcal{C}_{\rm ev} = \mathcal{P}_{\rm ev} R_r$. Based on Theorems 4.2 and 4.4, we can obtain the eavesdropping probability under both the SC and MRC eavesdropping, as given by

$$\mathcal{P}_{\rm ev}^{\rm sc} = 1 - \frac{(1+K_e)\, l^{-\alpha}\sigma_e^2}{(\mathbb{E}[\mathcal{L}])^{-\alpha_e}\sigma_r^2 + l^{-\alpha}\sigma_e^2} \times \exp\left(-\frac{K_e(\mathbb{E}[\mathcal{L}])^{-\alpha_e}\sigma_r^2}{(\mathbb{E}[\mathcal{L}])^{-\alpha_e}\sigma_r^2 + (1+K_e)l^{-\alpha}\sigma_e^2}\right);$$

$$\mathcal{P}_{\rm ev}^{\rm mrc} = 1 - \left(\frac{(N+K_e)\, l^{-\alpha}\sigma_e^2}{N(\mathbb{E}[l_i])^{-\alpha_e}\sigma_r^2 + (N+K_e)\, l^{-\alpha}\sigma_e^2}\right)^N \times \exp\left(-\frac{NK_e(\mathbb{E}[l_i])^{-\alpha_e}\sigma_r^2}{N(\mathbb{E}[l_i])^{-\alpha_e}\sigma_r^2 + (N+K_e)l^{-\alpha}\sigma_e^2}\right),$$

which are achieved by setting $\mathcal{C}_{\rm out}^{\rm sc}(l, \mathbb{E}[\mathcal{L}]) = 0$ in (4.4) and $\mathcal{C}_{\rm out}^{\rm mrc}(l, \mathbb{E}[l_i]) = 0$ in (4.7), and then evaluating $\mathcal{P}_{\rm ev} = \mathcal{P}_{\rm out}^{s}(0)$. This is because the definition of the secrecy outage probability is the probability at which the instantaneous secrecy capacity is lower than a threshold; see (4.3). To evaluate the effective eavesdropping rate here, the threshold can be set to zero.

For illustration convenience, we assume that $K_e = 0$ and $\sigma_e^2 = \sigma_r^2$. We can obtain the ratio of the eavesdropping probability (and hence the ratio of the effective eavesdropping rate) between the MRC and SC eavesdropping as

$$\frac{\mathcal{P}_{\rm ev}^{\rm mrc}}{\mathcal{P}_{\rm ev}^{\rm sc}} = \left(1 + \frac{l^{-\alpha}}{(\mathbb{E}[\mathcal{L}])^{-\alpha_e}}\right)\left[1 - \left(\frac{l^{-\alpha}}{l^{-\alpha} + (\mathbb{E}[l_i])^{-\alpha_e}}\right)^N\right].$$

The MRC eavesdropping can use up to N times the memory that the SC eavesdropping ideally uses. The SC eavesdropping can operate much longer, and sniff many more signals from the ground transmitter, than the MRC eavesdropping, given the finite memory per eavesdropper. In a plausible case, a distance or a received signal strength threshold can be set up. (The two thresholds are transferable). The distance threshold d_{th} can be set such that only the aerial eavesdroppers within d_{th} from the ground transmitter record and carry back their captured signals. As a result, the probability that none of the eavesdroppers record signals is $\mathcal{P}_{\text{th}} = \left(1 - \frac{d_{\text{th}}^3}{r^3}\right)^N$ based on the PDF of the shortest distance of all the eavesdroppers to the ground transmitter; see Appendix (C.3). The ratio of the effective eavesdropping rate between the MRC and SC eavesdropping becomes $\frac{\mathcal{P}_{\text{ev}}^{\text{mrc}}}{(1-\mathcal{P}_{\text{th}})\mathcal{P}_{\text{ev}}^{\text{sc}}}$. On the other hand, the average number of eavesdroppers recording the same piece of signals at any instant is $N_\varsigma = \frac{d_{\text{th}}^3}{r^3}N$, depending on the size of the region specified by the distance threshold d_{th}. As a result, the SC eavesdropping can potentially operate $\frac{d_{\text{th}}^3}{r^3}$ times as long as the MRC eavesdropping.

4.3 Numerical and Simulation Results

In this section, we conduct extensive simulations to validate our analysis for the secrecy rates of the ground transmission in the presence of multiple collaborative aerial eavesdroppers with 3D trajectories. Without loss of generality, we assume the same receiver noise powers at the ground receiver and the aerial eavesdroppers. $\theta = \frac{\pi}{2}$; unless otherwise specified. Other parameters are given in Table 3.1 with reference to [80, 191].

Figure 3.10 plots the ergodic secrecy capacity of the G2G link in the presence of N aerial eavesdroppers. Along with the analytical results from Theorems 4.1 and 4.3, Monte-Carlo simulation results are provided. We can see that the analytical results coincide with the simulation results, and provide an accurate approximation to the ergodic secrecy capacity. In Figure 3.10a, the ergodic secrecy capacity increases with r. We see that the ergodic secrecy capacity under SC eavesdropping is higher than that under MRC eavesdropping. This is due to the fact that the received SNR of the wiretapping channels is maximized by taking MRC. With the growth of N, the ergodic secrecy capacity decreases significantly, as shown in Figure 3.10b, while the threshold of r under which the ergodic secrecy capacity approaches zero, increases with N. The difference in the secrecy capacity between the two diversity schemes increases with N. In other words, the terrestrial radio link is vulnerable to aerial eavesdropping since the eavesdroppers can operate at a distance without being easily noticed, especially when there are multiple aerial eavesdroppers in the 3D space.

The impact of H_0 on the ergodic secrecy capacity is plotted in Figure 4.2, where we can find that the ergodic secrecy capacity grows with H_0 in the case presence SC and MRC eavesdropping. The reason for the growth is that the distances

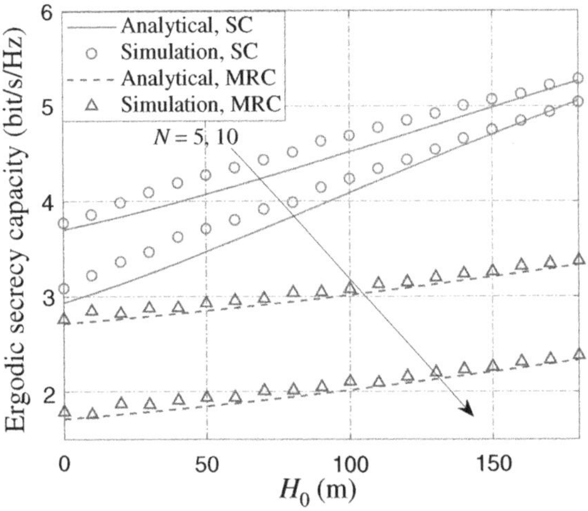

Figure 4.2 Ergodic secrecy capacity of aerial eavesdroppers conducting SC and MRC against the minimum altitude H_0, where $l = 50$ m and $r = 200$ m.

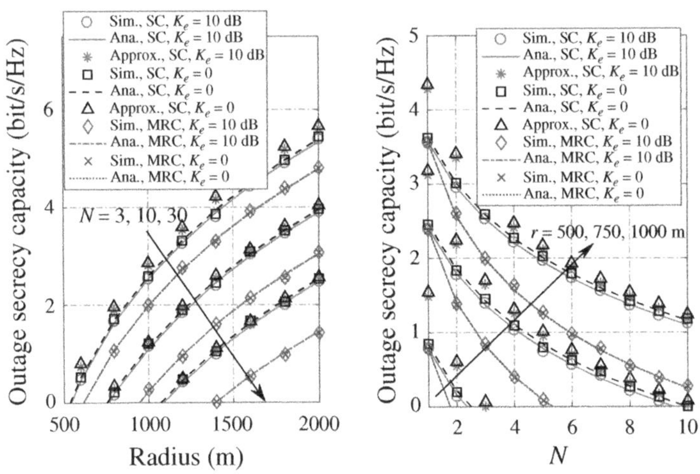

(a) Outage secrecy capacity vs. r. (b) Outage secrecy capacity vs. N.

Figure 4.3 Outage secrecy capacity of the eavesdroppers conducting SC and MRC vs. l and r, where $K_e = 0$, $\alpha = 3$, $\alpha_e = 2.8$, and $\epsilon = 0.1$.

between the ground transmitter and the aerial eavesdroppers become longer as H_0 increases.

Figure 4.3 plots the analytical and simulation results for the ϵ-outage secrecy capacity of the G2G link in the presence of N aerial eavesdroppers, as r and N

increase. We take different values for K_e and α_e. $K_e = 0$ is the case where there is no LoS path and the channel undergoes the Rayleigh fading. $K_e = 10$ dB is the case where there are LoS paths and non-LoS (NLoS) paths between the ground transmitter and the aerial eavesdroppers. The NLoS can result from the body blockage of the aerial eavesdroppers, and the reflection and scattering stemming from the surroundings. We see that the ϵ-outage secrecy capacity is higher under SC eavesdropping than under MRC eavesdropping. The difference in the outage secrecy capacity increases with N. This is consistent with the ergodic secrecy capacity. We also see that the ϵ-outage secrecy capacity is slightly lower when $K_e = 10$ dB than it is when $K_e = 0$. The reason is that the LoS avails between the ground transmitter and aerial eavesdroppers to facilitate eavesdropping. By comparing the analytical results of Theorems 4.2 and 4.4 with the simulation results, the accuracy of the analysis is confirmed. Also, the ϵ-outage secrecy capacity decreases as N increases, and increases with r.

Figure 4.4 plots the impact of the distance l and the radius of the aerial eavesdroppers' flying zone r on the cut-off densities of SC and MRC eavesdroppers. In the case of SC eavesdropping, we can see that the density only depends on l and is independent of r. In contrast, the density increases with r in the case of MRC eavesdropping. We also see that the density is lower under the MRC eavesdropping than it is under the SC eavesdropping. This is due to the fact that more eavesdroppers need to be dispatched to cover the 3D region to maximize their received SNR, if the G2G link is short and the transmit power of the link is low.

Figure 4.5 plots the ergodic secrecy capacity against P in the presence of non-negligible oscillator phase noises at the aerial eavesdroppers under different symbol intervals T_s (i.e., 0.1 and 0.5 μs). The ergodic secrecy capacity in the absence of the oscillator phase noises is also plotted for comparison purposes. We can see that the oscillator phase noises can substantially impair the eavesdropping capability of the aerial eavesdroppers, especially in high SNR regimes. This is because the eavesdropping SINR and, in turn, the ergodic capacity of the eavesdroppers asymptotically stabilize due to the oscillator phase noises and the resultant self-interference, as P increases. As a result, the ergodic secrecy capacity grows at the same rate as the ergodic capacity. This accounts for the consistent gap between them. The MRC eavesdropping has a larger gap than the SC eavesdropping, since the asymptotic SINR of the MRC eavesdropping is N times that of the SC eavesdropping, as $P \gg \sigma_e^2$. We also see that the impact of the oscillator phase noises is negligible in low SNR regimes. This is because the noises dominate over the oscillator phase noises at the aerial eavesdroppers in the low SNR regimes.

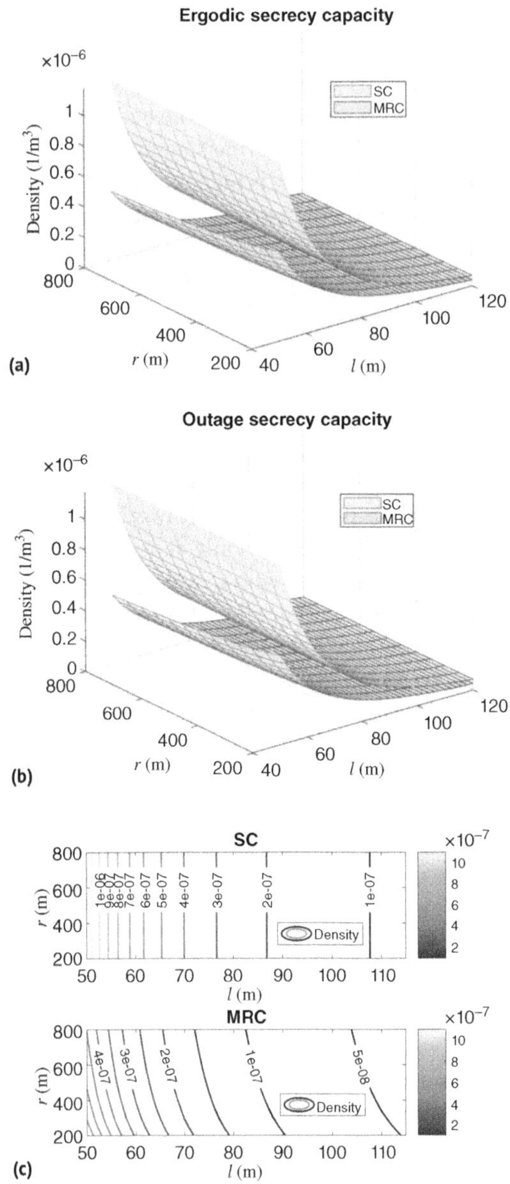

Figure 4.4 The density of aerial eavesdroppers vs. r and l, under both SC and MRC, where $\alpha = 3$, $\alpha_e = 2.8$, and $\epsilon = 0.1$. (c) provides the contours for (a), where the values, such as 2×10^{-7}, stand for the cut-off density of eavesdroppers (per cubic meter).

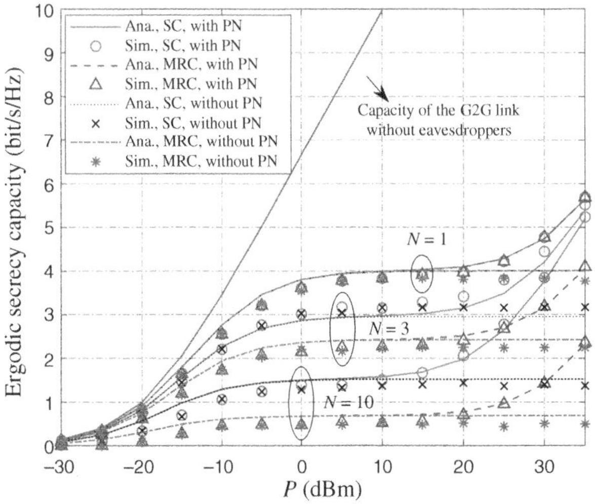

(a) Ergodic secrecy capacity vs. P ($T_s = 0.1\ \mu s$).

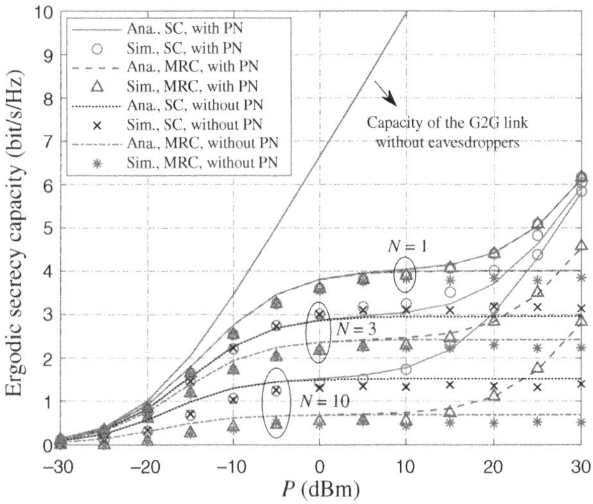

(b) Ergodic secrecy capacity vs. P ($T_s = 0.5\ \mu s$).

Figure 4.5 Ergodic secrecy capacity of the ground link vs. P in the presence and absence of oscillator phase noises at the aerial eavesdroppers, where $\sigma_r^2 = \sigma_e^2 = \sigma^2 = -80$ dBm, $l = 100$ m, $r = 500$ m, the path loss at a reference distance 1 m is 38.46 dB, and $\sigma_\xi^2 \triangleq 4\pi^2 f_c^2 c_\xi T_s$ (under $f_c = 2.4$ GHz, $T_s = 0.1$ and 0.5 µs, and $c_\xi = 4.7 \times 10^{-18}$ $(\text{rad} \cdot \text{Hz})^{-1}$).

Figure 4.6a plots the ratio of the eavesdropping probability (or in other words, the effective eavesdropping rate) between the SC and MRC eavesdropping, i.e., $\frac{\mathcal{P}_{\rm ev}^{\rm mrc}}{\mathcal{P}_{\rm ev}^{\rm sc}}$, against the number of aerial eavesdroppers, N. It is shown that $\frac{\mathcal{P}_{\rm ev}^{\rm mrc}}{\mathcal{P}_{\rm ev}^{\rm sc}}$ increases first, and then decreases and converges to 1, as N increases. Figure 4.6b plots the ratio of captured traffic between the MRC and SC eavesdropping, as the number of eavesdroppers increases. The ratio is $\frac{\mathcal{P}_{\rm ev}^{\rm mrc}}{(1-\mathcal{P}_{\rm th})\mathcal{P}_{\rm ev}^{\rm sc}} \times \frac{d_{\rm th}^3}{r^3} = \frac{\mathcal{P}_{\rm ev}^{\rm mrc}}{(1-\mathcal{P}_{\rm th})\mathcal{P}_{\rm ev}^{\rm sc}} \times (1-\mathcal{P}_{\rm th}^{\frac{1}{N}})$. We can see that the ratio declines with the growth of N,

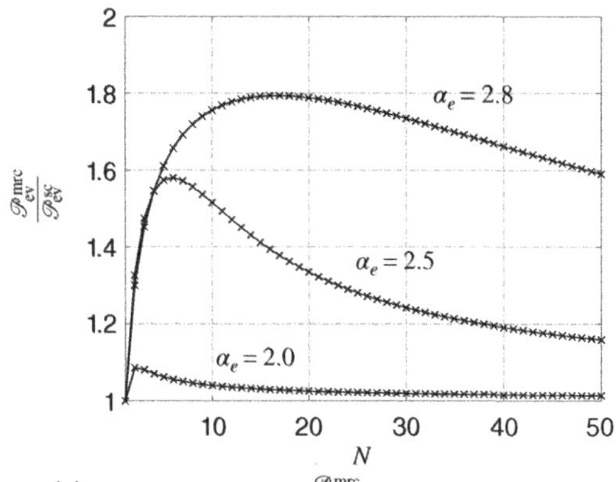

(a) $\frac{\mathscr{P}_{\rm ev}^{\rm mrc}}{\mathscr{P}_{\rm ev}^{\rm sc}}$ vs. N.

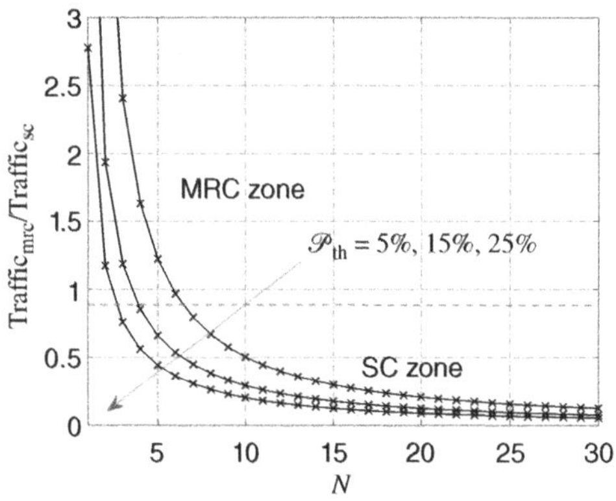

(b) $\frac{\text{Traffic}_{\rm mrc}}{\text{Traffic}_{\rm sc}}$ vs. N ($\alpha_e = 2.5$).

Figure 4.6 (a) The ratio of the eavesdropping probability between MRC and SC eavesdropping and (b) the ratio of captured traffic between MRC and SC eavesdropping, where $l = 100$ m, $r = 500$ m, and $\alpha = 3$.

and becomes lower than one, for example, when $N \geq 3$ and $\mathcal{P}_{\text{th}} = 15\%$. In other words, the SC eavesdropping is able to capture more traffic, due to its more efficient use of the memories of the eavesdroppers.

4.4 Discussion on AI Techniques and Noncooperative Eavesdroppers

4.4.1 AI Techniques Against Cooperative UAV Eavesdroppers

Eavesdropping by coordinated UAV swarms poses a serious threat to the confidentiality of terrestrial and aerial wireless communications. These aerial adversaries can dynamically adjust their positions, exploit LoS links, and collaborate to maximize interception capability. Traditional static defense mechanisms often fall short in such agile, high-mobility scenarios. To address this, AI offers a powerful suite of adaptive, learning-based techniques that enable proactive and intelligent countermeasures.

AI agents can be trained using deep reinforcement learning (DRL) to dynamically adjust the legitimate transmitter's position, power, or beamforming direction to reduce the risk of interception. For example, UAVs can learn optimal flight paths or transmission strategies that avoid high-risk regions based on the predicted positions of eavesdropping UAVs. Multi-agent DRL can further enable collaborative defense among multiple legitimate UAVs or base stations.

AI-enabled game-theoretic learning frameworks model the interaction between legitimate nodes and coordinated eavesdroppers as a dynamic, possibly repeated game. Techniques such as fictitious play or policy gradient-based zero-sum learning allow the system to adaptively learn defense strategies that anticipate and counter coordinated eavesdropping maneuvers.

Coordinated UAV eavesdroppers often form networked swarms. Graph neural networks (GNNs) can model their spatial relationships and learn patterns from historical mobility or communication behaviors. GNN-based models can be used to detect collaboration patterns, predict future positions, and proactively shape transmission strategies.

Given the distributed nature of UAV networks, federated learning allows individual UAVs to collaboratively train security models (e.g., eavesdropper classifiers or anomaly detectors) without sharing raw data. Online learning further enables real-time adaptation to new or evolving attack strategies, which is critical against intelligent adversaries.

4.4.2 Methods Against Noncooperative UAV Eavesdroppers

Eavesdropping by noncooperative UAVs presents a stealthy and unpredictable threat to wireless communications, particularly in open-air environments where LoS channels are common. Unlike coordinated attacks, these UAVs operate independently, without clear communication patterns or observable cooperation, making detection and defense more challenging. Addressing such threats requires a combination of signal-level protections, intelligent behavior modeling, and adaptive response mechanisms—where AI techniques can play a central role.

At the foundational level, physical layer security (PLS) methods offer robust defense by exploiting the characteristics of the wireless channel. For instance, artificial noise (AN) can degrade the signal quality at unintended receivers, directional beamforming can concentrate signal power toward legitimate users and away from suspicious regions, and cooperative jamming from friendly nodes can help mask the transmission from eavesdroppers. These methods can be further enhanced using AI to adapt to dynamic channel conditions and UAV movement.

Non-cooperative UAVs typically attempt to avoid detection while maintaining LoS to the transmitter. Machine learning (ML) models, such as anomaly detection algorithms or support vector machines, can be trained on historical RF signatures or spatial data to detect the presence of such UAVs based on irregular signal patterns, sudden fluctuations in channel state information (CSI), or unexpected motion in the UAV's airspace. AI techniques can enable context-aware security by integrating environment sensing, threat assessment, and policy adjustment. For instance, UAVs may learn to adjust their altitude, orientation, or antenna direction when flying near high-risk areas (e.g., open fields or urban rooftops). LSTM-based models can help in forecasting potential threats based on temporal movement patterns.

4.5 Conclusion

This chapter extended the analysis to scenarios where multiple aerial eavesdroppers cooperate by exploiting spatial diversity using techniques such as SC and MRC. Closed-form approximations for ergodic and ϵ-outage secrecy capacities were shown to be asymptotically tight as the number of eavesdroppers grows. Importantly, we derived the cut-off density of aerial eavesdroppers beyond which the secrecy of the terrestrial link is fully compromised. This density threshold depends solely on the terrestrial link range in the SC case and on both the link range and the spatial extent of the eavesdroppers' flight region under MRC.

The findings provide critical insights and practical guidelines for securing terrestrial wireless links in the presence of aerial adversaries. Notably, even independently flying UAVs with seemingly arbitrary trajectories can cooperate to perform highly effective eavesdropping. As the UAV density increases, the terrestrial link becomes increasingly exposed, underscoring the need for proactive regulatory and technical countermeasures. In particular, establishing UAV-restricted "no-fly zones" above sensitive ground infrastructure may be essential to prevent data breaches. Additionally, implementing precise timing synchronization mechanisms, such as handshaking protocols or the use of stable oscillators, can mitigate the impact of aerial eavesdropping, especially at high transmit power levels. Such measures can enable the legitimate link capacity to grow with power, while eavesdroppers remain constrained by oscillator-induced self-interference.

5

Joint Cruise Control and Data Collection of UAVs

5.1 Introduction

Mobile edge computing (MEC) has recently gained traction as an effective approach for bringing cloud-like services closer to edge Internet-of-Things (EdgeIoT) environments [59, 212, 318, 319]. By offloading resource-intensive tasks to nearby high-performance edge servers, MEC helps Internet-of-Things (IoT) devices overcome their inherent limitations in storage and processing power, thereby extending device lifespan and supporting demanding applications such as augmented and virtual reality [149, 221, 272]. As the number of IoT devices continues to grow, many EdgeIoT deployments are found in remote and unmonitored environments, such as agricultural fields, rural vineyards, and highway infrastructures, where they perform complex computational tasks like crop monitoring and adaptive irrigation management [13]. In these challenging settings, IoT devices often rely on ambient energy sources, such as solar panels, to replenish their batteries [156]. However, due to the sparse availability of traditional communication infrastructure in such regions, maintaining stable and reliable connectivity remains a significant challenge [152].

Despite their potential, several critical challenges and research gaps must be addressed to enable scalable, secure, and efficient UAV-IoT integration. UAVs are inherently energy- and resource-limited platforms, which restrict their flight time, coverage range, and communication duration. This limitation becomes more pronounced in large-scale IoT networks where continuous monitoring or frequent data relaying is required. There is a pressing need for energy-aware UAV scheduling, recharging strategies (e.g., via solar or wireless charging), and joint mobility-communication optimization. Coordinating a fleet of UAVs over dense or wide-area IoT deployments introduces significant scalability challenges. Maintaining reliable communication links, managing collision avoidance, and orchestrating task allocation become increasingly complex as the number of UAVs and

Artificial Intelligence for Unmanned Aerial Vehicles: Sensing, Communication, and Computing, First Edition. Shuyan Hu et al.

IoT nodes grows. Decentralized control algorithms and swarm intelligence remain underexplored in this context. The mobility of UAVs introduces a highly dynamic network topology, leading to fluctuating link quality, intermittent connectivity, and potential data loss. Designing robust protocols for routing, handover, and fault tolerance is critical, yet current solutions often assume static or semi-static scenarios.

UAV-assisted IoT systems are vulnerable to eavesdropping, spoofing, jamming, and unauthorized data collection. While physical layer security and authentication schemes are being developed, comprehensive, lightweight, and real-time security frameworks remain a key research gap—especially in multi-hop or edge-computing UAV scenarios. There is a lack of standardized communication protocols, interfaces, and architectures that enable seamless interoperability between UAVs, IoT devices, and terrestrial infrastructure. This impedes cross-platform deployment and integration with existing IoT ecosystems. Airspace regulations, safety requirements, and weather variability impose significant limitations on UAV deployment. These constraints are often neglected in algorithmic designs, leading to a gap between theoretical performance and real-world feasibility.

Figure. 5.1 presents an aerial EdgeIoT framework that utilizes the high mobility of UAVs to broaden the coverage of edge-enabled IoT systems [360]. In this setting, IoT devices, constrained by limited communication ranges, temporarily store computation tasks in local queues until a UAV is available for task offloading [75]. Equipped with lightweight edge servers, UAVs can dynamically move closer to spatially dispersed IoT devices, establishing high-quality line-of-sight (LoS) communication links that enhance channel conditions and facilitate rapid task transmission [155]. However, this setup introduces several practical challenges. When the UAV selects a particular device for task offloading, tasks queued on other devices risk being discarded if buffer capacities are exceeded or if they become outdated. Moreover, the UAV's continuous motion induces fluctuating channel conditions, which can degrade the offloading success rate and lead to task failures. These offloading inaccuracies contribute to buffer overflows and elevate the overall task missing rate (TMR). Given that the UAV typically lacks real-time visibility into the buffer states and channel qualities of all IoT devices, effectively scheduling task offloading, while simultaneously controlling the UAV's speed and heading, is critical to reducing TMR and ensuring efficient operation in aerial EdgeIoT systems.

This chapter aims to address the problem of jointly optimizing UAV navigation and task offloading decisions in real time, under the constraint that the UAV lacks prior knowledge of IoT devices' task queues, battery levels, and channel conditions. To maximize the number of successfully offloaded computation tasks, we formulate an optimization problem constrained by the devices' energy availability, their processing capacity, and the UAV's flight speed limits. To tackle the

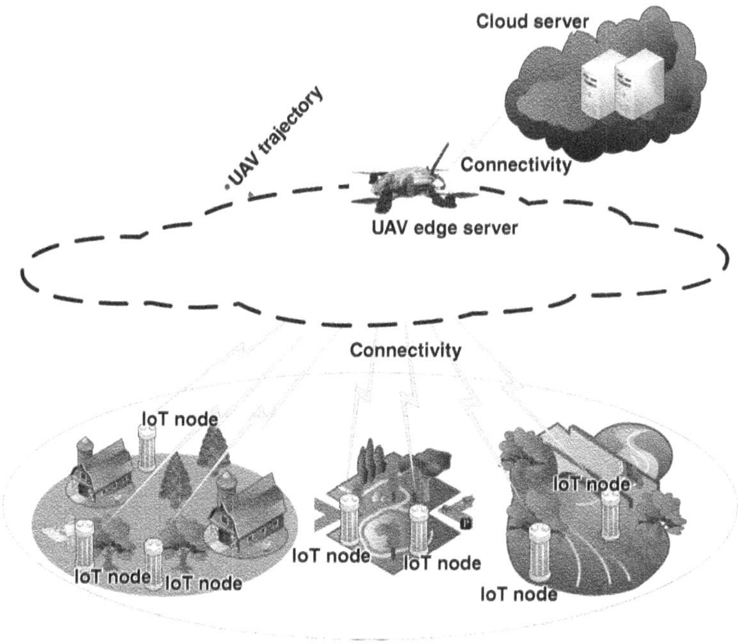

Figure 5.1 In an aerial EdgeIoT, the UAV carrying a lightweight edge server controls the trajectory and patrols over the area of interest. The IoT devices can offload their computation tasks to the UAV for processing or forwarding to the remote cloud server. *Source*: Hitec RCD USA, Inc.

uncertainty in network states, we employ an advantage actor-critic (A2C) framework to generate continuous control actions for the UAV, including trajectory adjustments and scheduling decisions. Notably, IoT devices deployed within the same geographic area often experience correlated environmental factors, such as solar or wind conditions, which can influence both their energy harvesting capabilities and task generation rates. Similarly, spatial proximity can lead to shared variations in wireless channel quality. To exploit these inherent correlations, we propose a graph neural network-assisted A2C model (GNN-A2C), where the GNN captures the underlying relational structure among devices and guides the training of the UAV's decision-making process within the A2C architecture.

The GNN models the system as a graph where nodes represent individual network states and edges capture the dependencies or relationships among them [295]. Within this framework, each node exchanges its hidden features with neighboring nodes through iterative message passing. This process continues over multiple steps, allowing information to flow across the graph until a stable representation is achieved for each node. These final node embeddings are then utilized to infer or forecast future network behavior. To handle the temporal dynamics in the UAV-to-ground communication links and the task generation patterns at IoT devices, the GNN-A2C framework incorporates a long short-term memory (LSTM) component. The LSTM leverages the learned GNN features to predict time-dependent variations in both communication and computational workloads.

This chapter is structured as follows. Section 5.2 reviews existing research on task offloading and UAV trajectory optimization in MEC. Section 5.3 defines the system model for the aerial EdgeIoT scenario. Section 5.4 introduces a new joint optimization strategy that combines UAV cruise control with task offloading to enhance computational efficiency. Due to the combinatorial nature of the problem and the lack of real-time visibility into network conditions, a novel GNN-A2C architecture is developed. This hybrid framework integrates a GNN to guide the training of the UAV's continuous control decisions within the A2C reinforcement learning model, leveraging latent correlations among network features to enhance decision-making. Section 5.5 details the design of the GNN-A2C framework.

In Section 5.6, the implementation setup and performance evaluation of GNN-A2C are discussed. The GNN-A2C framework can be implemented using Python 3.9 and Google TensorFlow on a 64-bit Ubuntu 18.04 Linux workstation. Simulation results confirm that our approach substantially lowers the TMR in aerial EdgeIoT scenarios while efficiently managing UAV navigation to execute offloaded computation tasks. Lastly, Section 5.7 concludes the chapter.

5.2 Existing Techniques

This section reviews recent work on task offloading and trajectory optimization in UAV-enabled MEC. In [270], a UAV equipped with an edge server is used to support ground-based IoT devices by processing their computation tasks. The study proposes a UAV placement strategy that selects optimal hovering locations over task-intensive regions to maximize service coverage. In [116], a bit-level resource management approach is investigated, where each ground device computes part of its data locally and offloads the remainder to the UAV. The proposed method jointly optimizes bit partitioning, transmission power levels for both UAV and devices, and UAV trajectory to minimize the ground device's energy consumption. Similarly, [108] explores a UAV-assisted MEC framework that jointly optimizes flight path planning, task offloading, and transmission scheduling to minimize the total system delay. Meanwhile, [349] employs a Poisson point process to model the ground device distribution and analyzes how UAV response time and computing resources can be provisioned to ensure latency guarantees, based on the probability of successful task transmission.

In [43], a UAV relies on terrestrial cellular base stations to offload its computational workload. By jointly optimizing the UAV's flight path and offloading schedule, considering its speed constraints and the computing capacities of base stations, the overall mission duration is minimized. Similarly, [340] investigates how trajectory planning in conjunction with task offloading can reduce both energy usage and mission completion time. To manage the non-convex nature of the problem, a path discretization approach is adopted, allowing the formulation to be separated into two tractable subproblems, which are addressed using successive convex approximation (SCA). In [352], UAV energy efficiency is further improved by jointly controlling transmission power, communication time, and offloading strategies. Given the complexity of the optimization, a decomposition framework is applied, breaking the problem into subproblems that are solved via Lagrangian duality and SCA techniques. Meanwhile, [263] explores how the UAV's route, task scheduling, and resource distribution can be collaboratively optimized to meet specific latency demands while minimizing energy expenditure.

In [157], a joint strategy involving transmit power control, task offloading decisions, and UAV trajectory design is proposed to enhance the UAV's energy efficiency. Given limited information about the ground devices, a mobility prediction model is incorporated to guide the UAV's path planning. In [322], the focus shifts to optimizing the UAV's discrete way point selection along with its allocation of communication and computational resources, aiming to minimize service delays for IoT devices. Meanwhile, [65] introduces a method for balancing task data distribution and UAV trajectory planning to promote fairness in energy consumption among ground devices.

In environments where the MEC system is not fully known, deep reinforcement learning (DRL) can be employed in [172] to dynamically plan the UAV's trajectory for servicing ground devices, while ensuring energy-efficient operation under quality-of-service constraints. In [348], deep Q-networks (DQN) and DRL techniques are used to jointly optimize the UAV's flight path and the proportion of tasks offloaded, aiming to minimize energy consumption for both the UAV and the IoT devices. In the context of wireless-powered sensor networks, studies such as [151, 153] apply deep deterministic policy gradient (DDPG) and DQN-based algorithms to coordinate UAV trajectories and allocate resources efficiently, to reduce buffer overflows at ground-based sensors.

Most existing studies concentrate on optimizing UAV trajectories and resource management to reduce mission duration, expand service coverage, or enhance energy efficiency, typically without accounting for the correlations among network states. In contrast, this work addresses a critical issue in aerial EdgeIoT systems: ineffective cruise control can lead to failed task offloading at selected IoT devices, while buffered tasks at unselected devices may become outdated and discarded. To tackle this, we propose a novel GNN-A2C framework, where the GNN captures underlying correlations in network conditions and guides the reinforcement learning process to improve UAV decision-making.

Particularly, GNN is a neural network utilizing graph structures to capture the dependence of graphs via message passing between the vertices of graphs [358]. For message passing, the vertices are connected by edges, where the hidden representations of the vertices are iteratively passed between neighboring vertices through the edges. The hidden representations are propagated throughout the graph iteratively until a fixed point is found. The final hidden representation is used to predict the features of the vertices. Although GNN is inspired by a convolutional neural network (CNN), which uses convolutional filters to reduce the number of parameters in fully connected deep neural networks, the convolution operations in the CNN are only applicable to grid-like image data in the Euclidean space. GNN redefines the convolution operation of graphs, generalizing the convolution operations performed in CNN to convolutions performed on arbitrarily structured data.

5.3 System Model

In this section, the system model of aerial EdgeIoT is presented. Table 5.1 lists the notations used in this chapter.

The studied aerial EdgeIoT consists of N ground devices, which offload computing tasks to a remote cloud server via the UAV, as depicted in Figure. 5.1. We consider that each ground device i can offload a part of its computation load, i.e., a computation task, to the UAV equipped with a portable edge server through a wireless link. The UAV patrols over the target field, while scheduling the ground devices for the task offloading. The battery level of device i ($\in [1, N]$) at time t is $b_i(t)$. Since the device's battery capacity is limited, $b_i(t) \leq E$ with E (in Joules) being the battery capacity of the ground device.

Table 5.1 Notation and definition.

Notation	Definition
N	Total number of IoT devices
$\mathcal{G}(\mathcal{V}, \mathcal{E}; \mathbf{w})$	A graph for modeling the network states
$\mathcal{V}, \mathcal{E}$	A vertex or an edge of the graph $\mathcal{G}(\mathcal{V}, \mathcal{E}; \mathbf{w})$
$\mathbf{f}$	Feature representation in GNN
K	Total number of layers that are composed in the graph
v_t	Instantaneous speed of the UAV
b_t^{UAV}	Battery level of the UAV
(x_t, y_t, z)	Location of the UAV
$\Lambda_i(t)$	Offloaded tasks of device i to the UAV

S_α, S_β	Network states
a_{S_α}	The action taken by the UAV at state S_α
i_{S_α}	The selected ground device at state S_α
$\mathcal{Q}$	Maximum number of tasks of the ground device
E	Battery capacity of the ground device
$b_i(t)$	Battery of device i
$q_i(t)$	Number of tasks generated at device i
$H_i(t)$	SNR received at the UAV from device i
ep	Number of training episodes
t_{learning}	Training time of each episode
μ	Discount value of the action-value function
$\nabla\Gamma(\vartheta)$	Policy gradient in A2C

It is assumed that the ground devices undergo a random task generation following the Poisson distribution. The tasks are buffered at the radio transmitter of the ground device and wait to be offloaded to the edge UAV, where the offloading of the buffered tasks follows the first-come-first-served discipline. Particularly, the number of tasks generated by device i at t is $q_i(t) \leq \mathcal{Q}$, where $\mathcal{Q}$ denotes the maximum number of tasks that can be buffered at the ground device. The buffer size of the device is finite. As a result, the newly generated tasks have to be dropped if $q_i(t) > \mathcal{Q}$, i.e., the buffer overflows.

Let (x_t, y_t, z) denote the position of the UAV at time t, and the UAV flies at a constant altitude of z meters [173]. With safety considerations, the instantaneous flight speed of the UAV, denoted by v_t, has to be no larger than the maximum speed $V_{\max}$, which depends on the maximum available thrust of the motors, the weight of the UAV, and the structural strength. Thus, we have $v_t \leq V_{\max}$. Δt is the time when the UAV moves from (x_t, y_t, z) to (x_{t+1}, y_{t+1}, z), respectively. The acceleration at (x_t, y_t, z) can be given by

$$\Delta v_t / \Delta t = (v_{t+1} - v_t)/\Delta t, \tag{5.1}$$

where

$$0 \leq \Delta v_t / \Delta t \leq V_{\max}/\Delta t. \tag{5.2}$$

Considering aeronautics, a smooth turn mobility is applied to the UAV, where the change of the heading direction follows an arc of a circle [148]. Suppose that θ_t is the turning angle of the UAV at time t, and (x_t^c, y_t^c, z) are the coordinates of the circle center at time t. Thus, we have

$$\theta_t = \arctan\left(\frac{y_{t+1} - y_t^c}{x_{t+1} - x_t^c}\right) - \arctan\left(\frac{y_t - y_t^c}{x_t - x_t^c}\right). \tag{5.3}$$

Notably, it is assumed that the UAV does not move backward, i.e., $\theta_t \neq 2\pi$.

Let b_t^{UAV} denote the battery level of the UAV at time t, which can be measured by onboard sensors in real-time. Suppose that the UAV has a solar panel to harvest energy and a rechargeable battery to store the energy. The harvested solar power of the UAV at time t can be given by [122]

$$\Delta b_t^{\text{UAV}} = \psi C_{\text{solar}} P_{\text{solar}} \exp\left(\frac{-\varkappa}{\cos \phi_t^{\text{solar}}}(1 - 2.2556 \times 10^{-5} z)^{5.2561}\right), \tag{5.4}$$

where $\psi \in (0, 1)$ and C_{solar} denote the charging efficiency and size of the solar panel, respectively. P_{solar} is the constant power intensity of the solar beams. $\varkappa > 0$ is the sum atmospheric extinction. ϕ_t^{solar} denotes the solar zenith angle at time t.

The propulsion energy of the UAV is given by [117]

$$\Delta b_t^{\text{UAV}'} = P_0\left(1 + \frac{3v_t^2}{\omega_t^2}\right) + P_0'\left(\sqrt{1 + \frac{v_t^4}{4v_0^4}} - \frac{v_t^2}{2v_0^2}\right)^{1/2} + \frac{1}{2}\xi_{\text{drag}}\rho_{\text{air}}\xi_{\text{rotor}}\xi'_{\text{rotor}}v_t^3, \tag{5.5}$$

where P_0 and P_0' are two constants. ω_t is the tip speed of the rotor blade. v_0 is the mean rotor induced velocity in hover. ξ_{drag} and ξ_{rotor} denote the fuselage drag ratio and rotor solidity, respectively. ρ_{air} and ξ'_{rotor} denote the air density and rotor disc area, respectively. Thus, we have

$$b_t^{\text{UAV}} + \Delta b_t^{\text{UAV}} - \Delta b_t^{\text{UAV}'} > b_0^{\text{UAV}}, \tag{5.6}$$

where b_0^{UAV} is the minimum safe battery level of the UAV.

5.4 Joint Optimization of Cruise Control and Task Offloading

In this section, joint optimization of cruise control and task offloading in aerial EdgeIoT is formulated to maximize the computation tasks offloaded from the IoT devices to the UAV. Given the coordinates of the UAV (x_t, y_t, z), the flying speed v_t can be decomposed to x and y components, i.e., v_t^x and v_t^y. The location of the UAV at time $t + 1$ can be given by

$$\begin{cases} x_{t+1} = x_t + v_t^x \cdot \Delta t \\ y_{t+1} = y_t + v_t^y \cdot \Delta t \end{cases}, \tag{5.7}$$

where Δt is the flight duration from t to $t + 1$ [69]. Assume that the location of ground device i is (x_i, y_i). Based on (5.7), the distance between device i and the UAV is

$$d_i(t) = \sqrt{(x_t - x_i)^2 + (y_t - y_i)^2 + z^2}. \tag{5.8}$$

Due to potential obstructions between the IoT device and the UAV, the propagation path yields a mixed LoS/non-LoS channel model [150], which contains two components

$$\begin{cases} \delta_i^{\text{LoS}}(t) = 20\log(f_c) + 20\log(4\pi/\upsilon_c) + 20\log(d_i(t)) + \eta_{\text{LoS}} \\ \delta_i^{\text{NLoS}}(t) = 20\log(f_c) + 20\log(4\pi/\upsilon_c) + 20\log(d_i(t)) + \eta_{\text{NLoS}} \end{cases}, \tag{5.9}$$

where f_c and υ_c denote the carrier frequency and the speed of light, respectively. η_{LoS} and η_{NLoS} stand for the excessive path loss of LoS and non-LoS, respectively. η_{LoS} and η_{NLoS} are two constants dependent on different applications [260].

Let $\Pr_i^{\text{LoS}}(t)$ denote the LoS probability, as given by

$$\Pr_i^{\text{LoS}}(t) = \frac{1}{1 + a\exp(-b[\varphi_i(t) - a])}, \tag{5.10}$$

where a and b are two Sigmoid function parameters. $\varphi_i(t)$ is the elevation angle between device i and the UAV at t. The SNR received at the UAV from device i, denoted by $H_i(t)$, is given by

$$H_i(t) = \sqrt{\frac{c_2^{-1}\text{In}(\frac{c_1}{\epsilon})(2^{\rho_i(t)} - 1)}{(H_i^{\text{LoS}}(t))^2}}, \tag{5.11}$$

where c_1 and c_2 are two constants, ϵ is the bit error rate (BER), and ρ_i is the modulation scheme of the IoT device. According to (5.9) and (5.10), we can obtain

$$H_i^{\text{LoS}}(t) = \Pr_i^{\text{LoS}}(t) \times \delta_i^{\text{LoS}}(t) + \Pr_i^{\text{NLoS}}(t) \times \delta_i^{\text{NLoS}}(t), \tag{5.12}$$

where $\Pr_i^{\text{NLoS}}(t) = 1 - \Pr_i^{\text{LoS}}(t)$.

According to (5.11), a regression model can map the SNR to the number of computation tasks successfully collected by the UAV edge server. The regression model is given by [147]

$$\Lambda_i(t) = \left(1 - \frac{1}{2}e^{\beta_1 - \beta_0 H_i(t)}\right)^{8(2f-l)}, \tag{5.13}$$

where β_0 and β_1 are two constants in the regression model. β_0 controls the shape of the regression curve and β_1 induces horizontal shifts of the curve. f and l denote the frame size and preamble size of a computation task offloaded to the UAV, respectively. We have $f > l$, as a frame must be longer than the preamble.

Therefore, maximizing the number of computation tasks offloaded to the UAV can be formulated as

$$\textbf{OPT}: \quad \max_{i_t,(x_t,y_t,z),(v_t^x,v_t^y)} \quad i_t\Lambda_i(t) \tag{5.14}$$

$$s.t.: (5.2), (5.6) \qquad 1 \leq i \leq N \tag{5.15}$$

$$0 \leq i_t \leq 1 \tag{5.16}$$

$$1 \leq t \leq T \tag{5.17}$$

$$q_i(t) \leq Q \tag{5.18}$$

where $i_t = 1$ indicates that device i is selected for task offloading at time t; otherwise, $i_t = 0$. Constraint $q_i(t) \leq Q$ indicates that the number of generated computation tasks at device i has to be smaller than the threshold Q.

5.5 Combinatorial GNN-enabled DRL

This section investigates the combinatorial learning architecture of GNN-enabled DRL. In the proposed structure, GNN extracts features and hidden connections of the network states for future state prediction. A2C utilizes the learning outcomes of GNN to train the action of the cruise control and task offloading.

5.5.1 Extracting Feature Correlation with GNN

To optimize jointly the cruise control and task offloading in practice (i.e., the **OPT** problem formulated by (5.14)~(5.18)), the UAV edge server needs to be aware of the *a-priori* knowledge of all the IoT devices, e.g., $q_i(t)$ and $H_i(t)$, across the entire time horizon. Moreover, collecting the real-time information of $q_i(t)$ and $H_i(t)$ during the flight can result in large communication overheads between the UAV and the IoT devices. Therefore, the **OPT** problem could only be optimized offline. To achieve optimal cruise control and task offloading online, we put forth the new combinatorial learning architecture, where GNN predicts the network states via modeling the dependencies between the states.

A network state consists of the battery levels of the IoT devices and the UAV at time t, i.e., $b_i(t)$ and b_t^{UAV}, the number of generated tasks at the IoT device $q_i(t)$, the channel condition $H_i(t)$, the location of the UAV, and the number of tasks of device i being offloaded to the UAV till time t, i.e., $D_i(t)$. Therefore, the network state, denoted by S_α, is given by

$$S_\alpha = \{b_i, q_i, H_i, D_i, b^{\text{UAV}}, (x, y, z)\}, \quad \forall i \in [1, N]. \tag{5.19}$$

The UAV edge server takes actions along the trajectory to control its heading and speed, while the scheduled IoT devices offload the computation tasks to the UAV. The action is given by

$$a_{S_\alpha} = \left\{ \left(x(S_\beta), y(S_\beta), z \right), \left(v_x(S_\alpha), v_y(S_\alpha) \right), i_{S_\alpha} \right\}, \tag{5.20}$$

where $(x(S_\beta), y(S_\beta), z)$ is the next location of the UAV. $(v_x(S_\alpha), v_y(S_\alpha))$ is the projection of the speed of the UAV on x- and y-axes. i_{S_α} indicates the selected IoT device at state S_α. We define an action space $\mathcal{A}$ that consists of all the actions the UAV can take to optimize cruise control and task offloading.

The correlation between the network states can be described by a graph $\mathcal{G}(\mathcal{V}, \mathcal{E}; \mathbf{w})$, where a vertex $\mathcal{V}$ in the graph represents one of the network states S_α and the presence of an edge $\mathcal{E}$ encodes the fact that the network states correlate with each other. The vertex space is denoted by $\mathcal{V}_\mathcal{G}$. $\mathbf{w}$ collects the weights of the edges in the graph.

The inputs of $\mathcal{G}(\mathcal{V}, \mathcal{E}; \mathbf{w})$ contain the vertices and edges of the graph, and the vertex representation matrix. Moreover, the GNN consists of many input, output, and hidden layers. Let K denote the total number of layers that are composed in $\mathcal{G}(\mathcal{V}, \mathcal{E}; \mathbf{w})$. In the k-th layer, $w_\mathcal{V}^k$ is a learnable vector of the weights of $\mathcal{V}$'s edges, where $w_\mathcal{V}^k \subset \mathbf{w}$. The hidden state of the vertex $\mathcal{V}$ can be given by

$$\mathbf{f}_\mathcal{V}^k = \Phi^k\left(\mathbf{f}_\mathcal{V}^{k-1} \oplus agt^k\left(\{\mathbf{f}_{\mathcal{V}',\mathcal{E}}^{k-1} : (\mathcal{V}, \mathcal{V}') \in \mathcal{E}^k\}_{\mathcal{E}^k \in \mathbb{R}^\mathcal{E}}\right); w_\mathcal{V}^k\right), \tag{5.21}$$

where $\oplus$ denotes the embedding summation operation; $\Phi^k(\cdot)$ is a nonlinear activation function (e.g., $\tanh(\cdot)$ or $ReLU(\cdot)$); $\mathbf{f}_\mathcal{V}$, $\mathbf{f}_\mathcal{E}$, and $\mathbf{f}_{\mathcal{V}'}$ are the representations of the vertex $\mathcal{V}$, edge $\mathcal{E}$, and neighbors of $\mathcal{V}$, respectively; $\mathcal{E}^k$ collects the edges at the k-th layer; $\mathbb{R}^\mathcal{E}$ denotes the hidden state dimension; and $agt^k(\cdot)$ is the aggregation function at the k-th layer, which maps the neighborhood information from different relations into a vector, e.g., mean aggregation and attention aggregation. $\mathbf{f}_\mathcal{V}^k$ can be initialized by $\mathbf{f}_\mathcal{V}^0 = \mathcal{V}$.

Based on (5.21), the graph generation loss, denoted by $\mathcal{L}_\mathcal{G}^k$, can be given by

$$\mathcal{L}_\mathcal{G}^k = \sum_{\mathcal{V} \in \mathcal{V}_\mathcal{G}} -\log(\Phi^k(\Omega(\mathbf{f}_\mathcal{V}^k))), \tag{5.22}$$

where Ω is a multilayer perceptron (with $\tanh(\cdot)$ in our work). The input of Ω at the k-th layer is the node embedding at the previous layer. The output is a scalar which is then fed into a nonlinear activation function $\Phi^k(\cdot)$.

5.5.2 GNN-A2C for Joint Cruise Control and Task Offloading

Figure 5.2 demonstrates the proposed GNN-A2C combinatorial learning structure, where network states are pretrained in GNN to predict the future states for training the UAV's actions in A2C, i.e., a_{S_α} in (5.20).

The reward of the A2C training, denoted by $R\{S_\beta | S_\alpha^\mathcal{G}, a_{S_\alpha^\mathcal{G}}\}$, measures the number of offloaded tasks to the UAV, as given by

$$R\{S_\beta | S_\alpha^\mathcal{G}, a_{S_\alpha^\mathcal{G}}\} = \min_{\forall i_{S_\alpha^\mathcal{G}} \in [1,N]} \Lambda_{i_{S_\alpha^\mathcal{G}}}(S_\alpha^\mathcal{G}), \tag{5.23}$$

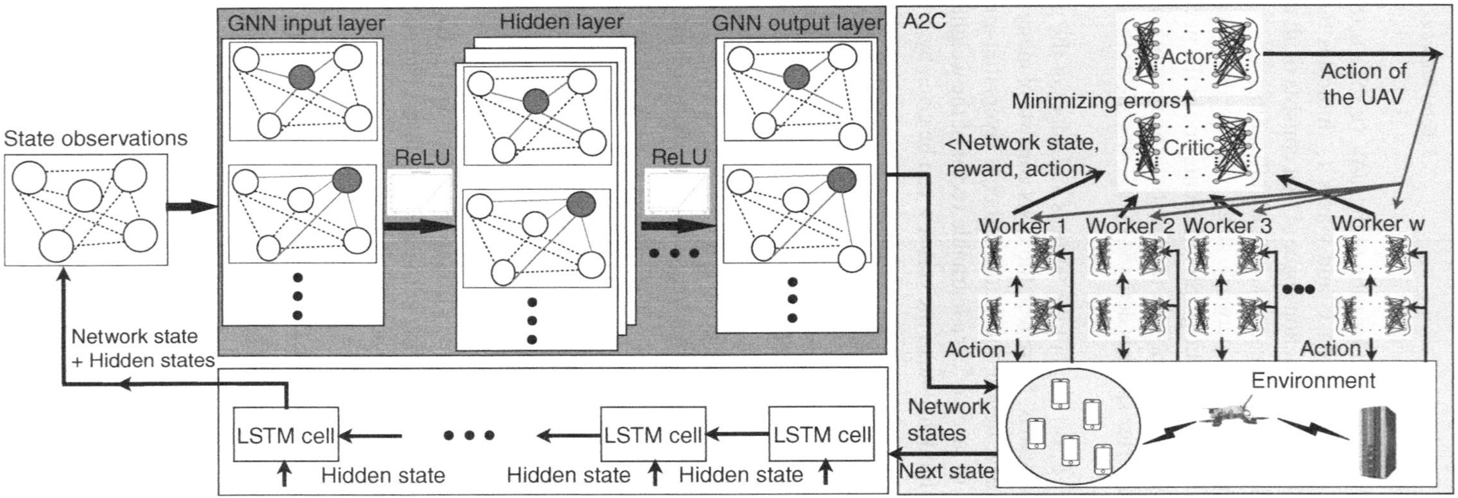

Figure 5.2 The combinatorial learning architecture based on GNN and A2C for aerial EdgeIoT. *Source*: Adapted from Li et al. [363]/IEEE.

where $S_\alpha^{\mathcal{G}}$ denotes the network state trained by the GNN. The action $a_{S_\alpha^{\mathcal{G}}}$ is carried out by the UAV and the network state transits from $S_\alpha^{\mathcal{G}}$ to S_β. In particular, the channel condition $H_i(t)$ in the network state $S_\alpha^{\mathcal{G}}$ indicates the required bandwidth of the selected IoT device, which can be affected by the transmission delay and jitter. According to the spatial and temporal channel conditions, GNN-A2C trains the cruise and speed of the UAV while selecting the IoT devices to maximize $R\{S_\beta | S_\alpha^{\mathcal{G}}, a_{S_\alpha^{\mathcal{G}}}\}$.

The actor neural network in the GNN-A2C approximates cruise control and task offloading rules and prioritizes data streams at the network state. In contrast to widely adopted value-based approaches, e.g., [196], GNN-A2C is a policy-based model-free structure, which directly parameterizes the policy $\pi\{S_\beta | S_\alpha^{\mathcal{G}}, a_{S_\alpha^{\mathcal{G}}}; \vartheta\}$. ϑ is updated by gradient ascents to approximate the expected reward $\mathbb{E}_\vartheta[\cdot]$. Therefore, the actor can be updated by the policy gradient, i.e.,

$$\nabla\Gamma(\vartheta) = \mathbb{E}_\vartheta[\nabla_\vartheta \log \pi\{S_\beta | S_\alpha^{\mathcal{G}}, a_{S_\alpha^{\mathcal{G}}}; \vartheta\} A(S_\alpha^{\mathcal{G}}, a_{S_\alpha^{\mathcal{G}}})], \tag{5.24}$$

where $A(S_\alpha^{\mathcal{G}}, a_{S_\alpha^{\mathcal{G}}})$ denotes the advantage function. The advantage function measures the difference between the action-value function $Q\{S_\beta | S_\alpha^{\mathcal{G}}, a_{S_\alpha^{\mathcal{G}}}\}$ and the state-value function $V(S_\alpha^{\mathcal{G}})$ [186]. The advantage function can be estimated by the temporal difference (TD) error, as given by

$$\begin{aligned} A(S_\alpha^{\mathcal{G}}, a_{S_\alpha^{\mathcal{G}}}) &= Q\{S_\beta | S_\alpha^{\mathcal{G}}, a_{S_\alpha^{\mathcal{G}}}\} - V(S_\alpha^{\mathcal{G}}) \\ &\approx R\{S_\beta | S_\alpha^{\mathcal{G}}, a_{S_\alpha^{\mathcal{G}}}\} + \mu V(S_\beta | S_\alpha^{\mathcal{G}}, a_{S_\alpha^{\mathcal{G}}}) - V(S_\alpha^{\mathcal{G}}), \end{aligned} \tag{5.25}$$

where μ is a discount value $(0 < \mu < 1)$.

The critic neural network in the GNN-A2C estimates the action-value function to evaluate the actions of the actor at each state. The loss function of the critic neural network, denoted by Γ_{loss}^c, defines the estimation error, as given by

$$\Gamma_{\text{loss}}^c = A(S_\alpha^{\mathcal{G}}, a_{S_\alpha^{\mathcal{G}}})^2. \tag{5.26}$$

At the updating step, the least squares TD is minimized to update the critic at each training step, where the state-value function is optimized by

$$a_{S_\alpha^{\mathcal{G}}}^* = \arg\min \Gamma_{\text{loss}}^c. \tag{5.27}$$

To implement the proposed GNN-A2C structure, Algorithm 1 is developed for the new joint UAV cruise control and task offloading in aerial EdgeIoT.

Algorithm 1: GNN-A2C for joint cruise control and task offloading in aerial EdgeIoT

1. Initialize: $\mathcal{G}(\mathcal{V}, \mathcal{E}; \mathbf{w})$, $\mathcal{V}_\mathcal{G}$, S_α, a_{S_α}, t_{learning}, $\mathcal{M}_{\text{replay}}$.
for episode $= 1, \cdots, T_{ep}$ **do**
  Map the network states S_α to the vertexes $\mathcal{V}$ of the graph.
  GNN:
  for Vertex $\mathcal{V} \in \mathcal{V}_\mathcal{G}$ **do**
    for $k = 1$ to K **do**
      For each vertex, $\mathbf{f}_\mathcal{V}^k \leftarrow$ (5.21).
      Calculate the graph generation loss, $\mathcal{L}_\mathcal{G}^k \leftarrow$ (5.22).
      Optimize $w_\mathcal{V}^k \in \mathbf{w}'$ to minimize $\mathcal{L}_\mathcal{G}^k$.
    end for
  end for
  Obtain $S_\alpha^\mathcal{G}$ according to the optimized graph $\mathcal{G}'(\mathcal{V}, \mathcal{E}; \mathbf{w}')$.
  Store the experience $< S_\alpha^\mathcal{G}, S_\beta, a_{S_\alpha}, R\{S_\beta | S_\alpha^\mathcal{G}, a_{S_\alpha}\} >$ to $\mathcal{M}_{\text{replay}}$.
  A2C:
  Sample random minibatches from $\mathcal{M}_{\text{replay}}$.
  $S_\alpha^\mathcal{G} \rightarrow$ the critic neural network, and calculate the state-value function $V(S_\alpha^\mathcal{G})$.
  The critic $\leftarrow R\{S_\beta | S_\alpha^\mathcal{G}, a_{S_\alpha^\mathcal{G}}\}$.
  $A(S_\alpha^\mathcal{G}, a_{S_\alpha^\mathcal{G}})$ is updated by (5.25).
  $S_\alpha^\mathcal{G}$ and $S_\beta \rightarrow$ the actor neural network.
  The policy gradient $\nabla\Gamma(\vartheta)$ is updated by (5.24). The actor is trained while the critic valuates (5.27) $\rightarrow a^*_{S_\alpha^\mathcal{G}}$.
  The UAV carries out the action $a^*_{S_\alpha^\mathcal{G}}$ in the environment.
  According to the training environment, the state observation S_α is renewed for GNN.
end for

5.6 Implementation and Performance Evaluation

5.6.1 Implementation of GNN-A2C

We implement the proposed GNN-A2C in Python 3.9 with Google Tensorflow, set up on a Linux workstation with 64-bit Ubuntu 18.04. The hardware for training the GNN-A2C has two Nvidia GPUs, one is GeForce GTX 1060 with 3 GB memory, and the other is GeForce RTX 2060 with 6 GB memory. The IoT devices are randomly distributed in a target area of 1000×1000 m^2. The initial position of the UAV is at

(600, 500) m. The highest flying speed is 15 m/s. The maximum transmit power of the IoT device is 100 mW, and the battery energy budget has 800 J.

The maximum number of iterations in GNN is set to 50, and the graph $\mathcal{G}(\mathcal{V}, \mathcal{E}; \mathbf{w})$ is trained in 1000 epochs for extracting the correlated features of the network states. In each iteration, the GNN is evaluated by a validation function every 20 epochs. The validation function provides the loss value of the training and the number of iterations at the convergence. The graph model that achieves the lowest cost is considered the best $\mathcal{G}(\mathcal{V}, \mathcal{E}; \mathbf{w})$, and is used to evaluate the testing data.

The A2C module contains two hidden layers separated from the same input tensor. Each hidden layer has 128 units. Policy gradients, weighted by advantages, define the policy loss. The actions of the UAV are trained in the A2C for 1500 steps, and the discount factor is 0.99. The loss function is calculated as the mean square error between the value estimates and returns, where the value function coefficient for the loss calculation is set to 0.5.

5.6.2 Performance Evaluation

5.6.2.1 Training of the GNN-A2C Framework

Figure 5.3 shows that the TMR of the proposed GNN-A2C framework gradually declines with the growing number of training episodes. This is because the GNN module works with LSTM to extract the feature correlation of the network states and predict the hidden network states, enriching the training environment of the A2C module. Hence, the action of the UAV is sufficiently trained by more state

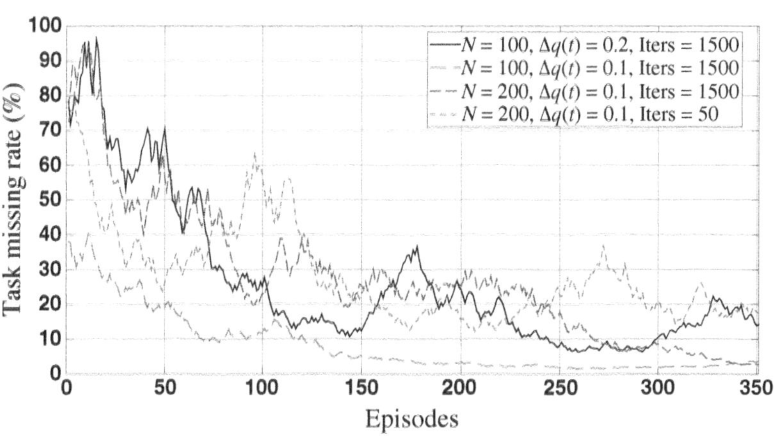

Figure 5.3 TMR in terms of training episodes.

observations S_α with the growing number of episodes. This leads to the fast convergence of the proposed GNN-A2C framework, where the TMR converges in about 150 episodes.

Given 200 nodes and that 10% of the IoT devices generate new tasks each time step (i.e., $\Delta q(t) = 0.1$), the TMR drops from 20% to 5% when the number of training iterations increases from 50 to 1500. Moreover, increasing the number of IoT devices that generate new computation tasks from 10% to 20% can slightly raise the TMR of GNN-A2C from 5% to 12%. This is due to the limited cruising speed of the UAV. Consequently, the IoT devices far away from the UAV may not get served in time.

Figure 5.4 shows the number of offloaded tasks and buffered tasks at each IoT device, where $N = 100$ and $Q = 400$. Each error bar is calculated with the standard deviation over 10 experiments. Thanks to the feature correlation extraction of the GNN-A2C, the UAV's flight speed and heading are trained with prior knowledge of the network state dynamics. This confirms that sufficient training of the UAV's cruise control enables all the IoT devices to timely offload their tasks to minimize the missing computation tasks.

5.6.2.2 Continuous Cruise Control

Figure 5.5 presents the flight trajectories of the UAV controlled by the proposed GNN-A2C framework, given the different number of IoT devices, $\Delta q(t)$ values, distribution patterns of the IoT devices, and learning iterations. Specifically, we can observe in Figure 5.5a and b that the UAV extends its flight coverage when $\Delta q(t)$ increases from 0.1 to 0.8. This is reasonable since a higher $\Delta q(t)$ value leads

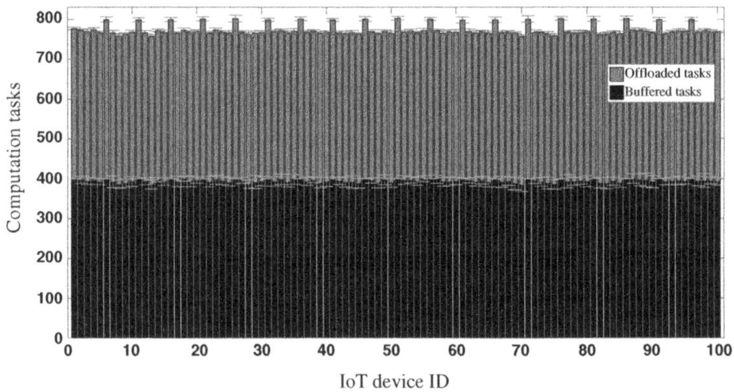

Figure 5.4 Number of offloaded tasks and buffered tasks at each IoT device, where $N = 100$ and $Q = 400$. Each error bar provides the standard deviation of 10 experiments.

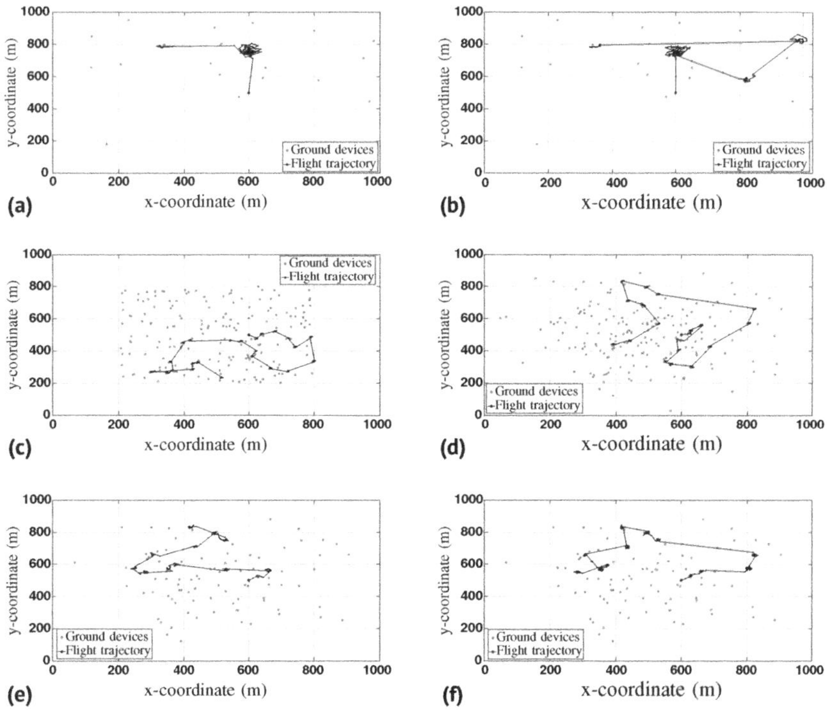

Figure 5.5 Flight trajectories of the UAV, which is controlled by the proposed GNN-A2C framework, given different numbers of IoT devices, $\Delta q(t)$ values, distribution patterns of the IoT devices, and training iterations. (a) $N = 20$ and $\Delta q(t) = 0.1$; (b) $N = 20$ and $\Delta q(t) = 0.8$; (c) $N = 200$ given uniform distribution; (d) $N = 200$ given normal distribution; (e) $N = 100$ and training iterations = 1500; (f) $N = 100$ and training iterations = 3000.

to more IoT devices producing computation tasks. Thus, the UAV adapts its cruise to approach the IoT devices to minimize the TMR.

In Figure 5.5c, the IoT devices are uniformly deployed in a 600 m × 600 m square area. In Figure 5.5d, the deployment follows the normal distribution, where the devices are dispersed in a 1000 m × 1000 m field. The performance validates that GNN-A2C effectively trains the cruise control of the UAV and the task offloading allocation according to the density of the task generation.

Furthermore, Figure 5.5e and f show the flight trajectories of the UAV when the training iterations of the GNN-A2C increase from 1500 to 3000. Once the GNN-A2C is sufficiently trained, the UAV alters its course to serve the devices with the full task buffers or high channel quality to minimize the TMR. This also cross-validates the performance in Figure 5.3.

5.6.2.3 Number of IoT Devices and Their Buffer Capability

Figure 5.6 shows the TMR in terms of different aerial EdgeIoT's scalability. We compare the proposed GNN-A2C framework with DQN- and DDPG-based trajectory planning strategies [139, 154], a pre-planned trajectory and random device selection (PTRS) strategy [254], and a channel-guided cruise control (CGC) strategy [205]. In general, the TMR grows with the number of IoT devices. When $N = 600$, GNN-A2C outperforms DDPG, DQN, CGC, and PTRS by about 16.7%, 43.5%, 63.9%, and 64.1%, respectively. This is because PTRS applies a predetermined flight trajectory of the UAV while the IoT devices are randomly selected to offload the computation tasks. Therefore, PTRS provides the worst. Although the DDPG enables the training of continuous actions (compared to the DQN-based strategy), the hidden states are hardly explored and predicted, resulting in insufficient action training.

Different from the existing solutions, the proposed GNN-A2C framework extracts the features of the network state observation. In contrast, the hidden states are predicted, which guides the actions in A2C to be sufficiently trained. Furthermore, GNN-A2C trains the actions of the UAV in a continuous domain, which optimally controls the real-time flight speed and heading of the UAV.

Figure 5.7 studies the TMR with respect to the maximum number of tasks that can be buffered at the IoT devices. In general, increasing the buffer sizes of the IoT devices reduces the TMR, since more generated computation tasks can be kept at an IoT device until the UAV visits the device. When $Q = 500$, the TMRs of the proposed GNN-A2C are 13.2%, 23.6%, 43.3%, and 54.8% lower than the strategies based on DDPG, DQN, CGC, and PTRS, respectively. The reason is that the GNN

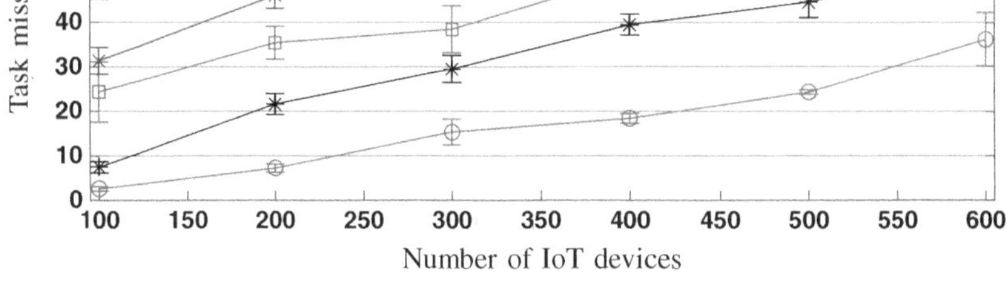

Figure 5.6 TMR vs. the aerial EdgeIoT's scalability N, where Q is 100. Each error bar provides the standard deviation of 10 independent experiments.

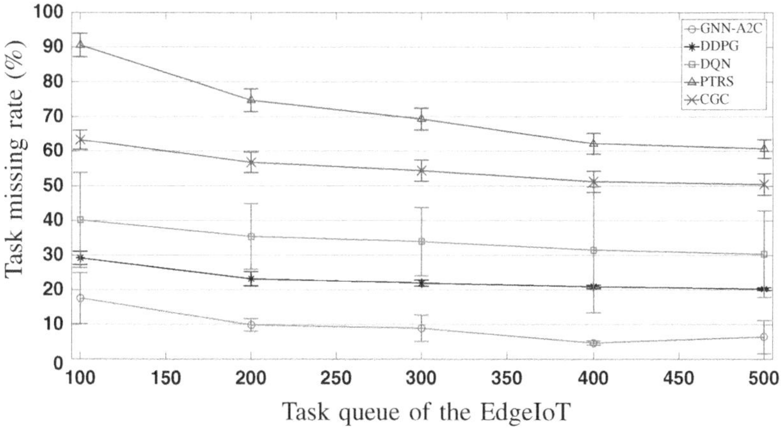

Figure 5.7 TMR vs. $\mathcal{Q}$, where N = 300. Each error bar shows the standard deviation of 10 independent experiments.

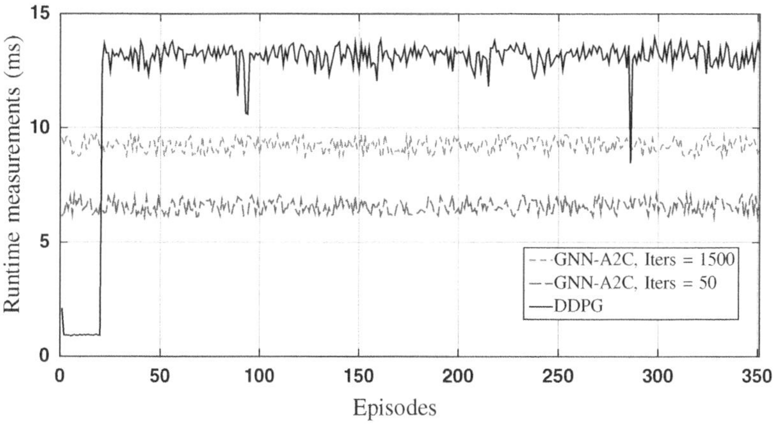

Figure 5.8 Runtime measurement of the GNN-A2C, where the number of training iterations is 50 or 1500.

enables the training of the cruise control and task offloading with observed network states and hidden states in the continuous action space.

5.6.2.4 Runtime of the GNN-A2C Framework

In Figure 5.8, the runtime of the GNN-A2C is measured after 50 or 1500 training iterations in comparison to the DDPG-based strategy. Particularly, the training time of the GNN-A2C with 1500 iterations is around 9 ms in each episode. When the number of training iterations decreases to 50, the training time drops to about 7 ms. Moreover, the DDPG-based strategy consumes 23 episodes to initialize its experience replay, consuming about 1.2 ms. Once the DDPG-based strategy starts the training, the runtime grows to 13 ms at each episode, which is slower than the GNN-A2C framework by at least 4 ms. The DDPG-based strategy trains the action with a random process for action exploration, which takes extra time. In contrast, the GNN-A2C predicts the future network states for training the UAV's actions in A2C at each episode, saving time on the random action exploration of the DDPG-based strategy.

5.7 Conclusion

In this chapter, a new joint optimization of UAV cruise control and task offloading allocation in aerial EdgeIoT was proposed to maximize the number of computation tasks offloaded to the UAV. Since the optimization contains a large solution space while the instantaneous network states are unknown to the UAV, a new deep graph-based DRL framework, i.e., GNN-A2C, was developed to take advantage of the GNN to supervise the training of UAV's actions in A2C, which explores the hidden network states based on the network feature correlation. Numerical studies showed that the GNN-A2C significantly reduces the TMR in aerial EdgeIoT while effectively controlling the flight cruise to process the computation tasks.

6

Radio and Video Surveillance with UAVs

6.1 Introduction

With their excellent flexibility, swift deployment, and wide coverage, UAVs have been extensively applied to search and rescue [14], landscape or power line inspection [359], bushfire monitoring [323], parcel delivery [269], data dissemination [334], relaying [146], eavesdropping [111, 260, 325], and mobile edge computing [313]. Equipped with sensors and cameras, UAVs have also been deployed to track and monitor mobile targets [119]. The UAVs can record the misbehaviors of the targets during tracking as evidence for forensic purposes [12]. However, the targets may notice the presence of the monitor, carry out countermeasures, change their behaviors, and potentially endanger public safety. In this sense, the covertness of the UAVs' operation is crucial, especially for public safety reasons.

Little to no existing studies have investigated the disguise of the UAV monitor during the tailing or surveillance mission. Most existing designs have required a UAV monitor to keep the target within its view for video surveillance and image-aided tracking. Image-aided control approaches have been proposed for a UAV monitor to track a mobile target [120, 121, 124, 354], calculate the 3D geological position of a mobile target [347], or detect a hidden stationary target [314]. In [354], geometric control was performed with nonlinear multi-ranking control, without requirement of a calculable thrust force. In contrast to traditional backstepping approaches, the geometric control could potentially simplify the execution of steps. Relying on the Lyapunov theory, the image-aided method was shown to converge. UAV-based target tailing was pursued in [237] and [320] to address visual obstruction in a city with high-rises. The route of a UAV monitor was designed to achieve the maximum detection rate, which is captured by the probability of keeping the target within its view. A time-varying zero-sum matching was developed in [215] to find the best moving direction for the monitor, who

Artificial Intelligence for Unmanned Aerial Vehicles: Sensing, Communication, and Computing, First Edition. Shuyan Hu et al.

attempts to have the target in the vicinity and sight. Dynamic programming was employed to achieve the lowest sum price about the target-monitor spatial separation. Of all the existing works on UAV-enabled target tailing, e.g., [16, 354], the design of the 3D UAV route was only addressed in [120]. The rest supposed that the monitoring UAV traveled at a fixed altitude and designed 2D routes to require the UAV to fly horizontally. A multi-objective optimization problem was formulated in [120] in pursuit of the least power consumption of a UAV monitor, and the largest number of visible targets. None of the above works have taken disguise into their designs.

Limited power supply to a UAV monitor could restrain tracking and monitoring missions. Most existing UAVs are powered by batteries, limiting the mission time and distance. The trajectory of a UAV monitor should be meticulously designed to elongate the mission, as mentioned in [120, 121, 124]. Realistic power usage models characterizing various flight statuses, including propulsion, thrust, and hanging, should be considered. Yet, the energy usage of the UAV monitor was simplified to a linear function of its velocity in [121] and [120], whereas the power usage of the hovering state was not considered in [124]. Solar power has been increasingly utilized in UAV platforms, e.g., [233]. Many latest UAVs can carry a payload of as much as 10 kg. It is plausible for such UAVs to be equipped with solar panels. The harvested solar power can potentially energize the UAVs' tracking and surveillance missions for sustainability and longevity. However, it is nontrivial to plan the trajectories of solar-powered UAVs due to the non-convexity of the power harvesting process [7].

This chapter focuses on two complementary studies on rotary-wing and fixed-wing solar UAVs, addressing the trade-offs between energy efficiency, dynamic constraints, and visual disguise for covert surveillance.

In the first study, we focus on UAV-on-UAV video tracking using rotary-wing UAVs. The core challenge lies in jointly optimizing propulsion power (horizontal flight), thrust power (vertical movement), and 3D trajectory to minimize energy consumption while the monitor disguises its intention of monitoring by adjusting its trajectory and heading. We solve a non-convex problem formulation incorporating solar power harvesting and rotor dynamics via successive convex approximation (SCA) and proximal difference-of-convex algorithm (PDCAE). We introduce an MPC framework that adapts to partially known target trajectories and further amplifies stealth by dynamically misaligning the monitor's heading with that of the target.

In the second study, we focus on tailing and video surveillance using fixed-wing UAVs. Fixed-wing dynamics introduce new constraints caused by air and load factor during the flight, which couple horizontal and vertical movements. In this study, a novel 3D propulsion model is considered, which captures aerodynamic forces (lift, drag, and thrust) and their impact on energy consumption. We develop a new algorithm that allows a solar-powered, fixed-wing UAV to carry out a disguised tailing and video surveillance mission, where the trajectory of the UAV is optimized to strike a balance between the multiple objectives of distance keeping, elevation variance, and power efficiency. In conjunction with receding horizon control, the proposed algorithm is applied online to deliver adaptive control of the UAV and refine the trajectory on the fly.

Extensive simulations corroborate the merits of our scheme over the baseline schemes with partial (i.e., only distance keeping or altitude changing) or no disguising. As compared with a rotary-wing UAV, the fixed-wing UAV shows its advantage of energy efficiency and sustainability to extend the surveillance mission. Such UAV-enabled visual monitoring can find critical implementations in domestic and civil security.

Notation: $\mathbb{R}^n$ denotes the n-dimensional Euclidean space; $\langle \cdot, \cdot \rangle$ and $|| \cdot ||$ stand for the inner product and Euclidean norm, respectively; dom F is the domain of function F; ∂F denotes the sub-differential of function F;[1] and ∇F stands for the gradient of a continuously differentiable function F. The notations used in the chapter are listed in Table 6.1.

Table 6.1 A list of notations and variables.

Notation	Description
$\mathcal{F}$	A set defining the feasible flight region of the monitor
T, N	Scheduling period and total number of time slots
δ	Duration of each time slot
(x_0, y_0, z_0)	Initial waypoint of the monitor
V_{hm}, V_{vm}	Maximum horizontal and vertical speeds of the monitor
P_0, P_1	Blade profile power and induced power
s, A	Rotor solidity and disc area
U_{tip}	Tip speed of the rotor blade
v_0	Mean rotor induce velocity
ρ, d_f	Atmospheric density and fuselage drag ratio
W	Body weight of the monitor
E_0	Amount of initial energy in the battery

(*Continued*)

1 For a convex function F defined on $\mathbb{R}^n$, its sub-differential at $x_0 \in \mathbb{R}^n$ is defined as $\partial F(x_0) := \{v \in \mathbb{R}^n | F(x) - F(x_0) \geq \langle v, x - x_0 \rangle, \forall x \in \mathbb{R}^n\}$.

Table 6.1 (Continued)

Notation	Description
η_0	Ratio of usable energy in the battery
η, S	Efficiency and size of a solar panel
P_i	Power intensity of solar beams
α	Sum atmospheric extinction
z_l	Minimum altitude of the monitor
D	Maximum 3D monitor-target distance
$c_1,\ c_2$	Coefficients for solar power approximation
$\mu_1,\ \mu_2$	Weights for disguising
t, ℓ	Index of each time slot and each iteration
(a_t, b_t, H)	3D waypoints of the target at time slot t
(x_t, y_t, z_t)	3D waypoints of the monitor at time slot t
Δ_t	Altitude change of the monitor between time slots t and $t-1$
$P_h^t,\ P_v^t$	Propulsion and thrust power of the monitor at time slot t
$P_s^t,\ \hat{P}_s^t$	Harvested solar power and its approximation at time slot t
ϑ_t	Solar zenith angle at time slot t
$d_t,\ \hat{d}_t$	2D and 3D monitor-target distances at time slot t
f_t	Performance of disguising at time slot t
q_t	Auxiliary variable at time slot t
$\lambda_{(\ell)}$	Extrapolation of all the variables at iteration ℓ
β_ℓ	Extrapolation parameter at iteration ℓ

6.2 System Model

In the considered system, a UAV with a 360-degree panoramic camera is employed to track and monitor a suspicious target UAV traveling at a fixed altitude of H (in meters).

Consider a solar-powered, fixed-wing UAV on a disguised tailing and video surveillance mission, where a target UAV flies at the altitude of H (in meters). The mission lasts T seconds, which is evenly split into N time slots with a duration of δ seconds per slot ($T = \delta N$).

We assume that the movement of the target is perfectly predictable; see Figure 6.1.

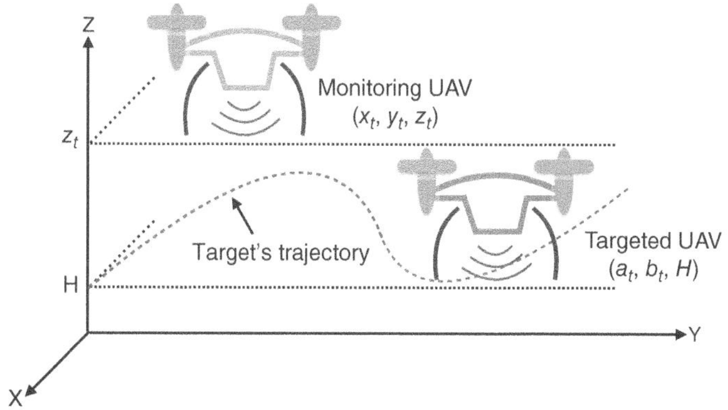

Figure 6.1 A UAV-enabled video surveillance system.

6.2.1 UAV Mobility

6.2.1.1 Rotary-wing UAV

We consider a 3D Cartesian coordinate system. The monitoring mission lasts a finite period of T seconds. We divide the period T evenly into N time slots, i.e., $t = 1, \dots, N$. The duration of a slot, denoted by δ, is so short that the position of the UAVs during each slot can be indicated as a single waypoint. The waypoints of the monitor, denoted by $(x_t, y_t, z_t), \forall t = 1, \dots, N$, are 3D. We suppose that the initial location of the monitor is $(x_0, y_0, z_0) = (0, 0, z_0)$, when the video surveillance starts, with $z_0 > H$. The maximum horizontal and vertical speeds of the monitor are V_{hm} and V_{vm}, respectively. The altitude change is Δ_t between two successive time slots $t-1$ and t. The monitor's mobility constraints, including its initial location and speed constraints, are given by [339, 255]

$$(x_1 - 0)^2 + (y_1 - 0)^2 \le (V_{hm}\delta)^2, \tag{6.1a}$$

$$\Delta_1 := |z_1 - z_0| \le V_{vm}\delta, \tag{6.1b}$$

$$(x_t - x_{t-1})^2 + (y_t - y_{t-1})^2 \le (V_{hm}\delta)^2, \ \forall t, \tag{6.1c}$$

$$\Delta_t := |z_t - z_{t-1}| \le V_{vm}\delta, \ \forall t, \tag{6.1d}$$

where $V_{hm}\delta$ and $V_{vm}\delta$ are the largest horizontal and vertical distances that the monitor can travel between two consecutive slots (i.e., two consecutive waypoints), respectively.

6.2.1.2 Fixed-wing UAV

Let $\boldsymbol{b}_t := (b_{xt}, b_{yt}, H)$ and $\boldsymbol{q}_t := (x_t, y_t, z_t)$ collect the 3D waypoints of the target and the monitor at time slot t, respectively, $t = 1, \dots, N$. Let $\boldsymbol{v}_t := (v_{xt}, v_{yt}, v_{zt})$ and

$\boldsymbol{a}_t := (a_{xt}, a_{yt}, a_{zt})$ collect the velocity and acceleration of the UAV monitor in the 3D space at time slot t, respectively.

The fixed-wing UAV monitor needs to obey the following mobility constraints [332]

$$\boldsymbol{q}_{t+1} = \boldsymbol{q}_t + \boldsymbol{v}_t\delta + \frac{1}{2}\boldsymbol{a}_t\delta^2, \ \forall t, \tag{6.2a}$$

$$\boldsymbol{v}_{t+1} = \boldsymbol{v}_t + \boldsymbol{a}_t\delta, \ \forall t, \tag{6.2b}$$

$$\|\boldsymbol{a}_t\| \leq a_{\max}, \ \forall t, \tag{6.2c}$$

$$\|\boldsymbol{v}_t\| \leq v_{\max}, \ \forall t, \tag{6.2d}$$

$$v_{\min} \leq \|\boldsymbol{v}_t\|, \ \forall t, \tag{6.2e}$$

$$v_{zt}/\|\boldsymbol{v}_t\| \leq \vartheta, \ \forall t, \tag{6.2f}$$

where $a_{\max}$ denotes the maximum acceleration of the UAV monitor; $v_{\min}$ and $v_{\max}$ are the lower and upper bounds of the UAV speed, respectively; and ϑ is the sine of the maximum pitch angle of the UAV monitor. The current location of the monitor depends on its previous location, current velocity, and acceleration, as captured in constraint (6.2a). The current velocity of the UAV monitor relies on its previous velocity and current acceleration, as captured in constraint (6.2b). Constraint (6.2f) makes sure that the UAV monitor ascends by a limited pitch angle. Constraints (6.2c)–(6.2e) are self-explanatory.

6.2.2 Visual Disguise

6.2.2.1 Rotary-wing UAV

Let (a_t, b_t, H) denote the 3D waypoints of the target at time slot t. The time-varying distances between the monitor and the target on the horizontal (x, y)-plane and in the 3D space are given by

$$d_t = \sqrt{(x_t - a_t)^2 + (y_t - b_t)^2}, \ \forall t; \tag{6.3}$$

$$\hat{d}_t = \sqrt{(x_t - a_t)^2 + (y_t - b_t)^2 + (z_t - H)^2}, \ \forall t. \tag{6.4}$$

We assume that the monitor always flies horizontally behind and above the target, i.e., $x_t \leq a_t$, $y_t \leq b_t$, $\forall t$,[2] and $z_t \geq z_l > H, \forall t$, where z_l provides the lower-bound altitude of the monitor. This is because the panoramic camera is typically installed at the bottom of the monitor. In order to successfully track the target,

---

2 Our proposed scheme can be readily applied to the problem without the requirements of $x_t \leq a_t$ and $y_t \leq b_t$, $\forall t$.

the monitor has to travel inside a *feasible flight region* (FFR) specified by the above constraints

$$\mathcal{F} := \{(x_t, y_t, z_t) | (\hat{d}_t)^2 \leq D^2, x_t \leq a_t, y_t \leq b_t, \\ H < z_l \leq z_t, \forall t\}, \tag{6.5}$$

where D specifies the distance for effective visual surveillance given the resolution of the panoramic camera at the monitor. The upper limit of the monitor's altitude is bounded by (6.4).

Different from the hypothesis test extensively considered in covert wireless communication, the monitor delivers covertness by visual disguise. The reason for the consideration of visual disguise is that optical cameras (e.g., including those at the target) can often achieve persistent observations if unobstructed, given the infinite maximum focusing distances of a typical camera lens [169].

We propose that the monitor disguises itself by keeping its distance from the target (at a higher altitude), e.g., (6.3), and constantly changing its altitude Δ_t to confuse the target, as considered in [121]. The constantly changing altitude of the monitor makes it difficult for the target to correctly focus its camera lens on the monitor for a clear view of the monitor. Keeping the monitor's distance from the target makes it hard for the target to get good image resolutions on the monitor, even if its camera is correctly focused.

6.2.2.2 Fixed-wing UAV

The time-dependent 3D distance, i.e., spatial separation, between the UAV monitor and the target is

$$d_t = \|\boldsymbol{q}_t - \boldsymbol{b}_t\|, \ \forall t. \tag{6.6}$$

By imposing $d_t \leq d_{\max}$, the monitor can keep the target in sight and achieve meaningful visual resolution on the target. $d_{\max}$ (in meters) is the maximum allowed spatial separation.

Suppose that the UAV monitor is always traveling at a higher altitude than the target, i.e., $z_t \geq z_l > H, \forall t$, in which z_l is the lowest allowed elevation of the monitoring UAV. This assumption stems from the fact that the 360° full-view surveillance camera is usually placed to the underbody of the UAV monitor. In light of [110], we consider that the monitoring UAV camouflages by maintaining a reasonable distance from the target, and continuously varying the elevation to bewilder the target. The continuously varying elevation of the monitor holds back the target from setting the correct focal length of its camera to the monitor, resulting in blurry sight. Keeping a reasonable monitor-target distance hinders the target from getting satisfying visual resolutions on the monitor, even though its camera could be reasonably concentrated. The *disguise* of the monitor is characterized by its horizontal distance to the target and its continuously varying elevation

$$f_t = \mu_1[(x_t - b_{xt})^2 + (y_t - b_{yt})^2] + \mu_2(z_t - z_{t-1})^2, \ \forall t \tag{6.7}$$

where μ_1 and μ_2 are two weighting coefficients to adjust the weights of the two aspects, respectively. The motivation of separating the disguising performance between the horizontal and vertical movement in (6.7) stems from the fact that the level flight and the ascending (or descending) of a fixed-wing UAV are typically controlled separately by controlling its rudder and elevator, respectively. Clearly, f_t is a convex function of (x_t, y_t, z_t, z_{t-1}) for any time slot t.

We further impose a constraint on the minimum target-monitor distance, as given by

$$d_{\min} \leq d_t, \ \forall t, \tag{6.8}$$

where $d_{\min}$ specifies the minimum distance to be kept from the target to help the monitor with its disguise.

6.2.3 UAV Propulsion and Thrust Powers

6.2.3.1 Rotary-wing UAV

For a rotary-wing UAV with a horizontal speed V_h^t, the propulsion power at time slot t, denoted by P_h^t, is given by [336]

$$P_h^t = P_0\left(1 + \frac{3{V_h^t}^2}{U_{tip}^2}\right) + P_1\left(\sqrt{1 + \frac{{V_h^t}^4}{4v_0^4}} - \frac{{V_h^t}^2}{2v_0^2}\right)^{\frac{1}{2}} + \frac{1}{2}d_f \rho s A {V_h^t}^3, \tag{6.9}$$

where P_0 and P_1 are the constant *blade profile power* and *induced power* in a hovering mode, respectively; U_{tip} is the tip speed of the rotor blade; v_0 denotes the average rotor-induced velocity in the hovering mode; d_f and s denote the fuselage drag ratio and rotor solidity, respectively; and ρ and A are the atmospheric density and rotor disc area, respectively.

When $V_h^t = 0$, (6.9) is the power consumption of the hovering state. We suppose that within each slot t, the UAV maintains a constant speed, as given by

$$V_h^t = \sqrt{(x_t - x_{t-1})^2 + (y_t - y_{t-1})^2}/\delta, \ \forall t. \tag{6.10}$$

By substituting (6.10) into (6.9), we find that the first and the third terms of (6.9) are jointly convex with respect to $(x_t, x_{t-1}, y_t, y_{t-1})$. However, the second term is neither convex nor concave.

The thrust power of vertical flight, i.e., altitude change, can be obtained by [255, equation (6)], [235, equation (7.12)]

$$P_v^t = W(z_t - z_{t-1})/\delta, \ \forall t, \tag{6.11}$$

where W is the body weight of the UAV. Clearly, P_v^t is jointly convex with respect to (z_t, z_{t-1}). The thrust power is positive for (altitude) ascending, and it is negative for (altitude) descending, since energy is saved thanks to gravity [354].

6.2.3.2 Fixed-wing UAV

We first derive the propulsion power of the fixed-wing UAV with a 3D trajectory. As illustrated in Figure 6.2, the fixed-wing UAV aloft undergoes four forces in total, namely, weight, drag, lift, and thrust at any instance [77].

- Weight (W): This is the force incurred by gravity with $W = mg$, in which m represents the UAV's mass, and $g \approx 9.8$ m/s^2 is the gravitational acceleration.
- Drag (D): This is the aerodynamic force in the direction of the airflow. At a zero wind speed, D operates in the reverse way of the UAV's heading.
- Lift (L): This is the force element in the direction perpendicular to the drag and lifting the UAV up.
- Thrust (F): This is the force generated by the UAV's engine, to surmount the drag and push the UAV forward and upward.

Given the UAV flying at a subsonic speed V, the drag is mathematically described as [77, equations (4.1), (4.2), (4.13), and (4.14)]

$$D = c_1 V^2 + \frac{c_2 \kappa^2}{V^2}, \tag{6.12}$$

where $\kappa = L/W$ is the load factor, and c_1 and c_2 are known coefficients relying on the air density, and the wing surface and span of the aircraft [332].

For straight-and-level flight (SLF) and banked level turn (BLT), the power models of the UAV are derived in [332, Appendix A]. In this chapter, we establish the power consumption for the ascending (and descent) of the UAV. We assume that

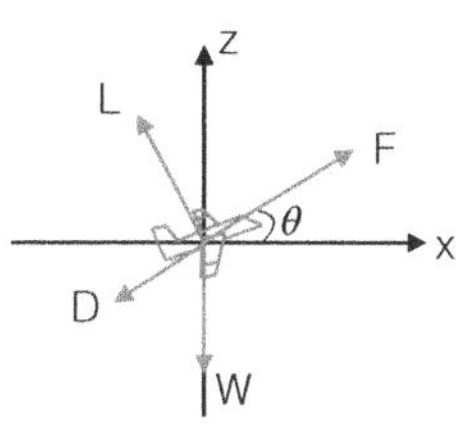

Figure 6.2 An illustration of the forces on an ascending aircraft, where W, D, L, and F stand for weight, drag, lift, and thrust, respectively.

the UAV takes straight ascending or descending (SAD), where the acceleration is in the same direction as the velocity, with no banked turn in any direction. It would be interesting to study in the future the scenario where the UAV ascends or descends spirally, i.e., with centrifugal accelerations (i.e., banked turns) on both the (x, y)-plane and the (y, z)-plane.

As shown in Figure 6.2, the acceleration a and the velocity V of the UAV monitor align with the thrust force F. The forces on the UAV obey the following equations

$$F - D - W\sin\theta = ma, \tag{6.13a}$$

$$L = W\cos\theta, \tag{6.13b}$$

$$\sin\theta = V_z/V, \tag{6.13c}$$

$$\cos\theta = \sqrt{V_x^2 + V_y^2}/V, \tag{6.13d}$$

where V_x, V_y, and V_z are the components of the velocity parallel to the x-, y-, and z-axes, respectively. We have $\kappa = \cos\theta$. The power required for SAD is given by

$$P = |F|V = \left| c_1V^3 + \frac{c_2\kappa^2}{V} + WV_z + maV \right|. \tag{6.14}$$

In (6.13c), $\sin\theta$ is positive when the UAV ascends, and negative when the UAV descends. This is intuitive, as ascending consumes more energy than descending due to gravity.

It is derived in [332, equation (30)] that the load factor of level flight (including SLF and BLT) is upper bounded and approximated by $\kappa = \sqrt{1 + a^2/g^2}$ (with $g \approx 9.8$ m/s^2 being the gravitational acceleration). Combining the values of κ for level flight and SAD, we develop an upper bound of the load factor for the 3D trajectory, which is given by $\kappa^2 = \cos^2\theta(1 + a^2/g^2)$. When $\cos\theta = 1$, κ^2 reduces to the case of level flight; when $\cos\theta < 1$, κ^2 serves as an upper bound for the SAD. To facilitate formulation and convexification of the optimization problem, we further relax the load factor κ^2 with its upper bound $\kappa^2 = \cos^2\theta(1 + a^2/g^2) \le \cos^2\theta + a^2/g^2$, which will be used in the ensuing derivations.

By substituting the upper bound of κ^2, the velocity $\boldsymbol{v}_t$ and acceleration $\boldsymbol{a}_t$ into (6.14), the propulsion power at time slot t is

$$\begin{aligned} P_v^t &= c_1\|\boldsymbol{v}_t\|^3 + \frac{c_2\kappa^2}{\|\boldsymbol{v}_t\|} + m(gv_{zt} + \boldsymbol{a}_t^T\boldsymbol{v}_t) \\ &\le c_1\|\boldsymbol{v}_t\|^3 + \frac{c_2(v_{xt}^2 + v_{yt}^2)}{\|\boldsymbol{v}_t\|^3} + \frac{c_2\|\boldsymbol{a}_t\|^2}{g^2\|\boldsymbol{v}_t\|} + m(gv_{zt} + \boldsymbol{a}_t^T\boldsymbol{v}_t). \end{aligned} \tag{6.15}$$

This provides the upper-bound propulsion power of the fixed-wing UAV under 3D movement.

The total energy consumption over time T is given by

$$\sum_{t=1}^{N} P_v^t \leq \sum_t c_1\|\boldsymbol{v}_t\|^3 + \frac{c_2(v_{xt}^2 + v_{yt}^2)}{\|\boldsymbol{v}_t\|^3} + \frac{c_2\|\boldsymbol{a}_t\|^2}{g^2\|\boldsymbol{v}_t\|} + mg(z_T - z_0) + \frac{m}{2}(\|\boldsymbol{v}_T\|^2 - \|\boldsymbol{v}_0\|^2), \tag{6.16}$$

where the initial elevation and speed of the UAV monitor z_0 and $\boldsymbol{v}_0$ are known in prior, and $m(\|\boldsymbol{v}_T\|^2 - \|\boldsymbol{v}_0\|^2)/2$ denotes the change of the UAV's kinetic energy [332].

6.2.4 Harvested Solar Power

Suppose that the monitor has a solar panel to harvest energy and a rechargeable battery to store energy. The battery initially had energy of E_0 (in Joules). The monitor needs to maintain a required minimum energy reserve $(1 - \eta_0)E_0$ at any moment for sustaining operations and dealing with emergencies. In practice, the monitor flies within a certain altitude range for aviation control and collision avoidance. In the scenario of tracking and monitoring for civilian purposes, the monitor typically flies lower than 1000 m. In this range of altitude, the harvested solar power of the monitor at each time slot t is given by [8]

$$P_s^t(z_t) = \eta S P_i \exp\left(-\frac{\alpha}{\cos\vartheta_t}(1 - 2.2556 \times 10^{-5} z_t)^{5.2561}\right), \tag{6.17}$$

where $\eta \in (0, 1)$ and S (in m^2) denote the efficiency and size of the solar panel, respectively; P_i (in Watts) is the constant power intensity of the solar beams before entering the atmosphere; $\alpha > 0$ is the sum atmospheric extinction; and $\vartheta_t \in [0, \pi/2]$ denotes the solar zenith angle at slot t. Note that ϑ_t is $\pi/2$ at sunrise (or sunset), and then gradually decreases to 0 at midday. The solar zenith angle changes negligibly within half an hour. Since the surveillance mission lasts much shorter than half an hour in the investigated case, ϑ_t is assumed to remain unchanged during the period of T.

Assume without loss of generality (w.l.o.g.) that the surveillance mission takes place at midday on a sunny day and therefore $\cos(\vartheta_t) = 1$. At each time slot t, the harvested solar power (6.17) is simplified as

$$P_s^t(z_t) = \eta S P_i \exp\left(-\alpha(1 - 2.2556 \times 10^{-5} z_t)^{5.2561}\right). \tag{6.18}$$

6.3 Trajectory and Power Optimization for Rotary-wing UAV

6.3.1 Problem Formulation

In this section, we pursue the optimal trajectory and power management of the solar-powered rotary-wing monitor, first offline by minimizing the total energy consumption and maximizing the disguising performance of the monitor. The offline scheme can be implemented if: (i) the target's route is perfectly predictable, e.g., based on historical data; or (ii) the target has only one possible trajectory to travel for the considered period (which is only part of the surveillance mission). The offline scheme can be extended for online operations, as will be discussed in Section 6.3.3.

The monitor can fly horizontally and vertically to adjust its position or posture for tracking, monitoring, and disguising. We use $\mu_1 d_t^2$ and $\mu_2 \Delta_t^2$ to measure the disguising performances of the monitor with regards to (horizontal) distance keeping and altitude changing at each time slot, respectively. Here, μ_1 and μ_2 (in Watts/m^2) are two nonnegative coefficients weighting the two aspects. The reason for decoupling the disguising measure between the horizontal and vertical controls is due to the fact that the propulsion and thrust of a UAV are typically controlled separately. The propulsion and thrust powers yield different models specified in (6.9) and (6.11), respectively. Note that $\mu_1 d_t^2$ is jointly convex in (x_t, y_t), and $\mu_2 \Delta_t^2$ is continuously differentiable and jointly convex in (z_t, z_{t-1}), $\forall t$.

Given the monitor's and the target's waypoints (x_t, y_t, z_t) and (a_t, b_t, H) at the t-th time slot, the propulsion and thrust powers of the monitor P_h^t and P_v^t, its harvested solar power P_s^t, disguising measure $f_t := \mu_1 d_t^2 + \mu_2 \Delta_t^2$, and mobility constraints (6.1a)–(6.1d), the problem of interest is cast as

$$\min_{\{x_t, y_t, z_t, \forall t\}} \sum_{t=1}^{N} (P_h^t + P_v^t - f_t)\delta \tag{6.19a}$$

$$\text{s.t.} \sum_{n=1}^{t} (P_h^n + P_v^n)\delta \le \sum_{n=1}^{t} P_s^n(z_n)\delta + \eta_0 E_0, \ \forall t, \tag{6.19b}$$

$$|z_t - z_{t-1} - (z_{t-1} - z_{t-2})| \le V_{vm}\delta, \forall t \ge 2, \tag{6.19c}$$

$$(x_t, y_t, z_t) \in \mathcal{F}, \ \forall t,$$

$$\text{(6.1a)–(6.1d)}, \tag{6.19d}$$

where (6.19b) is an energy harvesting causality constraint to ensure that the total energy consumed by a time slot does not exceed the total of the energies initially charged and progressively harvested by the time slot, while the monitor can still maintain a minimum energy reserve of $(1 - \eta_0)E_0$ Joules at any time slot for other functionalities; (6.19c) indicates that the altitude change between two consecutive slots does not exceed the maximum amount; and f_t is jointly convex with respect to (x_t, y_t, z_t, z_{t-1}), $t = 1, \dots, N$.

We note that the properties of video tracking are parameterized in (6.19a) and (6.19d). Specifically, (6.6) and (6.5), and in turn, (6.19d), indicate that the monitor needs to track and follow the target by keeping the target within its sight. The maximum visual range of the monitor can be specified in prior, depending on the optical characteristics of the monitor's camera lens (e.g., focal length, zoom capability, zoom magnification, etc.), the illumination condition of the environment [234], and the resolution requirement on the target.

The parameter f_t in (6.19a) specifies that the monitor needs to keep its distance from the target (as long as the target is within its sight), and also keep changing its altitude. As mentioned in Section 6.2.2, the constantly changing altitude of the monitor makes it difficult for the target to correctly focus its camera lens on the monitor for a clear view of the monitor. Keeping the monitor's distance from the target makes it hard for the target to get good image resolution on the monitor, even if correctly focused. By this means, the properties of video tracking are captured in problem (6.19).

Problem (6.19) is not convex, due to the non-convex term in P_h^t, the minimization of a concave function $-f_t$, and the non-convex element P_s^t in constraint (6.19b). Therefore, it is difficult to tackle (6.19) with standard convex solvers, such as the interior point method [38]. In the next section, we propose a new efficient method to convexify the non-convex terms in (6.19) and solve the problem with fast polynomial-time convergence.

6.3.2 Proposed Algorithm for Solar-powered Covert UAV-on-UAV Video Surveillance

In this section, we leverage the SCA [312] and PDCAE [284] to tackle the non-convexity of (6.19), and obtain a low-complexity solution for (6.19). The SCA technique successively approaches the global upper bound of the propulsion power P_h^t. The PDCAE replaces the concave part of the objective (6.19a), i.e., $-f_t$, with a linear function and solves the resultant convex problem. Non-trivial mathematical manipulations of (6.19) are involved to orchestrate SCA and PDCAE.

6.3.2.1 SCA-based Convexification

We start with the non-convex part of P_h^t, i.e., $\left(\sqrt{1+\frac{{V_h^t}^4}{4v_0^4}}-\frac{{V_h^t}^2}{2v_0^2}\right)^{\frac{1}{2}}$ in (6.9), by defining new slack variables $\{q_t \geq 0,\ \forall t = 1, \dots, N\}$

$$q_t^2 = \sqrt{1+\frac{{V_h^t}^4}{4v_0^4}}-\frac{{V_h^t}^2}{2v_0^2},\ \forall t, \tag{6.20}$$

which can be reorganized as

$$\frac{1}{q_t^2} = q_t^2 + \frac{{V_h^t}^2}{v_0^2},\ \forall t. \tag{6.21}$$

As a result, the second term on the right-hand side (RHS) of (6.9) can be substituted by the linear component $P_1 q_t$, with an additional new constraint (6.21).

For exposition, we now integrate the expression for V_h^t in (6.10) and define

$$\begin{aligned}\tilde{P}_h^t := P_0 + \frac{3P_0}{U_{tip}^2 \delta^2}\left[(x_t - x_{t-1})^2 + (y_t - y_{t-1})^2\right] + P_1 q_t \\ + \frac{d_f}{2\delta^3}\rho s A\left[(x_t - x_{t-1})^2 + (y_t - y_{t-1})^2\right]^{3/2}, \ \forall t.\end{aligned} \tag{6.22}$$

Here, $\tilde{P}_h^t$ is jointly convex in $(x_t, x_{t-1}, y_t, y_{t-1}, q_t)$. Given δ, problem (6.19) can be rewritten as

$$\min_{\{x_t, y_t, z_t, q_t, \forall t\}} \sum_{t=1}^{N} (\tilde{P}_h^t + P_v^t - f_t) \tag{6.23a}$$

$$\text{s.t.} \ \sum_{n=1}^{t} (\tilde{P}_h^n + P_v^n) \leq \sum_{n=1}^{t} P_s^n(z_n) + \eta_0 E_0/\delta, \ \forall t, \tag{6.23b}$$

$$\frac{1}{q_t^2} \leq q_t^2 + \frac{(x_t - x_{t-1})^2 + (y_t - y_{t-1})^2}{\hat{v}_0^2}, \ \forall t,$$

$$\text{(6.1a)–(6.1d), (6.19c), and (6.19d),} \tag{6.23c}$$

where $\hat{v}_0^2 = v_0^2 \delta^2$. Constraint (6.23c) is obtained by relaxing the equality in (6.21) with inequality. Yet, the equivalence still holds between problems (6.19) and (6.23). This is because, if (6.23c) holds with strict inequality for any t, we can always decrease the value of the related variable q_t to reduce the total energy consumption until (6.23c) is satisfied with the equality [341].

Problem (6.23) is still non-convex, since it still involves two non-convex constraints, i.e., (6.23b) and (6.23c). Constraint (6.23c) can be tackled with the SCA method [312] by evaluating the global lower bound of (6.23c) at a given local point. In particular, the left-hand side (LHS) of (6.23c) is a convex function of q_t, and the RHS is a jointly convex function of q_t and $(x_t, x_{t-1}, y_t, y_{t-1})$. Since the first-order Taylor expansion serves as the global lower bound of a convex function [38], we can obtain the following lower bound for the RHS of (6.23c)

$$\begin{aligned}q_t^2 + \frac{(x_t - x_{t-1})^2 + (y_t - y_{t-1})^2}{\hat{v}_0^2} \geq q_t^{(\ell)2} + 2q_t^{(\ell)}(q_t - q_t^{(\ell)}) \\ + \frac{2}{\hat{v}_0^2}\left[(x_t^{(\ell)} - x_{t-1}^{(\ell)})(x_t - x_{t-1}) + (y_t^{(\ell)} - y_{t-1}^{(\ell)})(y_t - y_{t-1})\right] \\ - \frac{1}{\hat{v}_0^2}\left[(x_t^{(\ell)} - x_{t-1}^{(\ell)})^2 + (y_t^{(\ell)} - y_{t-1}^{(\ell)})^2\right],\end{aligned} \tag{6.24}$$

where $q_t^{(\ell)}$, $x_t^{(\ell)}$, and $y_t^{(\ell)}$ are the respective values of the variables in the ℓ-th iteration of the SCA method.

We provide a lower bound for the concave function of the harvested solar power $P_s^t(z_t)$ by approximating it with a simple linear function, as given by

$$\hat{P}_s^t(z_t) := c_1 z_t + c_2, \tag{6.25}$$

where c_1 and c_2 are two scaling coefficients depending on the parameters of energy harvesting. We take the simple linear approximation of (6.25) to $P_s^t(z_t)$, rather than the first-order Taylor expansion as done in (6.24), since the first-order Taylor expansion provides an upper bound for a concave function, not a lower bound. Under our system and parameter settings, the approximation (6.25) serves as a tight lower bound for $P_s^t(z_t)$, as will be numerically verified in Section 6.5.1.

With the above mathematic manipulations and by letting $P_h := \sum_t \tilde{P}_h^t$, $P_v := \sum_t P_v^t$, and $f := \sum_t f_t$, problem (6.23) can be transformed to

$$\min_{\{x_t, y_t, z_t, q_t, \forall t\}} \quad (P_h + P_v - f) \tag{6.26a}$$

$$\text{s.t.} \ \sum_{n=1}^{t} (\widetilde{P}_h^n + P_v^n) \leq \sum_{n=1}^{t} \widehat{P}_s^n(z_n) + \eta_0 E_0 / \delta, \ \forall t, \tag{6.26b}$$

$$\frac{1}{q_t^2} \leq \frac{2}{\hat{v}_0^2} [(x_t^{(\ell)} - x_{t-1}^{(\ell)})(x_t - x_{t-1}) + (y_t^{(\ell)} - y_{t-1}^{(\ell)})(y_t - y_{t-1})]$$

$$- \frac{1}{\hat{v}_0^2} [(x_t^{(\ell)} - x_{t-1}^{(\ell)})^2 + (y_t^{(\ell)} - y_{t-1}^{(\ell)})^2]$$

$$+ q_t^{(\ell)2} + 2q_t^{(\ell)}(q_t - q_t^{(\ell)}), \ \forall t, \tag{6.26c}$$

$$q_t \geq 0, \ \forall t,$$

$$\text{(6.1a)–(6.1d), (6.19c), and (6.19d).} \tag{6.26d}$$

Given the global lower bound in (6.24) (and, in turn, the global upper bound of P_h^t) and the lower bound in (6.25), constraint (6.26b) tightens the original constraint (6.19b). When the constraints of (6.26) are satisfied, the constraints of the original problem (6.19) are satisfied; not the other way around. Therefore, the feasible solution region of (6.26) is a subset of the feasible solution region of (6.19), and the optimal value of (6.26) draws an upper limit to the optimal value of (6.19).

All the constraints of (6.26) are now convex. Yet, the problem is still not convex due to the non-convex part of the objective function (6.26a), i.e., $-f$, which prevents a direct use of any standard convex solvers. We propose to use the PDCAE to convexify (6.26a) and obtain a low-complexity (suboptimal) solution for (6.26) and, in turn, the original problem (6.19).

6.3.2.2 PDCAE-based Solution

The objective function of (6.26) has the same form as the classic DC problem

$$\min_{x\in\mathbb{R}^n} G_0(x) + G_1(x) - G_2(x), \tag{6.27}$$

where G_0 is a smooth convex function with a Lipschitz continuous gradient and a Lipschitz continuity modulus; G_1 is a proper closed convex function;[3] and G_2 is a continuous convex function. In this sense, (6.26) can be solved with the proximal DC algorithm [89], which, however, is slow to converge.

Developed in [284], PDCAE accelerates the proximal DC algorithm with an extrapolation technique. The extrapolation adds *momentum* terms that depend on the solutions in the previous iterations to update the variables during the current iteration [201]. The extrapolation has been widely used to speed up the proximal gradient algorithm and its variants for convex optimization problems [200].

We rewrite (6.26) into the standard form of (6.27). The mapping between (6.26) and (6.27) is as follows: $P_v := G_0$, $P_h := G_1$, and $f := G_2$, where $\mathbf{X} := x$ and $\lambda_{(\ell)} := \theta_{(\ell)}$. Since $f(\cdot)$ is a continuously differentiable function, we can obtain its gradient directly. Further let $\mathbf{X}_t := [x_t, y_t, z_t, q_t], \forall t$, and $\mathbf{X} := \{\mathbf{X}_t, \forall t\}$ collect the optimization variables at time slot t and over the entire time horizon, respectively; and $\mathcal{A}$ denote the feasible solution region of problem (6.26).

The PDCAE-based solver of problem (6.26) is summarized in Algorithm 2, where $\lambda_t^{(\ell)} := [\lambda_{xt}^{(\ell)}, \lambda_{yt}^{(\ell)}, \lambda_{zt}^{(\ell)}, \lambda_{qt}^{(\ell)}], \forall t$, and $\lambda_{(\ell)} := \{\lambda_t^{(\ell)}, \forall t\}$. $\lambda_{xt}^{(\ell)}$ denotes the extrapolation of the x-coordinate of the monitor at time slot t in the ℓ-th iteration,

$$\lambda_{xt}^{(\ell)} = x_t^{(\ell)} + \beta_{(\ell)}(x_t^{(\ell)} - x_t^{(\ell-1)}). \tag{6.28}$$

Similarly, $\lambda_{yt}^{(\ell)}$ and $\lambda_{zt}^{(\ell)}$ denote the extrapolations of the monitor's y- and z-coordinates, respectively; and $\lambda_{qt}^{(\ell)}$ the extrapolation of q_t.

The specific expressions in Step 5 of Algorithm 2 are given as follows.

$$\nabla P_v^t(\lambda_t^{(\ell)}) := \nabla P_v^t = \left[\frac{W}{\delta}, -\frac{W}{\delta}\right], \tag{6.29a}$$

$$\nabla f_t = \left[2\mu_1(x_t - a_t), 2\mu_1(y_t - b_t), 2\mu_2(z_t - z_{t-1}) - 2\mu_2(z_t - z_{t-1})\right], \tag{6.29b}$$

$$\begin{aligned}\langle \nabla P_v^t(\lambda_t^{(\ell)}) - \nabla f_t(\mathbf{X}_t^{(\ell)}), \mathbf{X}_t\rangle &= -2\mu_1(x_t^{(\ell)} - a_t)x_t \\ &-2\mu_1(y_t^{(\ell)} - b_t)y_t + \left(W/\delta - 2\mu_2(z_t^{(\ell)} - z_{t-1}^{(\ell)})\right)(z_t - z_{t-1}).\end{aligned} \tag{6.29c}$$

---

3 For an extended real-valued function $h : \mathbb{R}^n \to (-\infty, \infty)$, whose domain is denoted by dom $h = \{x \in \mathbb{R}^n : h(x) < \infty\}$, h is a proper function if it never equals $-\infty$ and dom $h \neq \emptyset$. Furthermore, h is a proper closed function if it is lower semicontinuous [284].

Algorithm 2 The PDCAE-based solution for problem (6.26)

1: **Initialization:** Input $\mathbf{X}_{(0)} \in \mathcal{A}$, $\beta_{(\ell)} \subseteq [0, 1), \forall \ell$. Set $\mathbf{X}_{(-1)} = \mathbf{X}_{(0)}$.
2: **for** $\ell = 0, 1, 2, \ldots$ **do**
3: Compute the gradients $\nabla f(\mathbf{X}_{(\ell)})$.
4: Apply the extrapolation technique to the previous two iterations: $\lambda_{(\ell)} = \mathbf{X}_{(\ell)} + \beta_{(\ell)}(\mathbf{X}_{(\ell)} - \mathbf{X}_{(\ell-1)})$.
5: Update the optimization variables $\mathbf{X}_{(\ell+1)} = argmin_{\mathbf{X} \in \mathcal{A}}\{\langle \nabla P_v(\lambda_{(\ell)}) - \nabla f(\mathbf{X}_{(\ell)}), \mathbf{X}\rangle + \frac{M}{2}||\mathbf{X} - \lambda_{(\ell)}||^2 + P_h(\mathbf{X})\}$ by the interior point method.
6: Update $\ell \leftarrow \ell + 1$.
7: **end for**

Algorithm 3 The overall algorithm for problem (6.19)

1: Approximate the harvested solar power $P_s^t(z_t)$ by (6.25).
2: **Initialization:** Input $\mathbf{X}_{(0)} \in \mathcal{A}$, $\beta_{(\ell)} \subseteq [0, 1), \forall \ell$. Set $\mathbf{X}_{(-1)} = \mathbf{X}_{(0)}$.
3: **for** $\ell = 0, 1, 2, \ldots$ **do**
4: Introduce slack variable $q_t, \forall t = 1, \cdots, N$ based on (6.21) and approximate the propulsion power P_h^t by (6.22) and (6.23c) with the SCA method.
5: Transform problem (6.19) to problem (6.26).
6: Implement Steps 3 to 5 of Algorithm 2 to obtain a suboptimal solution.
7: Update $\ell \leftarrow \ell + 1$.
8: **end for**

Recall that $P_v := \sum_t P_v^t$ and $f := \sum_t f_t$ in (6.26). We have $\nabla P_v = \sum_t \nabla P_v^t$, and $\nabla f = \sum_t \nabla f_t$ in Step 5.

A popular way to select and update the extrapolation parameters $\{\beta_{(\ell)}\}$ is to set $\bar{\beta}_{(-1)} = \bar{\beta}_{(0)} = 1$ and follow the recursive equations [284]

$$\beta_{(\ell)} = (\bar{\beta}_{(\ell-1)} - 1)/\bar{\beta}_{(\ell)}; \ \bar{\beta}_{(\ell+1)} = \frac{1}{2}\left(1 + \sqrt{1 + 4\bar{\beta}_{(\ell)}^2}\right). \tag{6.30}$$

We choose a fixed number $\bar{L}$, and reset $\bar{\beta}_{(\bar{L}-1)} = \bar{\beta}_{(\bar{L})} = 1$ every $\bar{L}$ iterations.

Given the same structure between problems (6.26) and (6.27) and by following the same lines as in [284], the proposed PDCAE-based algorithm can converge to a suboptimal solution for the original problem (6.19) within polynomial time.

6.3.2.3 Overview of the Proposed Algorithm

Algorithm 3 summarizes the proposed algorithm, which solves problem (6.19) by rewriting (6.19) as (6.26) given $\mathbf{X}_t^{(\ell)} = (x_t^{(\ell)}, y_t^{(\ell)}, z_t^{(\ell)}, q_t^{(\ell)})$, $\forall t = 1, \ldots, N$, and then running Algorithm 2 to solve (6.26) and obtain $\mathbf{X}_t^{(\ell+1)}$. Algorithm 3 repeats these steps until convergence, i.e., $\|\mathbf{X}_t^{(\ell+1)} - \mathbf{X}_t^{(\ell)}\| < \epsilon$, where ϵ is the required accuracy of convergence.

Problem (6.19) is challenging due to the non-convex propulsion power function (6.9), harvested solar power function (6.18), and the concave part $-f_t$ in the objective function. No known technique is able to obtain the globally or locally optimal solution to problem (6.19). As proved in [284], the objective of (6.26a) is non-increasing and lower bounded, and the proposed Algorithm 3 can converge efficiently to a suboptimal solution.

The computational complexity of Algorithm 3 is dominated by Step 6, and the step is further comprised of Steps 3 to 5 of Algorithm 2. In Algorithm 2, Step 3 involves computing the gradients with solutions obtained at the previous iteration, and Step 4 applies the extrapolation technique to the previous two iterations. Both of their complexities are $\mathcal{O}(N)$. In Algorithm 2, Step 5 updates the optimization variables with the PDCAE-based method. Since the PDCAE deals with a convex problem after convexification, transformation, and extrapolation, the computational complexity of Step 5 is primarily accounted for by the interior point method, which is $\mathcal{O}(N^{3.5})$ per iteration. Taking all the steps into account, the total complexity of Algorithm 3 is $\mathcal{O}(N^{3.5})$ per iteration. With the convergence accuracy ϵ, the overall computational complexity of Algorithm 2 is $\mathcal{O}(N^{3.5} \log \frac{1}{\epsilon})$, where $\mathcal{O}(\log \frac{1}{\epsilon})$ gives the number of iterations required for convergence.

We note that Algorithm 3 is suboptimal, due to the fact that the future trajectory of the target is hard to predict in practice and, more importantly, the considered problem (6.19) is non-convex. Nevertheless, the algorithm is convergent and stable with a polynomial time complexity, as discussed above, and can be used online to refine the trajectory on-the-fly, as will be described in Section 6.3.3.

Also note that Algorithm 3 is susceptible to the weather conditions. For example, the harvested solar energy can be substantially reduced if there are thick clouds, which can have an impact on the trajectory and heading of the monitor. Nevertheless, the algorithm can deal with this situation from the following two aspects. The first aspect is that Algorithm 3 plans the trajectory and heading of the monitor, adapting to the harvested solar energy. The impact of the cloud on solar power harvesting has been explicitly considered in the algorithm. In particular, the coefficient of atmospheric extinction α in (6.17) indicates that the intensity of the solar beam decreases in the presence of dense clouds. The second aspect is that the monitor is equipped with a rechargeable battery, which can supplement the harvested solar power when needed. In the presence of thick clouds, the monitor can withdraw energy from the battery, as the output of the battery is scheduled to power the monitor across the time slots, as captured in (6.19b). When the sky is clear, the battery can be recharged by any surplus solar energy. This is because the battery needs to be maintained at least a preconfigured minimum level for safety and emergency, as also specified in (6.19b). The battery can be either discharged or recharged, depending on the abundance of solar power and the complexity of the monitoring mission.

6.3.3 Extension to Online Implementation

In practice, it can be difficult for the monitor to predict precisely the entire trajectory of the target beforehand. Instead, the monitor estimates the future waypoints of the target, and adjusts its own trajectory accordingly. The proposed algorithm provides a modularized solution that can be a critical building block of online implementation. In this section, we generalize the proposed approach to online operations, where the predictions of future waypoints are based on the monitor's past observation and used to plan progressively the monitor's trajectory.

The online approach is an extension of the offline Algorithm 2, where Algorithm 2 serves as the dynamic method for the online control of the monitoring UAV, as shown in Figure 6.3. Specifically, at any instant τ, the monitor predicts the target's trajectory over the next N instants. Following the model predictive control (MPC) [43], the monitor can run Algorithm 2 on-the-fly to plan its trajectory from its current location for the next N instants (or steps). Only the first step of the N-step trajectory is implemented at instant $(\tau + 1)$ though. Then, the monitor predicts the target's trajectory over th upcoming N instants. Algorithm 2 is run again, starting from the new current location of the monitor, yielding a new control based on the new prediction. The prediction horizon of N instants (or steps) keeps shifting forward (by one at every instant). The details are provided in the following.

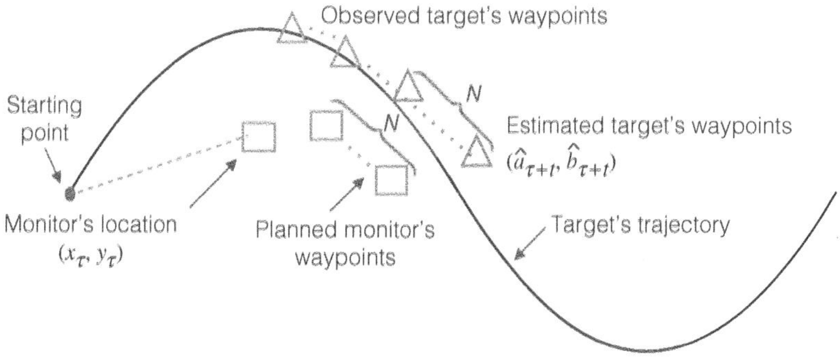

Figure 6.3 An illustration of the online scheme in the 2D space.

Suppose that the monitor keeps observing the locations (i.e., the coordinates) and headings of the target. At any (current) time slot τ, the monitor predicts the waypoints of the target for the upcoming N slots, i.e., $(\hat{a}_{\tau+t}, \hat{b}_{\tau+t})$, $t = 1, \ldots, N$ based on the past observation of the target. The monitor's current waypoint is (x_τ, y_τ, z_τ). As illustrated in Figure 6.4, the monitor plans its trajectory for the upcoming N slots (i.e., slots $\tau + 1, \ldots, \tau + N$), by solving the following problem at time slot τ

$$\min_{\{x_t, y_t, z_t, \forall t=1,\ldots,N\}} \sum_{n=\tau+1}^{\tau+N} (P_h^n + P_v^n - f_n) \tag{6.31a}$$

$$\text{s.t.} \sum_{n=1}^{\tau+t} (P_h^n + P_v^n) \leq \sum_{n=1}^{\tau+t} P_s^n(z_n) + \eta_0 E_0/\delta, \forall t, \tag{6.31b}$$

$$(x_{\tau+t} - x_{\tau+t-1})^2 + (y_{\tau+t} - y_{\tau+t-1})^2 \leq (V_{hm}\delta)^2, \ \forall t, \tag{6.31c}$$

$$|z_{\tau+t} - z_{\tau+t-1} - (z_{\tau+t-1} - z_{\tau+t-2})| \leq V_{vm}\delta, \ \forall t \geq 2, \tag{6.31d}$$

$$(x_{\tau+t}, y_{\tau+t}, z_{\tau+t}) \in \hat{\mathcal{F}}, \tag{6.31e}$$

where $\hat{\mathcal{F}}$ is the new FFR, as given by

$$\begin{aligned}\hat{\mathcal{F}} := \{&(x_{\tau+t}, y_{\tau+t}, z_{\tau+t}) | H < z_l \leq z_{\tau+t}, \ x_{\tau+t} \leq \hat{a}_{\tau+t}, \ y_{\tau+t} \leq \hat{b}_{\tau+t}, \\ &(x_{\tau+t} - \hat{a}_{\tau+t})^2 + (y_{\tau+t} - \hat{b}_{\tau+t})^2 + (z_{\tau+t} - H)^2 \leq D^2, \ \forall t\},\end{aligned} \tag{6.32}$$

and (6.31b) guarantees that the monitor can always maintain a minimum energy reserve of $(1 - \eta_0)E_0$ Joules in the battery at any time slot.

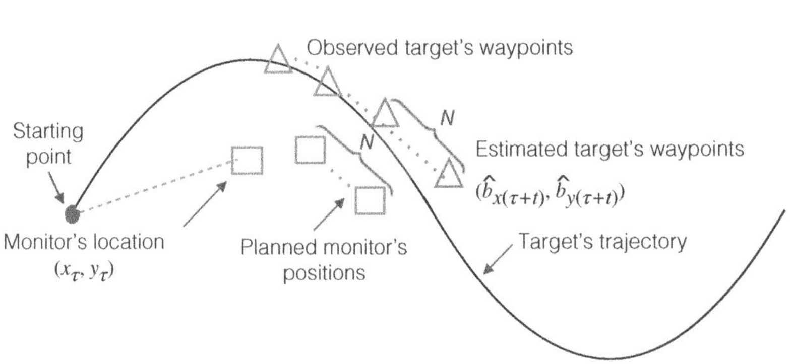

Figure 6.4 A bird's view illustration of the online control approach.

As with problem (6.19), the non-convex propulsion and solar power functions in (6.31) can be tightened linearly and iteratively with the SCA method, resulting in the following DC problem

$$\min_{\{x_t, y_t, z_t, \forall t=1,\dots,N\}} \sum_{n=\tau+1}^{\tau+N} (\widetilde{P}_h^n + P_v^n - f_n) \tag{6.33a}$$

$$\text{s.t.} \sum_{n=1}^{\tau+t} (\widetilde{P}_h^n + P_v^n) \leq \sum_{n=1}^{\tau+t} \widehat{P}_s^n(z_n) + \eta_0 E_0/\delta, \forall t, \tag{6.33b}$$

$$\frac{1}{q_t^2} \leq \frac{2}{\hat{v}_0^2}[(x_t^{(\ell)} - x_{t-1}^{(\ell)})(x_t - x_{t-1}) + (y_t^{(\ell)} - y_{t-1}^{(\ell)})(y_t - y_{t-1})]$$

$$- \frac{1}{\hat{v}_0^2}[(x_t^{(\ell)} - x_{t-1}^{(\ell)})^2 + (y_t^{(\ell)} - y_{t-1}^{(\ell)})^2]$$

$$+ q_t^{(\ell)2} + 2q_t^{(\ell)}(q_t - q_t^{(\ell)}), \ \forall t, \tag{6.33c}$$

$$q_t \geq 0, \ \forall t,$$

$$\text{(6.31c)–(6.31e)}. \tag{6.33d}$$

Problem (6.33) exhibits the same structure as (6.26). By replacing the actual waypoints of the target (a_t, b_t) in Algorithm 2 with the estimated waypoints $(\hat{a}_{\tau+t}, \hat{b}_{\tau+t})$, $t = 1, \dots, N$, problem (6.33) can be readily solved by using the PDCAE-based Algorithm 2. In this sense, the algorithm provides the so-called model, which is based on regularly updated observations and predictions, and can be solved in a structured manner.

Algorithm 4 summarizes the online implementation of the proposed approach by following the MPC framework. Under the MPC framework, the monitor solves (6.33) to produce a tentative trajectory of N waypoints for the upcoming N time slots, i.e., $\tau + 1, \dots, \tau + N$, at every slot $\tau = 1, 2, \dots$. The monitor only takes the trajectory at the next time slot, i.e., $\tau + 1$, and then solves (6.33) again to update the trajectory. By this means, the MPC can adapt quickly to new inputs and provide the control policy with consideration of future events, achieving a fast, flexible, and dynamic control. The impact of prediction errors can be mitigated by exploiting the online adjustability of the MPC and incorporating the error bounds into the control, thereby achieving an effective online control of the trajectory.

It is worth noting that the monitor can further enhance its disguise by regulating its horizontal heading to deviate from the heading of the target at each time slot (in addition to keeping the horizontal distance and changing the altitude at each time slot, as described in Sections 6.3.1 and 6.3.2). This can be implemented by controlling the slope of the monitor's trajectory on the (x, y)-plane, denoted by κ_τ, at any time slot τ. The slope κ_τ can change every slot, and deviate from the target's heading $\kappa_{ab,\tau}$ by at least a predefined nonnegative value $c_3 > 0$.

Algorithm 4 The proposed online algorithm for problem (6.31)

1: At time slot τ, the monitor observes the current location of the target and predicts the target's next N waypoints.
2: Approximate the harvested solar power $P_s^t(z_t)$ by (6.25).
3: **Initialization:** Set a feasible initial trajectory for the monitor, i.e., $\{(x_t^{(0)}, y_t^{(0)}, z_t^{(0)}), t = 1, \dots, N\}$, and initialize the auxiliary variables $\{\beta_{(0)}, q_t^{(0)}, \lambda_t^{(0)}, t = 1, \dots, N\}$.
4: **for** $\ell = 0, 1, 2, \dots$ **do**
5: Introduce slack variable $q_t, \forall t = 1, \dots, N$, approximate the propulsion power P_h^t by (6.22) and (6.23c), and tighten constraint (6.31b) with the SCA method.
6: Implement Steps 3 to 5 of Algorithm 2 to plan the monitor's trajectory for the upcoming N slots.
7: Update $\ell \leftarrow \ell + 1$.
8: **end for**

Considering the heading control of the monitor by increasing its slope, we have

$$\kappa_\tau := \frac{y_\tau - y_{\tau-1}}{x_\tau - x_{\tau-1}} \geq \frac{b_\tau - b_{\tau-1}}{a_\tau - a_{\tau-1}} + c_3, \ \forall \tau, \tag{6.34}$$

which is a linear inequality, and can be imposed as an additional constraint of (6.19) or (6.31). Given the linearity of (6.34), the convexification process of (6.19) or (6.31) is unaffected and Algorithm 2 remains effective with further improved disguising, as will be numerically validated in Section 6.5.1.

6.4 Trajectory and Power Optimization for Fixed-wing UAV

6.4.1 Problem Statement

We intend to minimize the energy usage of the fixed-wing UAV monitor and maximize its disguise throughout a disguised video surveillance mission period. The problem of interest is formulated as

$$\min_{\{\mathbf{q}_t, \mathbf{v}_t, \mathbf{a}_t, \forall t=1,\dots,N\}} \sum_{t=1}^{N} (P_v^t - f_t)\delta \tag{6.35a}$$

$$\text{s.t.} \sum_{n=1}^{t} P_v^n \delta \leq \sum_{n=1}^{t} P_s^n \delta + \eta_0 E_0, \ \forall t, \tag{6.35b}$$

$$d_{\min} \leq d_t \leq d_{\max}, \tag{6.35c}$$

$$z_l \leq z_t,$$

$$(6.2), \tag{6.35d}$$

where (6.35b) ensures that the UAV monitor can have a minimum battery level of $(1-\eta_0)E_0$ Joules at any slot in case of emergency.

Problem (6.35) is difficult to solve online due to the time-coupling in the 3D movement of the fixed-wing UAV, the non-convex power models P_v^t in (6.15), and the reliance on the correct prediction of the target trajectory. Traditional online algorithms are not effective in this case, as they would either result in a greedy optimization at the current slot or a coarse optimization over future slots with inaccurate predictions.

By exploiting receding horizon control (also known as MPC) [124], we decouple the problem between time slots and solve the problem online slot-by-slot to progressively improve the trajectory on-the-fly. In particular, at any time slot τ, the monitoring UAV forecasts the target's flight path over the upcoming N slots. Applying the receding horizon control, the UAV monitor optimizes the 3D flight path of the monitor for the forthcoming N time slots (or steps). At slot $(\tau+1)$, the monitor flies as planned for only a slot, and then forecasts the target's flight path again for the forthcoming N slots. The monitor adjusts its flight path according to its forecast of the target's trajectory. The forecast horizon of N slots keeps being shifted forward (by one slot at every step). The receding horizon control is known to provide excellent performance in real-world applications.

We assume that the UAV monitor keeps measuring and obtaining the positions and directions of the targeted UAV. At an arbitrary (present) slot, denoted by τ, the monitoring UAV forecasts the coordinates of the target for the next L slots ($L \le N$), i.e., $(\hat{b}_{x(\tau+t)}, \hat{b}_{y(\tau+t)})$, $t \le L$, according to the historical actions of the targeted object, e.g., by using Kalman filtering. The monitor's present coordinate is (x_τ, y_τ, z_τ). As shown in Figure 6.4, the monitoring UAV designs its flight path for the next L time slots (from $\tau+1$ up to $\tau+L$), by transforming problem (6.35) into the following problem and solving at slot τ

$$\min_{\{\boldsymbol{q}_t,\boldsymbol{v}_t,\boldsymbol{a}_t,\forall t=1,\dots,N\}} \sum_{t=1}^{L}(P_v^{\tau+t} - f_{\tau+t}) \tag{6.36a}$$

$$\text{s.t.} \sum_{n=1}^{\tau+t} P_v^n \le \sum_{n=1}^{\tau+t} P_s^n(z_n) + \eta_0 E_0/\delta, \forall t, \tag{6.36b}$$

$$\boldsymbol{q}_{\tau+t+1} = \boldsymbol{q}_{\tau+t} + \boldsymbol{v}_{\tau+t}\delta + \frac{1}{2}\boldsymbol{a}_{\tau+t}\delta^2, \ \forall t, \tag{6.36c}$$

$$\boldsymbol{v}_{\tau+t+1} = \boldsymbol{v}_{\tau+t} + \boldsymbol{a}_{\tau+t}\delta, \ \forall t, \tag{6.36d}$$

$$\|\boldsymbol{a}_{\tau+t}\| \le a_{\max}, \ \|\boldsymbol{v}_{\tau+t}\| \le v_{\max}, \ \forall t, \tag{6.36e}$$

$$\|\boldsymbol{q}_{\tau+t} - \hat{\boldsymbol{b}}_{\tau+t}\| \le d_{\max}, \ z_l \le z_t, \ \forall t, \tag{6.36f}$$

$$\max\{v_{\min}, v_{z(\tau+t)}/\vartheta\} \le \|\boldsymbol{v}_{\tau+t}\|, \ \forall t, \tag{6.36g}$$

$$d_{\min} \le \|\boldsymbol{q}_{\tau+t} - \hat{\boldsymbol{b}}_{\tau+t}\|, \ \forall t. \tag{6.36h}$$

The non-convexity of the UAV mobility, propulsion power, harvested solar power, and visual disguise makes problem (6.36) hard to address by off-the-shelf convex solvers, e.g., the interior point method. The 3D movement of the fixed-wing UAV has to be jointly optimized, since its horizontal and vertical propulsion powers are closely coupled in (6.16). Moreover, the online planning of the monitor's trajectory at different positions and time slots is correlated and interdependent. All this makes it intangible to control the 3D trajectory and minimize the power usage of the fixed-wing UAV monitor. In the next section, we convexify the problem and solve it efficiently.

6.4.2 SCA-based Per-slot Trajectory Planning

We employ the SCA [315] to deal with the non-convexity of problem (6.36) in pursuit of a computationally efficient solution for (6.36). The SCA progressively draws closer to the aforementioned upper limit of the propulsion power P_v^t, and replaces the concave component of (6.36a), i.e., $-f_t$, with a linear approximation to be readily tackled with convex programming.

By introducing slack variables $\zeta_t \leq \|\boldsymbol{v}_t\|$ and $\xi_t \leq \|\boldsymbol{v}_t\|^3$, $\forall t$, we can write the upper bound of P_v^t in (6.15) as

$$\hat{P}_v^t = c_1\|\boldsymbol{v}_t\|^3 + \frac{c_2(v_{xt}^2 + v_{yt}^2)}{\xi_t} + \frac{c_2\|\boldsymbol{a}_t\|^2}{g^2\zeta_t} + m(gv_{zt} + \boldsymbol{a}_t^T\boldsymbol{v}_t). \tag{6.37}$$

Problem (6.36) is transformed to

$$\min_{\{\boldsymbol{q}_t,\boldsymbol{v}_t,\boldsymbol{a}_t,\zeta_t,\xi_t,\forall t=1,\dots,L\}} \sum_{t=1}^{L}(\hat{P}_v^{\tau+t} - f_{\tau+t}) \tag{6.38a}$$

$$\text{s.t.} \sum_{n=1}^{\tau+t} \hat{P}_v^n\delta \leq \sum_{n=1}^{\tau+t} P_s^n\delta + \eta_0 E_0, \ \forall t, \tag{6.38b}$$

$$\zeta_{\tau+t} \leq \|\boldsymbol{v}_{\tau+t}\|, \ \forall t, \tag{6.38c}$$

$$\xi_{\tau+t} \leq \|\boldsymbol{v}_{\tau+t}\|^3, \ \forall t, \tag{6.38d}$$

$$\zeta_{\tau+t} \geq v_{\min}, \ \xi_{\tau+t} \geq v_{\min}^3, \ \forall t,$$

$$\text{(6.36b)–(6.36h)}. \tag{6.38e}$$

Despite the inequality constraints (6.38c) and (6.38d), the equivalence remains valid between problems (6.36) and (6.38). This is because, if (6.38c) and (6.38d) hold with tight inequality for an arbitrary t, we can always increase the values of ζ_t and ξ_t to suppress the energy consumption until (6.38c) and (6.38d) are satisfied with the equality [341].

Constraints (6.36g), (6.36h), (6.38c), and (6.38d), are non-convex. Nevertheless, the RHSs of the constraints are convex and differentiable in $\boldsymbol{v}_t$ or $\boldsymbol{q}_t$. They can be approached successively by using the first-order Taylor expansion at the local point attained during the ℓ-th iteration, as given by

$$\|\boldsymbol{v}_t\| \geq \|\boldsymbol{v}_t^{(\ell)}\| + \frac{(\boldsymbol{v}_t^{(\ell)})^T}{\|\boldsymbol{v}_t^{(\ell)}\|}(\boldsymbol{v}_t - \boldsymbol{v}_t^{(\ell)}) := \Gamma_{1t}, \tag{6.39a}$$

$$\|\boldsymbol{v}_t\|^3 \geq \|\boldsymbol{v}_t^{(\ell)}\|^3 + 3\|\boldsymbol{v}_t^{(\ell)}\|(\boldsymbol{v}_t^{(\ell)})^T(\boldsymbol{v}_t - \boldsymbol{v}_t^{(\ell)}) := \Gamma_{2t}, \tag{6.39b}$$

$$d_t \geq \|\boldsymbol{q}_t^{(\ell)} - \boldsymbol{b}_t\| + \frac{(\boldsymbol{q}_t^{(\ell)})^T}{\|\boldsymbol{q}_t^{(\ell)} - \boldsymbol{b}_t\|}(\boldsymbol{q}_t - \boldsymbol{q}_t^{(\ell)}) := \hat{d}_t, \tag{6.39c}$$

where $(\cdot)^T$ stands for transpose. The non-convex constraints (6.36g), (6.36h), (6.38c), and (6.38d), are now convexified by

$$\max\{v_{\min}, v_{z(\tau+t)}/\vartheta, \zeta_{\tau+t}\} \leq \Gamma_{1(\tau+t)}, \ \forall t, \tag{6.40a}$$

$$\xi_{\tau+t} \leq \Gamma_{2(\tau+t)}, \ \forall t, \tag{6.40b}$$

$$d_{\min} \leq \hat{d}_{\tau+t}, \ \forall t. \tag{6.40c}$$

We offer a lower limit for the concave solar power $P_s^t(z_t)$ through approximating the concave function by a linear expression, as given by

$$\hat{P}_s^t(z_t) := c_3 z_t + c_4, \tag{6.41}$$

where c_3 and c_4 are known scaling weights relying on the efficiency and amount of solar energy harvesting. Constraint (6.38b) and, in turn, (6.36b), can be convexified and tightened by

$$\sum_{n=1}^{\tau+t} \hat{P}_v^n \delta \leq \sum_{n=1}^{\tau+t} \hat{P}_s^n \delta + \eta_0 E_0, \ \forall t. \tag{6.42}$$

We do not perform the first-order Taylor expansion of the concave solar power $P_s^t(z_t)$, as it provides an upper bound for $P_s^t(z_t)$, not a lower bound. In the considered system, the approximation (6.41) offers a tight lower limit for $P_s^t(z_t)$ [112].

We proceed to convexify the concave part in the objective of (6.36), i.e., $-f_t$. Since f_t is convex, its global lower bound can be successively approached by plugging the first-order Taylor expansion at the local point attained during the ℓ-th iteration

$$\begin{aligned} f_t \geq\ & f_t(\boldsymbol{q}_t^{(\ell)}) + 2\mu_1(x_t^{(\ell)} - b_{xt})(x_t - x_t^{(\ell)}) + 2\mu_1(y_t^{(\ell)} - b_{yt})(y_t - y_t^{(\ell)}) \\ & + 2\mu_2(z_t^{(\ell)} - z_{t-1}^{(\ell)})\left[(z_t - z_t^{(\ell)}) - (z_{t-1} - z_{t-1}^{(\ell)})\right] \\ & := \hat{f}_t, \end{aligned}$$

where $\hat{f}_t$ is a linear function of $\boldsymbol{q}_t$.

Let $P_v = \sum_{t=1}^{L} \hat{P}_v^{\tau+t}$ and $f = \sum_{t=1}^{L} \hat{f}_{\tau+t}$. Since P_v is a convex function and $\hat{f}_t$ (therefore, f) is a linear function, their difference is a convex function. Problem (6.38) can be transformed into the following convex program

$$\begin{aligned} \min_{\{\mathbf{q}_t, \mathbf{v}_t, \mathbf{a}_t, \zeta_t, \xi_t, \forall t=1,\dots,L\}} \quad & P_v - f \\ \text{s.t.} \quad & \text{(6.36c)–(6.36f), (6.38e), (6.40), and (6.42).} \end{aligned} \tag{6.43}$$

The objective function and the constraints of (6.43) are all convex. Problem (6.43) can be readily tackled by standard convex solvers, e.g., the interior point method.

Given the global upper limit of P_v^t in (6.37) and the lower bounds in (6.39) and (6.41), constraints (6.40) and (6.42) tighten the original constraints (6.36g), (6.36h), and (6.36b), respectively. When the constraints of (6.43) are fully obeyed, the constraints of the initial problem (6.36) are fully obeyed; not the other way around. Hence, the feasible solution region of problem (6.43) is just part of that of problem (6.36). The minimum objective of (6.43) is an upper bound of the objective of (6.36). As proved in [341], the SCA-based approach can converge to at least a stationary point of problem (6.38). Therefore, integrating SCA into the MPC framework can generate a solution that is a stationary point at each optimization slot τ.

The proposed approach is summarized in Algorithm 5, which is computationally dominated by Step 9. Specifically, Steps 5 to 7 compute linear functions with a complexity of $\mathcal{O}(N)$. After a series of mathematical transformations, Step 9 tackles a convex program and refreshes the design variables by running the interior point method, whose complexity is $\mathcal{O}(N^{3.5})$ per round (or iteration). By summarizing every optimization procedure, the overall complexity of Algorithm 5 is $\mathcal{O}(N^{3.5})$ each round. Given the convergence precision ϵ, Algorithm 5 has a polynomial complexity of $\mathcal{O}\left(N^{3.5} \log \frac{1}{\epsilon}\right)$, wherein $\mathcal{O}\left(\log \frac{1}{\epsilon}\right)$ accounts for the number of rounds before convergence to a suboptimal solution to problem (6.36).

6.5 Numerical Results

This section presents MATLAB simulation results to corroborate the merits of the proposed approach. The common parameters concerning the UAV's power consumption and harvested solar power are given in Table 6.2.

6.5.1 Rotary-wing UAV

In this section, three baseline schemes are compared by setting the values of μ_1 and/or μ_2 to zero. We also include the extended schemes of moving direction (i.e., heading) regulation (labeled as "MDR") and MPC-based online optimization

Algorithm 5 The receding horizon control-based online approach for problem (6.36)

1: At an arbitrary time slot τ, the UAV monitor measures and obtains the present position of the target and forecasts its waypoints in the ensuing L slots.
2: Attain the lower bound on the harvested solar power $P_s^t(z_t)$ by a linear approximation in (6.41).
3: **Initialization:** Generate a feasible initial flight path for the monitor, i.e., $\{\boldsymbol{q}_t^{(0)}, \boldsymbol{v}_t^{(0)}, \boldsymbol{a}_t^{(0)}, t = 1, \ldots, L\}$, and input initial values for the slack variables $\{\zeta_t^{(0)}, \xi_t^{(0)}, t = 1, \ldots, L\}$.
4: **for** $\ell = 0, 1, 2, \ldots$ **do**
5: Approximate the propulsion power P_v^t by (6.37), (6.39), and (6.40) by the SCA technique.
6: Convexify constraints (6.36g), (6.36h), (6.38c), and (6.38d) by (6.39), and (6.40).
7: Linearize the concave part in the objective of (6.36), i.e., $-f_t$, with the SCA technique.
8: Transform problem (6.36) to problem (6.43).
9: Solve problem (6.43) with the interior point method and update the optimization variables $\{\boldsymbol{q}_t^{(\ell)}, \boldsymbol{v}_t^{(\ell)}, \boldsymbol{a}_t^{(\ell)}\}$ to design the UAV monitor's flight path for the forthcoming L slots.
10: Implement the first step of the L waypoints.
11: Update $\ell \leftarrow \ell + 1$.
12: **end for**

Table 6.2 The common parameters for UAV propulsion power, thrust power, and harvested solar power [132, 341].

Parameter	Value
UAV weight	4 kg
Atmospheric extinction, α	0.8978
Power intensity of solar beams, P_i	1367 W/m^2
Efficiency and size of solar panel, η, S	0.4, 0.5 m^2
Initial altitude of the target and monitor, H, z_0	100 m, 102 m
Coefficients for solar power approximation, c_1, c_2	0.0097, 165.83

(labeled as "Online") described in Section 6.3.3. The specific parameters concerning the rotary-wing UAV are given in Table 6.3. The total scheduling period is $T = 30$ s with each time slot of 0.2 s. The maximum monitor-target 3D distance is $D = 20$ m. The constant controlling the slope of the monitor's heading is $c_3 = 1$.

Table 6.3 The specific parameters for the rotary-wing UAV monitor's propulsion power, thrust power, and harvested solar power [132, 341].

Parameter	**Value**
Blade profile power and induced power, P_0, P_1	3.4 W, 118 W
Rotor solidity and disc area, s, A	0.03, 0.28 m^2
Tip speed of the rotor blade, U_{tip}	60 m/s
Mean rotor induce velocity, v_0	5.4 m/s
Atmospheric density and fuselage drag ratio, ρ, d_f	1.225 kg/m^3, 0.3
Maximum horizontal and vertical speed, V_{hm}, V_{vm}	30 m/s, 8 m/s
Ratio of usable energy in the battery, η_0	0.9

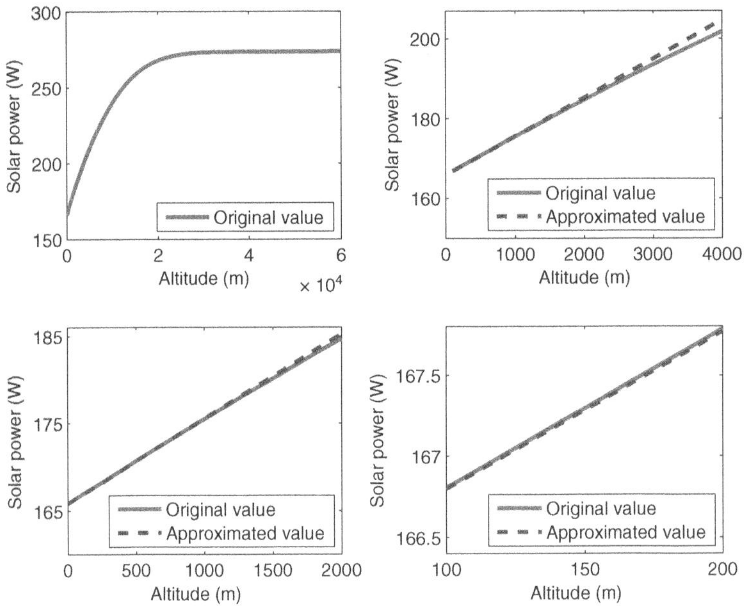

Figure 6.5 Original value and approximation of solar power.

The trajectory of the target is generated by $a_t = 10t$ and $b_t = 100\sin(t/5)$, unless otherwise stated. We set the initial trajectory of the monitor to be the same as the (predicted) target's trajectory. This initial trajectory is feasible, as it satisfies the UAV mobility and power consumption constraints.

We first show the tightness of the solar power approximation in (6.41). The scaling coefficients of the linear function c_1 and c_2 are given in Table 6.2. It can be seen from Figure 6.5 that the linear approximation is tight for an altitude below 4000 m and almost overlaps with the original function when the altitude is below 2000 m.

For the altitude typically ranging from 100 to 200 m in our application scenario, the approximation (6.41) provides a very tight lower bound for the original value, with a negligible approximation error of only 10^{-5} m.

Figure 6.6 shows the convergence speed of the proposed approach for problem (6.35), where $\mu_1 = 0.2$ and $\mu_2 = 0.1$. The red solid line is the original objective value in problem (6.35), the blue dash-dot (non-increasing) line is the total energy consumption of the monitoring UAV $(P_h + P_v)\delta$ [cf. (6.43)], and the black dashed line is the total disguising performance f. It is revealed in Figure 6.6 that the objective value can converge fast within 60 iterations. The monitor 3D trajectory is depicted in Figure 6.7. It can be observed that both the horizontal trajectory and the altitude control converge within 60 iterations. Horizontally, the monitor tails the target and surveils at a distance. Vertically, the monitor changes its altitude regularly, further deviating from the heading of the target. Bounded by constraint (6.35c), the monitor either ascends by the maximum height δV_{vm} that it reaches in the first slot and then stays at the altitude in the next slot, or first ascends and then descends by $\delta V_{vm}/2$ in the first two slots. This performance limitation averts a sharp ascent-then-descent action of the monitor at the full vertical speed, protecting the generator and preventing a speed loss.

Figure 6.8 depicts the 2D monitor-target distance and the relative 2D trajectory of the monitor observed by the target, which is treated as a reference point and marked by a blue square. The relative coordinates are $(x_t - a_t, y_t - b_t), \forall t$, and the target is at (0, 0). It can be seen from Figure 6.8 that the monitor always flies in the southwest region to the target at a distance of no more than $D = 20$ m . As observed by the target, the monitor exhibits random movement patterns.

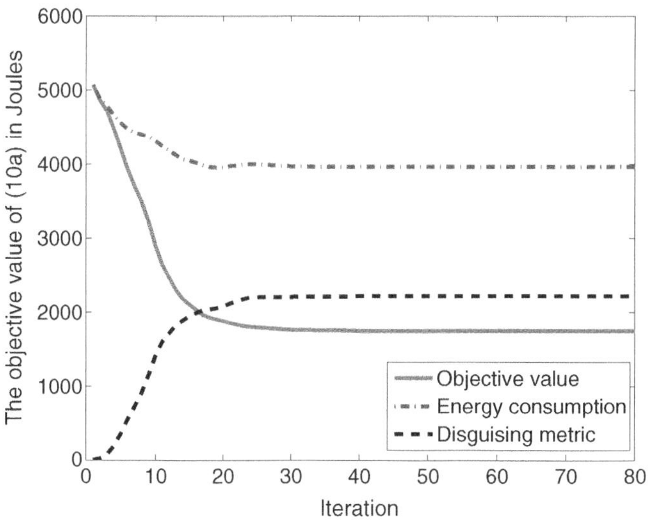

Figure 6.6 Convergence of the original objective values in (6.35a).

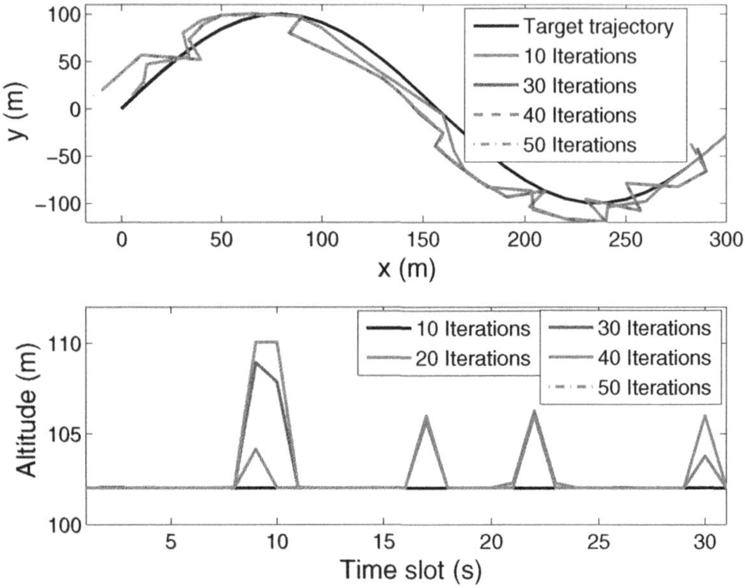

Figure 6.7 Convergence of the monitor 3D trajectory.

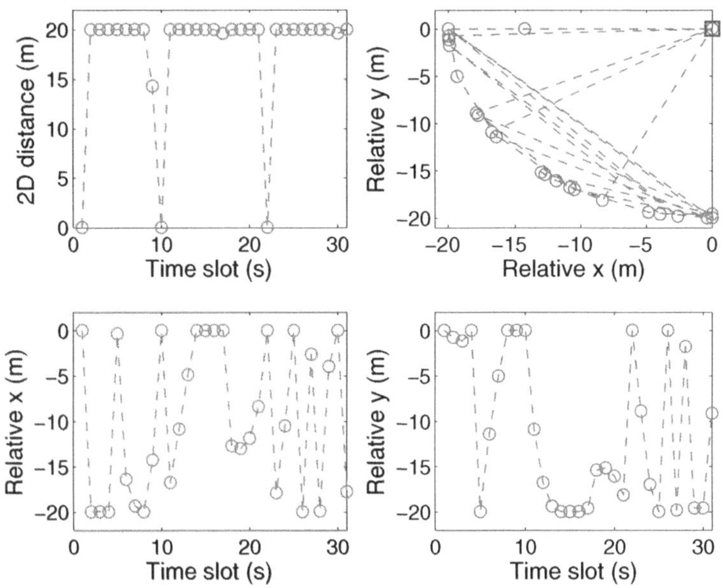

Figure 6.8 Relative monitor 2D trajectory observed by the target.

By changing the weights of the disguising performance μ_1 and μ_2, we observe their impact on the objective and the trajectory of the monitoring UAV. Figure 6.9 gives the objective values of problem (6.35) under different settings of μ_1 and μ_2. By comparing the performances under (0.1, 0.4) and (0.4, 0.1), we can conclude that keeping the horizontal distance is more efficient to reduce the objective value than changing the altitude. Yet, increasing the weights of disguising cannot enhance the performance indefinitely, as the monitor is also constrained by its maximum speed and the 3D distance to the target. Figure 6.10 depicts the corresponding monitor 3D trajectory of each scenario, where the blue solid line is the target trajectory. It is shown in Figure 6.10 that the altitude is more sensitive to the disguising performance than the horizontal distance. When the weight of altitude control rises from 0.2 to 0.4, the altitude changes even more frequently, which can cause the undesirable instability of the monitor and may trigger a speed loss. For (0.4, 0.1), the monitor always stays at the initial height, since the monitor-target horizontal distance is around 20 m, and there is no room for the monitor to rise or increase the vertical distance.

We also test the performance of the baseline schemes and the extended schemes. The proposed scheme runs with $\mu_1 = 0.2$ and $\mu_2 = 0.1$, which is labeled as "Proposed." The baseline schemes are the Proposed scheme with distance keeping only (labeled as "DKO"), i.e., $\mu_1 = 0.2$ and $\mu_2 = 0$, altitude changing only

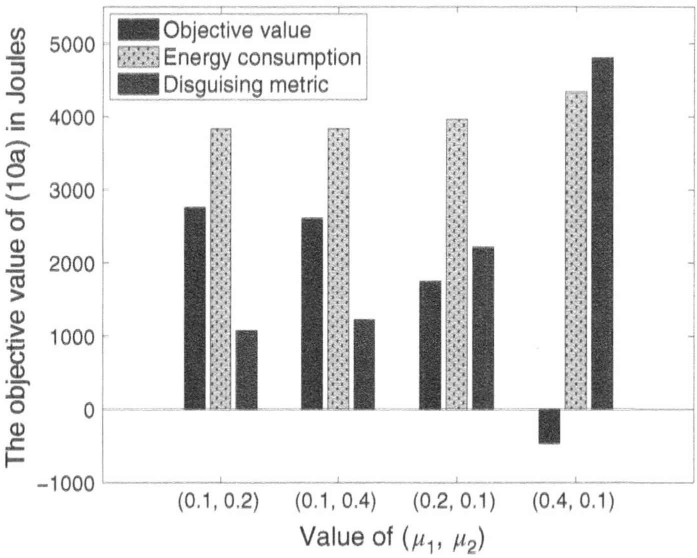

Figure 6.9 Objective values under different values of (μ_1, μ_2).

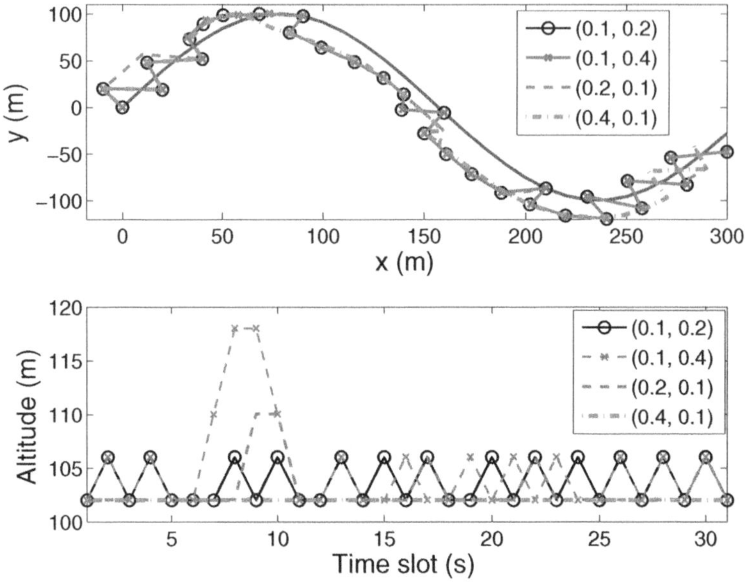

Figure 6.10 Monitor 3D trajectory under different values of (μ_1, μ_2).

(labeled as "ACO"), i.e., $\mu_1 = 0$ and $\mu_2 = 0.1$, and no disguising performance (labeled as "NDP"), i.e., $\mu_1 = \mu_2 = 0$, which essentially only minimizes the total energy consumption of the monitor. The extended schemes are the MDR and the Online schemes.

Figure 6.11 shows the objective values of the monitoring UAV, i.e., (6.35a), under different optimization schemes. Figures 6.12 and 6.13 depict the corresponding horizontal and vertical trajectories of the monitor, respectively. It is shown in Figure 6.11 that the Proposed scheme and the MDR scheme score the highest in terms of the disguising performance, followed by the DKO scheme. The Proposed scheme and the MDR scheme can thus significantly suppress the objective values. By comparing the DKO scheme and the ACO scheme, it is also seen that keeping distance is much more effective than changing altitude to enhance the disguising performance and reduce the objective value. On the other hand, the NDP scheme yields the least energy consumption, as it is actually formulated that way.

It is revealed in Figures 6.12 and 6.13 that the MDR scheme does a good job in keeping the distance from the target and changing the heading of the UAV at the same time on the horizontal plane. The Proposed scheme strikes the best balance between distance keeping and altitude changing, while the baseline schemes can only deal with either distance or altitude. Although the monitor lags behind the

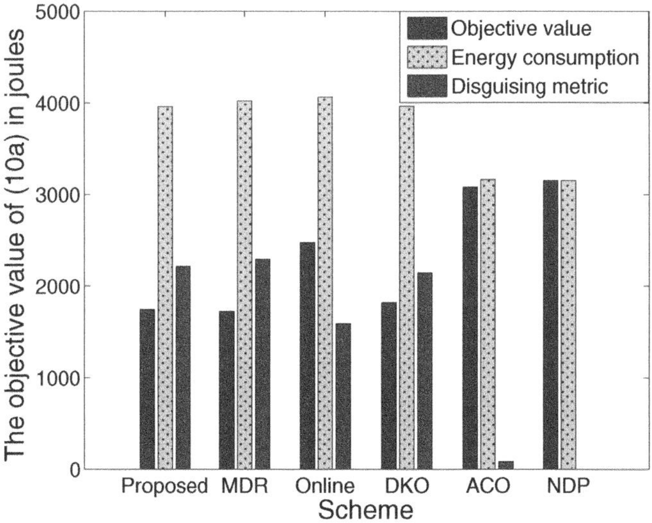

Figure 6.11 Objective values under different schemes.

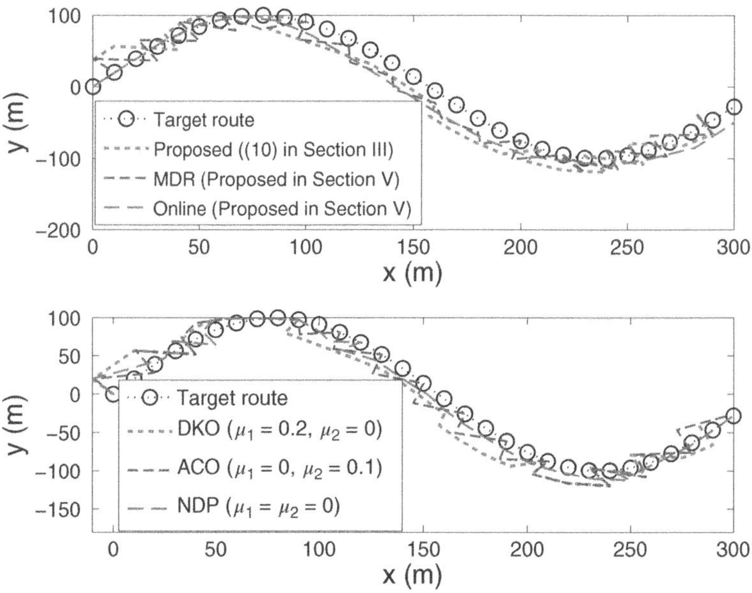

Figure 6.12 Monitor horizontal trajectory under different schemes.

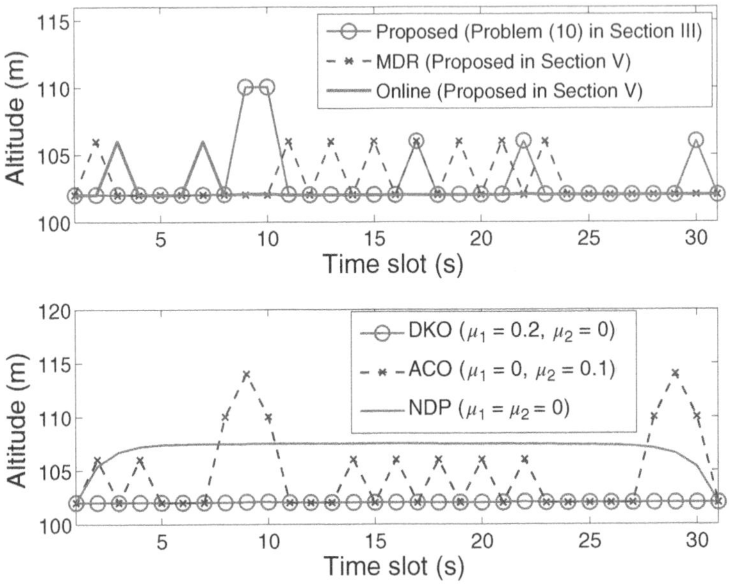

Figure 6.13 Monitor altitude under different schemes.

target under the Online scheme, its trajectory is close to and aligns with the target trajectory. The monitor flies to the same location projected on the (x, y)-plane with a few meters above the target, from time to time under the ACO scheme. The monitor always travels at the lowest height under the DKO scheme, and its horizontal trajectory aligns with the target trajectory from the 10th to the 18th second. Under the NDP scheme, the monitor stays close to the target horizontally and at a constant height for almost the entire scheduling period. All the aforementioned baseline schemes have their drawbacks, which put the target on alert and place the monitor at the risk of exposure. Besides, altitude changes are much more frequent than desired for the ACO scheme, which may incur the instability of the monitor and damage the generator. In a nutshell, these simulation results fully demonstrate and corroborate the merits of our proposed schemes.

Last but not least, to demonstrate the energy saving of the proposed framework under a practical model for propulsion power, we consider a baseline scheme (labeled as "DST") where the propulsion power is simply modeled as the total 2D distance traveled during the surveillance, i.e., $P'_h := \sum_t \sqrt{(x_t - x_{t-1})^2 + (y_t - y_{t-1})^2}$ [121]. With P'^2_h replacing the original $\sum_t P^t_h$ in (6.35), Figure 6.14 depicts the corresponding monitor trajectory and power consumption. It can be seen that the proposed schemes result in a more energy-efficient trajectory, and allow the monitor to consume less energy. A total energy of 1424 J can be saved with the proposed offline scheme.

It is also shown in Figure 6.14 that the online scheme is more energy-efficient than the DST scheme, given its objective of energy minimization (as opposed to the travel distance minimization objective of the DST scheme). Yet, the proposed online scheme consumes more energy than its offline counterpart, since the offline scheme assumes the *a-priori* knowledge of the target's trajectory throughout the video surveillance mission. On the other hand, the monitor keeps the target within its sight. As the target keeps changing its speed and heading, the monitor keeps adjusting its speed and heading accordingly. It can be seen in (6.9) that the propulsion power is a function of the UAV speed. Therefore, the time-varying propulsion power shown in Figure 6.14 also indicates that the UAV speed changes over time.

Figure 6.15 depicts the disguising performance and 3D monitor-target distance of the proposed scheme over time. It can be observed that the disguising metric decreases when the monitor flies near the target. The monitor keeps the maximum allowed 3D distance of $D = 20$ m for effective video surveillance from the target most of the time. Figure 6.16 plots the Pareto-front between the power consumption and the disguising metric by varying the values of μ_1 and μ_2 ($\mu_1 + \mu_2 = 1$). It is seen that the power consumption does not increase anymore when the disguising metric exceeds 4600 (m^2).

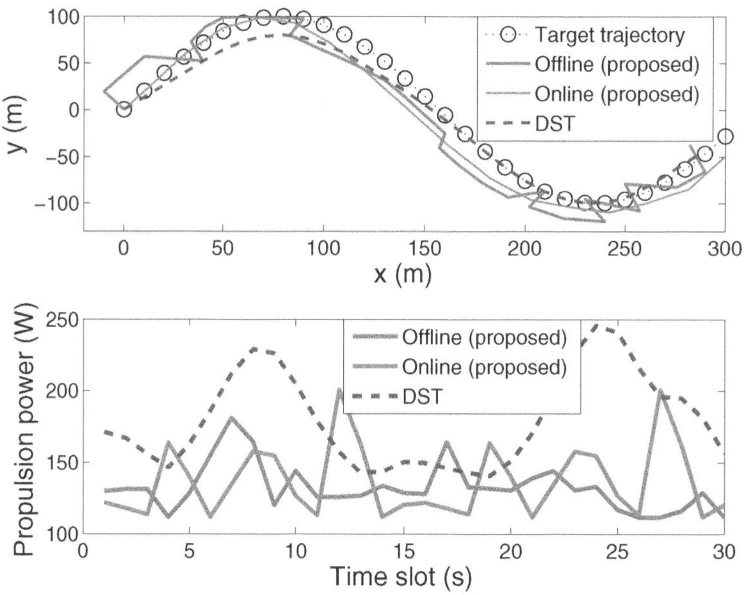

Figure 6.14 Energy saving of the proposed schemes.

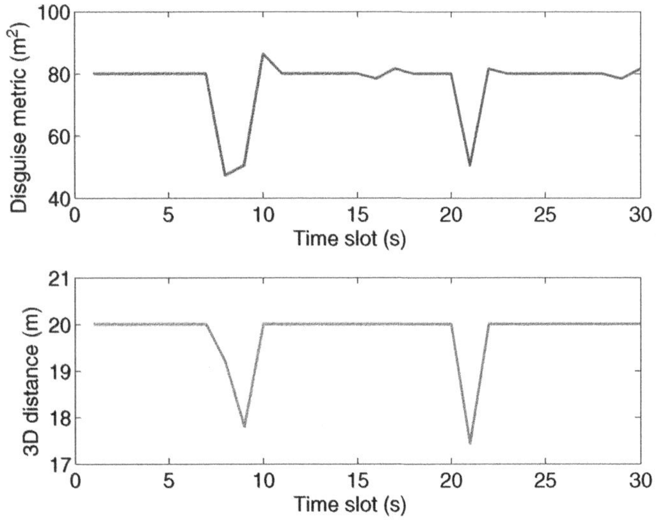

Figure 6.15 Disguising metric and 3D monitor-target distance.

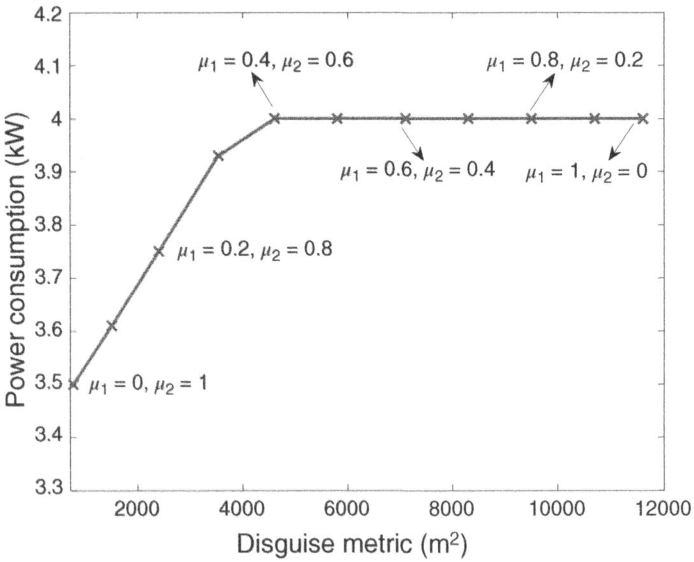

Figure 6.16 Power consumption versus disguising metric with $\mu_1 + \mu_2 = 1$.

6.5.2 Fixed-wing UAV

The specific parameters regarding the UAV monitor's energy usage and reaped solar energy are provided in Table 6.4. The overall mission duration is $T = 30$ s with a duration of 0.2 second per slot ($N = 150$). The minimum and maximum distances from the target to the monitor in the 3D space are $d_{\min} = 3$ m and $d_{\max} = 20$ m, respectively. The linear flight path of the target is simulated by $b_{xt} = 9t$ and $b_{yt} = 0$. At slot τ, the estimated target trajectory is generated by $(\hat{b}_{x(\tau+t)}, \hat{b}_{y(\tau+t)}) = (b_{x(\tau+t)}, b_{y(\tau+t)}) + \varepsilon_t, \forall t$, where ε_t is the estimation error following the zero-mean Gaussian distribution. The weighting coefficients of disguise are $(\mu_1, \mu_2) = (0.4, 0.6)$, unless otherwise stated. The initial trajectory of the monitor is set as $x_t = 8t$ and $y_t = 0$ for the first optimization slot $\tau = 1$. The optimized trajectory of the previous slot τ initializes the trajectory of the new slot $\tau + 1$.

We show the performance of the proposed MPC-based online approach. At each optimization slot τ, L steps of the trajectory are planned for the monitor by running Algorithm 4 until convergence. Figure 6.17 depicts the convergence of the online algorithm when $L = 50$. The red solid curve depicts the objective value of problem (6.36). The blue dash-dot curve depicts the sum energy usage of the monitoring UAV $P_v\delta$; see (6.43). The black dashed (non-decreasing) curve depicts the disguise measure f. It can be seen that the objective values (in Joules) converge fast within 10 iterations per slot τ, and become stable within five optimization slots. Figure 6.18 depicts the monitor trajectory by running the online approach for different numbers of iterations per slot τ when $L = 50$. It is revealed that the trajectory planning can achieve a desirable performance within 15 iterations per slot.

Figure 6.19 plots the 3D trajectory of the fixed-wing UAV monitor by the proposed online scheme with different numbers of planned waypoints L. It is observed that the trajectory is improved with the increasing number of planned waypoints per slot. Horizontally, the UAV follows and monitors the target with a reasonable spatial remoteness. Vertically, the monitor varies its elevation continuously, frequently deviating from the moving direction of the target. Figure 6.20 depicts the planned trajectory of the monitor at each optimization slot τ for $L = 50$ and

Table 6.4 Simulation parameters for the fixed-wing UAV monitor [132, 332].

Parameter	Value
Coefficients for the fixed-wing UAV, c_1 and c_2	9.26×10^{-4}, 2050
Lower and upper bounds of the speed, $v_{\min}$ and $v_{\max}$	4 m/s, 30 m/s
Maximum acceleration, $a_{\max}$	4 m/s^2
Sine of the maximum pitch angle, ϑ	0.7
Initial battery level, E_0	4000 J
Fraction of consumable energy in the battery, η_0	0.5

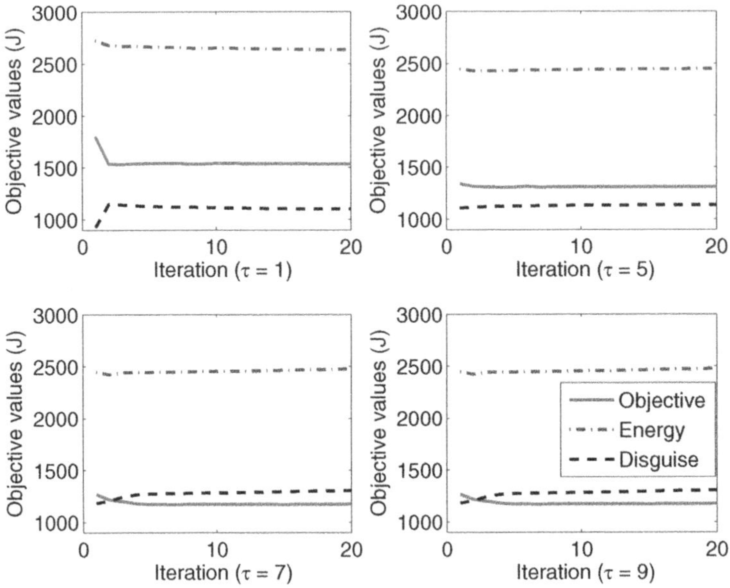

Figure 6.17 Convergence of the proposed online scheme at different optimization slot τ with $L = 50$.

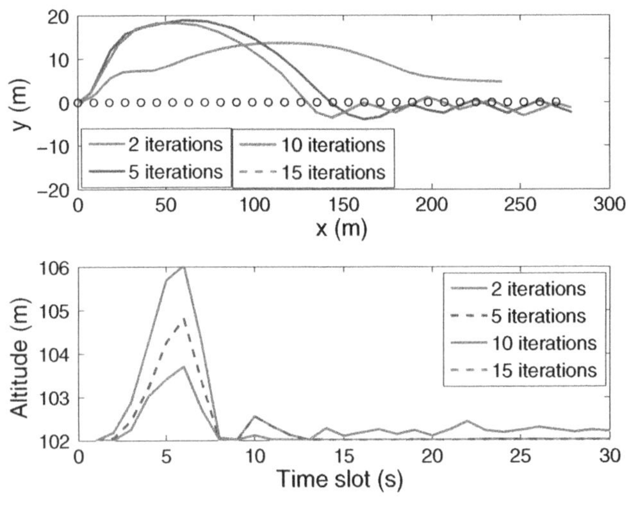

Figure 6.18 3D trajectory of the monitor by the proposed online scheme with different numbers of iterations when $L = 50$.

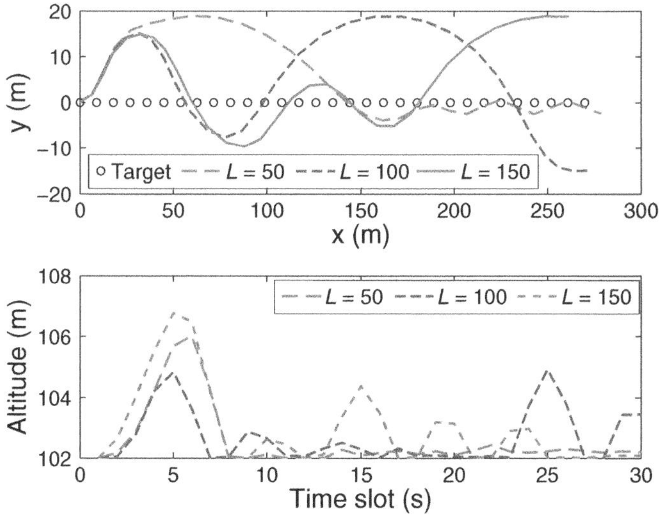

Figure 6.19 3D trajectory of the UAV monitor by different online schemes.

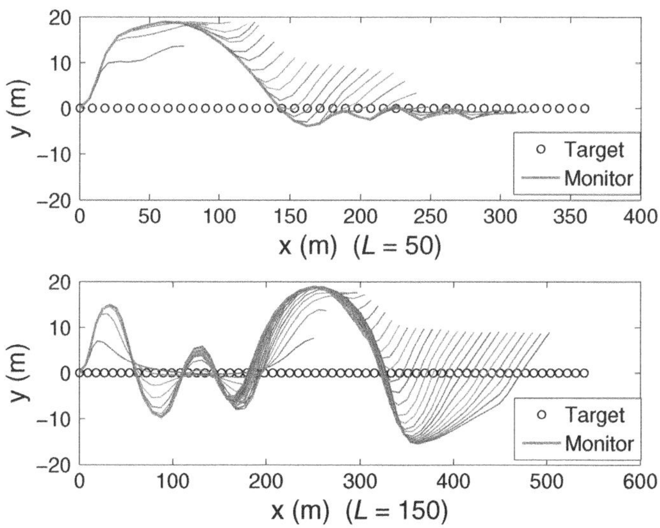

Figure 6.20 Horizontal trajectory of the UAV monitor at each optimization slot τ by the online scheme, where the black circle is the target trajectory, the green line is the optimized monitor trajectory, and the other trajectories in different colors are the planned L-step trajectories at each slot.

$L = 150$, respectively. It is seen that the trajectory is progressively refined over time by implementing one step at each slot. The objective values of the corresponding schemes are provided in Figure 6.21. By comparing the performances of $L = 50$ and 150, we see that the energy consumption is significantly reduced and the visual disguise is enhanced as a result of the trajectory planning over a longer period.

Figure 6.22 illustrates the distance from the target to the monitor on the (x, y)-plane, and the movement of the monitoring UAV in the view of the target by the online scheme with $L = 100$. The target is viewed as an anchor mark and highlighted by a green diamond. The relative waypoints are $(x_t - b_{xt}, y_t - b_{yt})$, $\forall t$. The target is located at (0, 0) in the figure. It is revealed in Figure 6.22 that the UAV monitor travels first northwestern, then southeastern to the target, at a reasonable 3D remoteness within the range of $[d_{\min}, d_{\max}] = [3, 20]$ m. In the view of the target, the UAV monitor demonstrates stochastic traveling patterns.

Figure 6.23 shows the battery level (in Wh) of the UAV monitor at each time slot during the surveillance mission, with and without solar energy reaping. It is observed that the sustainability of the monitor can be enhanced with solar energy reaping, and the residual energy in the battery is always above $(1 - \eta_0)E_0$ at any time slot for emergency. While the battery capacity can only support the UAV flight for 26 seconds without energy harvesting. The merits of solar energy harvesting are therefore corroborated.

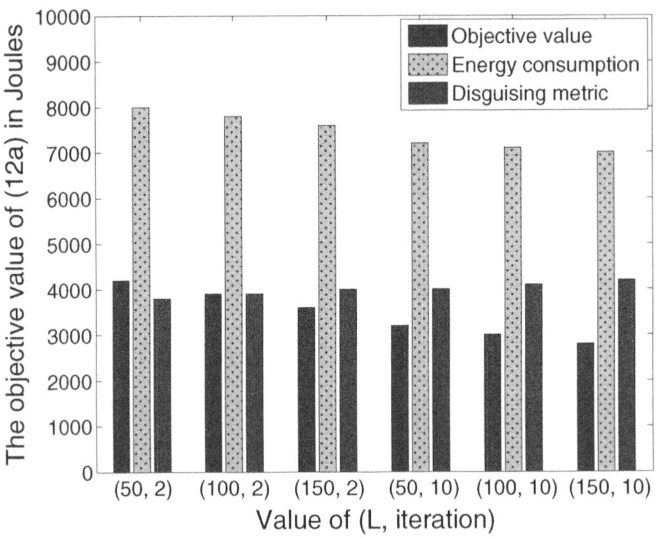

Figure 6.21 Objective values of the UAV monitor by different online schemes over $T = 30$ seconds.

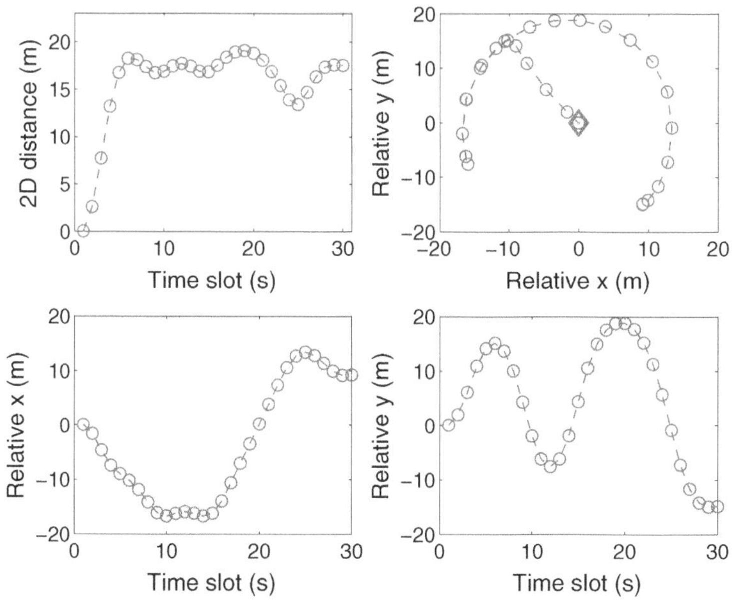

Figure 6.22 Relative 2D monitor trajectory in the view of the target by the online scheme with $L = 100$.

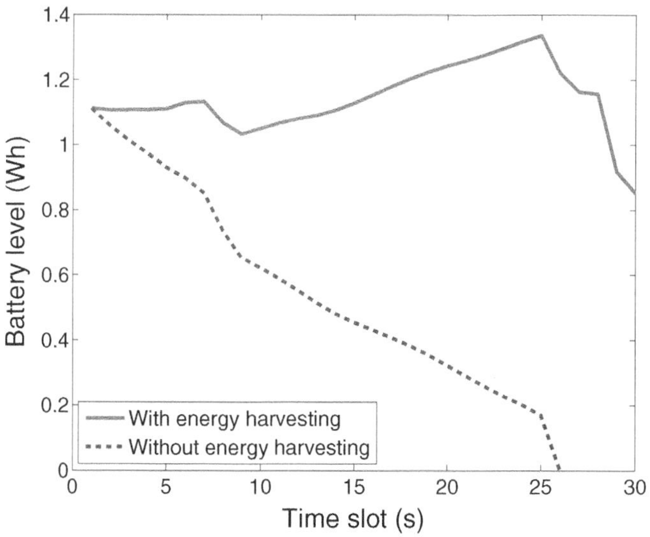

Figure 6.23 Battery level of the monitor with and without solar energy harvesting.

By varying the weighting coefficients of disguise (μ_1, μ_2), we show their influence on the objective value and the flight path of the monitoring UAV optimized online with the proposed Algorithm 5. Figure 6.24 depicts the 3D trajectory of the UAV monitor by Algorithm 5 and the baseline schemes (i.e., partial or no disguise) with $L = 100$. The baseline scheme without visual disguise is labeled "No Disguise" or "ND" for short. Compared to the ND scheme, the proposed algorithm strikes a good balance between distance keeping and elevation varying, and results in effective visual disguise. It is also shown in Figure 6.24 that the visual disguise is more susceptible to the elevation of the UAV monitor than the horizontal target-monitor spatial separation. When the weighting coefficient of the elevation variance is lifted from 0 to 1, the elevation varies increasingly frequently. Figure 6.25 presents the corresponding objective values of problem (6.36) in each scenario. By comparing $(\mu_1, \mu_2) = (0, 1)$ and $(1, 0)$, we can draw the conclusion that maintaining the horizontal target-monitor separation is more effective in suppressing the objective value than varying the height of the monitor. On the other hand, raising the weighting coefficients of disguise cannot improve the efficiency boundlessly, as the monitor is increasingly restrained by its maximum velocity and the 3D distance from the target. The energy consumption of the ND scheme is the lowest, as the monitor flies in the most energy-efficient way.

Figure 6.26 depicts the performance of the online algorithm when the monitor's estimation on the target trajectory gets increasingly inaccurate over time, i.e., the estimation error increases with time and is generated by $0.5t\varepsilon_t$, when

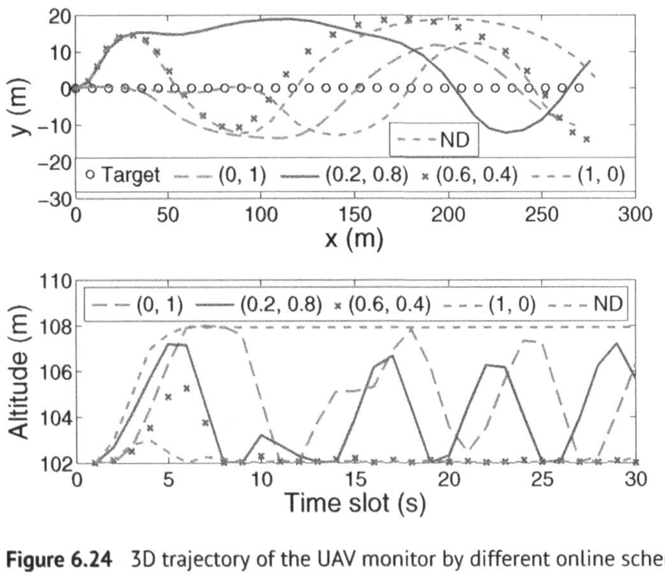

Figure 6.24 3D trajectory of the UAV monitor by different online schemes with $L = 100$.

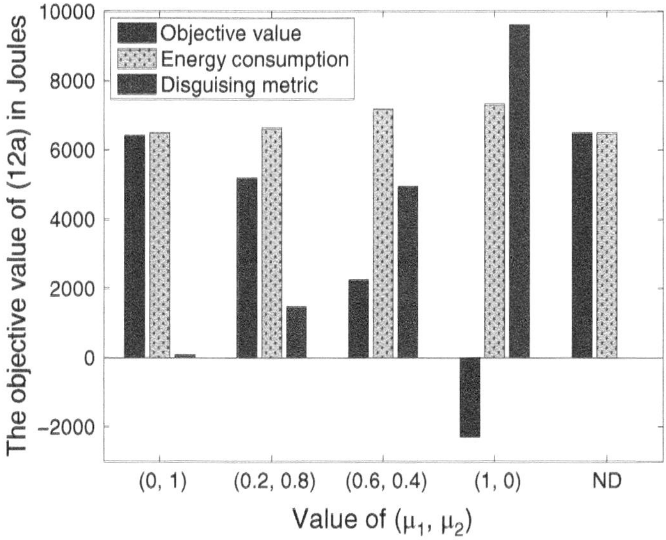

Figure 6.25 Objective values of the UAV monitor by different online schemes with $L = 100$.

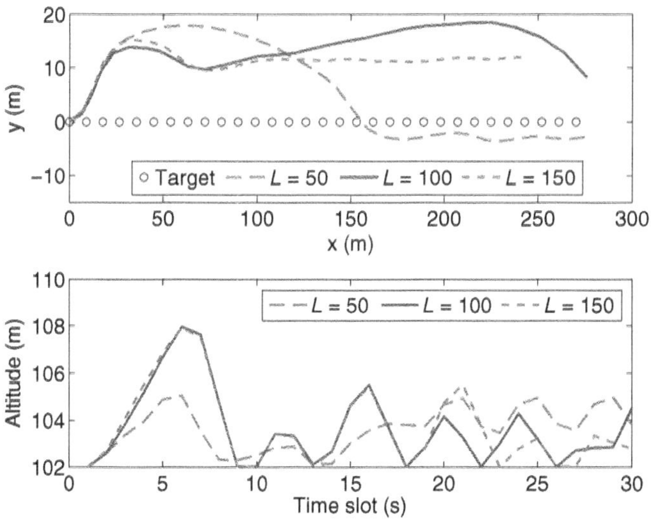

Figure 6.26 3D trajectory of the UAV monitor by the online scheme with an increasing estimation error over time when $(\mu_1, \mu_2) = (0.4, 0.6)$.

$(\mu_1, \mu_2) = (0.4, 0.6)$. It is unveiled that as the estimation error rises, the monitor takes fewer horizontal twists and turns, keeps a fair 2D distance from the target, travels in a relatively stable direction to the target on the (x, y)-plane, and changes its elevation more frequently, especially when $L = 150$. This is because the fixed-wing UAV changes its velocity continuously with no sudden or sharp movements due to its maneuverability and aerodynamics. Therefore, it seeks visual disguise in a direction that is more stable with less randomness to have robustness against the estimation error.

Next, we proceed to show the performance of the fixed-wing UAV monitor under the offline scheme, by running Algorithm 4 to solve problem (6.35) with perfect *a-priori* knowledge of the target's trajectory over the scheduling period of $T = 30$ seconds. Figure 6.27 depicts the convergence rate of the offline scheme with different values of (μ_1, μ_2). Figure 6.27 unveils that the objective of (6.35) becomes stable and converges quickly in less than 30 iterations. The objective values converge faster when $\mu_1 < \mu_2$. The 3D trajectory of the fixed-wing UAV monitor with $(\mu_1, \mu_2) = (0.4, 0.6)$ is delineated in Figure 6.28, where both the horizontal trajectory and the elevation variance become stable and converge in less than 20 iterations.

The 3D trajectories of the offline scheme and the baseline scheme are depicted in Figure 6.29. The corresponding objective values are provided in Figure 6.30.

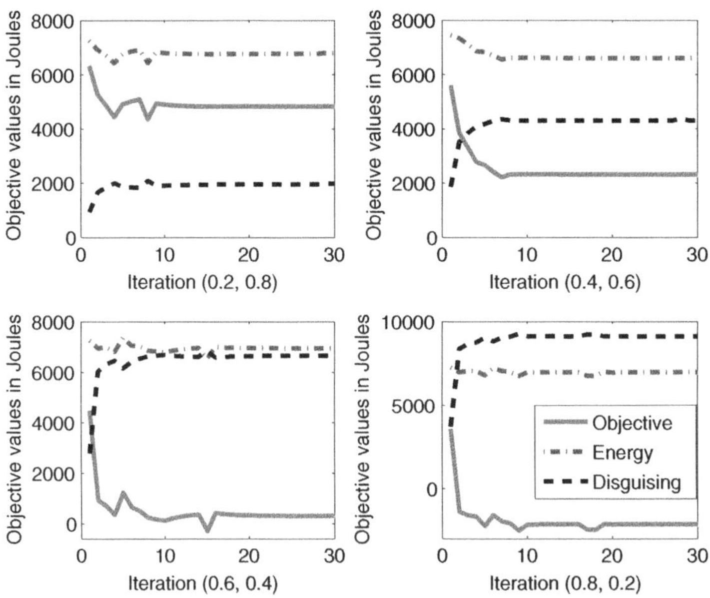

Figure 6.27 Convergence of the objective values in (6.35a) by the offline scheme with different (μ_1, μ_2).

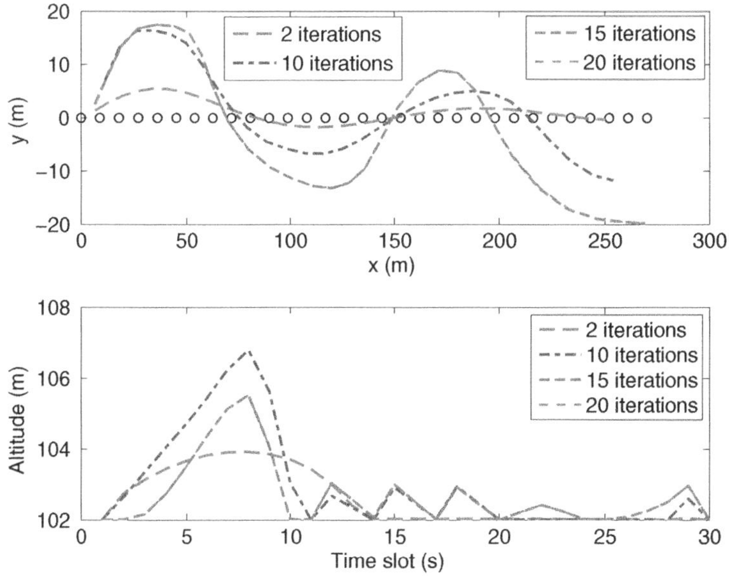

Figure 6.28 Convergence of the 3D monitor trajectory by the offline scheme with $(\mu_1, \mu_2) = (0.4, 0.6)$.

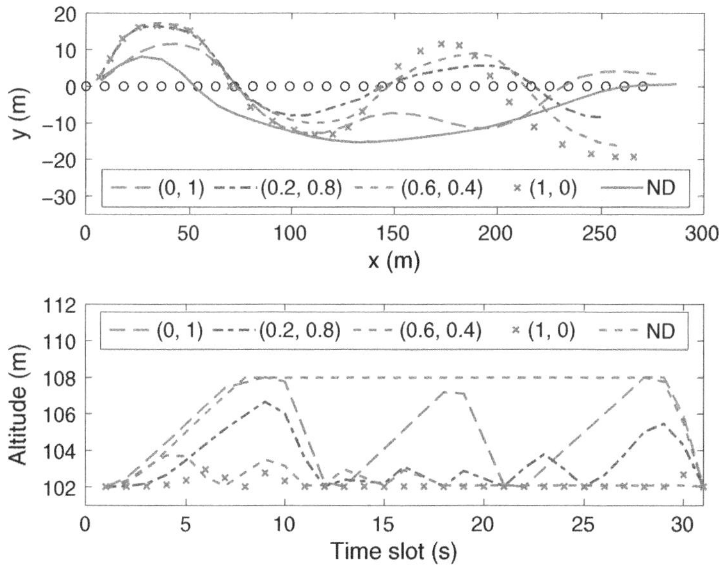

Figure 6.29 3D trajectory of the UAV monitor by the offline and baseline schemes.

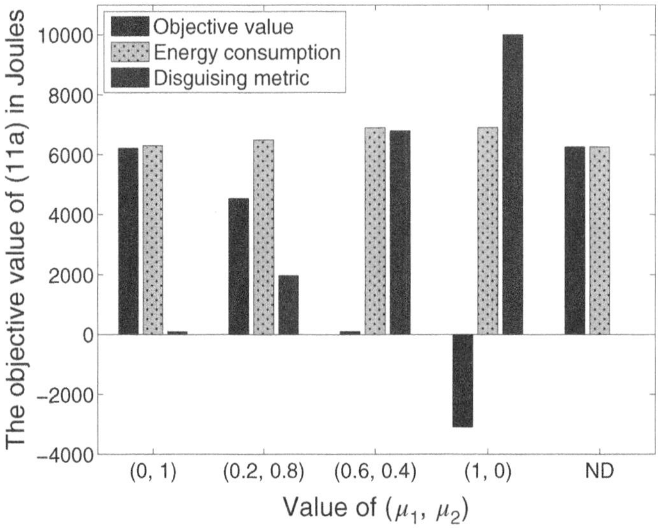

Figure 6.30 Objective values of the UAV monitor by different offline schemes.

By comparing these two figures with Figures 6.24 and 6.25, it is revealed that the online algorithm is as effective as the offline algorithm, but yields a slightly higher energy consumption and lower disguising performance, due to the estimation error of the target trajectory. By optimizing $L = 100$ waypoints per slot τ, the online algorithm can achieve a desirable performance as compared to the offline scheme.

We compare the performance of disguised video surveillance between the fixed-wing UAV and its rotary-wing counterpart, adjusting its trajectory online based on the receding horizon control model. For the convenience of readers, we provide the propulsion power model and the thrust power model of the rotary-wing UAV. The propulsion power for level flight is given by [112]

$$P^t_{lf} = P_0\left(1 + \frac{3{V^t_h}^2}{U^2_{tip}}\right) + P_1\left(\sqrt{1 + \frac{{V^t_h}^4}{4v^4_0}} - \frac{{V^t_h}^2}{2v^2_0}\right)^{\frac{1}{2}} + \frac{1}{2}d_f\rho sA{V^t_h}^3, \tag{6.44}$$

where V^t_h is the horizontal speed of the rotary-wing UAV at time slot t, given by $V^t_h = \sqrt{(x_t - x_{t-1})^2 + (y_t - y_{t-1})^2}/\delta$. The thrust power for vertical flight can be obtained by [110]

$$P^t_{vf} = W(z_t - z_{t-1})/\delta, \ \forall t. \tag{6.45}$$

Therefore, the power usage of the rotary-wing UAV at time slot t is $P_v^t = P_{lf}^t + P_{vf}^t$. The notations and values of the simulated parameters are provided in Table 6.5.

We plot the 3D trajectory of the rotary-wing monitor in Figure 6.31 by running the online Algorithm 5 with $L = 100$, and $(\mu_1, \mu_2) = (0.4, 0.6)$. It is observed that the rotary-wing monitor takes more and sharper horizontal turns, keeps a large horizontal distance from (and behind) the target, shows more randomness and better visual disguise on the (x, y)-plane, and does not change its altitude, as compared to the fixed-wing monitor (as plotted in Figure 6.29). While the rotary-wing monitor travels more smoothly by the offline scheme, it flies first behind and then ahead of the target, and always appears on the same side of the target. Figure 6.32

Table 6.5 The parameters of the rotary-wing UAV [336].

Parameter	Value
The weight of the rotary-wing UAV	4 kg
Blade profile power and tip speed, P_0 and U_{tip}	3.4 W, 60 m/s
Rotor induced power and velocity, P_1 and v_0	118 W, 5.4 m/s
Rotor solidity and disc area, s and A	0.05, 0.5 m^2
Air density and fuselage drag fraction, ρ and d_f	1.225 kg/m^3, 0.3

Source: Adapted from Zeng et al. [341].

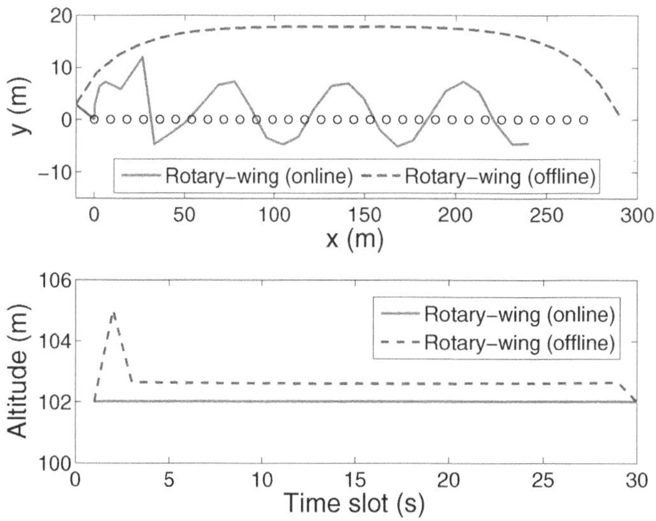

Figure 6.31 3D trajectory of the rotary-wing UAV monitor by the proposed schemes with $L = 100$ and $(\mu_1, \mu_2) = (0.4, 0.6)$.

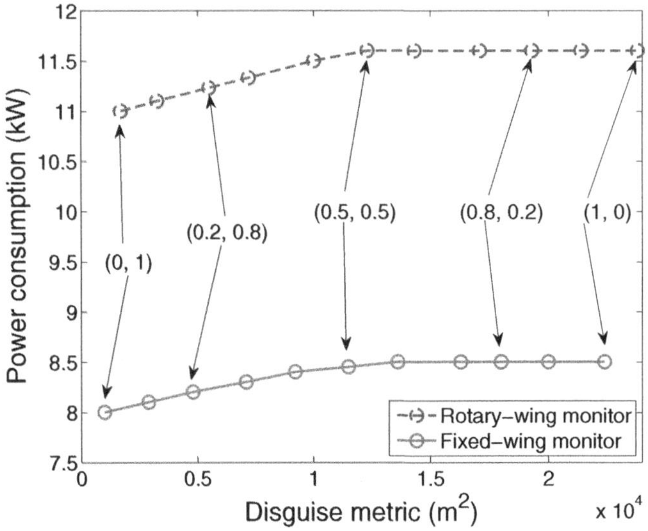

Figure 6.32 Power consumption versus disguising metric under different values of (μ_1, μ_2) with $\mu_1 + \mu_2 = 1$ for fixed-wing and rotary-wing UAV monitors running the online trajectory planning algorithm when $T = 60$ seconds.

shows the Pareto-front between the energy usage and the disguise measure by changing the numerical quantities of μ_1 and μ_2 ($\mu_1 + \mu_2 = 1$). Figure 6.32 shows that although the fixed-wing monitor scores a bit lower in disguise than the rotary-wing monitor, it is more energy-efficient and sustainable. The power consumption stops growing once the disguise measure surpasses 14,000 (m^2) for the fixed-wing monitor and 12,000 (m^2) for the rotary-wing monitor.

We show the performance of the proposed online scheme for the fixed-wing and rotary-wing monitors under different target trajectories, i.e., a sinusoidal trajectory and a random trajectory. It can be observed from Figures 6.33 and 6.34 that the monitors demonstrate good visual disguise under different target trajectories.

Last but not least, we show the performance of the proposed online scheme under a linear target trajectory when the mission duration is $T = 90$ seconds, and the maximum and minimum monitor-target distances are $d_{\max} = 100$ m and $d_{\min} = 20$ m. The reason is that the typical resolution of the cameras mounted on UAVs can recognize an object at a relatively large distance, e.g., over 100 m, when the cameras are reasonably concentrated. Figure 6.35 plots the 3D trajectory of the fixed-wing UAV, and the battery level of the UAV with and without energy harvesting (labeled as "W/EH" and "W/O EH," respectively). We see that the UAV monitor can effectively follow the target, while keeping a reasonable distance of

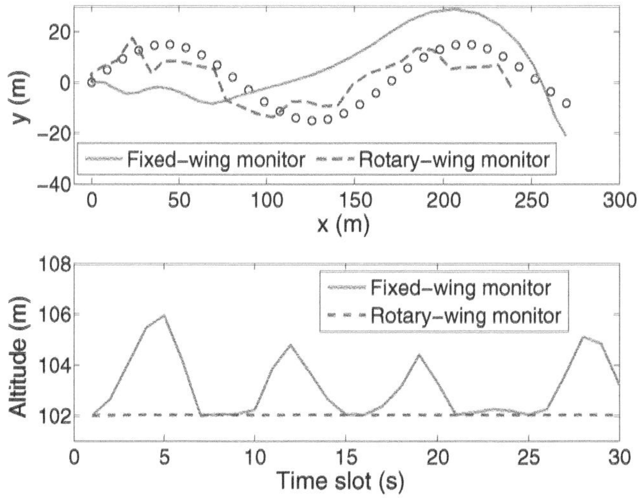

Figure 6.33 3D trajectory of the fixed-wing and rotary-wing monitors by the proposed online scheme under a sinusoidal target trajectory with $L = 100$ and $(\mu_1, \mu_2) = (0.4, 0.6)$.

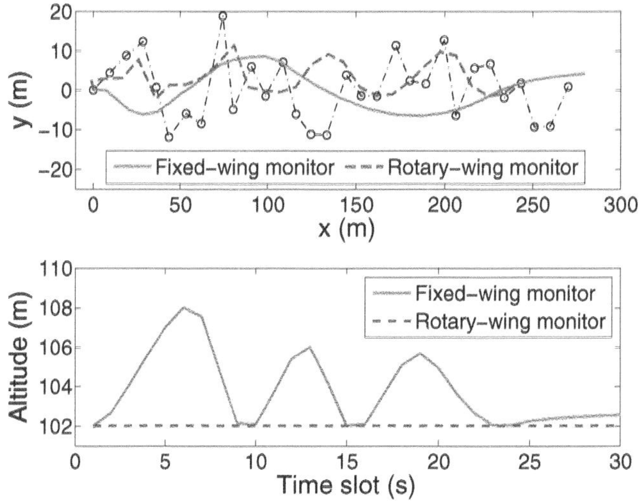

Figure 6.34 3D trajectory of the fixed-wing and rotary-wing monitors by the proposed online scheme under a random target trajectory with $L = 100$ and $(\mu_1, \mu_2) = (0.4, 0.6)$.

no larger than 100 m from the target. Moreover, the UAV can always maintain an energy reserve of no less than $(1 - \eta_0)E_0$ Joules in its battery with solar energy harvesting. In contrast, without energy harvesting, the battery energy can only support the UAV for 40 seconds of the mission.

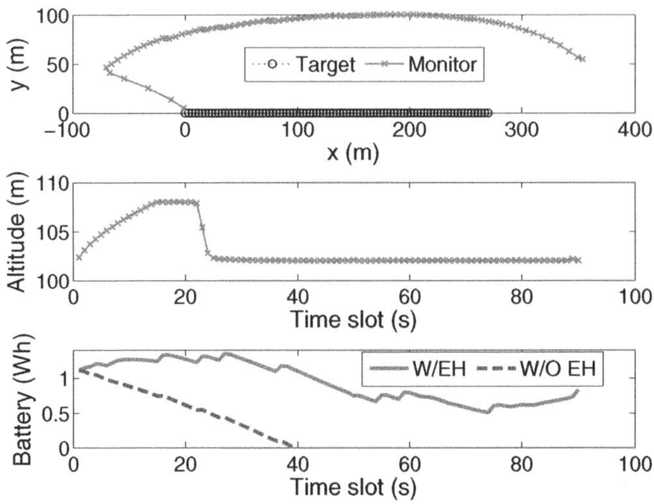

Figure 6.35 3D trajectory and battery level of the fixed-wing UAV monitor by the proposed online scheme under a linear target trajectory with $T = 90$ s, $d_{\max} = 100$ m, $L = 100$, and $(\mu_1, \mu_2) = (0.4, 0.6)$.

6.6 Authentication Management of UAVs

As UAV networks become increasingly integrated into critical applications such as surveillance, logistics, and public safety, securing these systems against unauthorized access and malicious activity has become a top priority. A foundational component of UAV network security is authentication management, which ensures that only legitimate UAVs can participate in network operations while detecting and excluding rogue or compromised nodes.

Given the constrained resources of UAVs, lightweight cryptographic protocols have been developed, such as ECC (Elliptic Curve Cryptography)-based mutual authentication, identity-based schemes, and certificateless signatures. Hierarchical authentication architectures, where a central ground station or edge node authenticates UAVs based on task domain or mission zone—have also been widely adopted to reduce computational overhead and latency in large-scale deployments.

Weight-based trust management systems are useful for dynamic trust assessment, where multiple behavioral metrics (e.g., signal consistency, response latency, and task compliance) are assigned quantifiable weights. These weights can be dynamically updated using statistical analysis or fuzzy logic to reflect evolving trust levels. This allows for flexible and context-sensitive decision-making in UAV acceptance or isolation.

Authentication decisions often rely on threshold-based models, where a UAV is deemed trustworthy if its aggregated trust score exceeds a predefined threshold. Adaptive thresholding can be adopted, where the threshold value is dynamically tuned based on environmental risk factors, network density, or mission criticality. Adaptive thresholds help balance security with availability, especially in rapidly changing or contested environments.

Another important issue is the tracking and analysis of UAV mobility patterns to detect malicious or abnormal behavior. Techniques include trajectory prediction using historical flight data, anomaly detection based on deviations from expected flight corridors, and RF fingerprinting to distinguish spoofed or impersonated UAVs. Some systems also use motion-based profiling and context-aware rules to flag suspicious UAVs in real time.

6.7 Conclusion

In the first study in this chapter, a new framework was proposed to exploit the flexibility of a solar-powered UAV for covert video surveillance by joint power management and 3D trajectory optimization. The SCA and PDCAE techniques were leveraged to convexify the optimization problem and obtain a low-complexity suboptimal solution. The proposed PDCAE-based solution can serve as the control model to operate online and achieve control-based refinement of the monitor's trajectory. Extensive numerical results corroborated the merits of the proposed scheme in maintaining covertness over baseline schemes with partial or no disguising.

In the second study in this chapter, a new scheme was developed to exploit the maneuverability of a solar-powered, fixed-wing UAV for disguised tailing and video surveillance collectively through energy management and trajectory optimization. The SCA technique was employed to convexify the investigated problem, and integrated with the MPC framework in pursuit of a computationally efficient, quality solution in an online fashion. As corroborated by extensive simulations, the proposed scheme outperformed the schemes with no or partial disguise, in terms of visual disguise. The fixed-wing UAV was also demonstrated to surpass its rotary-wing counterpart with regard to energy efficiency and sustainability.

7

Energy Efficiency for UAV-based Radio Sensing and Monitoring

7.1 Introduction

Featuring high flexibility, swift deployment, and wide coverage, UAVs have been extensively applied to activities, such as search and rescue in disaster areas, inspection of landscapes, and surveillance of forest fires. Recently, UAVs have found many use cases in wireless communication networks as cost-effective and on-demand aerial wireless platforms for areas without cellular coverage [293, 333, 334], or as flying mobile users within a cellular network [79, 192]. Cellular-connected UAVs can enhance connectivity, coverage, flexibility, and reliability of wireless communication networks [79, 192]. The UAVs are anticipated to engage significantly in the fifth-generation (5G) and beyond 5G (B5G) wireless networks, and provide new services such as real-time image transmission [190], caching and multicasting [304, 335], data dissemination or collection [202, 287, 291, 333], mobile relaying and edge computing [146, 313, 339], and wireless power transfer [194, 293, 302].

As the applications of the IoT continue to expand in the current and future wireless networks, many infrastructure-free wireless links (such as Bluetooth, Wi-Fi, and UAV-enabled transmission) have been established to support communications among IoT devices. Yet, these convenient networks can be abused for crimes and terrorism, if in the wrong hands. Therefore, authorized parties must surveil these suspicious communication links (see [23, 40, 107, 123, 179, 188, 290, 299, 300, 301, 338]). Optimization metrics for legitimate monitoring typically focused on maximizing the eavesdropping rate or the non-outage probability [123]. Spoofing schemes were proposed for a malicious transmission link to maximize the eavesdropping rate [338], or to intervene and change the communicated data [299]. For a suspicious communication link in [300], the largest achievable monitoring non-outage probability and comparative intercepting rate were obtained under delay-sensitive and delay-tolerant scenarios, respectively. Proactive jamming

Artificial Intelligence for Unmanned Aerial Vehicles: Sensing, Communication, and Computing, First Edition. Shuyan Hu et al.

schemes were developed to maximize the average monitoring rate for multi-input multi-output (MIMO) channels [40], relay networks [107], UAV-aided links [179], and with a deep-learning approach [188]. Methods of accurate detection, classification, and localization of unwanted UAVs in a no-fly zone were discussed in [23].

Most existing works considered a fixed ground node (GN) as the legitimate monitor, whose channel typically suffers from severe large-scale path loss and small-scale fading. Yet, the UAV-enabled monitor can enjoy a high-probability of line-of-sight (LoS) channels as its flying altitude rises. It is therefore easier for the UAV to obtain its channel gains with the GNs if their locations are known. Thanks to its flexibility, the UAV can dynamically adjust its positions for a better eavesdropping rate, e.g., by flying closer to the suspicious transmitter. Categorized by the power sources, there are two types of UAVs, namely, the tethered UAVs and the untethered UAVs [292]. A tethered UAV is linked with a ground control platform, and is powered stably through a cable or a wire. The lack of mobility has constrained tethered UAVs to a targeted area only [16, 181, 306, 311]. In particular, the horizontal positions of the UAVs were optimized in [181] to cover a set of GNs with the least possible number of UAVs. The optimal 3D deployment scheme of a UAV was developed in [16] to cover as many GNs as possible with a minimum transmit power budget.

By contrast, untethered UAVs are powered by a laser beam, onboard battery, and/or a solar panel. They can fly freely and enjoy full mobility in wide 3D space. Communication throughput was maximized for a laser-powered UAV in [207]. Despite their flexibility, the battery-powered UAVs have to revisit their home base repeatedly to refill their batteries during operations, due to the limited capacity of onboard batteries [25]. The optimal UAV trajectory and power management schemes were developed in [339] to obtain the largest achievable data rate of a relaying system, and in [335] to minimize the data dissemination time of a multicasting system. Since solar panels on the UAVs can harness and convert energy to electric power, supporting long endurance flights, solar-powered UAVs have also received great research interest. The optimal 3D trajectory optimization and resource assignment for a solar-powered UAV-aided communication system were developed in [255] to achieve the largest overall data rate in a fixed time horizon.

Apart from transmit power, the UAVs consume additional propulsion power to support hovering and moving activities. As a result, the energy management for UAV-enabled communications noticeably differs from that in current systems on the ground. The largest value of energy efficiency in bits/Joule was obtained in [332] for a fixed-wing UAV via trajectory optimization. Total (including communication and propulsion) energy usage of a rotary-wing UAV was minimized in [336] to satisfy the throughput requirement of each GN. Energy-constrained UAV trajectory was optimized in [230] to minimize the localization error of GNs.

We propose a simple model for a rotary-wing UAV-enabled monitoring system. The suspicious transmission link on the ground consists of one source (transmitting) node S and one destination (receiving) node D. When the UAV's channel condition is worse than that of node D, the UAV sends jamming signal to the latter as noise to degrade its channel for successful eavesdropping. The total jamming energy consumption of the UAV is minimized in a finite period via joint trajectory optimization and power allocation, based on the assumption of successful eavesdropping at each slot. By judicious reformulation, we transform this nonconvex optimization task into two separable subproblems, each of which is convex when the other set of variables is fixed. The alternating optimization method is leveraged to develop an efficient approach that is ensured to converge to a locally optimal solution. Based on such a solution, some useful insights are also drawn on the changing patterns of the UAV's trajectory and jamming policy. To achieve energy-efficient UAV operations in practice, we further consider a solar-powered rotary-wing UAV-enabled monitoring system by including the propulsion power consumption besides the jamming power. Capitalizing on the successive convex approximation (SCA) method, an efficient iterative approach is put forth to find a feasible solution fulfilling the Karush–Kuhn–Tucker (KKT) conditions. Numerical results demonstrate that with UAV trajectory optimization, the overall energy consumption can be greatly suppressed.

The rest of the chapter is organized as follows. Section 7.2 describes the system models. Section 7.3 develops an approach to the UAV trajectory design and jamming energy minimization, while Section 7.4 addresses the trajectory design and total (jamming and propulsion) energy management for a solar-powered UAV. Numerical results are provided in Section 7.5. This chapter is concluded in Section 7.6. The list of variables and notations is provided in Table 7.1.

Table 7.1 The list of variables and notations.

Notation	Description
H	UAV altitude
d	Distance between the source and destination
T, T_w	Scheduling period and total number of time slots
$\mathcal{T}$	A set collecting all the time slots
δ	Duration of each time slot
(x_0, y_0), (x_T, y_T)	Initial and final 2D coordinates of the UAV
V_m	Maximum speed of the UAV
β_0	Channel power gain at a reference distance
h_0	Channel power gain of the S-D link

(Continued)

Table 7.1 (Continued)

Notation	Description
E_0	Amount of initial energy in the battery
$E(H)$	Harvested solar power at each time slot
ϑ	Ratio of usable energy in the battery
P_0, P_1	Blade profile power and induced power
s, A	Rotor solidity and disc area
U_{tip}	Tip speed of the rotor blade
v_0	Mean rotor induce velocity
ρ, d_f	Atmospheric density and fuselage drag ratio
α, β_1	Atmospheric transmittance
β_c	Interception factor of clouds
F, Δ	Mean radiant power and scaling altitude
η, S	Efficiency and size of solar panel
L, U	Altitude of cloud
t	Index of each time slot
(x_t, y_t)	2D coordinates of the UAV at time slot t
P_j^t, P_m^t	UAV jamming and propulsion power at time slot t
h_1^t, h_2^t	Channel power gains of the S-U and U-D links at time slot t
γ_D^t, γ_U^t	Signal-to-interference-plus-noise ratio at the destination and the UAV at time slot t
u_t, w_t, q_t	Auxiliary variables at time slot t

7.2 System Models

Consider a point-to-point, frequency non-selective wireless communication link from a suspicious source node S to a suspicious destination node D, which are geographically set apart by d meters on the ground. An untethered UAV, traveling at a fixed altitude of H meters, serves as the legitimate monitor to eavesdrop on this link; see Figure 7.1.[1] The UAV can move forward horizontally or hover in the air. It can travel in the vicinity above the GNs to improve its eavesdropping performance. Suspicious nodes S and D have one antenna each, and the UAV operates

1 Although design freedom can be increased by further optimizing UAV's altitude, energy consumption as well as risks of instability and collision will rise. Therefore, rather than frequently adjusting altitude, it may be better for the UAV to fly at a fixed altitude and avoid vertical movement due to airspace regulation, collision avoidance, energy saving, and safety concerns.

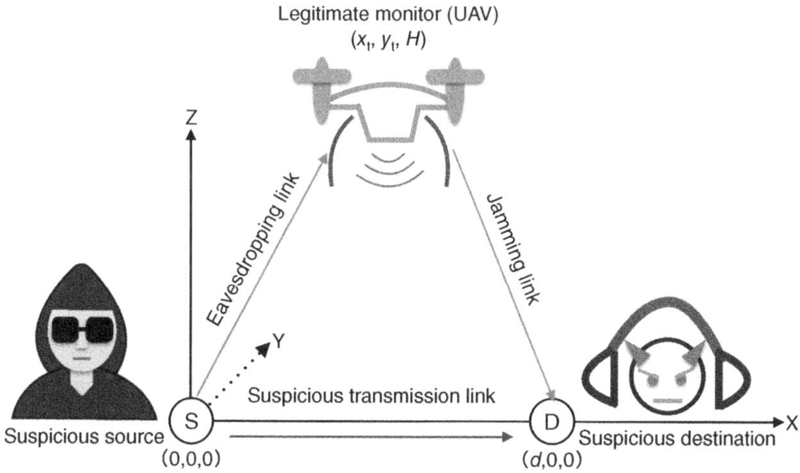

Figure 7.1 A UAV-enabled legitimate monitoring system.

with two antennas, one for monitoring and intercepting information from the S-D link (receiving) and the other for sending jamming signals to node D (transmitting). Therefore, the UAV can perform in a full-duplex state to jam and monitor simultaneously. Since its initial and final locations are given, the UAV's channel power gain can be worse than that of D at certain times. In this case, the UAV sends jamming signal to the latter as noise to degrade its channel for successful eavesdropping. We assume that the UAV can completely annul its self-interference from the transmitting antenna to the receiving antenna by adopting state-of-the-art analog and digital self-interference cancellation schemes [300].

7.2.1 UAV Mobility Model

Without loss of generality, we consider a 3D Cartesian coordinate system with nodes S and D located at $(0, 0, 0)$ and $(d, 0, 0)$, respectively. The UAV is deployed for the monitoring mission in a finite scheduling horizon of T seconds. We split the period T into T_w time slots given by $\mathcal{T} := \{1, \dots, T_w\}$; the duration of each slot is the same as δ. The slot length is selected to be short enough so that the UAV can be treated as static within each slot. Consequently, the time-varying coordinates of the UAV are given by $(x_t, y_t, H), \forall t \in \mathcal{T}$, with x_t and y_t being the UAV's x- and y-coordinates over time, respectively. The initial and final locations of the legitimate monitor are pre-defined and given by (x_0, y_0, H), and (x_T, y_T, H), respectively.[2]

2 The final location can be different from the initial location as the UAV may need to execute other missions after the monitoring, and thus lands on the predefined final location for a new start. The problem formulation with the same initial and final locations is a special case of our formulation.

The minimum traveling distance for the UAV to finish during the scheduling horizon T is thereby $d_{\min} = \sqrt{(x_T - x_0)^2 + (y_T - y_0)^2}$. Given the maximum speed of the UAV $\tilde{V}_m$, we let $\tilde{V}_m \geq d_{\min}/T$ so that at least one feasible trajectory can be found from the UAV's initial to final locations.[3]

Consequently, the UAV's mobile activity constraints, including its initial and final locations and speed constraints are given by [339]

$$(x_1 - x_0)^2 + (y_1 - y_0)^2 \leq V_m^2, \tag{7.1a}$$

$$(x_{t+1} - x_t)^2 + (y_{t+1} - y_t)^2 \leq V_m^2, \ \forall t \in \mathcal{T}, \tag{7.1b}$$

$$(x_T - x_{T-1})^2 + (y_T - y_{T-1})^2 \leq V_m^2, \tag{7.1c}$$

where $V_m := \tilde{V}_m \delta$ stands for the largest traveling distance of the UAV for each slot.

Remark 7.1 *(The choice of T_w):* In general, T_w is chosen such that the UAV can be treated as (quasi-) static within each time slot, observed from the ground. To guarantee a certain accuracy, the ratio of the largest traveling distance within each time slot $\tilde{V}_m \delta$ and the UAV altitude H can be restricted below a threshold, i.e., $\tilde{V}_m \delta / H \leq \varepsilon_m$, where ε_m is the given threshold and $\delta = T/T_w$. Then, the minimum number of time slots required for achieving the accuracy with a given ε_m can be obtained as $T_w \geq \tilde{V}_m T/(H\varepsilon_m)$. The optimization gets more precise with more discretized time samples, i.e., larger value of T_w. Yet, the computational complexity, given by $\mathcal{O}(T_w^{3.5})$, also increases significantly with the value of T_w. Therefore, the number of time slots T_w can be properly chosen in practice to balance between the accuracy and complexity [291].

7.2.2 Communication Channel Model

Malicious users of infrastructure-free wireless communication networks are more likely to appear in wide rural areas, where surveillance is overlooked. In open rural areas, the buildings and trees are sparsely distributed. LoS channels can be dominant even for communications between GNs. Therefore, we can suppose that the communication links between S, D, and UAV (i.e., node U) are all dominated by LoS channels, which can facilitate analysis on the structural properties of the optimal solution. The case with non-LoS channels will be addressed accordingly later. We further suppose that the Doppler effect resulting from the UAV's mobile activities is completely neutralized [163, 339]. The distance between S and D is fixed during the entire scheduling horizon, i.e., $d_{SD} = d$ meters. Hence, the

3 By considering the time for acceleration, the proposed maximum speed $\tilde{V}_m$ may be infeasible. However, in practice, the acceleration time could be very short and thus reasonably ignored, especially when the total flying period or distance is sufficiently long. From this perspective, we provide a lower bound for the maximum speed.

channel power gain of the suspicious link from S to D is constant and can be expressed as

$$h_0 = \frac{\beta_0}{{d_{SD}}^2} = \frac{\beta_0}{d^2}, \tag{7.2}$$

where β_0 stands for the channel power at the reference distance $d_0 = 1$ m. At each slot t, the channel power gain from S to U for legitimate eavesdropping follows the LoS model as

$$h_1^t = \frac{\beta_0}{{d_1^t}^2} = \frac{\beta_0}{x_t^2 + y_t^2 + H^2}, \quad \forall t \in \mathcal{T}, \tag{7.3}$$

where $d_1^t = \sqrt{x_t^2 + y_t^2 + H^2}$ is the link distance between S and U at slot t. Similarly, the channel power gain from U to D for jamming is

$$h_2^t = \frac{\beta_0}{{d_2^t}^2} = \frac{\beta_0}{(d - x_t)^2 + y_t^2 + H^2}, \quad \forall t \in \mathcal{T}, \tag{7.4}$$

where $d_2^t = \sqrt{(d - x_t)^2 + y_t^2 + H^2}$ is the separation distance between U and D at slot t.

Let P_x^t stand for the transmit power by S at time slot t, and P_j^t the jamming power from U to D to interfere with the channel at the suspicious receiver for a successful eavesdropping. The SINR at the suspicious receiver D is

$$\gamma_D^t = \frac{h_0 P_x^t}{h_2^t P_j^t + \sigma^2}, \quad \forall t \in \mathcal{T}, \tag{7.5}$$

where σ^2 is the variance of the additive white Gaussian noise (AWGN). On the other hand, the UAV can completely annul its self-interference from its jamming antenna to its receiving antenna. Hence, the SINR (which reduces to signal-to-noise ratio, SNR) of the legitimate eavesdropping channel at U is

$$\gamma_U^t = \frac{h_1^t P_x^t}{\sigma^2}, \quad \forall t \in \mathcal{T}. \tag{7.6}$$

Successful eavesdropping at the UAV requires $\gamma_U^t \geq \gamma_D^t$. The UAV can achieve this goal by dynamically adjusting its trajectory to fly close to the source node, and/or adjusting its jamming power to reduce the channel gain of the suspicious receiver D at each time slot, when the channel condition of the UAV is worse than that of D.[4]

4 When the channel condition of the legitimate monitor (the S-U link) is better than that of the suspicious receiver (the S-D link), eavesdropping is performed successfully without the UAV sending jamming signals to the receiver. However, when the S-U link suffers a worse channel condition than the S-D link, successful eavesdropping can be enabled through letting the UAV send jamming signals to the receiver to degrade its channel condition.

Note that the assumption of successful eavesdropping at each time slot is tenable and nontrivial. Malicious users of infrastructure-free wireless communication networks can also develop counter-eavesdropping measures to ensure secure transmissions on their behalf. One important method is to transmit secret information in cipher. In order to learn the pattern and decode the secret information, the legitimate agency treasures every bit of information. In this case, the cipher transmitted in each time slot is of equal importance for the legitimate agency to piece together the whole picture. Hence, it is of paramount importance that the eavesdropper can intercept information from the suspicious link successfully in every time slot.

Remark 7.2 *(Decoding the intercepted information):* we aim to investigate the fundamental performance limits of the physical layer approach for eavesdropping, and thus do not consider encryption for the suspicious link, which is a higher-layer technique and can be resolved as long as there are powerful computing resources. On the other hand, to avoid being monitored and tracked by legitimate parties, the suspicious link is very likely built, used, and discarded or destroyed in a day, which makes the link not complete or mature enough in terms of software and hardware to preserve privacy and security. We can thereby reasonably assume that the temporarily established infrastructure-free suspicious link is not vigilant against eavesdropping and does not employ any countermeasures such as signal encryption or anti-surveillance detection. From this perspective, the UAV can successfully decode the intercepted information from the suspicious link.

7.3 Legitimate Eavesdropping with Jamming

To ensure successful eavesdropping, the UAV may need to jam the transmission from S to D. For an untethered UAV without an incessant power supply, it is clear that we wish to minimize its overall jamming energy consumption. Building on the UAV's mobile activity constraints (7.1a)–(7.1c), together with the SINR expressions (7.5)–(7.6), the optimization task of interest can be formulated as

$$\min_{\{P_j^t\},\{x_t,y_t\}} \sum_{t\in\mathcal{T}} P_j^t\delta \tag{7.7a}$$

$$\text{s.t.}\ \frac{h_0 P_x^t}{h_2^t P_j^t + \sigma^2} \le \frac{h_1^t P_x^t}{\sigma^2},\ \forall t, \tag{7.7b}$$

$$(x_1 - x_0)^2 + (y_1 - y_0)^2 \le V_m^2, \tag{7.7c}$$

$$(x_{t+1} - x_t)^2 + (y_{t+1} - y_t)^2 \le V_m^2,\ \forall t, \tag{7.7d}$$

$$(x_T - x_{T-1})^2 + (y_T - y_{T-1})^2 \le V_m^2, \tag{7.7e}$$

$$P_j^t \ge 0,\ \forall t. \tag{7.7f}$$

Here, we aim to pursue the optimal jamming policy and trajectory design for the UAV. Note that the transmit power P_x^t by S can be canceled from both sides of the inequality constraints in (7.7b). This implies that the UAV does not need to know the P_x^t when making its jamming and trajectory decisions. This is of practical interest as the suspicious source is certainly reluctant to let the UAV know its transmit power value.

7.3.1 Proposed Solution

Problem (7.7) is not a convex program because of the non-convex constraints in (7.7b); hence, it cannot be dealt with by classic convex optimization methods. To make the problem more tractable, we introduce two slack variables $u_t := x_t^2 + y_t^2 + H^2$, and $w_t := (d - x_t)^2 + y_t^2 + H^2$, and rewrite (7.7) as

$$\min_{\{P_j^t, u_t, w_t\}, \{x_t, y_t\}} \sum_{t \in \mathcal{T}} P_j^t \delta \tag{7.8a}$$

$$\text{s.t. } x_t^2 + y_t^2 + H^2 - u_t \leq 0, \ \forall t, \tag{7.8b}$$

$$u_t - 2dx_t + d^2 - w_t \leq 0, \ \forall t, \tag{7.8c}$$

$$\frac{u_t w_t}{d^2} - w_t - P_j^t \beta_0 / \sigma^2 \leq 0, \ \forall t, \tag{7.8d}$$

$$w_t \geq H^2, \ \forall t, \tag{7.8e}$$

$$(7.7c)–(7.7f),$$

where (7.8d) results from (7.7b) by the following step

$$\frac{h_0}{P_j^t \beta_0 / w_t + \sigma^2} \leq \frac{\beta_0 / u_t}{\sigma^2}, \ \forall t. \tag{7.9}$$

Note that we change the "=" signs to "≤" signs in (7.8b) and (7.8c) to convexify those constraints. It can be justified that upon obtaining the optimal solution for (7.8), constraints (7.8b) and (7.8c) should always be met with equality, since otherwise, we can always decrease u_t and w_t, respectively, to improve the channel condition of the corresponding eavesdropping and jamming link, leading to smaller total jamming energy consumption. Therefore, problems (7.7) and (7.8) are equivalent.

Although problem (7.8) is not convex, it is easy to see that the problem becomes convex with regard to $\{P_j^t, x_t, y_t, u_t\}$ for fixed $\{w_t\}$, and it is also convex in $\{w_t\}$ for fixed $\{P_j^t, x_t, y_t, u_t\}$. For this reason, we resort to the alternating optimization method (a.k.a. block coordinate descent) to solve (7.8). The proposed algorithm is summarized in Algorithm 6. Since both subproblems are convex, the globally optimal solution for each of them can be obtained by standard convex optimization solvers, e.g., the interior point methods, in polynomial time [38]. The total jamming energy of the UAV is bounded above zero. For the proposed block coordinate descent method, the resultant total jamming energy is decreased in each iteration. Consequently, the proposed approach is ensured to converge to a locally optimal solution for problem (7.8). As problems (7.7) and (7.8) are equivalent, a locally optimal solution for (7.7) can be readily obtained.

Algorithm 6 Alternating optimization for problem (7.8)

1: **Initialize** $\{P_j^t(0), x_t(0), y_t(0), u_t(0)\}$, and set initial feasible values of $\{w_t(0)\}$ for problem (7.8).
2: **for** $m = 0, 1, 2, \ldots$ **do**
3: Obtain the optimal solution of $\{P_j^t(m+1), x_t(m+1), y_t(m+1), u_t(m+1)\}$ with $\{w_t(m)\}$ fixed.
4: Compute the optimal solution of $\{w_t(m+1)\}$ with $\{P_j^t(m+1), x_t(m+1), y_t(m+1), u_t(m+1)\}$ fixed.
5: Update $m = m + 1$.
6: **end for**

7.3.2 Structural Properties

To draw useful insights on the optimal trajectory optimization and jamming power allocation scheme, we analyze the structural properties of the optimal solution for the UAV-aided eavesdropping system.

Lemma 7.1 When the UAV is in the circular area of $\mathcal{A} := \{(x_t, y_t) | \sqrt{x_t^2 + y_t^2 + H^2} \leq d, \forall t\}$, it can eavesdrop successfully without jamming, i.e., $P_j^t = 0, \forall t$.

Proof. Lemma 7.1 can be proven by analyzing the characteristics of the transmit and eavesdropping rates. When the UAV is in the circular area of $\mathcal{A} := \{(x_t, y_t) | \sqrt{x_t^2 + y_t^2 + H^2} \leq d, \ \forall t\}$, the quality of the channel from S to U ($h_1^t = \beta_0/(x_t^2 + y_t^2 + H^2)$) is the same as or better than that from S to D ($h_0 = \beta_0/d^2$). It then readily follows that the UAV can eavesdrop successfully without jamming.

The circular area of $\mathcal{A}$ can be referred to as the jamming-free area. When the UAV is out of the range of $\mathcal{A}$, the channel quality from S to U is worse than that from S to D. In this case, the UAV can only eavesdrop successfully by degrading the SINR of the S-D link through jamming. The amount of the jamming power at each time slot increases with the UAV's distance to S.

Based on Lemma 7.1, it can be inferred that when both the initial and final locations of the UAV are inside $\mathcal{A}$, the optimal jamming policy is always zero, i.e., $P_j^{t*} = 0, \ \forall t$. As a result, the optimization problem (7.7) reduces to finding a feasible trajectory within the circular area of $\mathcal{A}$ with $P_j^t = 0, \ \forall t$, i.e.,

$$\begin{aligned} &\text{find } \{x_t, y_t\} \\ &\text{s.t. } x_t^2 + y_t^2 + H^2 \le d^2, \ \forall t, \\ &\quad (7.7\text{c})\text{–}(7.7\text{e}). \end{aligned} \tag{7.10}$$

Since problem (7.10) is convex, a classic convex solver can be leveraged to obtain the optimal solution, which is not necessarily unique.

Lemma 7.2 When the scheduling horizon T is larger than the minimum traveling time of the UAV $T_{\min} = d_{\min}/\tilde{V}_m$, the UAV will first fly toward the jamming-free area, then fly to its final location.

Lemma 7.2 is quite intuitive, as the UAV enjoys a better channel condition when it is closer to S. Based on Lemma 7.2, we can further characterize the changing patterns of the UAV's jamming policy.

Proposition 1 In general, the UAV's jamming power obeys the rule of first non-increasing, then non-decreasing. In some special cases, the jamming power either always non-increasing, or always non-decreasing.

Proof. When the UAV trajectory is fixed, (7.7) reduces to a jamming energy minimization problem

$$\begin{aligned} &\min_{\{P_j^t\}} \sum_{t \in \mathcal{T}} P_j^t \delta \\ &\text{s.t. } \frac{h_0 P_x^t}{h_2^t P_j^t + \sigma^2} \le \frac{h_1^t P_x^t}{\sigma^2}, \ \forall t, \\ &\quad P_j^t \ge 0, \ \forall t. \end{aligned} \tag{7.11}$$

For each time slot, the optimal solution for the jamming power is given by $P_j^{t*} = \max\{0, \frac{\sigma^2}{h_2^t}(\frac{h_0}{h_1^t} - 1)\}$, where $P_j^{t*} = 0$ when the UAV is in the jamming-free area of $\mathcal{A}$, and $P_j^{t*} = \frac{\sigma^2}{h_2^t}(\frac{h_0}{h_1^t} - 1) > 0$ when the UAV is outside $\mathcal{A}$. The latter can be rewritten into

$$P_j^{t*} = \frac{\sigma^2}{\beta_0 d^2}[(d - x_t)^2 + y_t^2 + H^2][(x_t^2 + y_t^2 + H^2) - d^2], \tag{7.12}$$

where $x_t^2 + y_t^2 + H^2 \ge d^2$. The projection of the jamming-free area on the ground is a circle centered at S (0, 0), with the radius of $\sqrt{d^2 - H^2}$. To observe how P_j^{t*}

changes with x_t outside $\mathcal{A}$, we let $y_t^2 = d^2 - H^2$ and take the first-order partial derivative of $P_j^{t^*}$ over x_t

$$\begin{aligned}\partial P_j^{t^*}/\partial x_t &= 4x_t^3 - 6dx_t^2 + 4d^2x_t \\ &= x_t[(2x_t - 3d/2)^2 + 7d^2/4].\end{aligned} \tag{7.13}$$

Clearly, the optimal jamming power $P_j^{t^*}$ increases with x_t when $x_t > 0$, and decreases with it when $x_t < 0$. The same pattern can be drawn from $P_j^{t^*}$ with respect to y_t. In one word, $P_j^{t^*}$ increases as the UAV flies away from S.

Now consider the following three cases.

- Case (i): Initial and final locations are both outside $\mathcal{A}$. When the UAV's traveling time is abundant, i.e., $T > T_{\min}$, it always seeks the trajectory that yields the least energy consumption. Therefore, the UAV first flies toward $\mathcal{A}$, then to its final destination. The jamming power experiences the process of first decreasing and then increasing. The same jamming policy applies when $T = T_{\min}$ and the line segment connecting the initial and final points goes through $\mathcal{A}$.
- Case (ii): Initial (or final) location is inside (or outside) $\mathcal{A}$, or vice versa. In the first scenario, the jamming power first decreases to zero, then stays constant till the eavesdropping mission is accomplished. The jamming power is always non-increasing. If we switch the initial and final locations, the jamming power then experiences a non-decreasing process.
- Case (iii): Both the initial and final locations are inside $\mathcal{A}$. The jamming power is always zero in this scenario.

Combining Cases (i)–(iii), the proposition follows.

Proposition 1 provides important insights on the optimal jamming policy of the UAV according to different initial and final locations. It shows that the UAV is willing to travel slowly inside the jamming-free area and even take detours to reduce the jamming power consumption. Such a strategy of the UAV is typically the consequence of minimizing the jamming energy only.

Remark 7.3 *(In and out of the jamming-free area):* The UAV usually stays in the home base, awaiting mission assignment, and is dispatched as a legitimate monitor once a suspicious link is detected. As the exact location of the suspicious link is not predictable, it is not likely that the UAV happens to be within the jamming-free area every time. Furthermore, by studying the UAV's trajectory with its initial and final locations in or out of the jamming-free area, we can provide more perspectives and insights for UAV trajectory design when it is assigned a mission of monitoring. This is why we consider a more general problem formulation, and the proposed solution applies to different scenarios, wherever the suspicious link is located.

7.3.3 Extension to Non-LoS Channels

If the suspicious transmission and legitimate monitoring links are located in an urban area, the channel between the suspicious source and destination experiences Rayleigh fading, which can be modeled as [342]

$$h_0^t = \beta_0 \xi_t d^{-\kappa}, \ \forall t, \tag{7.14}$$

where ξ_t is an exponentially distributed random variable with unit mean, and $\kappa \geq 2$ is the path loss exponent. The UAV-GN links can be formulated by considering the probabilities of both LoS and non-LoS (NLoS) channels, where the LoS probability at each time slot for the S-U ($j = 1$) or U-D ($j = 2$) link $p_{LoS,j}^t$ is given by [336]

$$p_{LoS,j}^t = \frac{1}{1 + C\exp(-D[\theta_j^t - C])}, \ \forall t. \tag{7.15}$$

Here the values of C and D is reliant on the propagation environment, and $\theta_j^t = \frac{180}{\pi}\sin^{-1}(H/d_j^t)$ is the elevation angle in degrees, which is closely related to the UAV's distance from the source node d_1^t or the destination node d_2^t. Thereby, the channel power gains of the UAV-GN links are given by [336]

$$h_j^t = p_{LoS,j}^t \beta_0 {d_j^t}^{-\kappa} + (1 - p_{LoS,j}^t)\zeta \beta_0 {d_j^t}^{-\kappa}, \ \forall t, \tag{7.16}$$

where $\zeta < 1$ is the extra reduction factor for the NLoS channel.

The Rayleigh fading in (7.14) does not affect the original problem (7.7), while the NLoS component in (7.16) renders problem (7.7) hardly tractable for existing solvers. To deal with it, we consider the case when $\kappa = 2$; then (7.16) can be approximated by [194]

$$h_j^t \approx \eta_1 {d_j^t}^{-2} + \eta_2, \ \forall t, \tag{7.17}$$

where η_1 and η_2 are two coefficients relying on the UAV altitude. Using the expressions in (7.17), the objective function and constraints for the NLoS scenario are generally in the same form as those in the original problem (7.7). Similar to problem (7.8), the variables in the NLoS scenario can be separated into three blocks, namely, $\{P_j^t, x_t, y_t\}$, $\{u_t\}$, and $\{w_t\}$, due to the product of $P_j^t u_t w_t$ invited in constraints (7.8d) by the NLoS component. The NLoS problem is convex regarding each block of variables when the other two blocks are fixed, and can thus be solved by the proposed block coordinate descent approach. Note that due to the approximation in (7.17), only a suboptimal solution can be obtained.

7.3.4 Generalization to Eavesdropping Non-outage Events

In this section, we propose a stochastic model for the eavesdropping system by considering Rayleigh fading for the suspicious S-D link, i.e., $h_0^t = \beta_0 \xi_t d^{-\kappa}$, $\forall t$ [cf. Equation (7.14)]. The UAV channels are all LoS, and successful eavesdropping is not required within each time slot anymore. Instead, we impose a constraint on non-outage probability to guarantee that the total successful eavesdropping events satisfy a certain threshold over time. We introduce the following indicator function I_t, $\forall t$ to denote the successful eavesdropping event of the UAV

$$I_t = \begin{cases} 1, & \text{if } \gamma_U^t \geq \gamma_D^t, \\ 0, & \text{otherwise,} \end{cases} \tag{7.18}$$

where $I_t = 1$ and $I_t = 0$ indicate eavesdropping non-outage and outage events, respectively.

The original problem is extended to the following form.

$$\min_{\{P_j^t\},\{x_t,y_t\}} \sum_{t\in\mathcal{T}} P_j^t \delta \tag{7.19a}$$

$$\text{s.t.} \sum_t I_t \geq p_{\text{non}} T_w, \tag{7.19b}$$

$$\text{(7.7c)–(7.7f),}$$

where $p_{\text{non}} \in [0, 1]$ is the eavesdropping non-outage probability and constraint (7.19b) guarantees that at least $100p_{\text{non}}\%$ of the total eavesdropping performances are successfully operated. Constraint (7.19b) is a relaxed (or generalized) version of constraints (7.7b), which can also take the form of non-outage probability

$$\mathbb{P}(\gamma_U^1 \geq \gamma_D^1, \dots, \gamma_U^{T_w} \geq \gamma_D^{T_w}) \geq p_{\text{non}}. \tag{7.20}$$

When $p_{\text{non}} = 1$, problem (7.19) specializes to the original problem (7.7). On the other hand, if $p_{\text{non}} = 0$, jamming is not needed at all, and the optimal value of the objective function $\sum_t P_j^t \delta$ is zero. In this case, problem (7.19) reduces to the feasibility problem of finding a trajectory constrained by the UAV's maximum speed with $P_j^t = 0$, $\forall t$. At optimality, jamming signals will be suppressed for at most $100(1 - p_{\text{non}})\%$ of the T_w time slots with worse S-U channels (or higher jamming power consumptions), and they will be sent, if necessary, in time slots with better S-U channels. As problem (7.19) is a relaxed one of the original problem (7.7), the optimal UAV trajectory for (7.7) is also an optimal one for (7.19), and the optimal jamming energy in (7.7) serves as an upper bound for that in (7.19). Problem (7.19) can be solved by first solving (7.7), then ranking the values of $\{P_j^{t*}\}_t$ from large to small and setting the top $100(1 - p_{\text{non}})\%$ to zero.

7.3.5 Extension to Two Suspicious Links

In this section, we extend the original problem (7.7) to include two suspicious links for the UAV to monitor simultaneously. The second pair of suspicious ground source and destination nodes, S_2 and D_2, are located at $(0, s_2)$ and (d, s_2), respectively, where s_2 is the given y-coordinate of the nodes. All communication links are assumed to be LoS for simplicity. We assume that the UAV has three antennas, with one of them for jamming and the other two for monitoring each link. Note that in this case, the jamming signal is sent to both links as long as eavesdropping is unsuccessful over one of the links.

The channel power gain for the S_2-D_2 link is the same as the S-D link, $h_0 = \beta_0/d^2$. The channel power gains for the S_2-U and U-D_2 links are given by

$$h_{21}^t = \frac{\beta_0}{x_t^2 + (s_2 - y_t)^2 + H^2}, \quad \forall t, \tag{7.21a}$$

$$h_{22}^t = \frac{\beta_0}{(d - x_t)^2 + (s_2 - y_t)^2 + H^2}, \quad \forall t. \tag{7.21b}$$

The SINR (or SNR) of the S_2-D_2 and S_2-U links are given by

$$\gamma_{D_2}^t = \frac{h_0 P_x^t}{h_{22}^t P_j^t + \sigma^2}, \quad \forall t, \tag{7.22a}$$

$$\gamma_{U_2}^t = \frac{h_{21}^t P_x^t}{\sigma^2}, \quad \forall t. \tag{7.22b}$$

Successful eavesdropping requires that $\gamma_{U_2}^t \geq \gamma_{D_2}^t$, and $\gamma_U^t \geq \gamma_D^t$ for both links. The new problem of interest can be formulated as

$$\min_{\{P_j^t\},\{x_t,y_t\}} \sum_{t\in\mathcal{T}} P_j^t \delta \tag{7.23a}$$

$$\text{s.t.} \quad \frac{h_0 P_x^t}{h_2^t P_j^t + \sigma^2} \leq \frac{h_1^t P_x^t}{\sigma^2}, \quad \forall t, \tag{7.23b}$$

$$\frac{h_0 P_x^t}{h_{22}^t P_j^t + \sigma^2} \leq \frac{h_{21}^t P_x^t}{\sigma^2}, \quad \forall t, \tag{7.23c}$$

$$\text{(7.7c)–(7.7f)}.$$

Problem (7.23) can be solved by following the same procedure summarized in Algorithm 6. The jamming-free area of the S_2-D_2 link is $\mathcal{A}_2 := \{(x_t, y_t) | \sqrt{x_t^2 + (s_2 - y_t)^2 + H^2} \leq d, \forall t\}$. When $|s_2| \leq 2d$, the common jamming-free area, i.e., the jamming-free area for problem (7.23) is the intersection of $\mathcal{A}$ and $\mathcal{A}_2$, which is essentially the intersection of two circles centered at $(0, 0)$ and $(0, s_2)$, respectively, both with a radius of d. When $|s_2| > 2d$, the common jamming-free area does not exist as $\mathcal{A} \cap \mathcal{A}_2 = \emptyset$. It is worth noting that the problem of two suspicious links can be further extended to address multiple suspicious links with different separating distances, or aerial (rather than ground) suspicious nodes with 3D optimization of the UAV trajectory.

In a nutshell, we address the problem of jamming energy minimization for a UAV-enabled monitoring system based on the assumption of sufficient power supply. We provide useful insights on the UAV trajectory design and reveal its impact on the jamming policy. However, in practice, such a trajectory design could result in a great cost (and waste) of propulsion power. Furthermore, it is not possible for untethered UAVs to possess an infinite power supply during flight. Motivated by this, we next investigate the energy optimization based on a more practical setting, by considering a finite power supply and propulsion power consumption at the UAV.

7.4 Energy Management for Solar-powered UAV

Compared with cables, laser beams, and onboard batteries, the solar-powered UAVs enjoy high flexibility and a long flight endurance in practical deployment. Apart from communication and jamming power, the UAV consumes additional propulsion power to maintain its airborne status and support its movement. Energy-efficient operation of the UAV needs to be achieved by considering propulsion energy management in system design [336].

Suppose that the UAV has a solar panel to harvest energy and an on-site battery to save energy. The UAV's battery is initially charged with an E_0 amount of energy, and it can consume ϑ portion of E_0 during the entire working period, and save the $(1 - \vartheta)$ portion for emergency during landing (to a prescribed platform or home base). The UAV can fly horizontally and adjust its positions dynamically to enhance the eavesdropping performance. We pursue the optimal trajectory design and energy management scheme of the UAV by minimizing the total jamming and propulsion energy consumption for the solar-powered rotary-wing UAV-enabled monitoring system.

7.4.1 UAV Propulsion Power Model

Besides transmit (jamming) power, the communication-related power also includes that for communication circuitry, information receiving and decoding, signal processing, etc. For simplicity, we suppose that such communication-connected power is a constant, represented by P_c in Watts (W) [109, 350]. The propulsion power, which typically depends on the UAV's speed, is essential to support the UAV's hovering and moving activities. For a rotary-wing UAV with speed V_t, the propulsion power at time slot t, denoted by P^t_m, is given by [230, 336]

$$P^t_m = P_0\left(1+\frac{3V_t^2}{U_{tip}^2}\right) + P_1\left(\sqrt{1+\frac{V_t^4}{4v_0^4}} - \frac{V_t^2}{2v_0^2}\right)^{\frac{1}{2}} + \frac{1}{2}d_f\rho sAV_t^3, \tag{7.24}$$

where P_0 and P_1 have fixed values and stand for the *blade profile power* and *induced power* under hovering mode, respectively, U_{tip} is the tip speed of the rotor blade, v_0 denotes the average rotor induced velocity in hover, d_f and s represent the fuselage drag ratio and rotor solidity, and ρ and A are the atmospheric density and rotor disc area, respectively. When $V_t = 0$, (7.24) corresponds to the power consumption of the hovering state. Figure 7.2 depicts the typical curve of P^t_m versus V_t. The parameters are set according to Table 7.2 [336]. It is revealed by Figure 7.2 that the UAV speed achieving the least power consumption, i.e., about 41.84 W, happens at approximately $V_e = 22.36$ m/s.

We suppose that within each slot t, the UAV maintains a constant speed, which is given by

$$V_t = \sqrt{(x_t - x_{t-1})^2 + (y_t - y_{t-1})^2}/\delta,\ \forall t. \tag{7.25}$$

By substituting (7.25) into (7.24), we find that the first and the third terms of (7.24) are jointly convex with respect to $(x_t, x_{t-1}, y_t, y_{t-1})$, whereas the second term is neither convex nor concave.

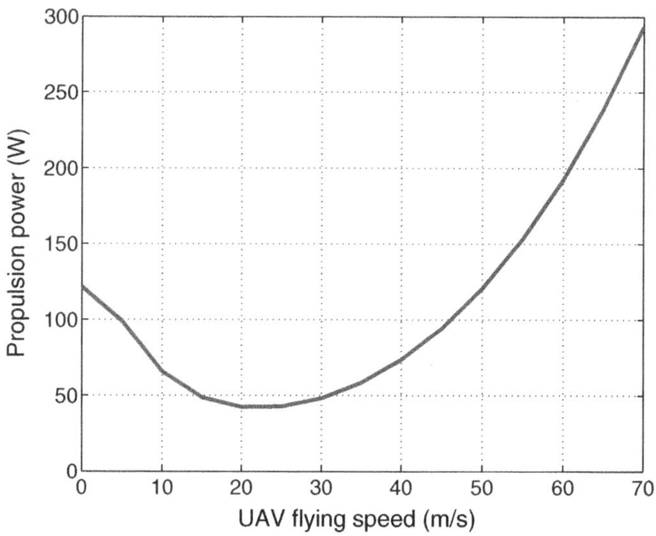

Figure 7.2 Propulsion power versus UAV speed.

Table 7.2 Parameters for propulsion power [336].

UAV weight	2 kg
Blade profile power and induced power, P_0, P_1	3.4 W, 118 W
Rotor solidity and disc area, s, A	0.03, 0.28 m^2
Tip speed of the rotor blade, U_{tip}	60 m/s
Mean rotor induce velocity, v_0	5.4 m/s
Atmospheric density and fuselage drag ratio, ρ, d_f	1.225 kg/m^3, 0.3

Source: Adapted from Zeng et al. [336].

7.4.2 Solar Power Model

Generally, the amount of the harvested solar power depends on the atmospheric transmittance and clouds. As higher altitude results in higher solar intensity, the atmospheric transmittance increases with altitude, which can be empirically approximated by the following equation at altitude H [142]

$$\phi(H) = \alpha - \beta_1 e^{-H/\Delta}, \tag{7.26}$$

where α is the largest possible amount of the atmospheric transmittance, β_1 is the extinguishing coefficient of the air, and Δ is the scaling altitude. On the other hand, the solar strength is diminished by cloud. The reduction of sunlight traveling through a cloud can be formulated as [255, 131]

$$\psi(d_c) = e^{-\beta_c d_c}, \tag{7.27}$$

where $\beta_c \geq 0$ is the interception factor of the cloud, and d_c represents the spatial length that the sunlight travels through the cloud. Overall, the electric generation power of the solar panels at height H is given by [142, 131]

$$E(H) = \begin{cases} \eta SF\phi(H)\psi(0), & H \geq U, \\ \eta SF\phi(H)\psi(U-H), & L \leq H < U, \\ \eta SF\phi(H)\psi(U-L), & H < L, \end{cases} \tag{7.28}$$

where $\eta \in (0, 1)$ and S (in m^2) denote the efficiency and size of the solar panel, respectively. Constant F is the mean radiant power on the ground, while U and L are the heights of the upper and lower limits of the cloud, respectively. Figure 7.3 illustrates the influence of UAV's altitude on the harvested solar power. The settings of the corresponding parameters are listed in Table 7.3 [131].

7.4.3 Problem Formulation

We aim to design the joint energy management and trajectory planning scheme for the solar-powered UAV-aided eavesdropping system by minimizing its total energy

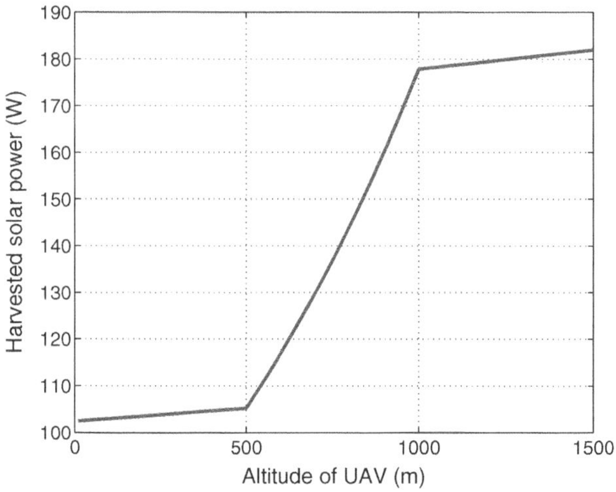

Figure 7.3 Harvested solar power versus UAV altitude.

Table 7.3 Parameters for solar power [131].

Atmospheric transmittance, α, β_1	0.8978, 0.2804
Interception factor of clouds, β_c	0.01
Mean radiant power and scaling altitude, F, Δ	1367 W/m^2, 8000 m
Efficiency and size of solar panel, η, S	0.4, 0.5 m^2
Altitude of cloud, L, U	500 m, 1000 m

Source: Adapted from Kokhanovsky [131].

consumption, including the jamming energy and the propulsion energy. Since the UAV flies at a fixed altitude, we can simply use E to denote the amount of solar power instead of $E(H)$. The problem is formulated as

$$\min_{\{P_j^t, x_t, y_t\}} \sum_{t\in\mathcal{T}} (P_j^t + P_m^t)\delta \tag{7.29a}$$

$$\text{s.t.} \sum_{i=1}^{t}(P_j^t + P_m^t + P_c)\delta \leq \sum_{i=1}^{t} E + \vartheta E_0, \ \forall t, \tag{7.29b}$$

$$\text{(7.7b)–(7.7f).}$$

Constraint (7.29b) is the energy-harvesting causality constraint, which is imposed to bound the total consumed energy up to the current time slot, not to exceed the harvested energy plus the battery capacity.

The minimum level of the initially stored energy $\underline{E}_0$ is chosen such that the UAV can finish the eavesdropping mission without harvested solar energy, following the shortest trajectory at a constant speed. In particular, given the line segment connecting its horizontal initial and final locations (x_0, y_0) and (x_T, y_T), the UAV travels at a fixed speed $\overline{V} = \sqrt{(x_T - x_0)^2 + (y_T - y_0)^2}/T$. With $\overline{V}$, we can obtain the total propulsion energy P_m and the UAV's coordinates (x_t, y_t) at each time slot. Based on the coordinates, we can further calculate its jamming power P_j^t according to (7.7b) per time slot. Then, we can readily obtain the value of $\underline{E}_0 = P_m + \sum_t (P_j^t + P_c)\delta$.

Remark 7.4 *(3D UAV trajectory design with altitude optimization):* 3D UAV trajectory design can be pursued by including the UAV altitude as an optimization variable $H_t, \forall t$. Considering problem (7.7) and the S-U channel condition $h_1^t = \frac{\beta_0}{d_1^{t\,2}} = \frac{\beta_0}{x_t^2 + y_t^2 + H_t^2}, \forall t$, the optimal altitude for the UAV is the lowest height within the regulated range that it can stay, since the UAV enjoys the best channel condition in this way, and there is no performance gain by increasing its altitude.

On the other hand, considering the model of harvested solar power in Chapter 7.4.2 [cf. (7.28)], a tradeoff can be observed between the UAV channel conditions and the amount of harvested energy [cf. (7.29)]. The UAV has to decide at each time slot whether to fly lower or higher to strike a balance between achieving better eavesdropping performance and harvesting more energy. With the UAV altitude included as an optimization variable $H_t, \forall t$, constraints (7.29b) become non-convex as the amount of harvested energy E_t is altitude-dependent and time-varying. It is difficult to convert (7.29b) to convex constraints due to the complicated expression of E_t, thus rendering the new problem hardly tractable for existing solvers. Furthermore, to the best of our knowledge, there is no general model to capture the power consumption incurred by both horizontal and vertical movements of the UAV, which in turn makes it difficult to pursue a joint 3D UAV trajectory design and power allocation. It will be an interesting direction to pursue in our future works with altitude optimization.

7.4.4 SCA-based Convexification and Solution

The problem (7.29) is not convex since it consists of the non-convex term $\left(\sqrt{1 + \frac{V_t^4}{4v_0^4}} - \frac{V_t^2}{2v_0^2}\right)^{\frac{1}{2}}$ in P_m^t, and the non-convex constraints (7.7b). The latter can be handled by leveraging the same method as in Section 7.3. With slack variables $\{u_t, w_t, \forall t\}$, (7.7b) can be replaced with the constraints (7.8b)–(7.8e).

To tackle the non-convexity with P_m^t, we first bring in slack variables $\{q_t \geq 0\}$ such that

$$q_t^2 = \sqrt{1 + \frac{V_t^4}{4v_0^4}} - \frac{V_t^2}{2v_0^2}, \ \forall t, \tag{7.30}$$

which is equivalent to

$$\frac{1}{q_t^2} = q_t^2 + \frac{V_t^2}{v_0^2}, \ \forall t. \tag{7.31}$$

The second term of (7.24) can thus be substituted by the linear component $P_1 q_t$, with the additional constraints (7.31). For the purpose of exposition, we now integrate the expression for V_t in (7.25) and let

$$\begin{aligned}\tilde{P}_m^t := P_0 + \frac{3P_0}{U_{tip}^2 \delta^2}\left[(x_t - x_{t-1})^2 + (y_t - y_{t-1})^2\right] + P_1 q_t \\ + \frac{d_f}{2\delta^3}\rho s A\left[(x_t - x_{t-1})^2 + (y_t - y_{t-1})^2\right]^{3/2}, \ \forall t.\end{aligned} \tag{7.32}$$

We can see that $\tilde{P}_m^t$ is now jointly convex with respect to $(x_t, x_{t-1}, y_t, y_{t-1}, q_t)$. With such a manipulation, problem (7.29) can be written as

$$\min_{\substack{\{P_j^t, q_t, u_t\} \\ \{x_t, y_t, w_t\}}} \sum_{t \in \mathcal{T}} (P_j^t + \tilde{P}_m^t)\delta \tag{7.33a}$$

$$\text{s.t.} \ \sum_{l=1}^{t} (P_j^l + \tilde{P}_m^l + P_c)\delta \le \sum_{l=1}^{t} E + \vartheta E_0, \ \forall t, \tag{7.33b}$$

$$\frac{1}{q_t^2} \le q_t^2 + \frac{(x_t - x_{t-1})^2 + (y_t - y_{t-1})^2}{\hat{v}_0^2}, \ \forall t, \tag{7.33c}$$

$$\text{(7.7c)–(7.7f), (7.8b), (7.8b)–(7.8e),}$$

where $\hat{v}_0^2 = v_0^2 \delta^2$.

Constraints (7.33c) are obtained from (7.31) by replacing the equations with inequalities. Yet, equivalence still holds between problems (7.29) and (7.33). To examine this, we assume that if any of the constraints in (7.33c) is met with strict inequality when achieving optimality for problem (7.33), we can decrease the related value of variable q_t to enable constraint (7.33c) met with equality, while reducing the total energy consumption (objective value) at the same time. Therefore, all constraints in (7.33c) are met with equality at optimality. The same equivalence also holds for constraints (7.8b) and (7.8c) as explained in Section 7.3. Hence, problems (7.29) and (7.33) are equivalent.

Problem (7.33) is still non-convex since it consists of the non-convex constraints in (7.33c). However, it can be tackled with the SCA method by calculating the global lower bounds at a given local point. In particular, for (7.33c), the left-hand-side (LHS) is a convex function in q_t, and the right-hand-side (RHS) is a jointly convex function regarding q_t and $(x_t, x_{t-1}, y_t, y_{t-1})$. Since the first-order Taylor

---

Algorithm 7 SCA-based method for problem (7.35)

1: **Initialization:** Find an initially feasible solution $\{P_j^t(0), x_t(0), y_t(0), q_t(0), u_t(0), w_t(0)\}$ for problem (7.35).
2: **for** $l = 0, 1, 2, \ldots$ **do**
3: Obtain the optimal solution of $\{P_j^t(l+1), q_t(l+1), u_t(l+1)\}$ with $\{q_t(l), x_t(l), y_t(l), w_t(l)\}$ fixed.
4: Compute the optimal solution of $\{x_t(l+1), y_t(l+1), w_t(l+1)\}$ with $\{P_j^t(l+1), q_t(l+1), u_t(l+1)\}$ fixed.
5: Update $l = l + 1$.
6: **end for**

---

expansion serves as the global lower bound of a convex function, we can obtain the following inequality for the RHS of (7.33c)

$$
\begin{aligned}
q_t^2 + \frac{(x_t - x_{t-1})^2 + (y_t - y_{t-1})^2}{\hat{v}_0^2} \geq\ & q_t^{(l)2} + 2q_t^{(l)}(q_t - q_t^{(l)}) \\
& + \frac{2}{\hat{v}_0^2}[(x_t^{(l)} - x_{t-1}^{(l)})(x_t - x_{t-1}) \\
& + (y_t^{(l)} - y_{t-1}^{(l)})(y_t - y_{t-1})] \\
& - \frac{1}{\hat{v}_0^2}[(x_t^{(l)} - x_{t-1}^{(l)})^2 + (y_t^{(l)} - y_{t-1}^{(l)})^2],
\end{aligned}
\tag{7.34}
$$

where $q_t^{(l)}$, $x_t^{(l)}$, and $y_t^{(l)}$ are the present values of the corresponding variables at the l-th iteration, respectively. By substituting the non-convex constraints (7.33c) with its lower bound at the l-th iteration acquired by (7.34), we can establish the following optimization problem

$$
\min_{\substack{\{P_j^t, q_t, u_t\} \\ \{x_t, y_t, w_t\}}} \sum_{t \in \mathcal{T}} (P_j^t + \tilde{P}_m^t)\delta \tag{7.35a}
$$

$$
\begin{aligned}
\text{s.t. } \frac{1}{q_t^2} \leq\ & q_t^{(l)2} + 2q_t^{(l)}(q_t - q_t^{(l)}) \\
& + \frac{2}{\hat{v}_0^2}[(x_t^{(l)} - x_{t-1}^{(l)})(x_t - x_{t-1}) + (y_t^{(l)} - y_{t-1}^{(l)})(y_t - y_{t-1})] \\
& - \frac{1}{\hat{v}_0^2}[(x_t^{(l)} - x_{t-1}^{(l)})^2 + (y_t^{(l)} - y_{t-1}^{(l)})^2],\ \forall t,
\end{aligned}
\tag{7.35b}
$$

$$
q_t \geq 0,\ \forall t, \tag{7.35c}
$$

(7.7c)–(7.7f), (7.8b)–(7.8e), (7.33b).

It can be justified that problem (7.35) is convex in $\{P_j^t, q_t, u_t\}$ for fixed $\{x_t, y_t, w_t\}$, and it is convex in $\{x_t, y_t, w_t\}$ for fixed $\{P_j^t, q_t, u_t\}$. Similarly, we can leverage the alternating optimization method to acquire the optimal values of one block of variables with the other fixed iteratively. The proposed algorithm is summarized in Algorithm 7.

In the proposed algorithm, each subproblem is a convex program, which can be efficiently tackled via classic convex optimization methodologies in polynomial time. It is worth noting that because of the global lower bounds in (7.34), when the constraints of problem (7.35) are fulfilled, those for the original problem (7.33) are also fulfilled; yet the reverse does not necessarily hold. Thereby, the feasible region of (7.35) is a subset of that for (7.33), and the optimal value of (7.35) draws an upper limitation to that of (7.33). By sequentially renewing the local point at each iteration through solving (7.35), our proposed approach is established for the non-convex optimization problem (7.33) or its original problem (7.29). Through the similar statements in [336] and [330], it is demonstrated that the proposed approach is ensured to converge to at least a solution that fulfills the KKT conditions of problem (7.33). A high-quality sub-optimal solution can therefore be obtained by our proposed algorithm with a computational complexity of $\mathcal{O}(T_w^{3.5})$ at a fast convergence speed, as will be corroborated by simulation results provided in Section 7.5.

7.5 Numerical Results

In this section, we provide numerical results for the proposed approaches. The reference channel power β_0 is set as 10^{-12}, the noise σ is set as -169 dBm/Hz, and the communication bandwidth is 10 MHz. The distance between S and D is $d = 200$ m, and the UAV flies at an altitude of $H = 100$ m. The maximum horizontal speed of the UAV is set as $\tilde{V}_m = 40$ m/s. The slot length is $\delta = 0.1$ seconds. The original capacity of the battery E_0 is 7×10^3 J. Parameters concerning the propulsion power and the harvested solar power are the same as in Tables 7.2 and 7.3. To evaluate the proposed optimal trajectory design and power allocation schemes, we test three pairs of coordinates for the initial and final locations of the UAV. The three test cases are: (i) JF (Jamming Free) scenario: both the initial and final locations are inside the jamming-free area of $\mathcal{A}$, namely, $(x_0, y_0) = (-50\text{ m}, -100\text{ m})$, and $(x_T, y_T) = (100\text{ m}, 140\text{ m})$; (ii) IF (Initial jamming Free) scenario: the initial location is inside $\mathcal{A}$ and the final location is outside $\mathcal{A}$, namely, $(x_0, y_0) = (-50\text{ m}, 0)$, and $(x_T, y_T) = (100\text{ m}, 350\text{ m})$; and (iii) NF (No jamming Free) scenario: both locations are outside $\mathcal{A}$, namely, $(x_0, y_0) = (300\text{ m}, 200\text{ m})$, and $(x_T, y_T) = (200\text{ m}, 400\text{ m})$. To further observe the UAV's behavior, we adopt three time horizons for each scenario, namely, $T = 10$, 30, and 60 seconds. Trajectory and power consumption of the UAV are depicted every second. Note that all pairs of coordinates are carefully selected such that at least one feasible trajectory can be found for the UAV in the shortest time horizon.

To corroborate the merits of our proposed scheme over existing schemes, we include the performances of the following baseline schemes for $T = 30$ seconds under the NF scenario.

- Low speed: The UAV travels straight from the initial location to the final location at a fixed speed (7.46 m/s).
- Fly half: The UAV flies straight to the final location at a constant speed (14.91m/s) during the first half of the period (15 seconds), and hovers at the destination for the rest of the period
- Two lines: The UAV first flies directly toward the point (200 m, 200 m), then flies to the final location, following the trajectory of two line segments at a constant speed of 10 m/s.
- Fly first: The UAV flies to the final location at approximately $V_e = 22.36$ m/s in the first 10 seconds, then hovers at the destination for the rest 20 seconds.
- Hover first: The UAV hovers above the initial location in the first 20 seconds, then flies to the final location for the rest 10 seconds at speed V_e.
- Round trip: The UAV first takes a round trip between the initial and final locations, then flies again to the destination, at speed V_e during the flight period.

Figure 7.4 depicts the UAV's trajectory designs (see Figure 7.4a) and jamming power allocations (see Figure 7.4b) for the simple system model in problem (7.7) under the NF scenario. The jamming-free area of $\mathcal{A}$ for the LoS links is illustrated by an orange dash-dot line. It can be observed from Figure 7.4 that when both the initial and final locations are outside $\mathcal{A}$, the UAV intends to fly toward $\mathcal{A}$ first, then travel to the final location. With sufficient traveling time ($T = 30$ and 60 seconds), the UAV first flies fast to $\mathcal{A}$, then takes a detour at a very low speed inside $\mathcal{A}$, and finally travels quickly to its final location. During this process, the jamming power first decreases, then stays at zero, and finally increases quickly in the last few time slots. This is consistent with the results in Lemma 7.2 and Proposition 1. We also include the performance of the UAV with NLoS links when $T = 30$ seconds in Figure 7.4 (labeled as "NLoS"). It can be seen that the UAV's trajectory does not vary much under this scenario, and that its jamming power does not change smoothly with its distance from the source due to the randomness introduced by the S-D link. To further validate the advantage of trajectory design on energy reduction, we examine three baseline schemes of the UAV under the NF scenario when $T = 30$ seconds, i.e., the "Low speed," "Fly half," and "Two lines" schemes.

Figure 7.5 shows the total energy consumption of the UAV under the NF scenario. It is unveiled in Figures 7.4 and 7.5 that the UAV consumes significantly

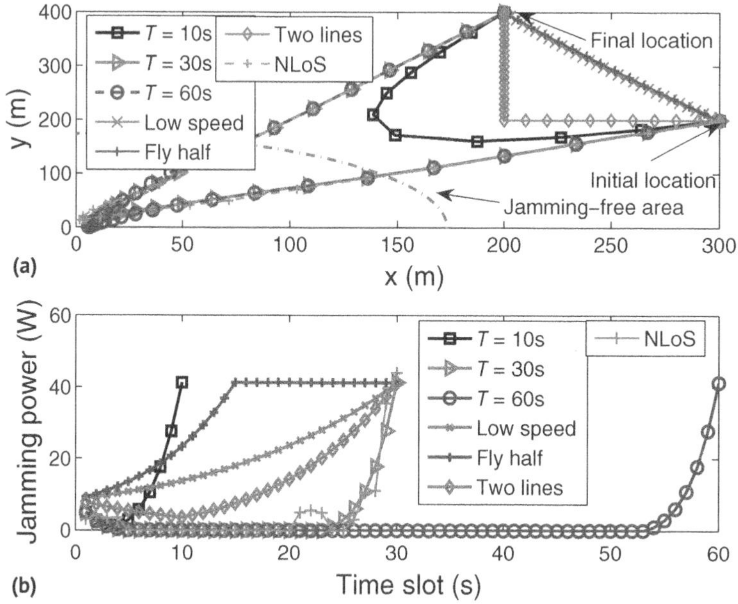

Figure 7.4 UAV trajectory designs and jamming power allocations under NF scenario.

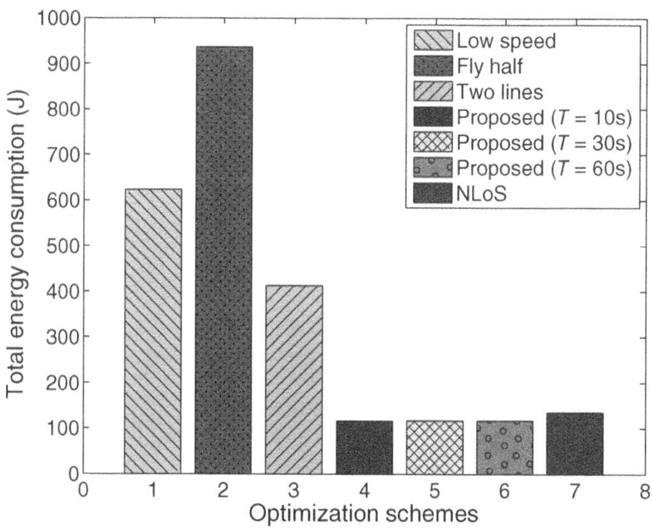

Figure 7.5 Total jamming energy consumption of the UAV under NF scenario.

more jamming energy without careful trajectory design. The overall energy consumption of the "Fly half" scheme is almost ninefold of that of our proposed scheme, since the UAV flies quickly to the destination and hovers there for a relatively long period. As the destination is far from the source node, the longer the UAV stays there, the more jamming energy it consumes. The "Two lines" scheme is the most energy-efficient among the baseline schemes as it amounts to a simple optimization of the trajectory. Under the same parameter setting, the UAV consumes more energy with NLoS links than with LoS links, since it experiences greater path loss with the former. Note that for the proposed scheme, the total jamming energy does not increase with the length of the scheduling period, since the UAV always spends the same duration outside $\mathcal{A}$.

Figures 7.6–7.10 depict the trajectory designs and energy management schemes of the UAV based on the system model proposed for problem (7.29) in Section 7.4, where harvested solar energy, propulsion power, and other circuit consumptions are considered. Specifically, Figures 7.6 and 7.7 demonstrate the convergence of the proposed approach for trajectory design and propulsion power assignment of the UAV under the JF scenario. Since the lines for the 50th and 60th iterations completely overlap, a fast convergence within 50 iterations can be readily observed in both figures. The UAV travels either inside the jamming-free area of $\mathcal{A}$ or at the edge of $\mathcal{A}$ and incurs no jamming power at any time, which corroborates Lemma 7.1. When $T = 10$ seconds, the UAV can only choose to finish the journey

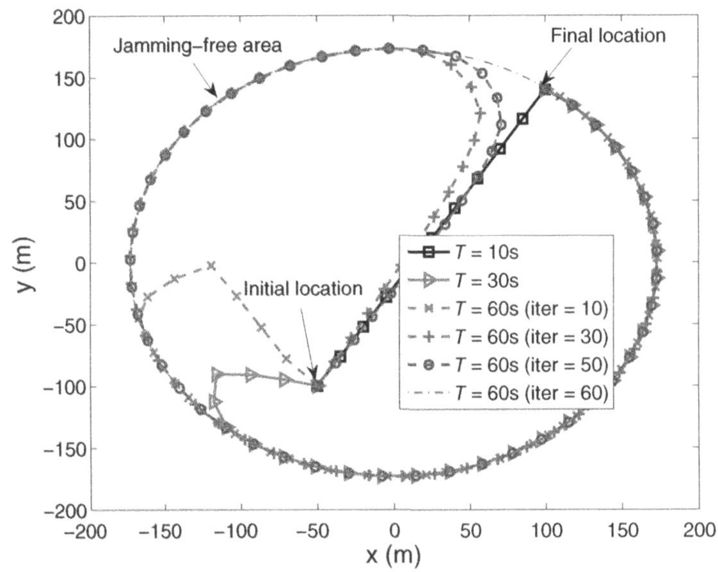

Figure 7.6 UAV trajectory designs under JF scenario.

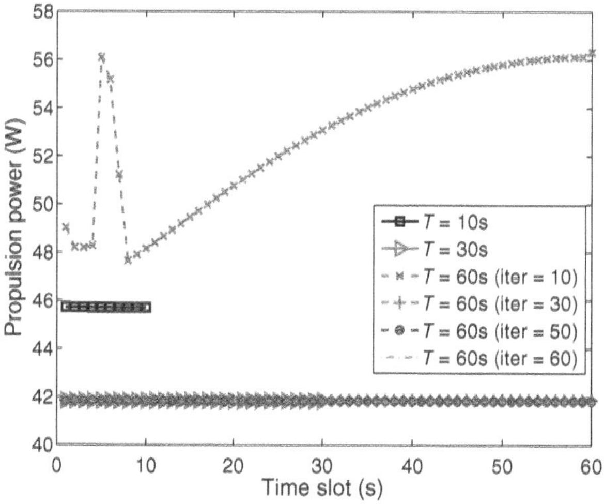

Figure 7.7 UAV power allocations under JF scenario.

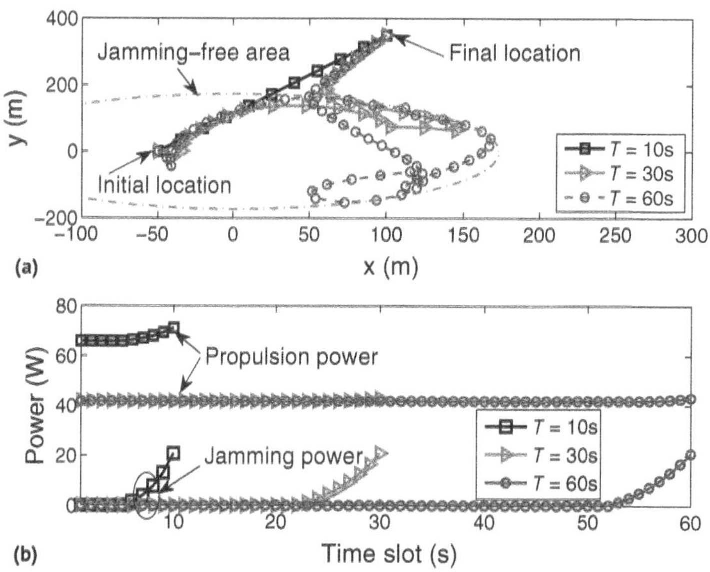

Figure 7.8 UAV trajectory designs and power allocations under IF scenario.

at the speed that incurs as little power consumption as possible. When the time horizon is sufficiently long ($T = 30$ and 60 seconds), the UAV carefully designs its trajectory and takes a detour inside $\mathcal{A}$ so that it can travel at the best energy-efficient speed V_e all the way.

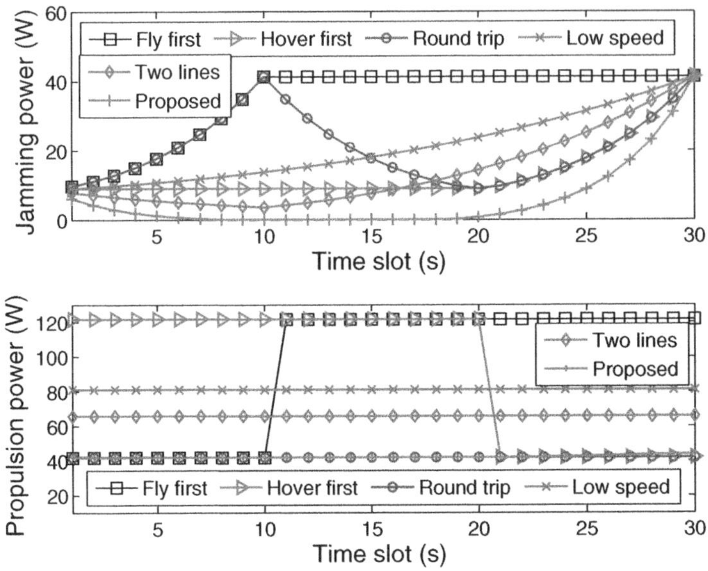

Figure 7.9 UAV power allocations for baseline schemes under NF scenario.

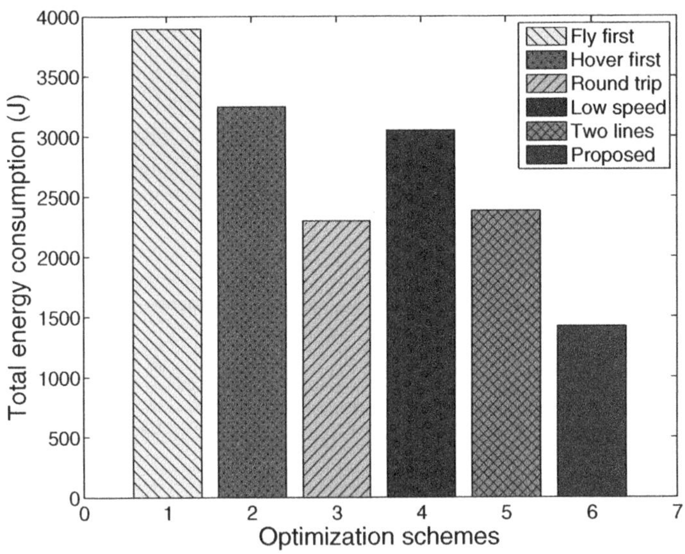

Figure 7.10 Total energy consumption of the UAV under NF scenario.

Furthermore, it is shown in Figure 7.8 that for the IF scenario when $T = 10$ seconds, the UAV has to travel straight from the initial location to the final location at a speed much faster than V_e, thus leading to a significant amount of propulsion power consumption at each slot. If there is surplus time, the UAV first travels inside $\mathcal{A}$ at a constant speed of V_e, which minimizes the propulsion power consumption. Then it flies to the final location, which is outside $\mathcal{A}$, in the last few time slots. This trajectory design enables the UAV to stay inside $\mathcal{A}$ for as long as possible, since the shorter the time it stays outside $\mathcal{A}$, the less jamming energy it consumes.

To fully demonstrate the influence and merits of delicate trajectory design for the UAV, we again compare with three baseline schemes where the UAV adopts different flying protocols for the same trajectory as the "Low speed" scheme when $T = 30$ s, i.e., the "Fly first," "Hover first," and "Round trip" schemes. To facilitate comparison, we also include the "Low speed," "Two lines," and our proposed schemes under the NF scenario. Figures 7.9 and 7.10 depict the jamming and propulsion power allocations at each time slot, and the total energy consumption of the UAV, respectively. Table 7.4 lists the respective jamming and propulsion energy consumptions for different schemes. It can be readily seen from Figure 7.9 that the UAV needs to send jamming signals at every time slot under the five baseline schemes, as it is always traveling outside $\mathcal{A}$. The propulsion power consumption for hovering triples that for traveling at speed V_e. The "Fly first" scheme incurs the largest energy consumption for both jamming and propulsion, due to its 20 seconds hovering at the farthest point from the suspicious source.

It is further revealed in Figure 7.10 that the total energy consumption of the "Round trip" scheme is the lowest among the baseline schemes, since it adopts a simple trajectory design with an energy-efficient speed. It is observed from Table 7.4 that the jamming energy consumption is the highest for the "Fly first" and "Round trip" schemes, while the propulsion energy consumption is the highest for the "Fly first" and "Hover first" schemes. Our proposed scheme consumes

Table 7.4 Respective jamming and propulsion energy consumptions for different schemes under NF scenario.

Optimization schemes	Jamming energy (J)	Propulsion energy (J)
Fly first	1042.9	2850.0
Hover first	396.3	2850.0
Round trip	1042.9	1255.2
Low speed	623.7	2427.5
Two lines	412.8	1968.6
Proposed	165.2	1255.2

the least jamming energy and propulsion energy. All of the baseline schemes consume more energy than our proposed scheme. In a nutshell, the UAV suffers significant waste of energy without careful trajectory optimization.

Remark 7.5 *(Mitigating the interference on other links):* When the suspicious link intentionally chooses to be located in a wild rural area to avoid surveillance by existing monitoring infrastructures, there would be few communication links in the vicinity, and the interference caused by jamming could thereby be reduced to the minimum level, which is negligible. In fact, it typically depends on the access scheme whether jamming suppresses the communications of other links. For instance, if the suspicious link occupies a certain frequency band all to itself, the UAV is able to send exclusive jamming signals to it, which will not affect other links. On the other hand, if serious communication degradation is reported by legitimate users within the neighborhood, the UAV can release the specific transmitted (and encrypted) information to these users so that they can decode the jamming signals and will not be interfered with. Note that the maximum jamming power of 40 W in Figures 7.4 and 7.9 is the worst-case value tested in the simulation. Yet in practice, the UAV is not usually that far away from the suspicious source and does not incur such a high jamming power consumption.

7.6 Conclusion

In this chapter, we addressed joint energy management and trajectory optimization for a rotary-wing UAV-enabled legitimate monitoring system. Building on a judicious (re-)formulation, we leveraged the alternating optimization and SCA methodologies to minimize the overall energy consumption of the UAV. Efficient algorithms were developed to compute the locally optimal solution or at least a feasible solution fulfilling the KKT conditions. We provided extensive numerical test results to justify the effectiveness of the proposed schemes. The proposed framework also inspires new directions for future research on security issues in UAV-aided wireless networks, such as wireless power transfer and/or mobile edge computing-based ones, especially with non-LoS channels and 3D trajectory planning.

8

Joint Sensing and Communications with UAVs

8.1 Introduction

Recent advancements in the 5G/B5G networks are increasingly enabling extensive background situation- and position-aware smart applications, including autonomous driving, distant medical care, and smart industry [112]. 5G/B5G networks are also envisioned to offer high-resolution sensing in support of these applications [138, 283]. For this reason, joint communication and radar sensing (JCAS) has been deemed as one of the essential technologies in 5G/B5G systems [52, 106, 282, 344]. By integrating radio communication and sensing into a single system, JCAS measures the reflections of communication signals to sense the location, velocity, and feature signal of targets and motions [114, 180, 250]. This is different from the widely accessible global positioning system (GPS), which enables every GPS receiver to locate itself by correlating the preambles/pilots from multiple (four or more) satellites [125, 248]. Neither of the GPS satellites nor ground receivers sense the environment or deliver situational awareness, as opposed to JCAS [344]. Cellular-connected UAVs have been increasingly considered for their operability and applicability to UAV operations over wide areas [325, 337]. Apart from being an aerial user of communication services, a cellular-connected UAV can potentially utilize the cellular signals from the corresponding BSs, and collect the echo signals from the ground (e.g., metallic) objects to perform radar sensing in areas covered by the BSs. Given the excellent mobility, the UAV can potentially form a bistatic SAR with its serving BS to sense the environment. While dedicated sensing or radar techniques may detect objects with even better resolution (e.g., because of their wider signal bandwidth and higher signal power), they inevitably bring radio radiation and pollution [344]. Nevertheless, the UAVs are passive receivers, and neither increases the radio footprint of the system nor produces radio pollution.

Artificial Intelligence for Unmanned Aerial Vehicles: Sensing, Communication, and Computing, First Edition. Shuyan Hu et al.

Figure 8.1 illustrates the new cellular-assisted radar sensing by a cellular-connected UAV, referred to as BS-UAV bistatic SAR sensing, where the reflections of the downlink cellular signals are picked up by the UAV and used to produce radio/radar images for object detection [166]. Since the UAV is typically power-constrained, it is critical to carefully plan its trajectory to deliver energy-efficient radar sensing missions and extend mission time. Existing works have only designed the trajectory to minimize the UAV propulsion energy for wireless communication services, such as [109, 110, 251, 335]. None has considered the energy consumption in the new scenario of cellular-assisted radar sensing with a cellular-connected UAV.

Combined with visual-based UAV tracking, UAVs can achieve better sensing and tracking performance. With eminent maneuverability, quick deployment, and expansive coverage, UAVs have been increasingly employed for nature resilience, disaster rescue, bushfire monitoring, mobile relaying, and eavesdropping [146]. Equipped with optical cameras, UAVs are increasingly utilized to visually tail and monitor moving targets [119]. Compared to a rotary-wing UAV, the energy efficiency and quietness make a fixed-wing UAV an excellent choice for surveillance applications. The suspicious targets can be aerial, e.g., misbehaved or malicious UAVs, due to the proliferation and easy accessibility of UAVs. A UAV monitor can record the targets' misbehaviors and provide evidence for forensics purposes, facilitating the visual-based surveillance for many Internet-of-Things (IoT) and security applications [113, 325]. While most existing drones are powered by batteries and restricted in terms of mission time and battery life. Renewable energy sources,

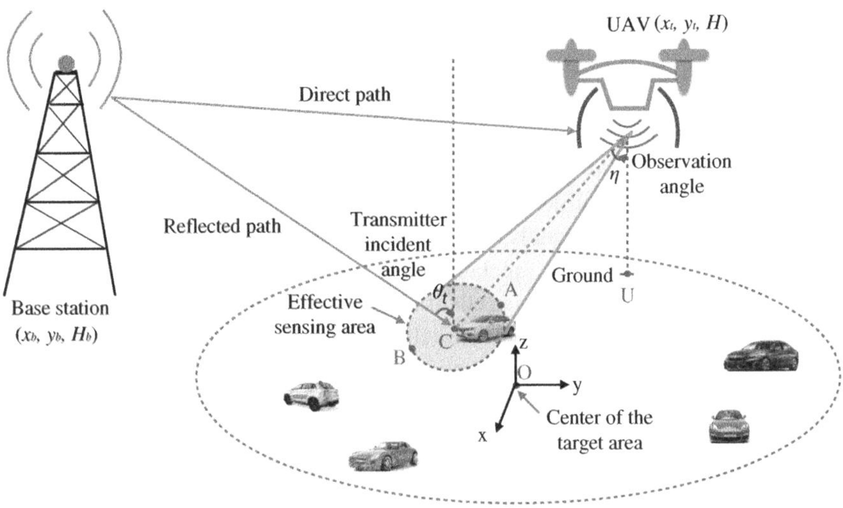

Figure 8.1 System model of JCAS by a BS-UAV bistatic SAR, where the UAV receives the downlink cellular signals from the BS and the reflections of the signals reflected by (metallic) objects. The reflection signals can be extracted and used to produce radio/radar images and detect objects.

such as solar power, are increasingly considered for UAVs. Solar panels can be readily installed on the top of fixed-wing UAVs to power visual-based tracking missions.

The trajectory of a UAV monitor needs to be carefully planned by taking into account not only the power usage models of the propulsion, thrust, and hanging of the UAV, but the energy harvesting process as well, as suggested in [109, 251, 332]. As far as we are concerned, there has been no investigation into the flight path planning of a solar-powered fixed-wing drone running a visual-based tailing and monitoring task. This is because the flight path plan of a fixed-wing drone is deemed nontrivial due to its relatively poor maneuverability, as compared to a rotary-wing drone.

8.1.1 Related Works

8.1.1.1 UAV-JCAS Systems

Existing studies of JCAS systems have typically focused on the beamforming design at the BSs, and have not considered the use of a UAV. These JCAS systems are, in essence, monostatic radars, which integrate the transmitter and receiver on the same platform. In [167], waveform design was optimized to confirm the produced signal to the expected sensing waveform subject to the SINR requirement for multiuser MIMO communications. In [168], globally optimal waveforms were devised for several expected sensing beam patterns to minimize multiuser interference. In [343], a multibeam technique was introduced for millimeter-wave (mmWave) JCAS with analog antenna arrays, and a fixed transmission subbeam was produced together with direction-changing scanning subbeams across various packets. In [248], a multi-metric waveform was designed for multiple-input single-output (MISO) communications to maximize the SINR at the users. On the other hand, a higher signal power is typically on demand for high-resolution sensing over long ranges, whereas the BSs usually have a limited transmit power. The sensing performance may considerably degrade if the targets are far away from the BS, because of the significant round-trip path loss attenuation of the echoed signals.

UAVs have been emerging as a new type of passive receiver in bistatic radar systems. Unlike a monostatic radar, a bistatic radar uses geographically separated antennas for signal transmission and reception. The German TanDEM-X project (initiated in June 2010) developed the first prototype of a space-borne bistatic radar, relying on twin low-Earth-orbiting (LEO) satellites traveling in near formation [97]. The high-resolution wide-swath was proposed to be the next-generation German space-borne radar platform for geological and geographical observation beyond 2030, by leveraging the formation movement of an active satellite and three passive satellites [29]. Mounted on mobile platforms, a bistatic SAR can perform HRWS sensing by making use of the relative movements between the SAR and the targets. The small-sized real antenna aperture can realize the capability of a larger aperture radar through data focusing and signal processing [276]. A clutter suppression and mobile target imaging strategy was developed in [351] for a GEO-LEO bistatic SAR system, which was robust against the quickly traveling object with the Doppler centroid ambiguity.

With a geosynchronous (GEO) SAR satellite as the transmitter, a UAV can receive the echo signals from terrestrial targets and achieve 2D imaging. The flight path of the UAV receiver was planned in [258] for such a GEO-UAV bistatic SAR system to sense rough terrains. A range model was developed in [257] to address the distinctiveness of the GEO-UAV bistatic SAR echo and the variations of the orbit positions of the GEO transmitter. A motion compensation algorithm was proposed in [279] for a mini-UAV-based bistatic SAR system to address the perturbation and spatial variance of the platform. However, none of the existing works have considered the BS-UAV bistatic SAR, as considered in this chapter. None of the existing studies have considered the energy consumption of a UAV-based bistatic SAR system.

8.1.1.2 Visual-based UAV

Many existing studies on visual-based UAV tracking have been focused on keeping its target within the UAV monitor's visual range, without considering the energy consumption and longevity of the surveillance missions. For instance, visual information-enhanced control mechanisms were developed for a UAV monitor to procure the 3D Cartesian coordinates of a moving object [247], or track a mobile object [354]. In [354], geometric control was operated along with nonlinear multi-rank control, suppressing the need for a mathematical model of the thrust force, to simplify the execution steps. UAV-enabled object tailing was pursued in [237] and [320] to deal with camera view occlusion in a city with tall buildings. The flight path of the drone monitor was devised to maximize the identification rate, which depends on the probability of having the object within sight. In [274], UAV-based target recognition and tracking were proposed by utilizing a smart gimbal with precise positioning and quick image analysis, where consensus-based target tailing, dynamic environment analyzing, and neural network-enabled target sensing were jointly optimized.

On the other hand, a target tracking system by UAVs equipped with gimbaled cameras was studied in [73]. The proposed system distributively incorporates a clustering algorithm, a sensor manager, and an optimal path planner to track multiple mobile targets. An integrated guidance and gimbal control coverage path planning approach was proposed in [209], in which the mobility and gimbal inputs of an autonomous UAV agent were jointly controlled and optimized to achieve full coverage of a given object of interest, according to a specified set of optimality criteria. None of these works considered the energy consumption of the UAV, and all of them relied on traditional control methods for UAV path planning.

Machine learning techniques, such as deep learning (DL) and deep reinforcement learning (DRL) [100–102, 112, 328, 329], have been found very useful in visual-based target tracking [183], to improve the performance of traditional control methods. For instance, the various features of deep convolutional neural networks (CNNs) were exploited in [182] to enhance the precision of video-based target tracking. The features of convolutional layers were interpreted to be a nonlinear corresponding term of a picture pyramid and are directly exploited to represent target objects. It was shown in [346] that basic two-layer CNNs could be sufficient to learn effective representations for video-based tailing, without training beforehand on a large dataset. An underwater target tailing task of an underwater vehicle was investigated in [278], where DRL was employed to resolve the constructed Markov decision task under uncertain hydrodynamics. The robustness of target detection and tracking was enhanced through radar and video camera data fusion in [177], especially under bad weather conditions. Radar acted as the main role, and the camera served as an assistant for real-time environment sensing. A dynamic occlusion-aware video-based tailing approach was developed in [67] to address the deformation and long-term occlusion of the target appearance during tracking. When the target is severely occluded or has been occluded for a long time, it is detected by a well-designed classifier chosen from a classifier group, based on an entropy minimization metric.

DL and DRL have been leveraged to UAV-enabled target tracking, to distinguish a target from its background, handle the object deformation, address the aspect ratio change (ARC) of captured images, and make coarse-to-fine tracking policies. Autonomous flight control was developed in [45] for path tailing missions in adversarial situations. A two-player zero-sum game was formulated, and the best UAV path was obtained by DL in real-time. Q-learning was utilized to optimize the UAV flight route for tailing a radio frequency (RF) target in a Rayleigh fading channel [55]. A UAV swarm was deployed to locate an RF mobile target in [53], where a constrained Markov decision process (MDP) was constructed to locate the target in the presence of channel uncertainties. Multi-agent reinforcement learning (RL) was utilized to coordinate the several UAVs, avoid excessive UAV flight paths, and perform real-time target tracking. In [353], DRL was employed by a UAV for vision-based target recognition and tracking to address the ARC problem of captured images and refine the boundaries of the bounding box, without addressing the UAV trajectory design during tracking. Yet, it is not straightforward to generalize these works to capture the power usage of the UAV monitors because the system model, UAV dynamics, and application scenario would be different when a specific UAV power consumption model is considered.

With the consideration of power usage (and especially solar power), the UAV control variables, i.e., acceleration, velocity, and trajectory, are coupled over time. The trajectory planning becomes a sequential decision process, which is challenging, attributed to its considerable state and action spaces and the requirement of real-time operation. UAVs were deployed to monitor or track mobile ground targets in [120, 121, 124]. A decentralized 3D navigation rule was developed in [120] to determine the best position of each UAV, and balance the UAV's power usage and the number of covered targets [144]. Dynamic programming was utilized in [121] to optimize the UAV path, which minimized the power usage and maximized the disguise performance of the surveillance. Receding horizon control was employed in [124] to generate an energy-efficient path for the solar-powered UAV. The power consumption of the UAV monitor was reduced to be linear with respect to its speed in [120] and [121], and the power requirement of hovering was overlooked in [124]. In [110] and [113], we investigated disguised visual-based target tracking by rotary- and fixed-wing UAVs, respectively. Difference-of-convex programming was employed to solve the non-convex trajectory design for rotary-wing UAVs [110]. Convex optimization was integrated with MPC to produce trajectories for fixed-wing UAVs [113].

None of the existing works fully investigated the energy-efficient 3D target tracking by a fixed-wing UAV with solar power harvesting capability, and compared the performances of ML methods with traditional control- or optimization-based algorithms.

This chapter focuses on two complementary studies. We first propose a novel framework for cellular-aided radar sensing with a cellular-connected UAV. Given a series of landmarks to be sensed/observed, the UAV's trajectory is optimized to minimize its propulsion energy while satisfying the range and azimuth resolutions of sensing, thereby delivering energy-efficient BS-UAV bistatic SAR sensing missions and extending mission durations. This trajectory planning problem is non-convex because of the non-convexity of its objective function and constraints. We convexify the problem and obtain a quality solution efficiently by utilizing successive convex approximation (SCA) and block coordinate descent (BCD).

Then, we present a new scheme for the online control of video-based UAV-on-UAV target tracking. Specifically, a solar-harvesting fixed-wing UAV installed with video cameras tails and visually surveils a suspicious UAV target. The monitoring UAV keeps the target inside its view, while staying away from the target to avoid raising the attention of the target. DRL is employed to train the trajectory of the monitor online.

The key contributions of the chapter are listed as follows.

- A novel BS-UAV bistatic SAR platform is proposed to reuse cellular downlink signals as excitation signals. A cellular-connected UAV receives the reflected signals and conducts radar sensing on landmarks.
- The range and azimuth resolutions of the BS-UAV bistatic SAR are analyzed. An effective sensing area of the BS-UAV bistatic SAR is established to effectively capture a landmark.

- The UAV's trajectory is optimized to minimize its propulsion energy and ensure that all landmarks are captured in the effective sensing areas with the required range and azimuth sensing resolutions. By applying the SCA and BCD methods, a new algorithm is developed to deliver energy-efficient trajectories with polynomial complexity.

Extensive simulations corroborate that, in terms of energy efficiency and energy usage fluctuations, the proposed trajectory planning algorithm outperforms a baseline scheme that minimizes the flight distance of the UAV on the cellular-aided radar sensing mission. The proposed algorithm can save up to 55% of the UAV's propulsion energy, as compared to the baseline scheme.

8.2 System Model

8.2.1 UAV-JCAS Systems

In the considered system, a BS is deployed to deliver downlink wireless services to cellular users. A cellular-connected UAV equipped with a side-looking SAR (receiver) flies and collects the echoes (or reflections) of the communication signals originating from the BS, to sense (metallic) objects and gain situational awareness; see Figure 8.1. The signals are buffered at the UAV, and offloaded and post-processed (at the BS) after the sensing mission. Given the knowledge of the communication signals, the direct path from the BS to the UAV can be canceled. The remaining echo signals can be used to sense and decide the number and locations of the objects. In this way, the UAV can perform the sensing task by utilizing the communication signals, without the need for dedicated sensing signals, and thereby reduces the radio fingerprint and carbon emission.

The proposed UAV-based sensing platform is insusceptible to inter-cell interference, since it does not need to detect or decode data signals. The transmissions from other BSs can serve as additional illuminating sources to enhance the reflection on the objects. The UAV may utilize the pilot signals of the BSs to extract the reflections of objects [83]. With the knowledge of the pilot signals, the collected reflection signals by the UAV can be correlated to cancel the direct paths from any BSs and extract the reflections by the objects to detect the objects. Alternatively, the UAV may store the captured signals, and only start to process the signals after it returns from the mission and has access to the transmitted signals of the BSs. Then, correlation can be conducted between the captured signals and the transmitted signals to extract the reflections and detect the objects. On the other hand, the proposed trajectory planning algorithm can be performed at the UAV with limited assistance from the BSs. It only requires the information about the landmarks prior to the trajectory planning. The inter-cell interference has little impact on the algorithm, when the algorithm is in operation.

8.2.1.1 Object Distribution

Consider a 2D ground region $\mathcal{A}$ with radius R. We assume that $\mathcal{A}$ is inside the coverage area of the BS. The number of illuminating (e.g., metallic) objects in $\mathcal{A}$, denoted by $N(\mathcal{A})$, follows the homogeneous spatial Poisson point process (SPPP). The reason is that the points are uniformly distributed in the circular area (with the same density λ). The probability density function (PDF) is given by [94]

$$\Pr(N(\mathcal{A}) = n) = \frac{\lambda|\mathcal{A}|^n}{n!} e^{-\lambda|\mathcal{A}|}, n = 0, 1, \dots , \tag{8.1}$$

where $\lambda > 0$ is the intensity, and $|\mathcal{A}|$ is the Lebesgue measure for the size of the region. The objects are uniformly randomly distributed in $\mathcal{A}$.

Let d_i, $i = 1, \dots , N(\mathcal{A}) - 1$, collect the Euclidean distances from an arbitrarily selected (and designated) object to the other $(N(\mathcal{A}) - 1)$ objects within the disk region with radius R; and assume that d_i, $i = 1, \dots , N(\mathcal{A}) - 1$, are independent and identically distributed (i.i.d.), non-negative random variables. Given the uniform stationary distribution of the objects in the disk region with radius R, the cumulative density function (CDF) of d_i, $i = 1, \dots , N(\mathcal{A}) - 1$ is given by [324][1]

$$\begin{aligned} F_{d_i}(l) &= 1 + \frac{2}{\pi}\left(\frac{l^2}{R^2} - 1\right)\cos^{-1}\left(\frac{l}{2R}\right) \\ &\quad - \frac{1}{\pi R}\left(1 + \frac{l^2}{2R^2}\right)\sqrt{1 - \frac{l^2}{4R^2}}, \quad 0 \le l \le 2R. \end{aligned} \tag{8.2}$$

Let $d_{\min}$ denote the shortest of the distances between two objects, i.e., $d_{\min} = \min_{i=1,\dots,N-1}\{d_i\}$. By exploiting order statistics, the CDF of $d_{\min}$ can be given by

$$\begin{aligned} F_{d_{\min}}(l) &= 1 - \left[1 - F_{d_i}(l)\right]^{N(\mathcal{A})-1} \\ &= 1 - \left[\frac{1}{\pi R}\left(1 + \frac{l^2}{2R^2}\right)\sqrt{1 - \frac{l^2}{4R^2}} - \frac{2}{\pi}\left(\frac{l^2}{R^2} - 1\right)\cos^{-1}\left(\frac{l}{2R}\right)\right]^{N(\mathcal{A})-1}. \end{aligned} \tag{8.3}$$

The resolution of the proposed UAV-based bistatic SAR platform is set to be finer than the expected shortest distance between two neighboring objects, i.e., $d_{\min}$, to reasonably distinguish any two different objects. For this reason, we derive the expectation of $d_{\min}$ based on the CDF in (8.3). The expectation is an integral

1 Given the uniform stationary distribution of the objects in the disk region with radius R, the PDF of d_i, $i = 1, \dots , N(\mathcal{A}) - 1$ is given by [324]

$$f_{d_i}(l) = \frac{2l}{R^2}\left(\frac{2}{\pi}\cos^{-1}\left(\frac{l}{2R}\right) - \frac{l}{\pi R}\sqrt{1 - \frac{l^2}{4R^2}}\right), 0 \le l \le 2R.$$

over l, the distance between any two objects in the circular coverage area of the BS with the radius of R. Here, $0 \leq l \leq 2R$. Then, the expectation of $d_{\min}$ can be given by

$$\begin{aligned}\mathbb{E}[d_{\min}] &= \int_0^{2R} l \, dF_{d_{\min}}(l) \\ &= -l\left[-F_{d_i}(l)\right]^{N(\mathcal{A})-1}\Big|_0^{2R} + \int_0^{2R} \left[1 - F_{d_i}(l)\right]^{N(\mathcal{A})-1} dl \\ &= \int_0^{2R} \left[\frac{1}{\pi R}\left(1 + \frac{l^2}{2R^2}\right)\sqrt{1 - \frac{l^2}{4R^2}} - \frac{2}{\pi}\left(\frac{l^2}{R^2} - 1\right)\cos^{-1}\left(\frac{l}{2R}\right)\right]^{N(\mathcal{A})-1} dl,\end{aligned} \tag{8.4}$$

where $F_{d_i}(0) = 1 - \frac{1}{\pi R}$ and $F_{d_i}(2R) = 1$.

8.2.1.2 UAV Mobility

In practice, the receiver of a bistatic SAR system flies at a constant altitude, as a stable flight can stabilize the accuracy of the SAR [355]. For this reason, we consider a horizontally flying UAV at a constant altitude, H (in meters), in line with practical implementations, and focus this chapter on 2D trajectory planning. Nevertheless, the algorithm developed in this chapter can be potentially extended to 3D trajectory planning with varying UAV altitudes H_t, $\forall t$, as will be discussed in Section 8.3.

The UAV executes the sensing mission for a fixed time horizon of T seconds, which is divided into T_w time slots indexed by t, and $t = 1, \dots, T_w$. The duration of a time slot t is δ seconds. The UAV's time-varying 2D coordinates are $\mathbf{q}_t = [x_t, y_t]^T$, $\forall t$, with the initial location $\mathbf{q}_0 = [x_0, y_0]^T$. Let V_t denote the UAV speed at time slot t, which is upper bounded by V_m. The UAV's mobility constraint is given by

$$\|\mathbf{q}_t - \mathbf{q}_{t-1}\| = \delta V_t \leq \delta V_m, \ \forall t. \tag{8.5}$$

For a rotary-wing UAV at a speed V_t, the propulsion power at time slot t, denoted by P_t, is shown by [336]

$$P_t = P_0\left(1 + \frac{3V_t^{\,2}}{U_{tip}^2}\right) + P_1\left(\sqrt{1 + \frac{V_t^{\,4}}{4v_0^4}} - \frac{V_t^{\,2}}{2v_0^2}\right)^{\frac{1}{2}} + \frac{1}{2}d_f \rho s A V_t^{\,3}, \tag{8.6}$$

where P_0 and P_1 stand for the fixed *blade* and *induced powers* when the aircraft hovers, respectively; U_{tip} denotes the rotor velocity; v_0 is the mean rotor velocity when the aircraft hovers; d_f and s stand for the fuselage drag fraction and rotor solidity, respectively; and ρ and A are the gaseous density and rotor disc area, respectively.

8.2.1.3 Airborne Bistatic Side-looking SAR

As shown in Figure 8.1, the effective sampling SAR sensing area of the UAV, denoted by $\mathcal{U}_t$, is an ellipsoid centered at point C as per slot t. We assume that the effective sensing area $\mathcal{U}_t$ is always to the right of the UAV, by considering a side-looking SAR [225, 231]. The major axis of the ellipsoid $\mathcal{U}_t$, L_{AB}, is perpendicular to the UAV's heading. Let $\alpha_t \in [0, 2\pi)$ denote the angle between the UAV's heading and the x-axis at time slot t. We have

$$\sin \alpha_t = \frac{y_t - y_{t-1}}{\delta V_t}; \quad \cos \alpha_t = \frac{x_t - x_{t-1}}{\delta V_t}. \tag{8.7}$$

Further, let $\eta \in (0, \pi/2)$ denote the observation angle of the side-looking SAR at the UAV, which is fixed over time. Given the horizontal coordinates of the UAV, i.e., $U = (x_t, y_t)$ at slot t, and $L_{\text{UC}} = H \tan \eta$, the coordinates of the center C, i.e., $\mathbf{q}_c^t = [x_c^t, y_c^t]^T$, can be written as

$$x_c^t = x_t + H \tan \eta \sin \alpha_t; \; y_c^t = y_t - H \tan \eta \cos \alpha_t. \tag{8.8}$$

To guarantee the sensing performance, C should always be inside the target area $\mathcal{A}$. We have $L_{\text{OC}} \leq R, \; \forall t$.

At any time slot t, the range resolution δ_r^t and the azimuth resolution δ_a^t of the BS-UAV bistatic SAR are given by [225, equation (5)], [187, equations (24) and (29)]

$$\delta_r^t = \frac{c}{B(\sin \eta + \sin \theta_t)}; \tag{8.9a}$$

$$\delta_a^t = \frac{\lambda_c H}{T_d V_t \cos \eta}, \tag{8.9b}$$

where c is the speed of light, λ_c is the wavelength, B is the transmit signal bandwidth of the BS, $\theta_t \in (0, \pi/2)$ is the incidence angle of the transmitted signal (from the BS) with respect to the point C, and T_d is the coherent SAR integration time [266]. Given the horizontal coordinates of the BS $\mathbf{q}_b = [x_b, y_b]^T$ and its height H_b, we have

$$\sin \theta_t = \frac{||\mathbf{q}_c^t - \mathbf{q}_b||}{\sqrt{||\mathbf{q}_c^t - \mathbf{q}_b||^2 + H_b{}^2}}. \tag{8.10}$$

In order to distinguish the objects, the resolution of the SAR has to be finer than the shortest expected distance between any two objects, $d_{\min}$, i.e., $\delta_r^t \leq d_{\min}$ and $\delta_a^t \leq d_{\min}$. This resolution requirement is used to detect and differentiate different objects by leveraging the typically narrow signal bandwidths of cellular systems and, in turn, the poor resolution of sensing [61]. It is worth mentioning that we do not consider object identification in this chapter, as it would require wider signal bandwidths to achieve better resolutions for detecting the shape of an object or even imaging the object [187]. Nevertheless, the proposed trajectory planning algorithm would still be relevant under a wider signal bandwidth and correspondingly a finer resolution requirement. The algorithm is generic and suitable for different resolution requirements.

Considering the characteristics of the side-looking SAR, the cosine of the UAV elevation angle is

$$\cos\eta \approx \frac{L_{\mathrm{AE}}}{W_g} \approx \frac{\theta_v R_m}{W_g} \approx \frac{\lambda_c R_m}{W_a W_g}, \tag{8.11}$$

where R_m is the slant range (distance) from the SAR antenna to the center of $\mathcal{U}_t$ (i.e., point C); W_a is the height of the SAR antenna, and W_g is the SAR range extent, i.e., the width of the ground swath covered by the SAR beam, to the right of the UAV; and the vertical beamwidth of the SAR is $\theta_v = \lambda_c/W_a$. The azimuth angular spread is $\theta_a = \lambda_c/L_a$, with L_a being the antenna length parallel to the direction of the UAV's heading. This spread is due to the interference of the waves emitted from and received by the dipoles of the antenna [57].

The effective sensing area is ellipsoidal, denoted by $\mathcal{U}_t$. The ellipsoidal effective sensing area is significantly smaller than the coverage area of the BS, $\mathcal{A}$, and is used to sample different parts of the coverage area for object detection. The major and minor axes of the ellipsoidal sensing area, $\mathcal{U}_t$, are $W_g = \frac{\lambda_c R_m}{W_a \cos\eta}$ and $\theta_a = \lambda_c/L_a$, respectively. We can obtain from Figure 8.2 that $R_m = H/\cos\eta$. Therefore, at any moment, the size of the effective sensing area $\mathcal{U}_t$, i.e., S, is

$$S = \frac{\pi}{4} \times \frac{\lambda_c}{L_a} \times \frac{\lambda_c H}{W_a \cos^2\eta}. \tag{8.12}$$

Given the range and azimuth resolutions of the SAR, and the center and size of the effective sensing area, the BS-UAV bistatic SAR can efficiently sense, detect,

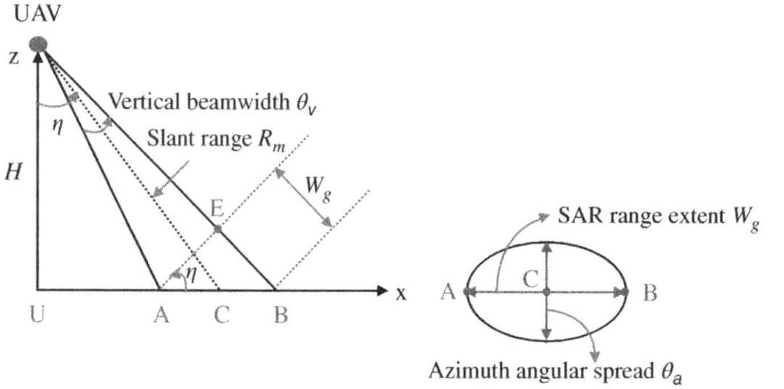

Figure 8.2 The (x, z)-plane view of the side-looking BS-UAV bistatic SAR effective sensing area, where the UAV heading is perpendicular to the chapter, and W_g is the SAR range extent.

and distinguish ground objects in the disk region while the UAV is flying. The details are provided in Section 8.3.

8.2.2 Visual-based UAV Tracking

8.2.2.1 Fixed-wing UAV Mobility

For ease of exposition, we describe the proposed method under a 2D setting; i.e., the UAV flies horizontally. Nevertheless, the method can be readily applied to the 3D scenario, as will be shown in Section IV. Let $\boldsymbol{b}_t := (b_{xt}, b_{yt}, h_0)$ and $\boldsymbol{q}_t := (x_t, y_t, h_1)$ denote the 3D coordinates of the target and monitor per slot $t\,(t = 1, \dots, N)$, respectively. $\boldsymbol{V}_t := (V_{xt}, V_{yt})$ and $\boldsymbol{A}_t := (A_{xt}, A_{yt})$ denote the monitor's speed and acceleration on the (x, y)-plane, respectively. Further, let $\boldsymbol{v}_t$ be the speed vector of the target at slot t. With a fixed-wing structure, the monitor obeys the following mobility constraints

$$\boldsymbol{q}_{t+1} = \boldsymbol{q}_t + \boldsymbol{V}_t\delta + \frac{1}{2}\boldsymbol{A}_t\delta^2, \ \forall t, \tag{8.13a}$$

$$\boldsymbol{V}_{t+1} = \boldsymbol{V}_t + \boldsymbol{A}_t\delta, \ \forall t, \tag{8.13b}$$

$$||\boldsymbol{A}_t|| \leq A_{\max}, \ \forall t, \tag{8.13c}$$

$$V_{\min} \leq ||\boldsymbol{V}_t|| \leq V_{\max}, \ \forall t, \tag{8.13d}$$

where $A_{\max}$ is the monitor's largest acceleration; and $V_{\max}$ and $V_{\min}$ are the maximum and minimum speeds of the monitor, respectively.

8.2.2.2 Power Consumption

Per time slot t, the UAV propulsion power consumption is modeled as [332]

$$P_v^t = c_1||\boldsymbol{V}_t||^3 + \frac{c_2}{||\boldsymbol{V}_t||}\left(1 + \frac{||\boldsymbol{A}_t||^2}{g^2}\right) + m\boldsymbol{A}_t^T\boldsymbol{V}_t, \tag{8.14}$$

where $(\cdot)^T$ stands for transpose; c_1 and c_2 are two known coefficients relying on the wing surface region and wing span ratio; m is the mass of the UAV monitor; $g \approx 9.8$ m/s^2 is the gravitational acceleration.

We note that the mobility model in (8.13) is generic and describes the motions of an object with a continuous and differentiable trajectory, such as a fixed-wing UAV, according to Newton's second law [232]. The model can capture and regulate the movement of UAVs with fixed wings, including their velocities, accelerations, and curvature radii. On the other hand, (8.14) provides the propulsion power of the fixed-wing UAV, which is nonlinear in the velocity and acceleration of the UAV (due to the aerodynamics) and makes it challenging to control and optimize the trajectory of the UAV.

Let P_s^t (in Watts) stand for the solar power harvesting capability of the UAV monitor. Suppose that the monitoring task is performed on the midday of a sunny or cloudy day. At slot t, the solar power collected is [7]

$$P_s^t(z_t) = \eta S P_i \exp\left(-\frac{\alpha}{\cos \vartheta_t}(1 - 2.2556 \times 10^{-5} z_t)^{5.2561}\right). \tag{8.15}$$

Here, z_t is the monitor's altitude; $\eta \in (0, 1)$ and S (in m^2) are the conversion efficiency and the solar panel size, respectively; P_i (in Watts) is the fixed power intensity of the solar beams before entering the clouds; $\alpha > 0$ is the total gaseous absorption; $\vartheta_t \in [0, \pi/2]$ stands for the solar zenith angle at slot t. At sunrise or sunset, $\vartheta_t = \pi/2$. At midday, $\vartheta_t = 0$. The change of ϑ_t is negligible within 15 minutes. We assume that $\vartheta_t = 0$ w.l.o.g.; i.e., the mission is at midday.

While some modern solar devices can capture energy in very low lighting conditions, such as solar cells based on Perovskite or amorphous silicon [49, 316], they cannot provide enough energy to power and sustain a typical UAV's flight. This is because these dim-light solar cells operate at low voltage and radiation conditions, yielding an output power in the magnitude of milliwatt (mW). In practice, dim-light solar cells are typically employed to support low-power devices, such as IoT sensors.

8.3 UAV-assisted JCAS

The UAV performs sensing at a series of landmarks at the specified time. The BS-UAV bistatic SAR satisfies the resolution requirements to distinguish objects. The 2D coordinates of the landmark to be sensed at the t-th time slot are $\mathbf{q}_g^t = [x_g^t, y_g^t]^T$, $\forall t$. We have $G \in \mathcal{U}_t$, $\forall t$.

Through antenna configuration, the ellipsoidal shape of $\mathcal{U}_t$ can be approximated by a circular area, i.e., $\theta_a = W_g$. The landmark of interest is inside the sensing area $\mathcal{U}_t$. Therefore, the distance between the landmark G and the center of the sensing area $\mathcal{U}_t$, i.e., point C, must not exceed the radius of $\mathcal{U}_t$ at any time slot t. With the coordinates of the center C and the diameter of $\mathcal{U}_t$, i.e., λ_c/L_a, we have

$$||\mathbf{q}_g^t - \mathbf{q}_c^t|| \leq \frac{\lambda_c}{2L_a}, \ \forall t. \tag{8.16}$$

The UAV flies at a constant altitude H, and detects objects. The UAV minimizes the overall propulsion energy consumption during the considered time horizon, while maintaining acceptable sensing resolution by reusing cellular communication signals. This can be formulated as

$$\min_{\{\mathbf{q}_t, V_t, \forall t\}} \sum_{t=1}^{T_w} P_t \delta \tag{8.17a}$$

$$\text{s.t.} \quad \frac{c}{B(\sin \eta + \sin \theta_t)} \leq d_{\min}, \ \forall t, \tag{8.17b}$$

$$\frac{\lambda_c H}{T_d V_t \cos \eta} \leq d_{\min}, \ \forall t, \tag{8.17c}$$

$$\|\mathbf{q}_t - \mathbf{q}_{t-1}\| \leq \delta V_t, \ \forall t, \tag{8.17d}$$

$$0 \leq V_t \leq V_m, \ \forall t, \tag{8.17e}$$

$$\|\mathbf{q}_g^t - \mathbf{q}_c^t\| \leq \frac{\lambda_c}{2L_a}, \ \forall t, \tag{8.17f}$$

where constraints (8.17b) and (8.17c) are the range and azimuth resolution requirements, respectively; (8.17d) and (8.17e) are the constraints on the UAV's mobility and maximum speed, respectively; and (8.17f) guarantees that the landmarks of interest are always captured at given time slots.

Problem (8.17) is non-convex because of the non-convexity of the objective (8.17a) and constraints (8.17b) and (8.17f). In what follows, we employ the SCA and BCD techniques to convexify (8.17), and attain a quality trajectory.

8.3.1 SCA-based Convexification

The convexification starts with the non-convex term of P_t in the objective (8.17a), i.e., $\left(\sqrt{1 + \frac{V_t^4}{4v_0^4}} - \frac{V_t^2}{2v_0^2}\right)^{\frac{1}{2}}$ in (8.6), by introducing new auxiliary variables $\{q_t \geq 0, \ \forall t = 1, \dots, T_w\}$

$$q_t^2 = \sqrt{1 + \frac{V_t^4}{4v_0^4}} - \frac{V_t^2}{2v_0^2}, \ \forall t, \tag{8.18}$$

which can be reorganized as

$$\frac{1}{q_t^2} = q_t^2 + \frac{V_t^2}{v_0^2}, \ \forall t. \tag{8.19}$$

Herein, the second component on the right-hand side (RHS) of (8.6) can be replaced by a linear element $P_1 q_t$, with a newly added constraint (8.19). As such, P_t is transformed into

$$\tilde{P}_t := P_0 + \frac{3P_0}{U_{tip}^2} V_t^2 + P_1 q_t + \frac{d_f}{2} \rho s A V_t^3, \ \forall t. \tag{8.20}$$

Here, $\tilde{P}_t$ is jointly convex in (V_t, q_t).

With a fixed δ, problem (8.17) becomes

$$\min_{\{\mathbf{q}_t, V_t, q_t, \forall t\}} \sum_{t=1}^{T_w} \tilde{P}_t \tag{8.21a}$$

$$\text{s.t.} \ \frac{1}{q_t^2} \leq q_t^2 + \frac{V_t^2}{v_0^2}, \ \forall t, \tag{8.21b}$$

$$(8.17\text{b})\text{–}(8.17\text{f}).$$

Constraint (8.21b) is attained by slackening the equality in (8.19) with inequality. Problems (8.17) and (8.21) are equivalent. The reason is that, if (8.21b) holds with inequality for any t, one could always diminish the value of the relevant variable q_t to save the overall power until (8.21b) takes equality [336].

Problem (8.21) is still non-convex due to the non-convex constraint (8.21b), and nevertheless can be resolved with the SCA approach [312] through determining the global lower limit of (8.21b) at a specified local point. Specifically, the left-hand side (LHS) of (8.21b) exhibits convexity in q_t, and the RHS exhibits convexity in $\{q_t, V_t\}$. Because the first-order Taylor expansion acts as a global lower limit of a convex function [38], we can attain the lower limit for the RHS of (8.21b), as given by

$$q_t^2 + \frac{V_t^2}{v_0^2} \geq q_t^{(\ell)2} + \frac{1}{v_0^2} V_t^{(\ell)^2} + 2q_t^{(\ell)}\left(q_t - q_t^{(\ell)}\right) + \frac{2}{v_0^2} V_t^{(\ell)}\left(V_t - V_t^{(\ell)}\right), \tag{8.22}$$

where $q_t^{(\ell)}$ and $V_t^{(\ell)}$ are the respective values of the variables at the ℓ-th iteration of the SCA method.

We proceed to convexify constraints (8.17b) and (8.17f), since the other constraints (8.17c)–(8.17e) are all convex by now. This starts by rewriting (8.17b) as

$$\sin\theta_t \geq \frac{c}{Bd_{\min}} - \sin\eta. \tag{8.23}$$

With $\sin^2\theta_t = 1 - \cos^2\theta_t$, we have

$$\frac{H_b^2}{\|\mathbf{q}_c^t - \mathbf{q}_b\|^2 + H_b^2} \leq 1 - \left(\frac{c}{Bd_{\min}} - \sin\eta\right)^2. \tag{8.24}$$

Introduce slack variables $w_t = \sin\alpha_t$ and $u_t = \cos\alpha_t, \forall t$. From (8.5) and (8.7), we have additional constraints on w_t and u_t

$$-1 \leq w_t \leq \frac{y_t - y_{t-1}}{\delta V_t}, \ \forall t, \tag{8.25a}$$

$$\frac{x_t - x_{t-1}}{\delta V_t} \leq u_t \leq 1, \ \forall t. \tag{8.25b}$$

Constraint (8.25) is non-convex because of the coupling between V_t and w_t (or u_t).

With w_t and u_t, (8.24) can be reorganized as

$$(x_t + w_t H \tan\eta - x_b)^2 + (y_t - u_t H \tan\eta - y_b)^2 \geq \frac{H_b^2}{1 - \left(\frac{c}{Bd_{\min}} - \sin\eta\right)^2} - H_b^2 := C_r, \ \forall t. \tag{8.26}$$

The LHS of (8.26) is convex in $\{\mathbf{q}_t, w_t, u_t\}$. The RHS of (8.26) is a constant and defined as C_r for illustration convenience, given the predetermined parameters H_b, c, B, $d_{\min}$, and η. We convexify (8.26) by determining the first-order Taylor expansion of its LHS at the local point attained during the ℓ-th iteration of the SCA:

Given $\{w_t^{(\ell)}, u_t^{(\ell)}\}$, (8.26) can be linearized in $\mathbf{q}_t$

$$\begin{aligned}&(x_t^{(\ell)} + w_t^{(\ell)} H \tan\eta - x_b)^2 + (y_t^{(\ell)} - u_t^{(\ell)} H \tan\eta - y_b)^2 \\ &+ 2(x_t^{(\ell)} + w_t^{(\ell)} H \tan\eta - x_b)(x_t - x_t^{(\ell)}) \\ &+ 2(y_t^{(\ell)} - u_t^{(\ell)} H \tan\eta - y_b)(y_t - y_t^{(\ell)}) \geq C_r, \ \forall t.\end{aligned} \tag{8.27}$$

Given $\{\mathbf{q}_t^{(\ell)}, V_t^{(\ell)}, q_t^{(\ell)}\}$, (8.26) can be linearized in $\{w_t, u_t\}$

$$\begin{aligned}&(x_t^{(\ell)} + w_t^{(\ell)} H \tan\eta - x_b)^2 + (y_t^{(\ell)} - u_t^{(\ell)} H \tan\eta - y_b)^2 \\ &+ H \tan\eta(x_t^{(\ell)} + w_t^{(\ell)} H \tan\eta - x_b)(w_t - w_t^{(\ell)}) \\ &+ H \tan\eta(y_t^{(\ell)} - u_t^{(\ell)} H \tan\eta - y_b)(u_t - u_t^{(\ell)}) \geq C_r, \forall t.\end{aligned} \tag{8.28}$$

As done to (8.26), constraint (8.17f) can be rewritten as

$$(x_t + w_t H \tan\eta - x_g^t)^2 + (y_t - u_t H \tan\eta - y_g^t)^2 \leq \left(\frac{\lambda_c}{2L_a}\right)^2, \forall t, \tag{8.29}$$

which is convex in $\{\mathbf{q}_t, w_t, u_t\}$.

As a result, problem (8.21) is transformed into

$$\min_{\{\mathbf{q}_t, V_t, q_t, w_t, u_t, \forall t\}} \sum_{t=1}^{T_w} \tilde{P}_t \tag{8.30a}$$

$$\text{s.t. } \frac{1}{q_t^2} \leq q_t^{(\ell)2} + \frac{1}{v_0^2} V_t^{(\ell)^2} + 2q_t^{(\ell)} \left(q_t - q_t^{(\ell)}\right) + \frac{2}{v_0^2} V_t^{(\ell)} \left(V_t - V_t^{(\ell)}\right), \forall t, \tag{8.30b}$$

(8.17c)–(8.17e), (8.25), (8.27)–(8.29),

where (8.30b) tightens the original constraint (8.21b) by using a lower bound of its RHS. Since problem (8.30) is a tightened version of problem (8.21), the feasible region of (8.30) is also the feasible region of (8.21); not the other way around.

8.3.2 Block Coordinate Descent (BCD)

Problem (8.30) is still non-convex due to the non-convexity of constraint (8.25). We apply the BCD to optimize $\{\mathbf{q}_t, V_t, q_t\}$ and $\{w_t, u_t\}$ in an alternating manner, because the variables V_t and $\{w_t, u_t\}$ are coupled in the non-convex constraint (8.25)

Given fixed $\{w_t, u_t\}$, problem (8.30) is reduced to the following convex problem

$$\min_{\{\mathbf{q}_t, V_t, q_t, \forall t\}} \sum_{t=1}^{T_w} \tilde{P}_t \tag{8.31a}$$

$$\text{s.t. } w_t^{(\ell)} \leq \frac{y_t - y_{t-1}}{\delta V_t}, \forall t, \tag{8.31b}$$

$$\frac{x_t - x_{t-1}}{\delta V_t} \leq u_t^{(\ell)}, \forall t, \tag{8.31c}$$

$$(x_t + w_t^{(\ell)} H \tan\eta - x_g^t)^2 + (y_t - u_t^{(\ell)} H \tan\eta - y_g^t)^2 \leq \left(\frac{\lambda_c}{2L_a}\right)^2, \forall t, \tag{8.31d}$$

(8.17c)–(8.17e), (8.27), (8.30b).

Given fixed $\{\mathbf{q}_t, V_t, q_t\}$, problem (8.30) becomes a feasibility checking problem to find the values of $\{w_t, u_t\}$ that satisfy the following convex constraints

$$\text{find } \{w_t, u_t, \forall t\} \tag{8.32a}$$

$$\text{s.t.} \; -1 \leq w_t \leq \frac{y_t^{(\ell)} - y_{t-1}^{(\ell)}}{\delta V_t^{(\ell)}}, \; \forall t, \tag{8.32b}$$

$$\frac{x_t^{(\ell)} - x_{t-1}^{(\ell)}}{\delta V_t^{(\ell)}} \leq u_t \leq 1, \; \forall t, \tag{8.32c}$$

$$(x_t^{(\ell)} + w_t H \tan \eta - x_g^t)^2 + (y_t^{(\ell)} - u_t H \tan \eta - y_g^t)^2 \leq \left(\frac{\lambda_c}{2L_a}\right)^2, \; \forall t, \tag{8.32d}$$

(8.28).

Problems (8.31) and (8.32) can be solved by existing convex tools, such as MATLAB CVX.

When considering 3D trajectory planning, we can potentially decouple the optimizations of the 2D horizontal trajectory and the altitude by using alternating optimization techniques. Given fixed altitudes of the UAV H_t, $\forall t$, the 2D trajectory can be designed using the proposed Algorithm 8. Given fixed 2D trajectory (and auxiliary variables), i.e., $\{\mathbf{q}_t, V_t, q_t, w_t, u_t, \forall t\}$, the varying UAV altitude $H_t, \forall t$, can be potentially optimized by solving a similar subproblem to (8.32) by replacing w_t and u_t with H_t in (8.32a), and H with H_t in (8.32d). The subproblem can be convexified and solved in the same way as problem (8.32), as described in Algorithm 8. The horizontal trajectory and the altitude can be optimized in an alternating manner until convergence. The detailed design and validation of the 3D trajectory planning will be presented in our subsequent study.

8.3.3 Algorithm Summary and Complexity Analysis

Algorithm 8 summarizes the steps proposed in this section to convexify and solve efficiently problem (8.17). The algorithm is computationally dominated by Steps 6 and 7, as Steps 3 to 5 compute linear functions with a complexity of $\mathcal{O}(N)$. After a series of mathematical manipulations and transformations, the original problem (8.17) is recast as (8.30) with a convex objective and non-convex constraints, and further decomposed into two subproblems, i.e., problems (8.31) and (8.32), to convexify the non-convex constraints. Problems (8.31) and (8.32) are convex programs, given their convex objective functions and constraints (i.e., all the constraints are either linear or quadratic). In this sense, Steps 6 and 7 each solve a convex program and update the optimization variables, i.e., by running the interior point method with the complexity of $\mathcal{O}(N^{3.5})$ per iteration [199]. The overall computational complexity of Algorithm 8 is $\mathcal{O}(N^{3.5})$ per iteration. Interested readers can refer to [38] for a detailed introduction to the interior point method and convex optimization in general.

Algorithm 8 The proposed trajectory planning algorithm for the cellular-aided radar sensing to problem (8.17)

1: **Initialization:** Generate a feasible initial flight path for the UAV, i.e., $\{\mathbf{q}_t^{(0)}, V_t^{(0)}, q_t^{(0)}, t = 1, \dots, T_w\}$, and input initial values for the slack variables $\{w_t^{(0)}, u_t^{(0)}, t = 1, \dots, T_w\}$.
2: **for** $\ell = 1, 2, \dots$ **do**
3: Approximate the propulsion power P_t by (8.20), (8.21b), and (8.22) by the SCA technique.
4: Introduce slack variables w_t and $u_t, \forall t$, and bound them by constraint (8.25).
5: Convexify constraint (8.17b) by (8.27) and (8.28), and rewrite constraint (8.17f) into a convex one (8.29).
6: Given fixed $\{w_t^{(\ell-1)}, u_t^{(\ell-1)}\}$, solve problem (8.31) by the interior point method to update the optimization variables $\{\mathbf{q}_t^{(\ell)}, V_t^{(\ell)}, q_t^{(\ell)}\}$ to design the UAV trajectory.
7: Given $\{\mathbf{q}_t^{(\ell)}, V_t^{(\ell)}, q_t^{(\ell)}\}$, solve problem (8.32) by the interior point method to update the slack variables $\{w_t^{(\ell-1)}, u_t^{(\ell-1)}\}$.
8: Update $\ell \leftarrow \ell + 1$.
9: **end for**

Let $\mathbf{x}$ and $\mathbf{x}^*$ denote a feasible solution and the optimal solution to problem (8.30), respectively. Also, let P denote the objective value of problem (8.30). As established in [198], given the convergence precision ϵ of the algorithm, i.e., $|P(\mathbf{x}) - P(\mathbf{x}^*)| \le \epsilon$, the interior-point method takes $\mathcal{O}\left(\log \frac{1}{\epsilon}\right)$ iterations before convergence. Therefore, Algorithm 8 has a polynomial complexity of $\mathcal{O}\left(N^{3.5} \log \frac{1}{\epsilon}\right)$.

8.4 Problem Formulation and DDPG-based Solution for Visual-based UAV

The UAV monitor is installed with a video camera to spot and identify the target of interest, which requires the representations of various visual variables, such as resolution and viewpoint [357]. To successfully tail the target, the UAV monitor needs to follow the target within a specific resolution range specified by $[p_{\min}, p_{\max}]$ and minimize the energy consumption during the mission time. Here, $p_{\min}$ and $p_{\max}$ are the smallest and largest perceived width/height of the target in pixels, respectively. Based on the resolution requirement, we can use the triangle similarity to determine the maximum and minimum allowed monitor-target distances, $d_{\max}$ and $d_{\min}$ (in meters), as given by [105]

$$d_{\max} = \frac{W_T \times F}{p_{\min}}, \tag{8.33a}$$

$$d_{\min} = \frac{W_T \times F}{p_{\max}}, \tag{8.33b}$$

where W_T is the actual width/height of the target, and F is the focal length of the onboard video cameras. The monitor optimizes its trajectory online by controlling its acceleration, to keep a certain distance $d_t = \|\boldsymbol{q}_t - \boldsymbol{b}_t\|$, $\forall t$ from the target

$$d_{\min} \leq d_t \leq d_{\max}, \ \forall t. \tag{8.34}$$

Let E_0 (in Joules) denote the energy level of the monitor's battery at the beginning of the mission. Suppose the monitor needs to maintain at least $(1 - \eta_0)E_0$ Joules at every moment in the battery for other functions and contingency plans, i.e.,

$$\sum_{n=1}^{t} P_v^n \delta \leq \sum_{n=1}^{t} P_s^n \delta + \eta_0 E_0, \ \forall t. \tag{8.35}$$

The battery constraint (8.35) is important, without which the UAV monitor could coarsely design its trajectory, consume all the battery energy at the beginning, and fail for the rest of the mission (as shown by simulations in Section 8.5.1). The studied problem is cast as

$$\min_{\{(\boldsymbol{q}_t - \boldsymbol{b}_t), \boldsymbol{v}_t, \boldsymbol{a}_t, \forall t\}} \sum_{t=1}^{N} P_v^t, \tag{8.36}$$

$$\text{s.t. } (8.13), (8.34), (8.35).$$

Here, we propose to optimize the relative position of the UAV monitor with reference to the target, i.e., $(\boldsymbol{q}_t - \boldsymbol{b}_t)$, as opposed to designing the monitor trajectory, i.e., $\boldsymbol{q}_t$. This helps avoid the infeasibility incurred by optimizing the monitor trajectory directly. Specifically, the range and distribution of the relative distance and velocity between the drone monitor and the target are expected to be reasonably consistent across different time slots for an effective tracking mission. In contrast, the range and distribution of the absolute positions of the UAV monitor would change dramatically over time. Moreover, the optimization of the relative distance and speed contributes to the reduction of the state space; or the absolute position and velocity of the target would be part of the state as the monitor's observations. The reduced state space is conducive to the training and convergence of the proposed DRL model.

DDPG is known to effectively tackle problems with continuous action spaces [244, 162]. In contrast, traditional DRL methods, such as deep Q-learning, perform poorly and incur divergence under a continuous action space. Based on (8.36), a novel DRL-based method is developed to compute the instantaneous acceleration of the monitor, which is the *action* of the designed control system. The proposed tracking problem has continuous action and state spaces, and can be solved by designing a DDPG-based method, as described in the next section. While we primarily focus on video input from onboard cameras in this paper, it is worth noting that the proposed DRL-based control mechanism for UAV tracking can be adapted to utilize alternative distance measurement techniques, such as lidar and radar, to determine the direction and distance of a target.

8.4.1 Markov Decision Process (MDP)

The UAV propulsion power consumption, i.e., the objective function in (8.36), is a non-convex function of the optimization variables, i.e., the UAV acceleration, velocity, and waypoint. In addition, these variables are tightly coupled in (8.36) and in the constraints (8.13) and (8.35) over time. As a result, the problem is non-convex and challenging to tackle.

On the other hand, the task is a sequential decision process and can be interpreted as an MDP, since the environment (or state) is observable and the future state only depends on the present state [27]. Specifically, the UAV's position at the $(t+1)$-th time slot depends on its position, velocity, and acceleration at the t-th slot; see (8.13a). The UAV's velocity at the $(t+1)$-th time slot depends on its velocity and acceleration at the t-th slot; see (8.13b). However, it is difficult to procure the optimal policy online for this MDP problem, due to the continuous state space of the problem [189]. The problem cannot be optimally solved using traditional control or optimization methods that would rely on accurate predictions of the target's trajectory, which may not be possible in practice.

We start by constructing an MDP for the considered target tracking problem and defining the state, action, and reward of the drone monitor per time slot [214].

- *State Space* $\mathcal{S}$: The current system state $s_t \in \mathcal{S}$ consists of the relative position, $\boldsymbol{q}_t - \boldsymbol{b}_t$, and the relative velocity, $\boldsymbol{V}_t - \boldsymbol{v}_t$, of the UAV monitor with regards to the target.
- *Action Space* $\mathcal{A}$: Define $\mathcal{A} := \{a_t, \forall t = 1, \dots, N\}$ to gather all potential actions. The current action a_t is the acceleration value of the UAV monitor, $\boldsymbol{A}_t$, constrained by (8.13c)–(8.13d). The values of the action are constrained to the range of $[-1, \ 1]$ (m/s^2). Given the initial location and velocity of the monitor, its future waypoints $\boldsymbol{q}_t$ and velocities $\boldsymbol{V}_t$ are decided by the accelerations, i.e., by (8.13a) and (8.13b).
- *Policy:* A policy, denoted by $\pi : \mathcal{S} \to \mathcal{A}$, is a mapping from the state space, $\mathcal{S}$, to the action space, $\mathcal{A}$. In other words, given a state $s \in \mathcal{S}$, the policy determines a distribution $\pi(a|s) = \Pr(a_t = a | s_t = s)$ over state $s \in \mathcal{S}$.

- *Experience:* Define $e_t = (s_t, a_t, r_t, s_{t+1})$ as an experience, which is stored in a replay memory $\boldsymbol{R}$.
- *Reward* r_t: The reward function offers non-negative rewards at each time slot (or step) if the monitor-target distance is within $[d_{\min}, d_{\max}]$ and the onboard battery level is no lower than $(1 - \eta_0)E_0$ Joules; or incurs penalties otherwise. The reward function is defined as

$$r_t = \underbrace{1 + C_{d1} \tanh(d_t - d_{\min}) + C_{d2} \tanh(d_{\max} - d_t)}_{\text{for distance-keeping}} + \underbrace{C_e \tanh\left(\eta_0 E_0 + \sum_{n=1}^{t} P_s^n \delta - \sum_{n=1}^{t} P_v^n \delta\right)}_{\text{for battery constraint}}, \tag{8.37}$$

where C_{d1}, C_{d2}, and C_e $(C_{d1}, C_{d2}, C_e \leq 1)$ are configurable coefficients that can be tuned during the learning process; and "tanh(·)" is used to scale the rewards and ensure that all the rewards are in similar magnitudes.

The state space and reward function are particularly designed for problem (8.36). In particular, we optimize the relative position of the monitor to the target, as opposed to directly designing the monitor's trajectory. This is because the range and distribution of the relative distance and velocity are expected to be reasonably consistent across different time slots for an effective tracking mission. In contrast, a direct design of the monitor's trajectory would suffer from a substantially changing value range and distribution of the absolute positions of the UAV monitor. On the other hand, the state $\boldsymbol{s}_t$ consists of the relative distance and velocity, the two of which can have substantially different ranges and distributions. A state vector can also be sparse. We adopt the max-abs normalization [98] to preprocess the state. The max-abs normalization does not destroy the original data distribution, and can rescale the value range of the data. Moreover, we perform the max-abs normalization separately for the relative distance and the relative velocity, to avoid the loss of state information due to substantial differences in values.

In practice, the target motion estimation using a camera can be influenced by the time-varying monitor-target distance, illumination condition, background characteristics, and target appearance. Yet, some commercially available UAVs can achieve a high-resolution view and an accurate target motion estimation with a camera array, such as Intel RealSense D435i. Yet, this type of camera may suffer from temperature change and noisy data when used outdoors. The accuracy of their distance estimation can be as good as within 0.28 m. We assume that the motion estimation of the target is reasonably accurate using such cameras and visual-aided methods.

In the reward, the "tanh(·)" function is selected due to its smooth and bounded nature. It can normalize each of the penalty terms, barrier the reward function, stabilize training, and facilitate the convergence of the algorithm. C_{d1}, C_{d2}, and C_e can be fine-tuned to account for the relative importance or priority of each of the penalty terms. Moreover, the reward function can be applied under cloudy weather conditions. For a particular time slot n, P_s^n follows (8.15) when the UAV monitor is exposed under the sun and harvests the solar power. P_s^n takes zero when the UAV monitor flies under a cloud and its harvested solar power diminishes.

Moreover, the shape and size of the target in an image depend on the image-capturing direction, and can change during the tracking process. The proposed algorithm has the potential to maintain a consistent view of the target object's shape and size by keeping the image capturing direction/angle within a predefined range relative to the target's movement direction while maintaining a safe distance between the drone and the target. This can be potentially achieved by introducing a new term in the reward function defined in (8.37) that incorporates the image capturing angle. The new term can provide a reward when the angle is within the specified range, or a penalty when it falls outside of it.

The agent senses the present state s_t, executes a legitimate action a_t, receives a reward r_t, and evolves to state s_{t+1}. A policy $a_t = \pi(s_t)$ projects s_t to a_t. The agent chooses the strategy that achieves the largest accumulated reward $R_t = \sum_{n=t}^{N} \gamma^{n-t} r_t$ with $\gamma \in (0, 1)$ denoting the discount coefficient. With the determined state s_t, action a_t and the strategy $\pi(\cdot)$, R_t is evaluated by an action-value function, i.e., the Q-function, as given by

$$Q_\pi(s_t, a_t) = \mathbb{E}_\pi[R_t | s_t, a_t], \tag{8.38}$$

The action-value function, $Q_\pi(s_t, a_t)$, meets the following Bellman Expectation Equation

$$Q_\pi(s_t, a_t) = \mathbb{E}_{r_t, s_{t+1} \sim \mathcal{E}} \left[r_t + \gamma \mathbb{E}_{a_{t+1} \sim \pi} \left[Q_\pi(s_{t+1}, a_{t+1}) \right] \right]. \tag{8.39}$$

Here, $\mathcal{E}$ represents the interacting environment. When this given policy, denoted by $\mu_\pi : \mathcal{S} \to \mathcal{A}$, is deterministic, (8.39) can be rewritten as

$$\begin{aligned} Q_{\mu_\pi}(s_t, a_t) = \mathbb{E}_{r_t, s_{t+1} \sim \mathcal{E}} \big[& r_t \\ & + \gamma Q_{\mu_\pi}(s_{t+1}, \mu_\pi(s_{t+1})) \big]. \end{aligned} \tag{8.40}$$

In general, it is difficult to apply an RL algorithm directly to handle the MDP or obtain the Q-value, $Q(s_t, a_t)$, because of the continuous state and action spaces, and the uncertain target trajectory. We design a DDPG-based algorithm to control the UAV monitor's trajectory, as described in the following.

8.4.2 Actor-critic Framework-based DDPG

In the DDPG-based network, four DNN approximators are used, including training-actor and training-critic networks, and target-actor and target-critic networks, as illustrated by Figure 8.3. The training-actor network, represented by $\mu\,(s_t; \theta_a)$, approaches the strategy of the UAV monitor and generates actions. θ_a stands for the parameters of the training-actor network. The training-critic network, denoted by $Q_\mu(s_t, a_t; \theta_c)$, estimates the action-value function of the actions [244]. θ_c stands for the model parameters of the training-critic network. The target-actor network, represented by $\mu'\,(s_t; \theta'_a)$, and the target-critic network, represented by $Q'_{\mu'}(s_t, a_t; \theta'_c)$, generate the target Q-value for learning the training-actor and training-critic networks. θ'_a and θ'_c stand for the model parameters of the target-actor and target-critic networks, respectively.

Relying on the actor-critic setting, the DDPG network abides by the deterministic policy gradient (DPG) theorem [244] to refresh θ_a, θ_c, θ'_a, and θ'_c, and improve the actions. The adoption of the target network (containing the target-actor and target-critic networks) helps address an issue of oscillating operation stemming from employing only a training network [104]. We consider fully CNNs (FCNNs) with two hidden layers for each training-actor, training-critic, target-actor, and target-critic network.

The UAV monitor (i.e., the agent) passes its current state s_t to the training-actor network. In the DPG theorem [244], the training-actor network yields explicitly the present tactics by definitely mapping a state into an action. The training-actor network approaches the strategy function of the agent and chooses an action a_t.

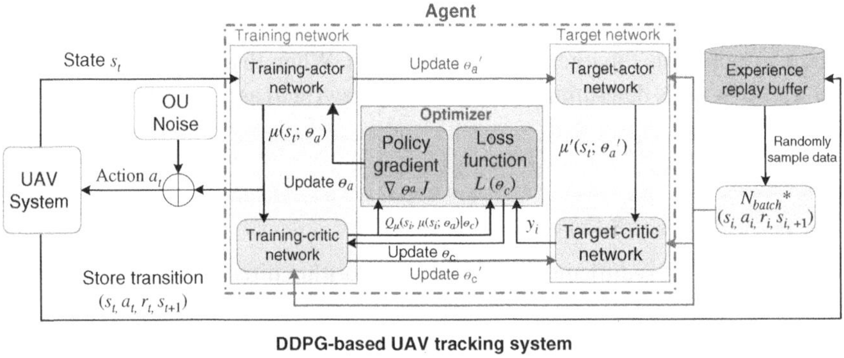

Figure 8.3 The architecture of the proposed DDPG-based UAV tracking system, where the training network and the target network each comprise an actor network and a critic network. The experience replay buffer provides batches of samples of state transitions to train and update the networks.

Exploration noises are attached to the action to balance new action exploration and known action exploitation. The output action is given by

$$a_t = \mu\,(s_t; \theta_a) + \mathcal{N}_t, \tag{8.41}$$

where $\mathcal{N}_t$ is an Ornstein–Uhlenbeck (OU) noise process that generates temporally correlated exploration in the physics domain to improve exploration efficiency [265]. As a result of the action a_t, the monitor is rewarded with r_t and evolves to state s_{t+1}. It also stores (s_t, a_t, r_t, s_{t+1}) in $\boldsymbol{R}$.

The training-critic network evaluates the action-value function of the executed action a_t, that is, $Q_\mu\,(s_t, \mu(s_t; \theta_a); \theta_c)$. By taking a recorded transition (s_i, a_i, r_i, s_{i+1}) at random from $\boldsymbol{R}$, the action-value function generated at the training-critic network is approximately evaluated as $Q_\mu(s_i, \mu(s_i; \theta_a); \theta_c)$.

We assume a probability distribution of θ_a, denoted by $J(\theta_a)$, for policy estimation. To improve the strategy fastest, the training-actor network is refreshed based on the gradient of $J(\theta_a)$. The gradient is given by [244]

$$\nabla_{\theta_a} J(\theta_a) = \mathbb{E}_{s\sim\rho^\mu}\left[\nabla_{\theta_a}\mu(s_t; \theta_a)\nabla_a Q_\mu(s_t, a; \theta_c)|_{a=\mu(s_t;\theta_a)}\right], \tag{8.42}$$

where ρ^μ stands for a discounted state distribution of strategy $\mu(s_t; \theta_a)$ [162]; $\nabla_{\theta_a}\mu(s)$ provides the gradient of the training–actor network $\mu(s)$ regarding the parameter θ_a; $\nabla_a Q_\mu(s_t, a; \theta_a)$ stands for the gradient of $Q_\mu(s_t, a; \theta_a)$ regarding action a.

By randomly drawing N_{batch} sampled historical transitions from $\boldsymbol{R}$, $\nabla_{\theta_a} J(\theta_a)$ is approximated by

$$\nabla_{\theta_a} J(\theta_a) \approx \frac{1}{N_{\text{batch}}} \sum_{i=1}^{N_{\text{batch}}} \left[\nabla_{\theta_a}\mu(s_i)\nabla_a Q_\mu(s_i, a; \theta_c)|_{a=\mu(s_i)}\right]. \tag{8.43}$$

The model parameter of the training-actor network, θ_a, is refreshed based on the gradient ascent [259]

$$\theta_a \leftarrow \theta_a + \frac{\eta_a}{N_{\text{batch}}} \sum_{i=1}^{N_{\text{batch}}} \left[\nabla_{\theta_a}\mu(s_i)\nabla_a Q_\mu(s_i, a; \theta_c)|_{a=\mu(s_i)}\right]. \tag{8.44}$$

Here, η_a stands for the learning rate of the training-actor network.

The training-critic network is refreshed by achieving the minimum loss, as given by

$$L(\theta_c) = \mathbb{E}_{s_t\sim\rho^\mu, a_t\sim\mu(s_t;\theta_a)}\left[\left(Q_\mu\,(s_t, a_t; \theta_c) - y_t\right)^2\right]. \tag{8.45}$$

Here, $y_t = r_t + \gamma Q'_{\mu'}\,(s_{t+1}, \mu'\,(s_{t+1}; \theta'_a)\,; \theta'_c)$ is the target Q-value yielded by the target network according to (s_t, a_t, r_t, s_{t+1}). Here, the parameters of the target-actor and target-critic networks, θ'_a and θ'_c, provide declined copies of θ_a and θ_c, respectively.

With the N_{batch} stochastically sampled evolutions, the loss function, $L(\theta_c)$, is approximately evaluated by

$$L(\theta_c) \approx \frac{1}{N_{\text{batch}}} \sum_{i=1}^{N_{\text{batch}}} \left[\left(Q_{\mu} \left(s_i, \mu(s_i; \theta_a); \theta_c \right) - y_i \right)^2 \right]. \tag{8.46}$$

Here, $y_i = r_i + \gamma Q'_{\mu'} \left(s_{i+1}, \mu' \left(s_{i+1}; \theta'_a \right); \theta'_c \right)$ approximates the target Q-value from the target network according to N_{batch} samples drawn randomly from the replay memory. Differentiating $L(\theta_c)$ with respect to θ_c, we obtain the gradient as

$$\begin{aligned} \nabla_{\theta_c} L(\theta_c) \approx \frac{2}{N_{\text{batch}}} \sum_{i=1}^{N_{\text{batch}}} & \left[\left(Q_{\mu} \left(s_i, \mu(s_i; \theta_a); \theta_c \right) - y_i \right) \right. \\ & \left. \cdot \nabla_{\theta_c} Q_{\mu}(s_i, \mu(s_i; \theta_a); \theta_c) \right]. \end{aligned} \tag{8.47}$$

The parameter of the training-critic network, θ_c, is refreshed utilizing the stochastic gradient descent method [259].

The target-actor and target-critic networks evolve from the training-actor and training-critic networks based on the following rule

$$\begin{aligned} \theta'_a &\leftarrow \tau_a \theta_a + (1 - \tau_a) \theta'_a, \\ \theta'_c &\leftarrow \tau_c \theta_c + (1 - \tau_c) \theta'_c, \end{aligned} \tag{8.48}$$

where τ_a and τ_c are decaying rates for the training-actor and training-critic networks, respectively.

The proposed DDPG-based UAV-target tailing approach is summarized in Algorithm 9. The algorithm contains model initialization, model training, and model updates. We first randomly initialize the four networks, including the training-actor and training-critic networks, target-actor and target-critic networks, and the experience replay buffer. The relative locations and velocities of the UAV monitor and target are the input state of the algorithm. Given the present state of the UAV monitor, the action is produced through model training. The model parameters are updated with historical transitions from the replay memory until the maximum cumulative reward is obtained. Finally, the monitor executes the action, changes its state, and decides whether to stop the training.

8.4.3 Complexity and Optimality

To evaluate the time-complexity of Algorithm 9, we consider the time-complexity of an FCNN with two hidden layers, and the mini-batch gradient ascent (8.43). The time-complexity of the FCNN with two hidden layers is $T_{\text{fc}} = \mathcal{O}\left(\sum_{l=1}^{2} F_{\text{in}}^{l} \cdot F_{\text{out}}^{l}\right)$, where F_{in}^{l} and F_{out}^{l} are the input and output sizes of fully connected layer l [236].

The mini-batch gradient ascent has the time complexity of $T_{\text{bgd}} = \mathcal{O}\left(N_{\text{batch}}/\epsilon_0\right)$, where ϵ_0 is the accuracy requirement to terminate the iterations [129]. Therefore, the time-complexity of Algorithm 9 is $T_{\text{fc}} + T_{\text{bgd}} = \mathcal{O}\left(\sum_{l=1}^{2} F_{\text{in}}^{l} \cdot F_{\text{out}}^{l} + N_{\text{batch}}/\epsilon_0\right)$.

The optimality of the proposed DDPG-based algorithm can be established by proving that the DDPG algorithm satisfies the conditions under the optimality of the policy gradient (PG) methods. It was established in [33] that the optimality gap of PG is bounded by

$$L(\pi_\theta) - \min_{\pi\in\Pi} L(\pi) \leq -\frac{\kappa_\rho}{1-\gamma} \min_{\upsilon\in\Theta}\left[c\langle\nabla_\theta L(\theta), \upsilon - \theta\rangle + \frac{u}{2}\|\upsilon - \theta\|^2\right], \quad (8.49)$$

Algorithm 9 DDPG-based UAV-based tracking

Initialization:

Randomly initialize training-actor network μ and training-critic network Q_μ with weights θ_a and θ_c, target-actor network μ' and target-critic network $Q'_{\mu'}$ with weights $\theta'_a \leftarrow \theta_a$ and $\theta'_c \leftarrow \theta_c$, and $\boldsymbol{R}$.
for episode = 1, ..., T_{ep} **do**
    Initialize the relative locations and velocities between the UAV monitor and the target as the initial state s_0.
    for timestep = 1, ..., T_s **do**
        Select an action $a_t^1 = \mu\left(s_t; \theta_a\right) + \mathcal{N}_t$.
        Operate action a_t^1, procure reward r_t and transits to next state s_{t+1}.
        Store current evolution (s_t, a_t, r_t, s_{t+1}) in $\boldsymbol{R}$.
        Stochastically draw N_{batch} sampled historical evolutions (s_i, a_i, r_i, s_{i+1}) from $\boldsymbol{R}$.
        Refresh the target Q-value: $y_i = r_i + \gamma Q'_{\mu'}\left(s_{i+1}, \mu'\left(s_{i+1}; \theta'_a\right); \theta'_c\right)$.
        Calculate the loss function:
        $L(\theta_c) \approx \frac{1}{N_{\text{batch}}} \sum_{i=1}^{N_{\text{batch}}} \left[\left(Q_\mu\left(s_i, \mu(s_i; \theta_a); \theta_c\right) - y_i\right)^2\right]$.
        Update the critic network by minimizing the loss function:
        $\min_{\theta_c} \frac{1}{N_{\text{batch}}} \sum_{i=1}^{N_{\text{batch}}} \left[\left(Q_\mu\left(s_i, \mu(s_i; \theta_a); \theta_c\right) - y_i\right)^2\right]$.
        Update the actor network by the sampled policy gradient:
        $\nabla_{\theta_a} J(\theta_a) \approx \frac{1}{N_{\text{batch}}} \sum_{i=1}^{N_{\text{batch}}} \left[\nabla_{\theta_a}\mu(s_i)\nabla_a Q_\mu(s_i, a; \theta_c)|_{a=\mu(s_i)}\right]$;
        $\theta_a \leftarrow \theta_a + \eta_a \nabla_{\theta_a} J(\theta_a)$.
        Update the target-actor and target-critic networks:
        $\theta'_a \leftarrow \tau_a\theta_a + (1-\tau_a)\theta'_a$, $\theta'_c \leftarrow \tau_c\theta_c + (1-\tau_c)\theta'_c$.
    end for
end for

---

in which $L(\cdot)$ is the loss function. κ_ρ is the concentrability coefficient defined for the class of cost-to-go functions $\mathcal{J}_\Theta = \{J_{\pi_\theta} : \theta \in \Theta\}$, which is the smallest scalar to satisfy

$$\|J - J^*\|_{1,\rho} \leq \frac{\kappa_\rho}{1-\gamma}\|J - TJ\|_{1,\rho}, \forall J \in \mathcal{J}_\Theta, \tag{8.50}$$

where $T : \mathcal{J} \to \mathcal{J}$ is the Bellman optimality operator

$$(TJ)(s) \triangleq \min_{a \in \mathcal{A}} \left[g(s,a) + \int J(s')P(ds'|s,a)\right]. \tag{8.51}$$

The optimality gap holds when the following conditions are met.

Condition 1: For any $\theta \in \Theta$, the functions $\bar{\theta} \mapsto \mathcal{B}(\bar{\theta}|\eta_{\pi_\theta}, J_{\pi_\theta})$ and $\bar{\theta} \mapsto \mathcal{B}(\theta|\eta_{\pi_{\bar{\theta}}}, J_{\pi_\theta})$ are continuously differentiable on an open set containing θ.

Condition 2: For each $\pi \in \Pi_\Theta$, there exists $\pi^+ \in \Pi_\Theta$, such that $\mathcal{B}(\pi^+|\eta_\pi, J_\pi) = \min_{\pi' \in \Pi} \mathcal{B}(\pi'|\eta_\pi, J_\pi)$.

Condition 3: For any $\pi \in \Pi_\Theta$, the function $\mathcal{B}(\theta|\eta_\pi, J_\pi)$ has no suboptimal stationary points.

Condition 4: For each $\pi \in \Pi_\Theta$, there exist constants $c > 0$ and $u \geq 0$, such that the function $\mathcal{B}(\theta|\eta_\pi, J_\pi)$ satisfies the following property

$$\begin{aligned} &\min_{\theta' \in \Theta} \mathcal{B}(\theta'|\eta_\pi, J_\pi) \geq \mathcal{B}(\theta|\eta_\pi, J_\pi) \\ &+ \min_{\theta' \in \Theta}\left[c\langle \nabla \mathcal{B}(\theta|\eta_\pi, J_\pi), \theta' - \theta\rangle + \frac{u}{2}\|\theta' - \theta\|^2\right]. \end{aligned} \tag{8.52}$$

Conditions 1 and **3** imply that the loss function L is continuously differentiable with $\nabla L(\theta) = \nabla_{\bar{\theta}} \mathcal{B}\left(\bar{\theta}|\eta_{\pi_\theta}, J_{\pi_\theta}\right)|_{\bar{\theta}=\theta}$, $\theta \in \Theta$ presents a stationary point of $L(\cdot)$ if $L(\pi_\theta) = L(\pi^*)$ [33, Lemma 6], and $\mathcal{B}\left(\theta|\eta_{\pi_\theta}, J_{\pi_\theta}\right) = \min_{\bar{\theta} \in \Theta} \mathcal{B}\left(\bar{\theta}|\eta_{\pi_\theta}, J_{\pi_\theta}\right)$. **Condition 4** is the gradient-dominance property of the loss function, which guarantees quick global convergence of first-order approaches, even with non-convex objectives.

We confirm that the loss function in (8.46) satisfies the four conditions for the following reasons:

- The UAV propulsion power in (8.14), the harvested solar power in (8.15), and the monitor-target distance in (8.34) are all continuously differentiable functions defined in a closed set, hence satisfying **Conditions 1** to **3**.
- The state and action spaces of problem (8.36) are compact and lie in a closed set, as the states and actions are constrained by the UAV's highest acceleration and velocity, thus meeting **Condition 2**.
- The distance keeping constraint and battery constraint in (8.35) are lower- and upper-bounded linear, quadratic, or smooth functions that are continuously differentiable and satisfy the gradient-dominance property. Hence, **Condition 4** is satisfied.

- Moreover, the training network consists of feedforward FCNNs. Their activation functions are twice continuously differentiable, for example, Rectified Linear Units (ReLUs) and Sigmoid, hence satisfying **Conditions 1** and **2**.

Therefore, the solution obtained by the proposed method is bounded by a finite optimality gap.

The proposed DRL-based method offers the advantage of both offline training and online testing. Offline training boasts computational efficiency, quicker convergence, and optimal utilization of available data. While the initial parameter setting during offline training is important, online testing enables the agent to adapt and refine its policy in real-time. Specifically, online testing facilitates continuous improvement of the policy by evaluating its performance and making adjustments on-the-fly-based on real-time feedback. To this end, the combination of offline training and online testing empowers the DRL-based method to effectively address the challenges of dynamic environments.

8.5 Simulation and Experiment

8.5.1 Simulation Results for UAV-JCAS Systems

This section provides the simulation results of the proposed trajectory planning algorithm for BS-UAV bistatic SAR using MATLAB. The entire sensing period lasts $T = 300$ s with each time slot of $\delta = 0.5$ s, unless stated otherwise. The default setting of the landmarks (and associated sensing time) is a series of points with the first 15 of them, i.e., when $t \leq 150$ s, along the y-axis starting from (0, 0), and the rest along the x-axis starting from $\mathbf{q}_g^{16}$. The adjacent landmarks are apart by an equal distance. The UAV initial location is $\mathbf{q}_0 = [-1000, -500]^T$ m. The BS is located at $\mathbf{q}_b = [1000, -1000]^T$ m, with the height $H_b = 50$ m. The other parameters concerning the BS-UAV bistatic SAR sensing performance and the UAV propulsion power are provided in Table 8.1.

We note that the standard sub-6 GHz band has been extensively deployed to provide broad 5G coverage, where the antennas are typically horizontally or quasi-horizontally oriented in the sub-6 GHz band [61]. The transmission bandwidth is typically narrow in the sub-6 GHz band, e.g., up to 100 MHz, and consequently, the range resolution could be poor [2]. Nevertheless, a bandwidth of 150 MHz can be achieved in a standard sub-6 GHz band by using carrier aggregation techniques [130]. For instance, China Mobile deployed a 4G and 5G concurrent integrated network supporting a transmission bandwidth of up to 160 MHz at the 2.6 GHz band in 2019 [1]. On the other hand, the 5G new radio (NR) has also specified the use of mmWave bands. The propagation of mmWave signals can be limited within a close range of the BS, subject to antenna orientation (i.e., downtilt) [61]. Nevertheless,

Table 8.1 The parameters of the BS-UAV bistatic SAR [187, 336].

Parameter	Value
Transmission bandwidth, B	150 MHz
Wavelength, λ_c	0.1 m
Coherent integration time, T_d	1 s
SAR observation angle, η	$\pi/4$
Minimum object distance, $d_{\min}$	20 m
UAV weight and altitude, H	2 kg, 1000 m
Maximum UAV speed, V_m	50 m/s
Blade profile power and tip speed, P_0 and U_{tip}	3.4 W, 60 m/s
Rotor induced power and velocity, P_1 and v_0	118 W, 5.4 m/s
Rotor solidity and disc area, s and A	0.02, 0.5 m^2
Air density and fuselage drag fraction, ρ and d_f	1.225 kg/m^3, 0.3

the signals can still be reflected by objects (e.g., with smooth metallic surfaces) or the edges of objects [86, 126], and captured by the UAV. Given the quasi-optical property of mmWave signals [85], the ground reflections (and reflections by other non-smooth surfaces, e.g., walls, vegetation, etc.) are expected to be substantially weaker, resulting in strong contrast to manifest objects [85].

We also note that no existing studies have addressed the problem of energy-efficient trajectory design for the considered UAV-based bistatic SAR system, as discussed in Section 8.1.1. In other words, no existing algorithm is directly comparable to the proposed algorithm. With due diligence, we come up with a new baseline scheme for the considered problem, which minimizes the total flight distance of the UAV without considering the energy consumption of the UAV, i.e., $\min_{\mathbf{q}_t} \sum_t \|\mathbf{q}_t\|$. The baseline has a convex objective function, while its constraints can be convexified in the same way as done in Algorithm 8.

Figure 8.4a shows the expected shortest distance between any two objects in the area, $d_{\min}$, versus the radius of the given area $\mathcal{A}$, R (in meters), when there are $N = 800,\ 1000$, or 1200 objects in $\mathcal{A}$. It can be seen that $d_{\min}$ increases with R, given the number of objects in the area; i.e., the objects are more sparsely distributed when the area is larger. When $R = 5000$ m, the expected shortest distance between any two objects in the area $d_{\min}$ is about 20 m. Therefore, we choose the value of $d_{\min} = 20$ m. Figure 8.4b depicts the convergence of the proposed scheme when the landmarks are separated by 30 and 200 m under two settings.

It can be seen that the objective function of problem (8.30), i.e., (8.30a), can quickly converge within about only 20 iterations. When the two neighboring landmarks are apart for a longer distance, the UAV consumes less energy and shows

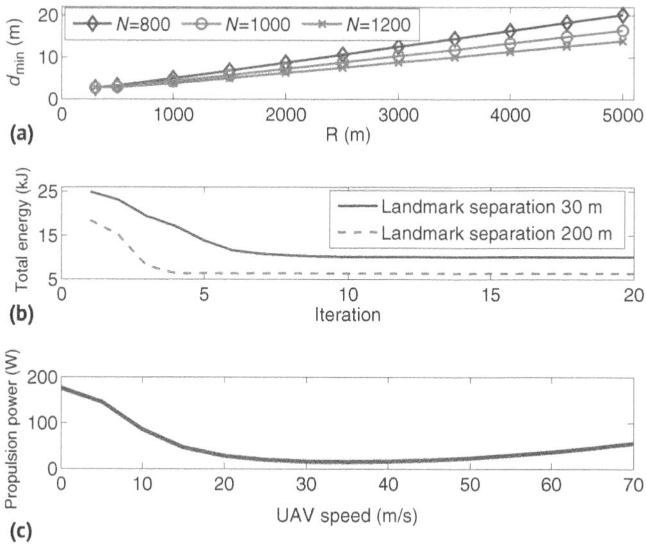

Figure 8.4 (a) Plots the expected shortest distance between any two objects in the area, d_{min}, vs. the radius of the given area, R; (b) plots the convergence of the objective value (8.30a) under the default setting; and (c) plots the UAV propulsion power vs. speed.

better energy efficiency. This is because the UAV can move faster, and the propulsion power first decreases and then increases with the increasing speed of the UAV, as shown in Figure 8.4c. Figure 8.4c plots the UAV power consumption P_t in (8.6) by varying the instantaneous speed of a rotary-wing UAV from 0 to 70 m/s. The UAV's propulsion power is independent of its trajectory and heading. The UAV consumes the least power when its speed is about 35 m/s, validating the results in Figure 8.4b.

Figure 8.5 shows the per-slot energy consumption of the rotary-wing UAV conducting the BS-UAV bistatic SAR sensing under the proposed and baseline schemes, where the default setting of the landmarks is considered, i.e., with an equal distance of 30 or 200 m between adjacent landmarks. It is revealed that the minimization of the flight distance does not necessarily lead to the minimization of the UAV's energy consumption. When the landmarks are separated by 30 m, the proposed scheme can save 55% of the total energy, as compared to the baseline.

It is interesting to notice that the baseline scheme is better than the proposed algorithm in terms of per-slot UAV power consumption when $t \leq 50$ seconds and the spacing between adjacent landmarks is 30 m. This is because the objective of the proposed algorithm, i.e., (8.17a), is to minimize the total energy consumption. As a result, the algorithm has the UAV fly slowly away from a landmark (while still keeping the landmark within its effective sensing area) at the beginning of the

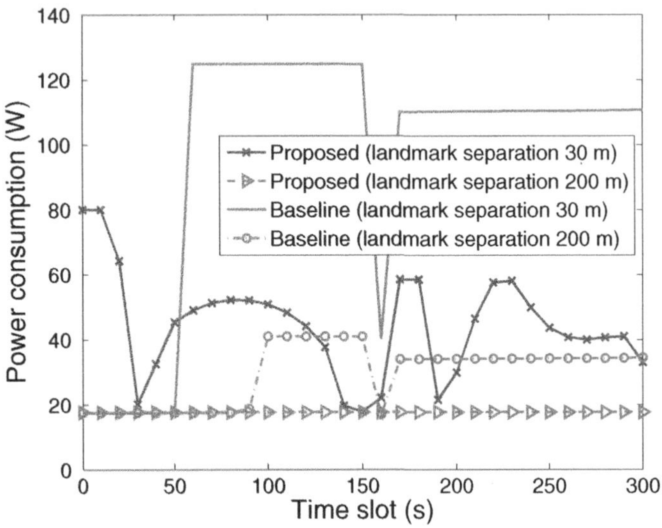

Figure 8.5 The UAV per-slot energy consumption by the proposed and baseline schemes when the landmarks of interest are separated by 30 and 200 m, under the default setting.

sensing mission to benefit the later stage of the mission in Figure 8.5. The UAV has to restrain its speed at the beginning of the mission at the cost of high energy consumption, as will be shown in Figure 8.6. On the other hand, the baseline scheme requires the UAV to fly fast toward the landmarks in the beginning and then slow down to minimize the total flight distance to sense all landmarks. As a result, the baseline scheme consumes less energy than the proposed algorithm at the beginning of the mission. Nevertheless, the proposed algorithm is much more energy-efficient when the entire mission is considered.

Figures 8.6 and 8.7 plot the UAV trajectory (the red dashed line with triangle markers) and its sensing performance under the proposed and baseline schemes. The black and magenta circles provide some examples of the effective sensing areas at different time slots. The green triangle shows the location of the BS. We see in Figures 8.6 and 8.7 that the UAV can always capture the landmarks within its sensing area at the time slots required. Under the proposed schemes, the UAV seeks an energy-efficient flight path, and is better adapted to the distribution of the landmarks. In contrast, under the baseline schemes, the UAV first catches up with the nearest landmark at its highest speed, and then flies at a relatively stable speed to satisfy the resolution requirements. This can cause drastic changes in the instantaneous power consumption of the UAV under the baseline method, as already shown in Figures 8.5.

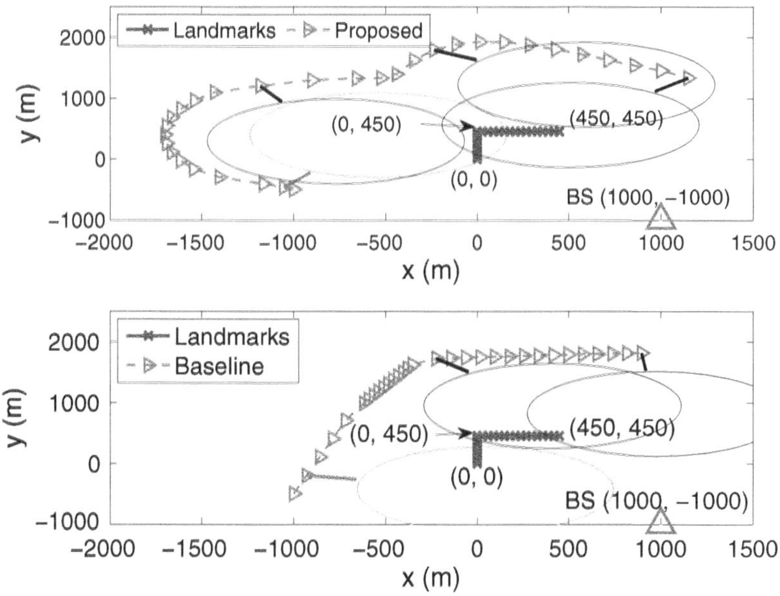

Figure 8.6 The UAV trajectory by the proposed and baseline schemes, where the landmarks of interest are separated by 30 m, and $H = 1000$ m.

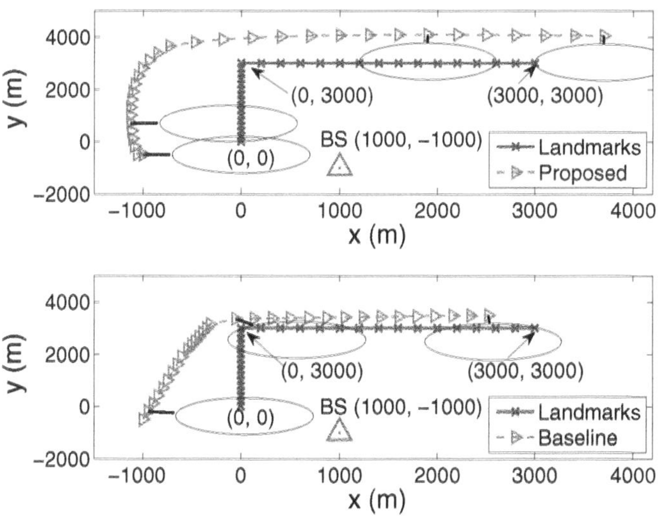

Figure 8.7 The UAV trajectory by the proposed and baseline schemes, where the landmarks of interest are separated by 200 m, and $H = 1000$ m.

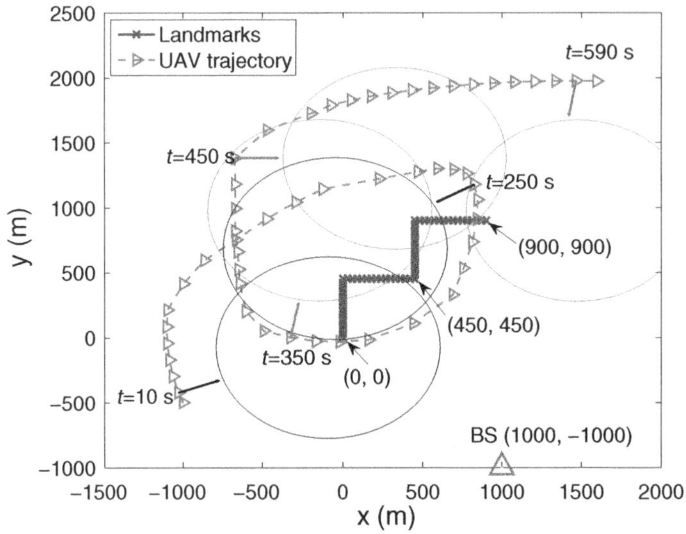

Figure 8.8 The UAV trajectory by the proposed scheme, where the landmarks of interest are separated by 30 m, $H = 1000$ m, and $T = 600$ s.

Figure 8.8 plots the UAV trajectory when $T = 600$ s. The landmarks are separated by 30 m and distributed in a staircase (that is, two street blocks situated along a diagonal). The effective sensing area of the BS-UAV bistatic SAR is indicated by the black, magenta, or green circles, when $t = 10$ s, 250 s, 350 s, 450 s, and 590 s, respectively. It is shown in Figure 8.8 that the BS-UAV bistatic SAR can always capture the landmarks in its sensing coverage at the required time slots.

Figure 8.9 plots the UAV trajectory when the landmarks are separated by 100 m and distributed in a square (e.g., a closed loop around a triangular street block), $T = 600$ seconds, and $H = 1000$ m (Figure 8.9a) or 1200 m (Figure 8.9b). The effective sensing area of the BS-UAV bistatic SAR is shown when $t = 10$, 170, 300, 320, 450, and 530 seconds, under the UAV elevation of $H = 1000$ m, and when $t = 10$, 220, 440, and 580 seconds, under the UAV elevation of $H = 1200$ m. It is observed that the UAV intends to fly around the landmarks for both effective sensing and energy saving. The UAV trajectory of the corresponding baseline scheme is shown in Figure 8.9b when $H = 1200$ m; see the green dashed line with circle markers. The effective sensing area of the UAV is illustrated when $t = 250$ s. We see that the UAV takes a much shorter flight path around the landmarks to satisfy the sensing resolution requirements, which is at the cost of a much higher propulsion energy consumption, as shown in Figure 8.10.

Figure 8.10 plots the per-slot energy consumption of the proposed scheme when the landmarks are distributed in a staircase (i.e., the two diagonally situated street

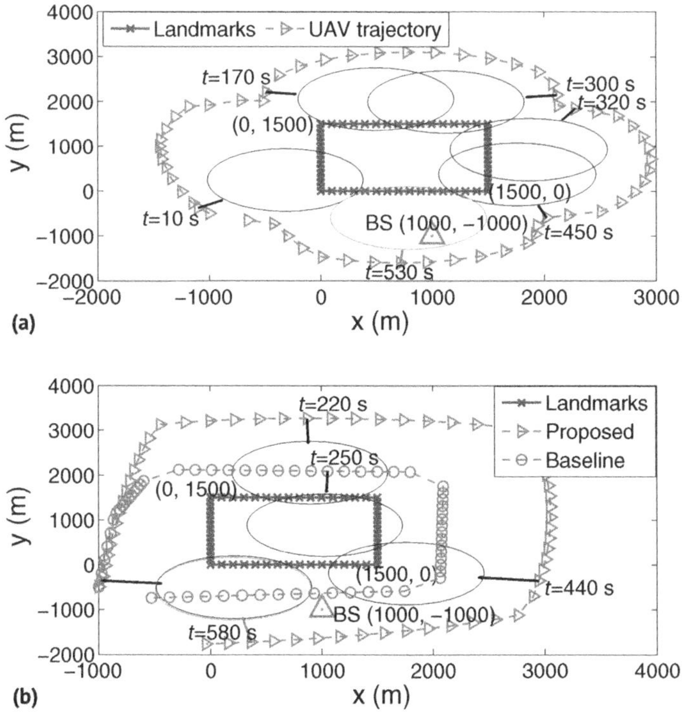

Figure 8.9 The UAV trajectory by the proposed scheme, where the landmarks of interest are separated by 100 m and $T = 600$ s. (a) $H = 1000$ m; (b) $H = 1200$ m.

blocks) or a square (i.e., the closed loop of a triangular street block), with the spacing of 30 and 100 m, respectively. Here, $T = 600$ seconds, and $H = 1000$ or 1200 m. We see that the UAV consumes less energy when the landmark spacing or the UAV altitude is larger, as the UAV has more flexibility to design an energy-efficient trajectory and meet the sensing requirements. The per-slot energy consumption of the baseline approach is also plotted for the landmarks arranged on the square with the spacing of 100 m, and the UAV elevation is $H = 1200$ m. We see that the proposed scheme consumes dramatically lower energy than the baseline, and prevents drastic fluctuations in the per-slot consumption of the UAV propulsion energy.

Figure 8.11 plots the UAV trajectory and per-slot energy consumption of the proposed and baseline schemes, When the spacing between neighboring landmarks obeys a random uniform distribution within [0, 100] m. The averaged results of 20 independent realizations of the landmark locations are plotted. While the UAV

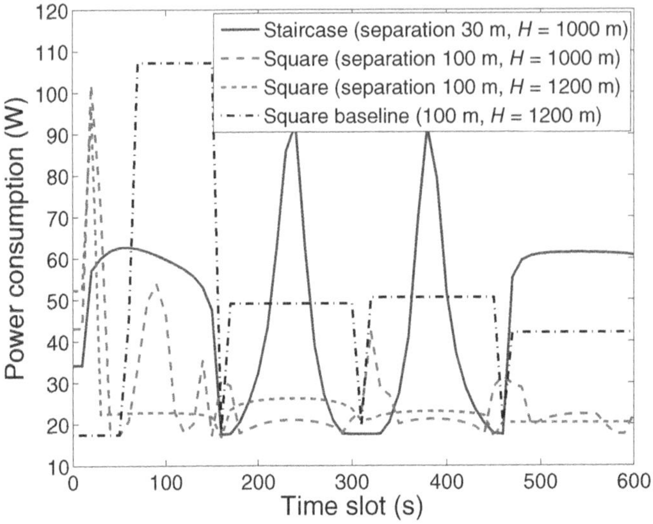

Figure 8.10 The UAV per-slot energy consumption by the proposed scheme when the landmarks of interest are distributed like a staircase (separated by 30 m, $H = 1000$ m), and a square (separated by 100 m, $H = 1000$ and 1200 m), and $T = 600$ s.

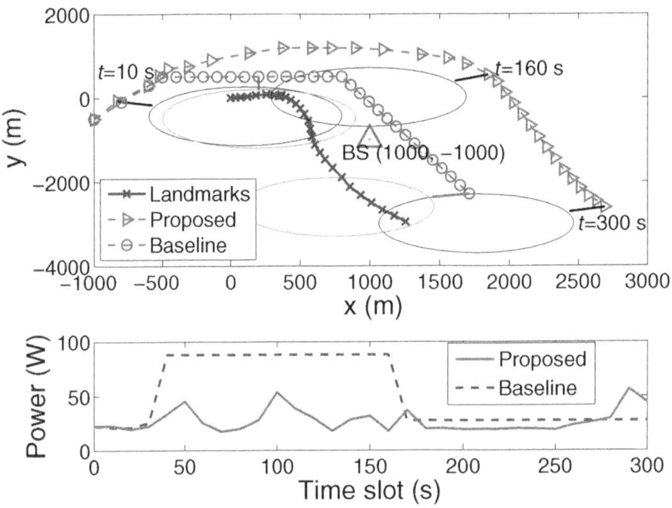

Figure 8.11 The UAV trajectory and per-slot energy consumption by the proposed and baseline schemes when the landmarks of interest are distributed randomly, $H = 1000$ m and $T = 300$ seconds.

Table 8.2 The range resolutions of the BS-UAV bistatic SAR under different transmission bandwidths, where the transmitter incident angle is $\theta_t = \pi/3$.

Transmission bandwidth, B (MHz)	50	100	150	200
Range resolution, δ_r (m)	3.81	1.91	1.27	0.95

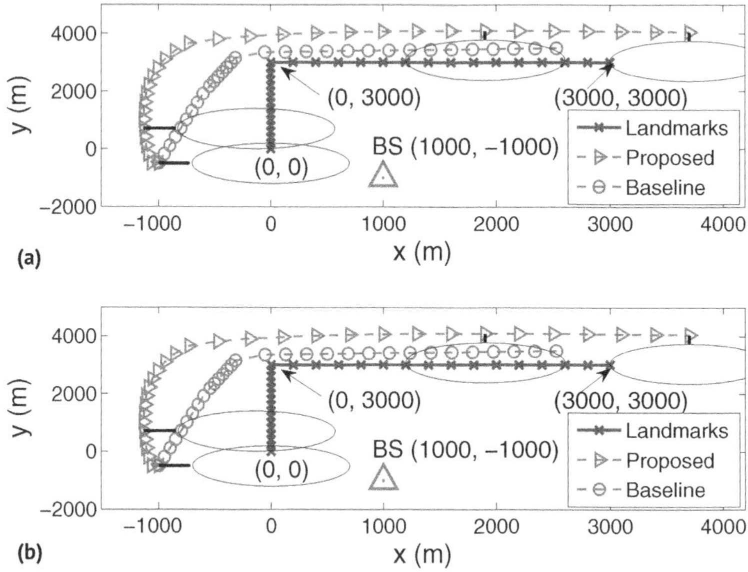

Figure 8.12 The UAV trajectory by the proposed and baseline schemes when the landmarks of interest are separated by 200 m, and the transmission bandwidth is 50 and 100 MHz under the default setting. (a) $B = 50$ MHz; (b) $B = 100$ MHz.

flies a longer distance under the proposed algorithm than it does under the baseline approach, as shown in the top of the figure, the energy consumption is substantially lower and undergoes much smaller fluctuations under the proposed algorithm, as shown in the bottom of the figure. This again corroborates the merits of the proposed scheme over the baseline scheme in terms of energy efficiency and energy consumption fluctuation.

Last but not least, we examine the performance of the proposed BS-UAV bistatic SAR under different transmission bandwidths B in the sub-6 GHz band and the mmWave band. Table 8.2 shows the range resolutions of the SAR, where B is 50 MHz, 100 MHz, and 150 MHz at the sub-6 GHz band, and 200 MHz at the mmWave band. Here, θ_t is assumed to be $\pi/3$. We see that the resolutions are always better than the expected shortest distance between any two objects, $d_{\min}$; in other words, the SAR can effectively distinguish ground objects. Figure 8.12 shows the UAV trajectory by the proposed and baseline schemes, where the landmarks are separated by 200 m, and the transmission bandwidth B is 50 and 100 MHz. The other settings are consistent with those in Figure 6. By comparing Figure 8.12 to Figure 8.6, we see that the UAV trajectory has similar patterns while satisfying the sensing accuracy.

8.5.2 Experiment Results for Visual-based UAV

This section provides Python experimental results to corroborate the merits of the proposed method. We include the reward value during learning to show the convergence of the DRL algorithm. We test the learned policy under different target trajectories and compare it with an MPC-based approach to further justify the advantage of DRL methods over traditional control- and optimization-based approaches. We also consider the scenarios with and without the battery constraint on the UAV monitor to show the effect of energy-aware target tracking. The details of the schemes, scenarios, and simulation results are provided in the ensuing sections.

8.5.2.1 Experiment Settings

The actor networks are implemented by FCNNs with two hidden layers, and their learning rates are 10^{-4}. The first layer has 32 neurons. The second layer has 64 neurons. The output layer of the actor networks takes the $\tanh(\cdot)$ activation function that bounds the output actions within $[-1, 1]$. The critic networks use the FCNNs with two hidden layers, and their learning rates are 10^{-3}. Both hidden layers use the ReLU activation function. The first layer has 32 neurons. The second layer has 64 neurons. The OU noise is added to the actor policy, sampled from the OU process with parameters $\mu = 0$, $\theta = 0.15$, and $\sigma = 0.25$. The training of the policy network is done on a server with an Nvidia Tesla P100 SXM2 16GB GPU. Table 8.3 collates the hyperparameters.

We include the performance of the proposed scheme under a practical parameter setting to verify its effectiveness for real-world applications. In particular, we set the power conversion efficiency and size of the solar panel as $\eta = 0.2$ and $S = 0.5$ m^2, respectively [46]. We consider a UAV of 10 kg inclusive of a battery [331]. Typically, a solar panel of 1 m^2 weighs about 2.8 kg [362]. Therefore, when the UAV is equipped with a solar panel of 0.5 m^2, its weight is 11.4 kg. Interested readers can refer to [242] for an in-depth study on the solar power management system of a UAV. The other parameters regarding the monitor's power and the solar power are listed in Table 8.4.

The maximum monitor-target distance for effective target detection depends on the focal length of the UAV camera, where a longer focal length results in a larger distance. It was experimentally demonstrated in [274] that a UAV-based visual tracking system can detect and recognize a moving target within a distance of

Table 8.3 Summary of hyperparameters.

Parameters	Descriptions	Values
γ	Discounting factor for future reward	0.99
η_c	Learning rate for critic network	0.001
η_a	Learning rate for actor network	0.0001
τ_c	Decaying rate for critic network	0.001
τ_a	Decaying rate for actor network	0.001
	Buffer size for experience replay	10^6
T_{ep}	Number of episodes	2000
T_s	Total number of steps in each episode	4000
N_{batch}	Number of experiences in the mini-batch	64

Table 8.4 The parameters for fixed-wing UAV propulsion power and harvested solar power.

Parameter	Value
UAV weight	10 kg [331]
Coefficients, c_1, c_2	9.26×10^{-4}, 2250 [332]
Maximum/minimum speed, $V_{\max}, V_{\min}$	100 m/s, 10 m/s
Maximum acceleration, $A_{\max}$	8 m/s^2
Gaseous absorption, α	0.8978 [131]
Solar power intensity, P_i	1367 W/m^2 [131]
Efficiency and size of solar panel, η, S	0.2, 0.5 m^2 [46]
Ratio of battery energy reserve, η_0	0.1
Elevation of the target and monitor, h_0, h_1	10 m, 10 m
Battery capacity, E_0	750 Joules

40 meters. In light of [274], we set the maximum and minimum monitor-target distances to 40 and 2 m, respectively.

8.5.2.2 Results of Policy Learning

In the training process, the average reward for the i-th training episode, denoted by $\bar{r}_i$, is calculated by taking the average of the step reward from the first episode up to the i-th episode, i.e., $\bar{r}_i = \frac{1}{i}\sum_{j=1}^{i} r_j, i = 1, \dots, T_{ep}$, where r_j is the step reward for the j-th training episode; see (9.40).

The proposed model is trained using the "random-linear" target trajectory (along the x-axis) for each episode. We assume that the target travels along the x-axis at an initial velocity of $\boldsymbol{v}_0 = [15, 0]$ m/s. A time-varying acceleration

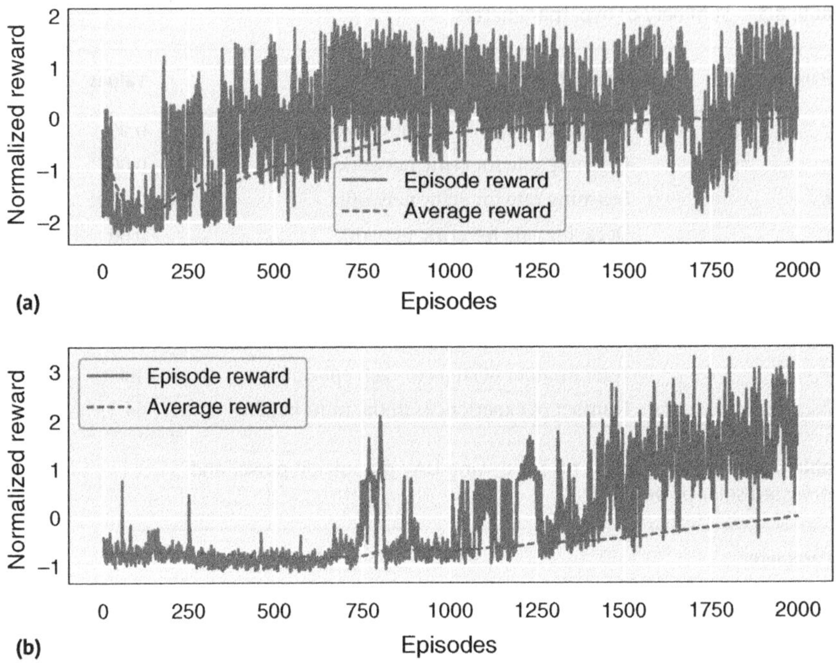

Figure 8.13 Episode and average rewards with and w/o the battery constraint. (a) Reward with battery constraint; (b) Reward without battery constraint.

$\boldsymbol{B}_t := (B_{xt}, B_{yt}), \forall t$, is randomly chosen from $[-1, 1]$ for each time step, i.e., $\boldsymbol{B}_t \in U(-1, 1)$, where $U(\cdot)$ denotes the uniform distribution. The UAV monitor has an initial velocity of $\boldsymbol{V}_0 = [15, 15]$ m/s. The total scheduling period is $T = 120$ s with 0.05 s per slot. The performance of the drone monitor without (w/o) the battery constraint is examined as a baseline.

Figure 8.13a plots both the per-episode and the average rewards of our algorithm. For comparison, Figure 8.13b plots the per-episode reward and the average reward w/o the battery constraint. Figure 8.13 demonstrates that both the performances of our proposed algorithm with and w/o the battery constraint improve over time. With the battery constraint, the proposed algorithm performs better and learns much faster than w/o the battery constraint, and reaches the maximum reward at the 1067th episode. The average reward gradually increases and reaches its peak around the 1400th episode, as the UAV control policy changes with a randomly generated target trajectory in each episode. Furthermore, the reward function (9.40) takes an important role in the learned policy. The parameters C_{d1}, C_{d2}, and C_e have to be carefully tuned to allow the UAV monitor to learn the desired behavior.

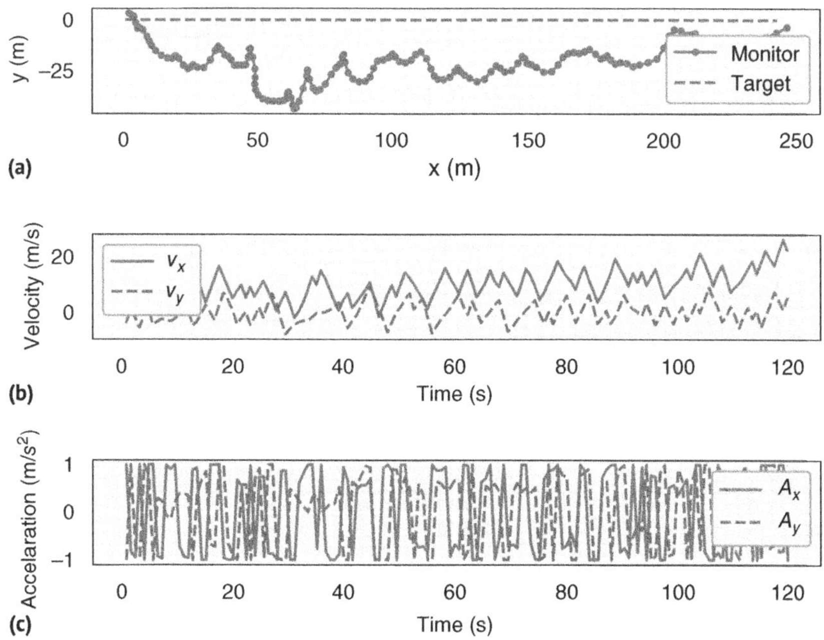

Figure 8.14 Trajectory, velocity, and acceleration of the best control policy for the UAV monitor (Episode 1067). (a) Trajectories of the UAV monitor and the target; (b) Velocity over time; (c) The output action (Acceleration).

Figure 8.14 plots the trajectory, velocity, and acceleration of the control policy learned at the 1067th episode for the UAV monitor over time under the "random-linear" target trajectory. Figure 8.15 shows the corresponding monitor-target distance, energy consumption of the monitor, and the residual battery level over time. As shown in both figures, the UAV monitor can quickly adapt its control policy (i.e., the acceleration) to the random target trajectory, effectively track the target and keep a reasonable distance from it. The UAV monitor can maintain the required amount of energy in the battery for future use.

8.5.2.3 Test Results of Learned Policy

The network parameters obtained at the 1067th training episode are selected to test the learned policy. With the learning results from Section 8.5.2.2, we test the proposed energy-efficient DDPG-based tracking algorithm under three different types of target trajectories over 100 episodes, namely, the "random-linear," "random," and "sinusoidal" target trajectories, as shown in Figures 8.16–8.18. The model parameters of the "random-linear" target trajectory are set up consistently with those in Section 8.5.2.2. The UAV monitor has an initial velocity of $\boldsymbol{V}_0 = [15, 15]$ m/s. Each episode consists of 6000 time slots (steps). No exploration noise is added during the testing process.

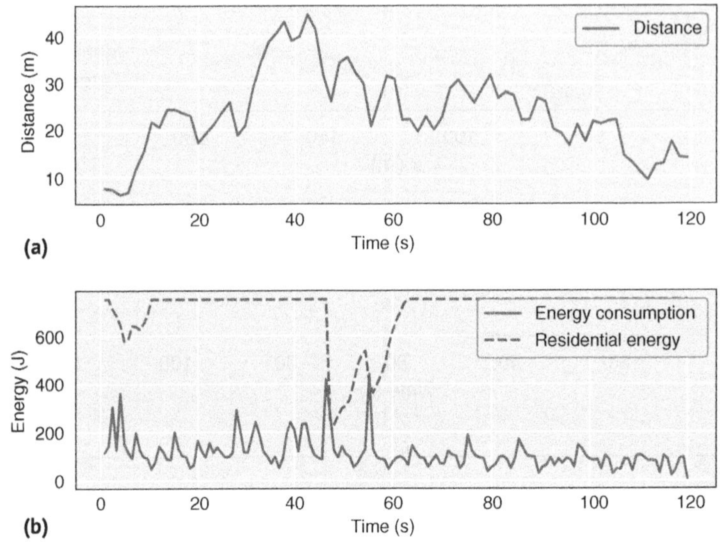

Figure 8.15 Monitor-target distance, energy consumption, and battery level of the best control policy for the UAV monitor (Episode 1067). (a) Distance between the UAV monitor and the target over time; (b) Power consumption over time.

As a state-of-the-art online trajectory plan technique, the MPC-based online optimization combines rigorous and efficient convex optimization-based trajectory generation with control-based bias/error mitigation (cf. [124, 238] using numerical heuristics to generate trajectories). The MPC-based online optimization approach is implemented every time slot τ by first predicting the target trajectory $\{\boldsymbol{b}_{\tau+t}, \forall t = 1, \dots, N\}$. As typically considered, the Kalman filtering technique is adopted to forecast the target trajectory relying on the past observations [19]. Then, the MPC plans the monitor trajectory for N steps ahead. Nevertheless, only the first step of the devised trajectory is executed. The forecast horizon keeps being shifted forward at each (new) time slot $\tau + 1$.

In Figures 8.16–8.18, the tested trajectories of the UAV monitor are indicated by blue solid lines under DDPG and orange dashed-dot lines under MPC, and the target trajectories are indicated by green dashed lines. It is seen that the new algorithm can successfully follow the target under all three target trajectories, even when the time-varying acceleration is randomly chosen from $[-3, 3]$; see Figures 8.16 and 8.17. The trajectories of the UAV and target do not match perfectly (or overlap); see Figures 8.16 and 8.18. This is because the UAV keeps the target in its sight for effective tracking and maintains a reasonable distance from the target, rather than heading directly towards the target.

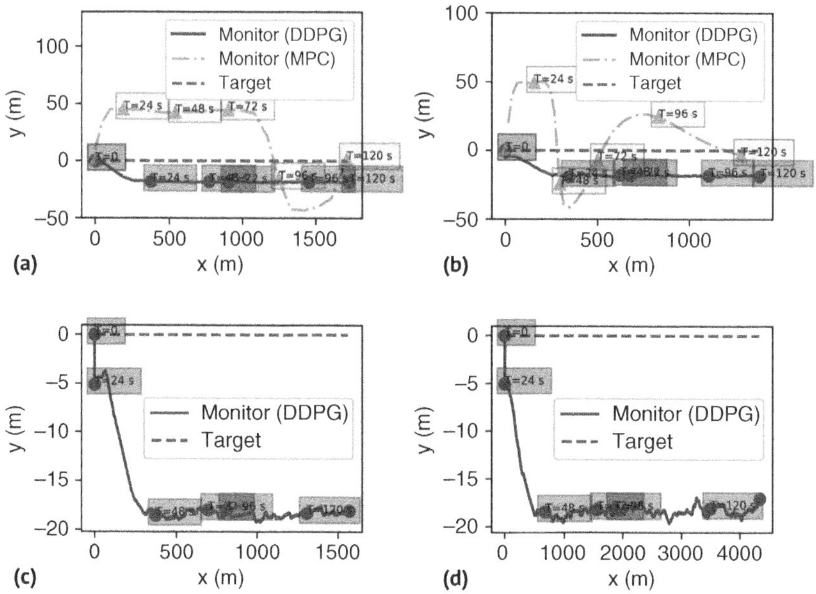

Figure 8.16 Linear target trajectory (along the x-axis) with accelerations randomly chosen from $[-1, 1]$ and/or $[-3, 3]$ m/s^2. (a) $\eta = 0.2$, $S = 0.2$, and $B_{xt} \in [-1, 1]$; (b) $\eta = 0.2$, $S = 0.2$, and $B_{xt} \in [-3, 3]$; (c) $\eta = 0.1$, $S = 0.4$, and $B_{xt} \in [-1, 1]$; (d) $\eta = 0.1$, $S = 0.5$, and $B_{xt} \in [-3, 3]$.

In Figures 8.16–8.18, we consider $\eta = 0.1$ and $S = 0.4$ or 0.5 m^2 to show the effectiveness of the DDPG-based algorithm under different energy harvesting capabilities, and understand the impact of the energy conversion rate and solar panel size on the tracking. The UAV monitor with a stronger energy conversion capability η and a larger size of solar panel S can have a higher probability of successful tracking.

We consider $\eta = 0.2$ and $S = 0.5$ m^2 when plotting the MPC-based algorithm. We do not include $\eta = 0.1$ (and $S = 0.4$ or 0.5 m^2) because the MPC-based algorithm is prone to infeasibility under this setting. Specifically, the MPC plans the UAV monitor's trajectory based on the estimation of the target's trajectory over longer time horizons. The estimation can be poor at the beginning of tracking, due to the lack of past observations. As a consequence, the trajectory planning can be heavily penalized, leading to a need for excessively high energy for drastic adjustments of the trajectory in the first several decision rounds. When η is small, i.e., $\eta = 0.1$, the energy available is likely to fail to meet the need, and the MPC-based approach can become infeasible.

Figure 8.19 plots the monitor battery level under three types of target trajectories under the DDPG and MPC algorithms, when $(\eta, S) = (0.2, 0.5)$. We see from Figures 8.16–8.19, that the UAV takes different flight paths under the MPC scheme, and consumes more energy than under the DDPG-based algorithm. This

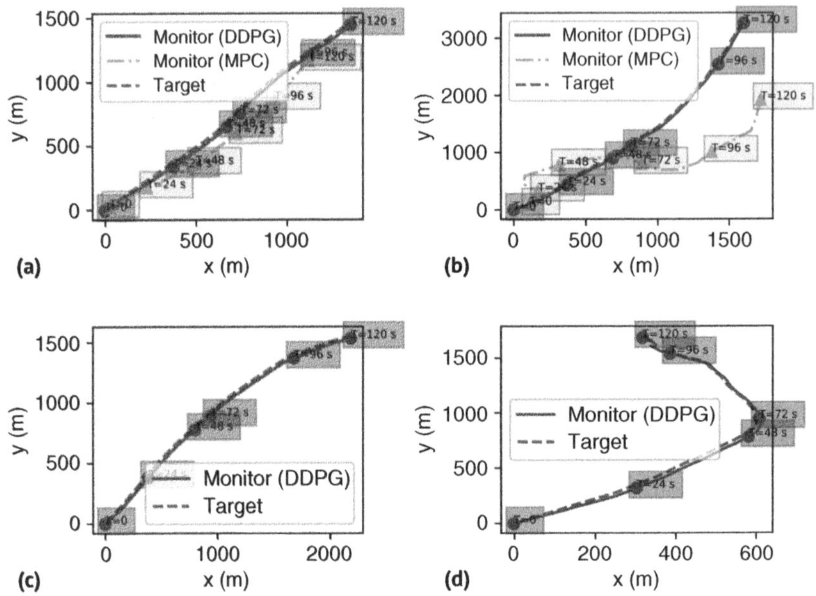

Figure 8.17 Random target trajectory with accelerations randomly chosen from $[-1, 1]$ m/s^2 and/or $[-3, 3]$ m/s^2. (a) $\eta = 0.2$, $S = 0.5$, and $(B_{xt}, B_{yt}) \in [-1, 1]$; (b) $\eta = 0.2$, $S = 0.5$, and $(B_{xt}, B_{yt}) \in [-3, 3]$; (c) $\eta = 0.1$, $S = 0.4$, and $(B_{xt}, B_{yt}) \in [-1, 1]$; (d) $\eta = 0.1$, $S = 0.4$, and $(B_{xt}, B_{yt}) \in [-3, 3]$.

is due to inaccurate predictions of the target trajectory, and the suboptimal control policy resulting from the SCA.

Figures 8.20 and 8.21 plot the changes in the monitor's battery level and the monitor-target distance during tracking, respectively, where the efficiency and size of the solar panel are $(\eta, S) = (0.1, 0.4)$. Figure 8.20 shows that, when $(\eta, S) = (0.1, 0.4)$, the monitor's battery maintains the initially charged level of 750 J under the "random" target trajectories with $\boldsymbol{B}_t \in U(-1, 1)$. This indicates that the harvested solar energy supports the UAV monitor effectively. In the rest of the scenarios studied in Figure 8.20, the battery level drops below 500 J, indicating the insufficiency of the harvested solar energy for UAV propulsion. Figure 8.21 shows that the average monitor-target distance increases quickly to about 20 m first, and then stabilizes around 20 m during tracking.

We show that the proposed scheme can be readily extended from the 2D UAV target tracking to the 3D. In this case, with slight abuse of notation, the constraints of 3D mobility in (1) concern 3D variables, including the UAV waypoint $\boldsymbol{q}_t := (x_t, y_t, z_t)$, velocity $\boldsymbol{V}_t := (V_{xt}, V_{yt}, V_{zt})$, and acceleration $\boldsymbol{A}_t := (A_{xt}, A_{yt}, A_{zt})$. In addition, the maximum pitch angle constraint has to be satisfied when the drone is ascending or descending, i.e., $V_{zt}/\|\boldsymbol{V}_t\| \leq \vartheta$, where ϑ is the sine value of the largest

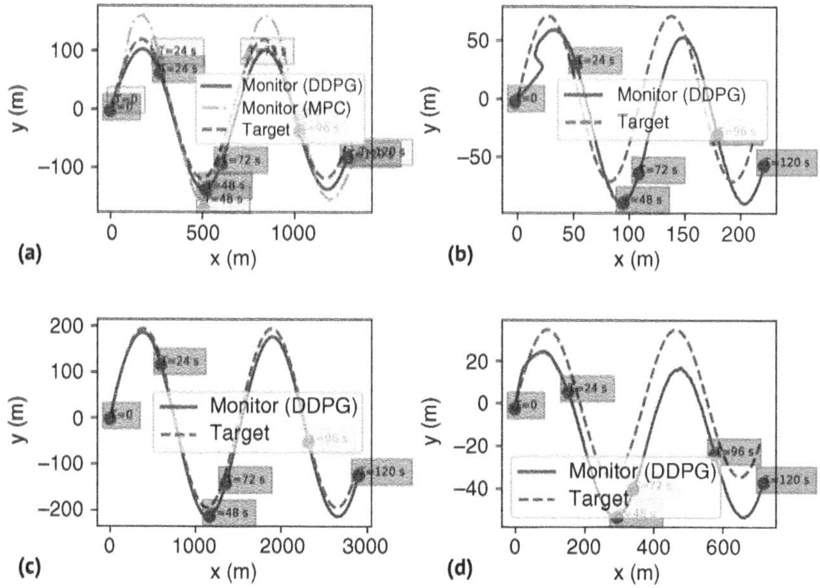

Figure 8.18 Sinusoidal target trajectory with accelerations randomly chosen from $[-1, 1]$ m/s^2. (a) $\eta = 0.2$, $S = 0.5$; (b) $\eta = 0.2$, $S = 0.4$; (c) $\eta = 0.1$, $S = 0.5$; (d) $\eta = 0.1$, $S = 0.4$.

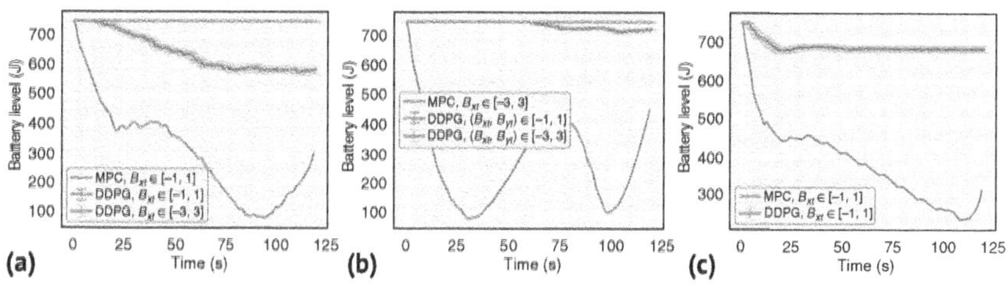

Figure 8.19 Monitor battery level under three types of target trajectories by the DDPG and MPC algorithms when $\eta = 0.2$ and $S = 0.5$. (a) Random-linear trajectory; (b) Random trajectory; (c) Sinusoidal trajectory.

allowable pitch angle of the drone. The propulsion power usage for the UAV 3D flight is [298]

$$P_v^t = c_1\|\boldsymbol{V}_t\|^3 + \frac{c_2(V_{xt}^2 + V_{yt}^2)}{\|\boldsymbol{V}_t\|^3} + \frac{c_2\|\boldsymbol{A}_t\|^2}{g^2\|\boldsymbol{V}_t\|} + \frac{2c_2A_{zt}}{g\|\boldsymbol{V}_t\|} + m(gV_{zt} + \boldsymbol{A}_t^T\boldsymbol{V}_t). \tag{8.53}$$

We plot the 3D UAV trajectory and the corresponding battery level under the proposed scheme in a 3D scenario in Figures 8.22 and 8.23, where the solar panel efficiency and size are set to $(\eta, S) = (0.2, 0.5)$. The UAV acceleration is 3D and

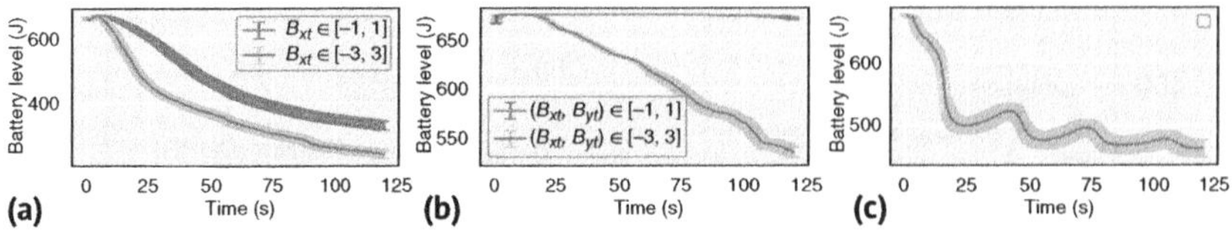

Figure 8.20 Monitor battery level under three types of target trajectories when $\eta = 0.1$ and $S = 0.4$ m^2. (a) Random-linear trajectory; (b) Random trajectory; (c) Sinusoidal trajectory.

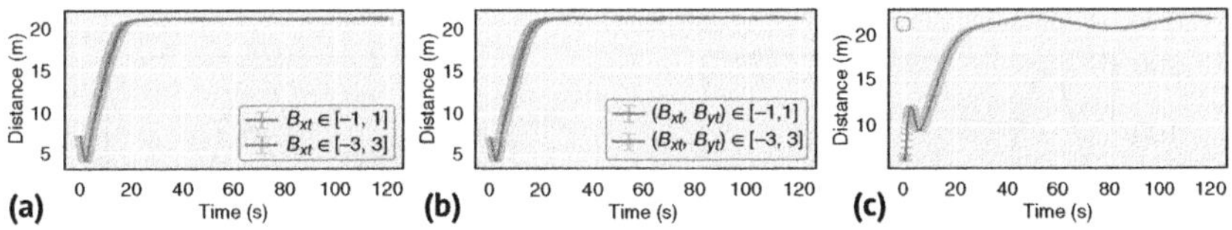

Figure 8.21 The distance-keeping performance of the UAV monitor under three types of target trajectories when $\eta = 0.1$ and $S = 0.4$ m^2. (a) Random-linear trajectory; (b) Random trajectory; (c) Sinusoidal trajectory.

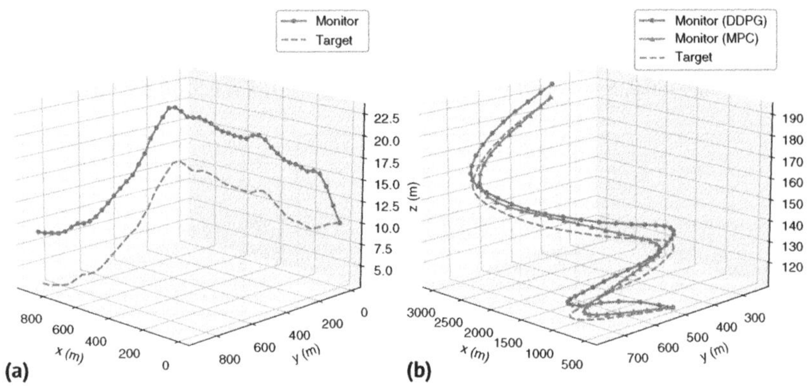

Figure 8.22 A demonstration of the 3D trajectories generated, when $(\eta, S) = (0.2, 0.5)$. (a) 3D Random trajectory; (b) 3D Sinusoidal trajectory.

can change continuously along each of the x-, y-, and z-axes under the proposed method. By default, we set the acceleration within $[-1, 1]$ m/s^2 along each of the x-, y-, and z-axes. It is seen that the proposed DDPG-based algorithm outperforms the MPC-based algorithm under the sinusoidal target trajectory, providing more effective tracking and better energy efficiency.

We note that controlling a fixed-wing UAV through its 3D acceleration is feasible due to the inherent connection between acceleration and the key aerodynamic forces involved in flight. By influencing the UAV's acceleration, one indirectly

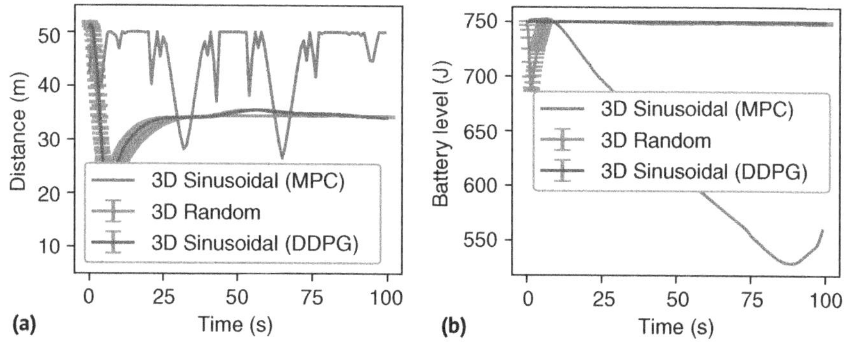

Figure 8.23 Monitor-target distance and battery level under 3D random and sinusoidal trajectories of the UAV monitor. (a) Monitor-target distance; (b) Monitor battery level.

manages the interplay of thrust, lift, drag, and weight, which are fundamental to flight dynamics. The throttle control, linked to the propulsion system, governs the UAV's longitudinal acceleration by adjusting thrust. Elevator control influences the pitch, affecting the balance between lift and weight for changes in altitude. Ailerons control roll, adjusting the distribution of lift between wings, facilitating turns. Rudder control manages yaw, enabling coordinated turns by manipulating the aircraft's heading. These control inputs collectively shape the forces and moments acting on the UAV. The ability to control a fixed-wing UAV through its 3D acceleration can leverage these aerodynamic principles to achieve responsive flight control.

8.5.2.4 Comparison with Learned Policy w/o Battery Constraint

As discussed in Section 8.5.2.2, our DDPG-based energy-efficient algorithm with the battery constraint helps improve the learning efficiency of the UAV monitor. Now, we further test the tracking capability of the proposed algorithm either with or w/o the battery constraint, using the learned policy in Section 8.5.2.2. Different sizes of the solar panel are considered to gauge the impact of the inflow energy volume on the proposed DDPG-based algorithm and its resulting UAV trajectory. A typical flexible, lightweight solar panel is about 3 kg/m^2.

Tables 8.5 and 8.6 list the successful tracking probability (STP) of the proposed DDPG-based algorithm with and w/o the battery constraint (8.35) under the "random-linear" and "random" target trajectories, respectively, over 100 testing episodes. We see that the proposed algorithm offers a higher STP with the battery constraint (8.35) than it does w/o the battery constraint, under both target trajectories. This is due to the fact that, with the battery constraint (8.35), the UAV monitor carefully designs its trajectory to maintain the minimum energy reserve. In contrast, without (8.35), the UAV monitor consumes all battery energy at an early stage and fails the mission.

Table 8.5 Comparison of tracking capability between the DDPG-based algorithms with and w/o battery constraint under the "random-linear" target trajectory.

Random-linear trajectory					
Initial velocity (UAV)	Acceleration (Target)	Efficiency of solar panel η	Size of solar panel S	Successful tracking probability (%)	
				with battery constraint	w/o battery constraint
$V_0 = [15, 0]$	$[-1, 1]$	0.2	0.5	100	95
			0.4	100	93
			0.35	98	91
		0.1	0.5	87	65
			0.4	60	30
			0.35	32	11
	$[-3, 3]$	0.2	0.5	75	58
			0.4	72	57
			0.35	67	38
		0.1	0.5	61	30
			0.4	43	9
			0.35	12	6
$V_0 = [25, 0]$	$[-1, 1]$	0.2	0.5	100	100
			0.4	100	100
			0.35	100	100
		0.1	0.5	100	99
			0.4	99	95
			0.35	83	82
	$[-3, 3]$	0.2	0.5	96	86
			0.4	93	83
			0.35	91	77
		0.1	0.5	82	61
			0.4	57	39
			0.35	32	17

The UAV monitor has higher STPs when the solar panel has a higher efficiency η or a larger size S, since the monitor can harvest more solar energy and improve flexibility in trajectory planning. The STP also increases with a decrease in the target randomness and/or an increase in the monitor's initial velocity, i.e., $\boldsymbol{B}_t \in U(-1, 1)$ and/or $\boldsymbol{V}_0 = [25, 0]$ m/s, as compared to $\boldsymbol{B}_t \in U(-3, 3)$ and/or

$\boldsymbol{V}_0 = [15, 0]$ m/s. This is because the monitor enjoys higher flexibility in the trajectory design when $\mathbf{B}_t \in U(-1, 1)$ and/or $\mathbf{V}_0 = [25, 0]$ m/s. In addition, Table 8.6 also shows that when $(\eta, S) = (0.1, 0.3)$, the STP of the monitor is zero. This failed target tracking is the result of an energy deficiency, as the monitor consumes all energy and falls before catching up with the target.

Table 8.7 summarizes the STP of the UAV monitor by the MPC-based online control with and w/o the battery constraint under the "random" target trajectory with $\boldsymbol{B}_t \in [-3, 3]$, over 100 realizations of the random target trajectories. By comparing Tables 8.6 and 8.7, we see that the new DDPG-based algorithm outperforms the MPC-based control in terms of successful target tracking. This is because the feasibility of the SCA-based convexification in the MPC-based control can be impacted or restrained by the initial trajectory of the monitor, resulting in less effective UAV control and a higher probability of mission failure.

Table 8.6 Comparison of tracking capability between the DDPG-based algorithms with and w/o battery constraint under the "random" target trajectory.

Random trajectory				
Acceleration (target)	**Efficiency of solar panel η**	**Size of solar panel S**	**Successful tracking probability (%)**	
			with battery constraint	**w/o battery constraint**
[−1, 1]	0.2	0.5	100	100
		0.4	100	100
		0.35	100	100
		0.3	100	90
	0.1	0.5	100	95
		0.4	100	57
		0.35	98	6
		0.3	49	0
[−3, 3]	0.2	0.5	100	97
		0.4	99	90
		0.35	95	84
		0.3	87	67
	0.1	0.5	92	53
		0.4	83	20
		0.35	55	9
		0.3	49	0

Table 8.7 Comparison of tracking capability between the MPC-based control algorithms with and w/o battery constraint under the "random" target trajectory.

Random Trajectory			
Efficiency of solar panel η	**Size of solar panel S**	**Successful tracking probability (%)**	
		with battery constraint	**w/o battery constraint**
0.2	0.5	93	88
	0.4	82	76
	0.35	77	65
	0.3	75	53
0.1	0.5	67	21
	0.4	63	10
	0.35	50	0
	0.3	45	0

By comparing the simulation results in [110, Figures 9, 10, and 13], it is observed that the rotary-wing UAV monitor maintains the maximum monitor-target distance for most of the time, as opposed to a moderate monitor-target distance kept by a fixed-wing counterpart, as shown in Figure 8.21. In addition, the UAV monitor with rotary wings consumes much more energy than its counterpart with fixed wings. This again corroborates the merits of the proposed DDPG-based algorithm over the existing MPC-based control.

We also implement the proposed algorithm in a popular, computer-based robotic simulator, i.e., CoppeliaSim robot simulator, to demonstrate the effectiveness of real-world UAV tracking. CoppeliaSim can be regarded as a software-in-the-loop (SITL) verification tool and has been broadly used to demonstrate novel designs and concepts of UAV applications [223]. We connect Python with the CoppeliaSim simulator and implement our proposed method by training the DDPG model embedded in CoppeliaSim using the reward function (8), the relative monitor-target distance-based state space, and the acceleration-based action space. The CoppeliaSim simulator provides the instantaneous locations of the target and monitor, based on which the relative target-monitor distance is obtained to evaluate (8).

The flight trajectories of the UAV and the target in the CoppeliaSim simulator are available at https://drive.google.com/drive/folders/1Qoiq_uWEzG0ATP77diZcI1R0DHOCkKFb, with a screenshot provided in Figure 8.24. A small simulation area is displayed due to a small default acceleration range of [−0.1, 0.1] m/s^2 in

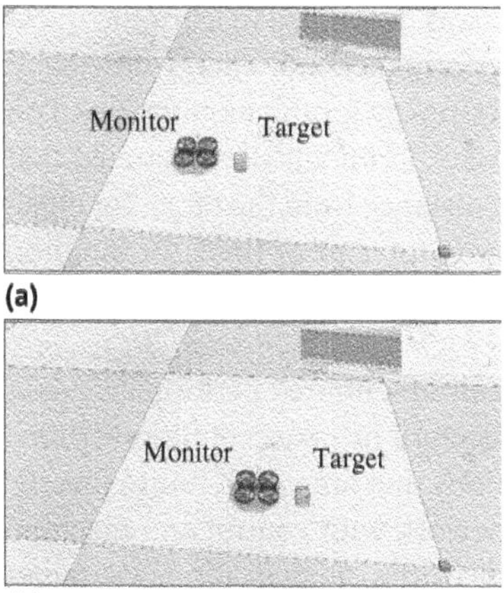

Figure 8.24 Snapshots from the CoppeliaSim simulator. The recordings of the CoppeliaSim simulations are at https://drive.google.com/drive/folders/1Qoiq_uWEzG0ATP77diZcl1R0DHOCkKFb. (a) $t = 0$; (b) $t = 20$.

the simulator. Nevertheless, the simulation test can be proportionally projected onto large areas if larger acceleration ranges, e.g., $[-1, 1]$ m/s^2, are considered. We will extend and test the proposed scheme for various application scenarios and hardware-in-the-loop (HITL) and/or SITL simulators, such as Gazebo and AirSim, in our future works.

8.6 Conclusion

This chapter proposed a novel framework of cellular-aided radar sensing with a BS-UAV bistatic SAR platform and a novel resolution framework for the online control of UAV-on-UAV visual-based tracking performed by a solar-powered UAV monitor with fixed wings.

In the first study, the trajectory of the UAV was optimized to minimize the propulsion energy while satisfying the range and azimuth resolutions of sensing. The trajectory planning problem was convexified and solved efficiently by utilizing the SCA and BCD methods. Extensive simulations revealed that, in terms of energy efficiency and effective consumption fluctuation, the proposed trajectory planning algorithm is superior to its alternative that minimizes the flight distance of a cellular-aided sensing mission. The energy saving can be as large as 55% for the UAV, by running the proposed algorithm. In the future, we will extend the BS-UAV bistatic SAR system to a fixed-wing UAV-based platform, where a more sophisticated propulsion energy model and dynamic/mobility model will be taken into account. In the second study, a DRL framework was developed to learn the tracking trajectory on-the-fly, balancing tracking performance and energy efficiency. DDPG was tailored to learn the optimal control strategy for the monitor based on its continuous state and action spaces. Extensive experiments demonstrated that our DDPG-based algorithm can keep a required distance from the target for effective visual tracking, while satisfying the battery constraint of the UAV monitor. The new algorithm outperformed traditional control and optimization methods in power efficiency and tracking accuracy.

The proposed DRL-based approach could result in a limited performance gain when the target has a drastically changing trajectory or takes countermeasures against the monitor. Interesting research directions that are worth future investigation include an integrated optimization framework with control and machine learning techniques, and transformer-based learning models for UAV target tracking.

9

Reconfigurable Intelligent Surface-assisted UAVs

9.1 Introduction

Unmanned aerial vehicles (UAVs) have emerged as transformative platforms for wireless communication and surveillance [253], leveraging their mobility and line-of-sight (LoS) advantages to address critical challenges in wireless systems and IoT networks [32, 146, 170]. On the one hand, RISs have been increasingly considered to create programmable radio propagation environments for efficient wireless communications [44, 145, 252]. It is particularly useful in radio surveillance. This is because the RISs can be deployed unnoticeably on building facades, roadside billboards, and indoor walls, and passively augment the radio propagation environments without catching the targets' attention, thereby creating a covert surveillance channel to a seemingly unlikely monitor, such as a UAV farther away or with the LoS obstructed. In this sense, it is prudent and necessary to consider RISs when designing UAV-based radio surveillance to develop future-proof surveillance solutions [327].

On the other hand, RISs were developed as a means of creating programmable wireless transmission environments [118]. These surfaces, made up of passive reflecting units with reprogrammable phase shifts [44], can be placed on building surfaces, and are expected to be an integral part and effective enhancer for the IoT. By configuring the phase shifts, the RIS-reflected signals are added constructively to the direct signal to improve signal quality or destructively to reduce interference [252]. The performance gain of RIS-assisted wireless communication was experimentally validated in [261].

For surveillance, a key challenge arises from the joint UAV navigation and RIS configuration for radio surveillance. One reason is that a fixed-wing UAV is preferred for surveillance applications for its energy efficiency and quietness [332]. However, the fixed-wing UAV can have a much more rigid mobility/dynamic

Artificial Intelligence for Unmanned Aerial Vehicles: Sensing, Communication, and Computing, First Edition. Shuyan Hu et al.

model than its rotary-wing counterpart [110]. Another reason is that a large number of RIS reflecting elements would incur prohibitive overhead and complexity for radio channel acquisition. Let alone an uncooperative suspicious Tx [145]. A predictive reconfiguration of the RIS may even be needed when a fixed-wing UAV is employed for surveillance, as the UAV has to keep flying to stay afloat (or it stalls) [115]. Furthermore, the waypoints of the fixed-wing UAV can be nested in the exponents of the array responses of the RIS-reflected channels [159], making existing optimization tools, such as successive convex approximation (SCA) [109], inapplicable to the joint UAV navigation and RIS configuration.

Radio surveillance is important to detect malicious users, intercept information, prevent attacks, and enhance security in wireless communication networks. An alternate-jamming-aided wireless physical-layer surveillance protocol was developed in [253], where two devices work cooperatively to eavesdrop on a suspicious point-to-point radio link. Radio surveillance performance was quantified in [4], where a monitor captures subtle changes in received signal strength to infer sound vibrations. By analyzing the phase and amplitude of CSI, radio biometric signatures were extracted for human surveillance and reidentification in [20]. Yet, none of these works considered the use of proliferating UAVs and RISs. Some recent works examined radio- and video-based UAV surveillance, e.g., [110, 111, 147, 325, 326]. The eavesdropping rate was maximized for a UAV monitor in [111, 147]. The energy consumption of the UAV was minimized for disguised video surveillance in [110]. None of the works has attempted to exploit the capability of RISs to augment radio environments and facilitate radio surveillance.

For anti-jamming applications, captured radio signals can be employed to decipher malicious messages, understand signal signatures, integrate information, and gather intelligence [51]. On the one hand, UAVs have been extensively applied to surveillance applications. The potential of RISs for anti-jamming applications has not been widely studied [176, 294]. In [288], a collective active and passive beamforming design was formulated to lower transmit power by optimizing the continuous phase shifts of an RIS. The result was later extended to discrete phase shifts in [289]. The RIS-assisted secure transmission was studied in [210, 303, 308], where fast RL and deep RL (DRL) were used to design the active and passive beamformers. For instance, in [303], jamming signals were modulated and reflected by an RIS to reduce the eavesdropping data rate.

In the context of UAV-borne IoT platforms, RIS-assisted jamming rejection has yet to be addressed. The design of an RIS-assisted jamming rejection UAV-borne IoT platform faces the following challenges. First, the received data rate at the UAV is a non-convex function of the UAV's flight path and the RIS phase shifts. Second, the UAV flight path waypoints are embedded in the exponents of the communication link, making the trajectory optimization intractable for existing convex tools, such as SCA. In addition, the large number of RIS reflecting units can cause prohibitive overhead and complexity for radio channel estimation, acquisition, and reconfiguration.

In a different yet relevant context, some studies have examined general RIS-assisted UAV communications [159, 160, 175, 219]. The authors of [159] jointly optimized UAV flight path and RIS passive beamformers to achieve the largest average rate of terrestrial users. In [219], a UAV and RIS were configured to deliver ultra-reliable and low-latency commands among terrestrial IoT devices using non-convex optimization. UAVs and RISs have been considered to assist with communications in a vehicular network [8], IoT [305], and wireless powered communication network [208]. For instance, the coverage probability, bit error rate, block error rate, and throughput of a vehicular network were analyzed in [8]. Alternating optimization was employed in [208] to maximize the sum rate of energy harvesting-powered IoT nodes. Moreover, UAV-mounted RISs have also been employed to assist communication or data collection in a multi-user network [245, 264, 356]. The sum rate of non-orthogonal multiple-access was maximized in [245] by block coordinate descent (BCD) and Riemannian conjugate gradient. The average throughput was analyzed in [264]. The number of UAVs was minimized by optimizing the RIS deployment and assignment by integer linear programming in [356].

The practicality and advantages of RIS-assisted UAV communications were experimentally corroborated by computer simulations in [159, 160, 175, 219, 245, 264, 356]. A UAV-borne RIS was designed for communication enhancement using experimental results from UAV hovering in [285]. On the other hand, an RIS is typically controlled via a field programmable gate array (FPGA) [226]. The execution delays of FPGA units are in nanoseconds [58]. Therefore, FPGA-controlled RISs can achieve a phase shift change with a delay of microseconds. Consider a UAV's typical speed of about 30 m/s. The reconfiguration of an RIS is fast enough to cater to the time-varying UAV communication channels. It is feasible to introduce RIS into UAV communications.

This chapter presents two studies based on the DRL architecture. In the first study, a new problem of RIS-assisted, UAV-based radio surveillance is formulated to jointly optimize the UAV trajectory and RIS configuration for monitoring a suspicious ground link. The problem is a sequential decision process with a non-convex, non-smooth objective, and cannot be solved by traditional optimization techniques. A new DRL framework is designed to allow the UAV to learn its trajectory and the RIS configuration only based on its observed transmit rate of the suspicious Tx, while the latest twin-delayed DDPG (TD3) model is tailored to implement the DRL model, coping with the continuous state and action spaces and learning efficient policies. The action and reward of the DRL capture the UAV's mobility constraint and maximize the eavesdropping rate.

In the second study, we put forth a new DRL-based architecture for the flight path planning and jamming cancellation of UAV-borne IoT platforms. A fixed-wing UAV is considered. Our architecture utilizes an RIS, which dynamically modifies the wireless transmission environment to mitigate jamming power and enhance intended signals to the UAV. To accomplish this, we devise a new DDPG model and its enhancement, TD3, to allow the UAV to learn its flight path and the RIS configuration based solely on its received data rate, eliminating the need for the CSI involving the RIS and jammer in the flight path training. This presents a significant practical advantage, as the estimation of CSI involving an RIS is complex and may not be able to be performed in real-time.

9.2 DRL-driven RIS-assisted Radio Surveillance

9.2.1 System Model

As illustrated in Figure 9.1, we consider an RIS-assisted, UAV-based surveillance system, where a fixed-wing UAV is employed to monitor and wiretap a suspicious radio link on the ground with the assistance of a benign RIS. The suspicious Tx and

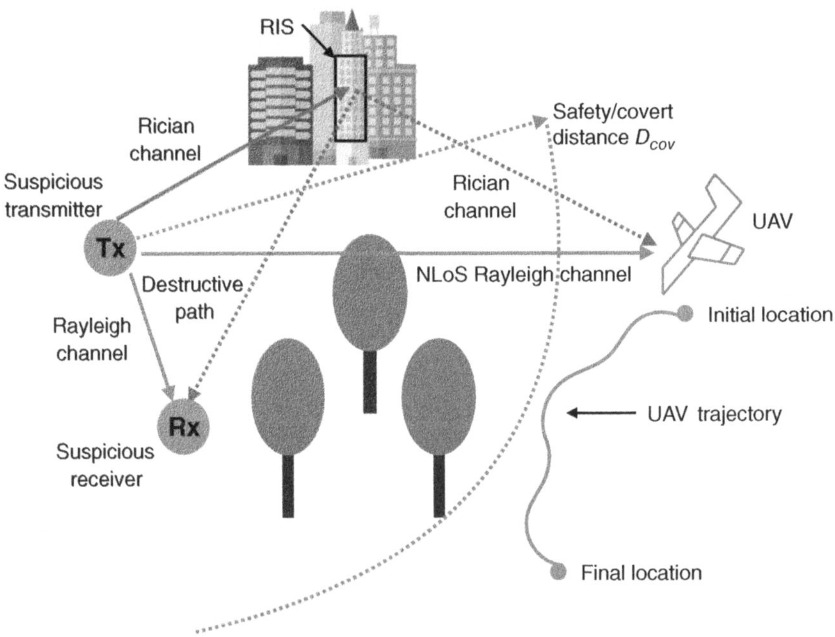

Figure 9.1 An RIS-assisted UAV-based radio surveillance system, where a fixed-wing UAV monitors a suspicious ground link at a distance. The RIS strengthens the UAV channel and/or weakens the suspicious link.

Rx are apart by d m. The UAV and the suspicious Tx/Rx all have a single antenna for illustration convenience. The RIS is equipped with a smart controller to receive instructions from the UAV for its configuration [195].

9.2.1.1 Threat Model

The threat model considered is that the suspicious Tx transmits messages that potentially contain plans or evidence of crimes to its intended Rx in a point-to-point communication mode. This can circumvent typical wiretapping techniques. In this case, radio surveillance is critical to detect crime, determine transmitter identities, or gather evidence for criminal prosecution. The suspicious Tx can be equipped with anti-reconnaissance capability, e.g., cameras. Therefore, it is important for the monitor, i.e., a UAV in the context of this chapter, to stay unnoticed by keeping its distance from the Tx. For this reason, we propose the RIS-assisted, UAV-based surveillance system, where an RIS augments the radio propagation environment to allow the UAV to eavesdrop on the suspicious link and keep its distance.

On the other hand, the UAV needs to maintain an active communication link with the RIS to send configurations to the RIS in the proposed system. However, this may be susceptible to several security attacks, e.g., launched by the Tx or its other accomplices, including: (i) Eavesdropping, where an attacker eavesdrops on the UAV-RIS link to obtain the UAV's credentials; (ii) Replay attack, where the attacker replays the UAV's transmissions to mislead the RIS; (iii) Spoofing, where the attacker pretends to be the UAV by encoding its messages with the UAV's identity and credentials; and (iv) Message forgery, where the attacker sends forged messages following the spoofing to take control of the RIS.

A countermeasure to these attacks is strong data encoding and encryption at the UAV and RIS to increase the difficulty in deciphering the UAV-RIS link. The UAV and RIS can frequently and regularly update their encryption keys, coordinated by a trusted third party, e.g., a cellular operator. The UAV can timestamp its messages to prevent replay attacks. By preventing eavesdropping and replay attacks, the UAV can protect its identity and credentials. This makes the spoofing and message forgery unlikely. In this chapter, we assume that these security measures are in place. The UAV can have a secure and reliable communication link with the RIS.

9.2.1.2 UAV Mobility Model

The UAV operates for a finite scheduling horizon of T seconds, which is divided into T_w time slots indexed by t, i.e., $t = 1, \dots, T_w$. The duration of a slot is $\delta = T/T_w$, which is short enough that the UAV can be treated as stationary within each slot. The UAV flies from a known initial location $\boldsymbol{q}_0 = [x_0, y_0, z_0]^T$ to a known final location $\boldsymbol{q}_F = [x_F, y_F, z_F]^T$. Consider a 3D Cartesian coordinate system. The UAV's time-varying 3D coordinates are $\boldsymbol{q}_t = [x_t, y_t, z_t]^T, \forall t$. Let $\boldsymbol{V}_t = [V_{xt}, V_{yt}, V_{zt}]^T$ and $\boldsymbol{A}_t = [A_{xt}, A_{yt}, A_{zt}]^T$ collect the velocity and acceleration of the UAV at time slot t, respectively. The UAV obeys the following mobility constraints [251]

$$\boldsymbol{q}_{t+1} = \boldsymbol{q}_t + \boldsymbol{V}_t\delta + \frac{1}{2}\boldsymbol{A}_t\delta^2, \ \forall t, \tag{9.1a}$$

$$\boldsymbol{q}_F - \boldsymbol{q}_{T_w} \leq d_0, \tag{9.1b}$$

$$\boldsymbol{V}_{t+1} = \boldsymbol{V}_t + \boldsymbol{A}_t\delta, \ \forall t, \tag{9.1c}$$

$$|A_{i,t}| \leq A_{\max}, i = x, y, z, \ \forall t, \tag{9.1d}$$

$$V_{\min} \leq \|\boldsymbol{V}_t\| \leq V_{\max}, \ \forall t, \tag{9.1e}$$

$$z_{\min} \leq z_t \leq z_{\max}, \ \forall t, \tag{9.1f}$$

where d_0 is the radius of the desired region centered at the final location; $A_{\max}$ is the maximum acceleration of the UAV; $V_{\max}$ and $V_{\min}$ are the maximum and minimum speeds of the UAV monitor, respectively; $z_{\max}$ and $z_{\min}$ are the maximum and minimum allowed altitudes of the UAV, respectively.

The UAV dynamic model in (9.1) is physically meaningful. Specifically, the UAV's current location depends on its velocity, acceleration, and previous location, as captured in constraint (9.1a). The UAV's final location needs to be within the proximity of the prespecified destination, as indicated by (9.1b). The UAV's current velocity relies on its previous velocity and current acceleration, as stated in (9.1c). The UAV's acceleration must not exceed a maximum value, as specified in (9.1d). The UAV's speed is upper bounded by its maximum speed, and it also needs to be lower bounded to avoid the loss of speed (or lift) and fall, as specified in (9.1e). Last but not least, (9.1f) makes sure that the UAV ascends or descends by a limited altitude.

9.2.1.3 RIS Configuration

The RIS is installed on the facade of a building, which is on the x-z plane and aligns with the x-axis. The RIS has a uniform rectangular array (URA) of $N = N_x N_z$ reflecting elements. The UAV can use out-of-band signaling to deliver the configurations of the reflecting elements to the controller of the RIS, thereby remotely configuring the RIS.

Consider a practical RIS model, where the reflection amplitude of a reflecting element depends on its phase shift. Let $\boldsymbol{\Theta}_t := \text{diag}(\upsilon_n^t e^{j\theta_1^t}, \dots, \upsilon_N^t e^{j\theta_N^t})$ be the reflection coefficient matrix of the RIS at time slot t. Here, $\theta_n^t = \theta_{(n_x-1)N_z+n_z}^t \in [-\pi, \pi), n = 1, \dots, N$, is the phase shift of the n-th reflecting element located at the n_z-th row and the n_x-th column of the RIS, i.e., $n = (n_x - 1)N_z + n_z$; $\upsilon_n^t \in [0, 1]$ is the corresponding amplitude; and $j = \sqrt{-1}$. The amplitude υ_n^t is a function of the phase shift θ_n^t, as given by [3]

$$v_n^t = f(\theta_n^t) = (1 - v_{\min})\left(\frac{\sin(\theta_n^t - \phi) + 1}{2}\right)^{\alpha_0} + v_{\min}, \tag{9.2}$$

where $v_{\min} \geq 0$ is the minimum amplitude; $\phi \geq 0$ denotes the difference between $-\pi/2$ and the phase shift $\theta_{\min}$ that corresponds to the minimum amplitude $v_{\min}$, i.e., $\phi = |-\pi/2 - \theta_{\min}|$ and $v_{\min} = f(\theta_{\min})$; and $\alpha_0 \geq 0$ controls the steepness of the function curve. $v_{\min}$, ϕ, and α_0 are three constants depending on the specific circuit implementation. When $v_{\min} = 1$ ($v_n^t = 1$) or $\alpha_0 = 0$, (9.2) is the widely adopted phase shift model with a constant amplitude of 1. The first element of the RIS is at the bottom right corner of the RIS, and its coordinates are $\boldsymbol{q}_{\mathrm{R}} = [x_{\mathrm{R}}, y_{\mathrm{R}}, z_{\mathrm{R}}]^T$.

9.2.1.4 Channel Model

The suspicious Tx and Rx nodes are at $(0, 0, 0)$ and $\boldsymbol{q}_{\mathrm{Rx}} = [d, 0, 0]^T$ on the ground (i.e., the x-y plane). The distances of the Tx-UAV link $d_{\mathrm{Tx,U}}^t$, Tx-RIS link $d_{\mathrm{Tx,R}}$, RIS-UAV link $d_{\mathrm{R,U}}^t$, and RIS-Rx link $d_{\mathrm{R,Rx}}$ are

$$d_{\mathrm{Tx,R}} = \|\boldsymbol{q}_{\mathrm{R}}\|,\ d_{\mathrm{Tx,U}}^t = \|\boldsymbol{q}_t\|,\ \forall t, \tag{9.3a}$$

$$d_{\mathrm{R,Rx}} = \|\boldsymbol{q}_{\mathrm{R}} - \boldsymbol{q}_{\mathrm{Rx}}\|,\ d_{\mathrm{R,U}}^t = \|\boldsymbol{q}_t - \boldsymbol{q}_{\mathrm{R}}\|,\ \forall t. \tag{9.3b}$$

9.2.1.5 Tx-Rx Link

Consider a Rayleigh fading channel for the Tx-Rx link. The channel power gain of the Tx-Rx link is

$$h_{\mathrm{Tx,Rx}} = \sqrt{\rho d_{\mathrm{Tx,Rx}}^{-\kappa_1}}\tilde{h}, \tag{9.4}$$

where ρ is the path loss at the reference distance $d_0 = 1$ m with the path loss exponent $\kappa_1 > 2$; and $\tilde{h}$ is the random scattering component modeled by a zero-mean, unit-variance circularly symmetric complex Gaussian (CSCG) random variable.

9.2.1.6 Tx/Rx-RIS Link

The channel power gains of the Tx-RIS and RIS-Rx links, denoted by $\boldsymbol{h}_{\mathrm{Tx,R}}$ and $\boldsymbol{h}_{\mathrm{R,Rx}} \in \mathbb{C}^{N\times 1}$, are

$$\boldsymbol{h}_{\mathrm{Tx,R}} = \underbrace{\sqrt{\rho d_{\mathrm{Tx,R}}^{-\kappa_2}}}_{\text{path loss}}\underbrace{\left(\sqrt{\frac{\beta_1}{1+\beta_1}}\boldsymbol{h}_{\mathrm{Tx,R}}^{los} + \sqrt{\frac{1}{1+\beta_1}}\boldsymbol{h}_{\mathrm{Tx,R}}^{nlos}\right)}_{\text{array response \& small-scale fading}}; \tag{9.5a}$$

$$\boldsymbol{h}_{\mathrm{R,Rx}} = \underbrace{\sqrt{\rho d_{\mathrm{R,Rx}}^{-\kappa_3}}}_{\text{path loss}}\underbrace{\left(\sqrt{\frac{\beta_2}{1+\beta_2}}\boldsymbol{h}_{\mathrm{R,Rx}}^{los} + \sqrt{\frac{1}{1+\beta_2}}\boldsymbol{h}_{\mathrm{R,Rx}}^{nlos}\right)}_{\text{array response \& small-scale fading}}, \tag{9.5b}$$

where κ_2 and κ_3 are the path loss exponents for the Tx-RIS and RIS-Rx links, respectively; β_1 and β_2 are the Rician factors; and $\boldsymbol{h}^{nlos}_{\mathrm{Tx,R}}$ and $\boldsymbol{h}^{nlos}_{\mathrm{R,Rx}}$ are the non-LoS (NLoS) components with the variables independently drawn from the zero-mean, unit-variance CSCG distribution.

In (9.5a) and (9.5b), $\boldsymbol{h}^{los}_{\mathrm{Tx,R}}$ and $\boldsymbol{h}^{los}_{\mathrm{R,Rx}}$ are the LoS components

$$\boldsymbol{h}^{los}_{\mathrm{Tx,R}} = \left[1, \dots, e^{-j\frac{2\pi d_x}{\lambda}(N_x-1)\phi^x_{\mathrm{Tx,R}}}\right]^T \otimes \left[1, \dots, e^{-j\frac{2\pi d_y}{\lambda}(N_y-1)\phi^y_{\mathrm{Tx,R}}}\right]^T, \tag{9.6a}$$

$$\boldsymbol{h}^{los}_{\mathrm{R,Rx}} = \left[1, \dots, e^{-j\frac{2\pi d_x}{\lambda}(N_x-1)\phi^x_{\mathrm{R,Rx}}}\right]^T \otimes \left[1, \dots, e^{-j\frac{2\pi d_y}{\lambda}(N_y-1)\phi^y_{\mathrm{R,Rx}}}\right]^T, \tag{9.6b}$$

where d_x and d_y are the antenna spacings in the directions of the x- and y-axes, respectively; $\phi^x_{\mathrm{Tx,R}} = x_{\mathrm{R}}/d_{\mathrm{Tx,R}}$ and $\phi^y_{\mathrm{Tx,R}} = y_{\mathrm{R}}/d_{\mathrm{Tx,R}}$ are the spatial frequencies corresponding to the angles-of-arrivals (AoAs) from the Tx to the RIS, and $\phi^x_{\mathrm{R,Rx}} = (d - x_{\mathrm{R}})/d_{\mathrm{R,Rx}}$ and $\phi^y_{\mathrm{R,Rx}} = y_{\mathrm{R}}/d_{\mathrm{R,Rx}}$ are the spatial frequencies corresponding to the AoDs from the RIS to the Rx in the directions of the x- and y-axes, respectively [178].

9.2.1.7 Tx/RIS-UAV Link

The PLoS channel can be constructed by considering the probabilities of the LoS and NLoS channels, where the LoS probability of the Tx/RIS-UAV link for a given time slot, $p^{los,t}_b$, $b \in \{\mathrm{Tx}, \mathrm{R}\}$, is given by [194]

$$p^{los,t}_{b,\mathrm{U}} = \frac{1}{1 + C_{b,\mathrm{U}} \exp(-D_{b,\mathrm{U}}[\theta^t_{b,\mathrm{U}} - C_{b,\mathrm{U}}])}, \ \forall b, t. \tag{9.7}$$

Here, $C_{b,\mathrm{U}}$ and $D_{b,\mathrm{U}}$ rely on the propagation environment, and $\theta^t_{b,\mathrm{U}} = \sin^{-1}(\Delta z^t_{b,\mathrm{U}}/d^t_{b,\mathrm{U}})$ is the elevation angle in radii, where $\Delta z^t_{b,\mathrm{U}}$ is the altitude difference between the Tx or RIS and the UAV.

The channel power gains of the Tx-UAV and RIS-UAV links, denoted by $h^t_{\mathrm{Tx,U}}$ and $\boldsymbol{h}^t_{\mathrm{R,U}} \in \mathbb{C}^{N\times 1}$, $\forall t$, are

$$\begin{aligned} h^t_{\mathrm{Tx,U}} &= p^{los,t}_{\mathrm{Tx,U}} \sqrt{\rho (d^t_{\mathrm{Tx,U}})^{-\kappa_4}} \\ &+ (1 - p^{los,t}_{\mathrm{Tx,U}}) \sqrt{\rho (d^t_{\mathrm{Tx,U}})^{-\kappa_4}}\, h^{nlos,t}_{\mathrm{Tx,U}}, \end{aligned} \tag{9.8a}$$

$$\boldsymbol{h}_{\mathrm{R,U}}^{t} = p_{\mathrm{R,U}}^{los,t}\sqrt{\rho(d_{\mathrm{R,U}}^{t})^{-\kappa_5}}\boldsymbol{h}_{\mathrm{R,U}}^{los,t} + (1 - p_{\mathrm{R,U}}^{los,t})\sqrt{\rho(d_{\mathrm{R,U}}^{t})^{-\kappa_5}}\boldsymbol{h}_{\mathrm{R,U}}^{nlos,t}, \tag{9.8b}$$

where κ_4 and κ_5 are the path loss exponents of the Tx-UAV and RIS-UAV links, respectively; and $h_{\mathrm{Tx,U}}^{nlos,t}$ and $\boldsymbol{h}_{\mathrm{R,U}}^{nlos,t}$ are the NLoS components with the variables independently drawn from the zero-mean, unit-variance CSCG distribution.

In (9.8b), $\boldsymbol{h}_{\mathrm{R,U}}^{los,t}$ is the array response, as given by

$$\boldsymbol{h}_{\mathrm{R,U}}^{los,t} = \left[1, \ldots, e^{-j\frac{2\pi d_x}{\lambda}(N_x-1)\phi_{\mathrm{R,U},x}^{t}}\right]^{T} \otimes \left[1, \ldots, e^{-j\frac{2\pi d_y}{\lambda}(N_y-1)\phi_{\mathrm{R,U},y}^{t}}\right]^{T}, \quad \forall t, \tag{9.9}$$

where $\phi_{\mathrm{R,U},x}^{t} = (x_t - x_{\mathrm{R}})/d_{\mathrm{R,U}}^{t}$ and $\phi_{\mathrm{R,U},y}^{t} = (y_t - y_{\mathrm{R}})/d_{\mathrm{R,}}^{t}$ are the spatial frequencies corresponding to the AoDs from the RIS to the UAV along the x- and y-axes, respectively.

It is worth pointing out that the UAV does not require the CSI of the channels to and/or from the RIS, i.e., $\boldsymbol{h}_{\mathrm{Tx,R}}$, $\boldsymbol{h}_{\mathrm{R,Rx}}$, and $\boldsymbol{h}_{\mathrm{R,U}}$, to produce its trajectory and configure the RIS in this chapter. Instead, the UAV only observes the transmit rate of the Tx and measures its own received SNR to learn the UAV trajectory and RIS configuration. This consideration is of practical value because the RIS-reflected channels are difficult to estimate without the cooperation of the Tx. Let alone that the UAV has no access to the CSI of the Tx-Rx link [44].

9.2.1.8 Passive Beamforming for Radio Surveillance

Under the proposed DRL framework, the UAV only assesses the resolvability of its received signals, and adjusts its flight and the RIS in an attempt to keep the signals resolvable at the UAV. (The received signals are always resolvable at the intended receiver, since the Tx can adjust its transmit rate, adapting to the channel capacity between itself and the Rx.) In other words, the Tx-UAV link needs to sustain a higher channel capacity than the Tx-Rx link, so that the UAV can capture distinguishable and resolvable signal constellations useful for signal detection or cyber analysis.

A channel outage occurs when the UAV cannot resolve the signal constellations intended for the Rx (because the channel capacity of the Tx-UAV channel is lower than the transmit rate of the Tx-Rx link). To accomplish radio surveillance and prevent a channel outage, the UAV maneuvers its flight and remotely configures the RIS to enhance its own channel from the Tx, or degrade the channel between the Tx and Rx, or both. By adjusting the RIS configuration in real-time, the UAV can maintain a high channel capacity for the Tx-UAV link and prevent channel outage, while also ensuring that the Rx can still receive the desired signal.

We design that, when the Tx-UAV link has a lower gain than the Tx-Rx link, the RIS adjusts its phase shifts to weaken the Tx-Rx channel and lower the transmit rate of the Tx-Rx link until the UAV can recognize the signal constellations of the Tx node with reasonable resolution. When the Tx-UAV link has a higher gain than the Tx-Rx link, the RIS enhances the Tx-Rx channel as long as the received signals remain resolvable at the UAV. This allows the Tx to send more data to the Rx, and the UAV can shed more insights into the interaction between the Tx and Rx.

The eavesdropping data rate of the UAV at time slot t can be evaluated as

$$R_{\mathrm{U}}^{t}=\begin{cases}R_{\mathrm{Rx}}^{t}=\log_2(1+\gamma_{\mathrm{Rx}}^{t}), & \text{if } \gamma_{\mathrm{U}}^{t}\geq\gamma_{\mathrm{Rx}}^{t};\\ 0, & \text{otherwise.}\end{cases} \tag{9.10}$$

Here, the SNRs at the UAV, γ_{U}^{t}, and the Rx, γ_{Rx}^{t}, are

$$\gamma_{\mathrm{U}}^{t}=\frac{P_t}{\sigma_{\mathrm{U}}^{2}}|h_{\mathrm{Tx,U}}^{t}+(\boldsymbol{h}_{\mathrm{R,U}}^{t})^{H}\boldsymbol{\Theta}_t\boldsymbol{h}_{\mathrm{Tx,R}}|^{2},\ \forall t; \tag{9.11a}$$

$$\gamma_{\mathrm{Rx}}^{t}=\frac{P_t}{\sigma_{\mathrm{Rx}}^{2}}|h_{\mathrm{Tx,Rx}}+\boldsymbol{h}_{\mathrm{R,Rx}}^{H}\boldsymbol{\Theta}_t\boldsymbol{h}_{\mathrm{Tx,R}}|^{2},\ \forall t, \tag{9.11b}$$

where P_t is the transmit power of the Tx, and σ_{U}^{2} and σ_{Rx}^{2} are the variances of the additive white Gaussian noise (AWGN) at the UAV and the Rx, respectively.

The UAV can practically evaluate its SNR, i.e., γ_{U}^{t} and the transmit rate over the suspicious Tx-Rx link, i.e., R_{Rx}^{t}.[1] Specifically, the UAV can estimate the CSI-at-the-Rx (CSIR) based on the pilot of the Tx in the same way as the Rx, especially when the Tx, Rx, and UAV are equipped with multiple antennas [74]. With the CSIR, the UAV can estimate its SNR, γ_{U}^{t}, and recover the signal constellations to estimate the transmit rate over the suspicious Tx-UAV link, R_{Rx}^{t}, if the signal constellations are resolvable at the UAV, i.e., $\gamma_{\mathrm{U}}^{t}\geq\gamma_{\mathrm{Rx}}^{t}$. Otherwise, $R_{\mathrm{U}}^{t}=0$, as shown in (9.10); the UAV needs to adjust the RIS and maneuver its flight until the constellations become resolvable at the UAV.

The new model can be extended under imperfect CSIR (i.e., at the intended Rx and the UAV). Define the effective channels from the suspicious Tx to the UAV and the Rx at time slot t as $h_{\mathrm{eff,U}}^{t}=h_{\mathrm{Tx,U}}^{t}+(\boldsymbol{h}_{\mathrm{R,U}}^{t})^{H}\boldsymbol{\Theta}_t\boldsymbol{h}_{\mathrm{Tx,R}}^{t}$ and $h_{\mathrm{eff,Rx}}^{t}=h_{\mathrm{Tx,Rx}}^{t}+(\boldsymbol{h}_{\mathrm{R,Rx}}^{t})^{H}\boldsymbol{\Theta}_t\boldsymbol{h}_{\mathrm{Tx,R}}^{t}$, respectively. A minimum mean-square error (MMSE) estimator is often adopted at the Rx [74]. The distribution of channel estimation errors is typically agnostic to the channel distribution. The estimation errors, denoted by $\Delta h_{\mathrm{eff,U}}^{t}$ and $\Delta h_{\mathrm{eff,Rx}}^{t}$, are Gaussian, i.e., $\Delta h_{\mathrm{eff,U}}^{t}$, $\Delta h_{\mathrm{eff,Rx}}^{t}\in\mathcal{CN}(0,\epsilon_{\mathrm{eff}})$,

1 The transmit rate R_{Rx}^{t} depends on the SNR at the Rx, i.e., γ_{Rx}^{t}; see (9.10). Yet, the UAV does not need to explicitly know γ_{Rx}^{t} in the proposed system.

where $\mathbb{E}[|\Delta h^t_{\text{eff,Rx}}|^2] = \mathbb{E}[|\Delta h^t_{\text{eff,U}}|^2] = \epsilon_{\text{eff}}$ only depends on the noises at the Rx/UAV and is independent of the RIS configuration $\boldsymbol{\Theta}_t$ and the channels. The SNRs/signal-to-interference-plus-noise ratios (SINRs) at the Rx and the UAV in (9.10) can be rewritten as

$$\gamma^t_{\text{U}} = \frac{P_t|\hat{h}^t_{\text{eff,U}}|^2}{P_t\mathbb{E}[|\Delta h^t_{\text{eff,U}}|^2] + \sigma^2_{\text{U}}}, \ \forall t; \tag{9.12a}$$

$$\gamma^t_{\text{Rx}} = \frac{P_t|\hat{h}^t_{\text{eff,Rx}}|^2}{P_t\mathbb{E}[|\Delta h^t_{\text{eff,Rx}}|^2] + \sigma^2_{\text{Rx}}}, \ \forall t, \tag{9.12b}$$

where $\hat{h}^t_{\text{eff,U}} = h^t_{\text{eff,U}} + \Delta h^t_{\text{eff,U}}$ and $\hat{h}^t_{\text{eff,Rx}} = h^t_{\text{eff,Rx}} + \Delta h^t_{\text{eff,Rx}}$ are the estimated effective channels.

At time slot t, the eavesdropping data rate of the UAV is

$$R^t_{\text{U}} = \begin{cases} R^t_{\text{Rx}} = \log_2\left(\dfrac{P_t|\hat{h}^t_{\text{eff,Rx}}|^2}{P_t\epsilon_{\text{eff}} + \sigma^2_{\text{Rx}}}\right), & \text{if } \gamma^t_{\text{U}} \geq \gamma^t_{\text{Rx}}; \\ 0, & \text{otherwise,} \end{cases} \tag{9.13}$$

which shows that the channel estimation errors (at the Rx) would result in reduced SNR/SINR and an increased requirement for the transmit power of the Tx. The UAV could underestimate the transmit rate due to less accurately recovered signal constellations. The UAV would configure the RIS to further degrade the Tx-Rx link and/or enhance the Tx-UAV link until the signal constellations become resolvable at the UAV, as in the case of perfect CSIR.

9.2.2 Problem Formulation

We maximize the total amount of data that the UAV eavesdrops on the Tx node during the mission of T_w slots, which is a long-term goal over the entire mission. The UAV also needs to keep its distance from Tx to avoid being noticed by Tx. The problem is written as

$$\max_{\{\mathbf{q}_t, \boldsymbol{\Theta}_t, \forall t\}} \sum_{t=1}^{T_w} R^t_{\text{U}} \tag{9.14a}$$

$$\text{s.t.} \ -\pi \leq \theta^t_n < \pi, \ \forall n, t, \tag{9.14b}$$

$$\hat{d}^t_{\text{Tx,U}} \geq D_{\text{cov}}, \ \forall t, \tag{9.14c}$$

(9.30a)–(9.30f),

where $\hat{d}^t_{\mathrm{Tx,U}}$ is the distance between the UAV and Tx on the horizontal plane; and D_{cov} is the distance that the UAV keeps horizontally from the Tx to stay unnoticed to the Tx (or the UAV is likely to be noticed if flying over the Tx).[2]

In existing works, a no-fly zone has been typically formulated as a polyhedron or a circle on the 2D plane [158]. Those works typically created offline trajectory plans to bypass no-fly zones using traditional optimization techniques, such as Lagrangian dual-based method [328], BCD [72], and SCA [109]. In contrast, problem (9.14) requires a 3D *cylindrical* no-fly zone that takes a different shape from the popular polyhedron or circular no-fly zones on the 2D plane (for 2D UAV trajectory design) considered in the literature [158].

Moreover, problem (9.14) is fundamentally different from the existing works. It is more than only creating a no-fly zone, as the UAV needs to keep its distance from the Tx (i.e., staying outside the cylindrical no-fly zone) and maximize its eavesdropping rate on the Tx by jointly designing its trajectory and configuring an RIS. Problem (9.14) has never been studied in the existing literature.

Alternating optimization and BCD can be applied to problem (9.14) to decouple the optimization variables of UAV trajectory and RIS coefficients, and yield two subproblems that can be solved in an alternating manner, as discussed in [176, 294]. However, neither of the subsequent subproblems of (9.14) can be solved using traditional optimization techniques popular in UAV or RIS optimizations, such as semidefinite programming (SDP), Lagrange-dual method, SCA, and manifold optimization (MO). One reason is that the objective function in (9.14a) is a non-smooth, non-convex, piecewise function of the UAV trajectory and RIS phase shifts, which cannot be tackled by the existing optimization techniques, e.g., SDP or SCA. Another reason is that the large number of RIS reflecting elements incurs prohibitive overhead and complexity for radio channel acquisition. Let alone that the CSI of the Tx-Rx link is unavailable to the UAV in practice.

To further corroborate that it is not possible to obtain the globally optimal solution to problem (9.14) within polynomial time, we simplify the problem by considering a 1-bit resolution (discrete) phase shift at the RIS, where each RIS element

---

2 Maximizing the UAV's eavesdropping rate is equivalent to minimizing the user's secrecy rate, which is, however, different from maximizing the user's secrecy rate in the presence of a UAV eavesdropper. This is because the eavesdropping rate given in (9.13) is different from the secrecy rate given by

$$R^t_{\mathrm{S}} = \begin{cases} \log_2(1+\gamma^t_{\mathrm{Rx}}) - \log_2(1+\gamma^t_{\mathrm{U}}), & \text{if } \gamma^t_{\mathrm{U}} \le \gamma^t_{\mathrm{Rx}}; \\ 0, & \text{otherwise.} \end{cases}$$

More importantly, maximizing a user's secrecy rate in the presence of a UAV eavesdropper would not be able to optimize the trajectory of the UAV. As a result, the state and action spaces, and reward designs would be different.

enjoys a phase shift of 0 or π, and an adjustable amplitude that is a function of θ_n^t, i.e., $v_n^t(\theta_n^t)$ with $|v_n^t(\theta_n^t)| \leq 1$. Given the optimal UAV trajectory, $\{\boldsymbol{q}_t^*, \forall t\}$, the subproblem of discrete phase shift optimization can be cast as

$$\text{(P.DPS)} \max_{\{\boldsymbol{\Theta}_t, \forall t\}} \sum_{t=1}^{T_w} R_{\mathrm{U}}^t \tag{9.15a}$$

$$\text{s.t. } \theta_n^t \in \{0, \pi\}, \ \forall n, t, \tag{9.15b}$$

$$\sum_{n=1}^{N} |v_n^t(\theta_n^t)| \leq N, \ \forall t, \tag{9.15c}$$

where the optimization of $\boldsymbol{\Theta}_t$ is a binary decision problem with a sum modulus constraint.

This subproblem (9.15) is in the same form as the classic NP-hard 0/1 knapsack problem (0/1 KP), where the modulus constraint of each RIS unit is relaxed to the sum modulus constraint (9.15c) to conform to the constraint of the 0/1 KP. Yet, the objective function (9.15a) is non-convex, while the objective of the 0/1 KP is linear. As a result, solving problem (9.15) takes no less time than solving the 0/1 KP. It is known to be difficult to obtain the optimal solution to the 0/1 KP within polynomial time. Let alone problem (9.15). Moreover, problem (9.15) cannot be solved using SDP and SCA due to its non-smooth, non-convex, piecewise objective function, and the unavailability of perfect and instantaneous CSI of all channels.

Problem (9.14) is much more complex than problem (9.15), as the former has more optimization variables (i.e., the UAV trajectory and RIS phase shift) and temporally coupled constraints (i.e., the UAV mobility and dynamics). It is not possible to obtain the optimal solution to problem (9.14) with a polynomial time-complexity. Nevertheless, the problem has a long-term goal that indicates a sequential decision process suitable for DRL. For this reason, we propose to utilize DRL to solve problem (9.14), as elaborated on in the next section.

9.2.3 TD3 Framework for UAV Navigation and RIS Configuration

This section presents a new TD3-based method under the actor-critic framework to solve problem (9.39). Our method learns to configure the RIS and navigate the UAV (including heading and acceleration) based on its observed transmit rate of the suspicious link, suppressing the need for precise CSI, including the Tx-Rx link and the RIS reflecting channels.

9.2.3.1 State, Action, and Reward

The UAV trajectory (i.e., waypoints) is a Markov decision process (MDP). The RIS configuration depends on the instantaneous position of the UAV. We can interpret problem (9.39) as an MDP with its state, action, and reward defined below.

State Space $\mathcal{S}$: At time slot t, the system state $s_t \in \mathcal{S}$ is made of the relative position of the UAV with regards to its final location, $\boldsymbol{q}_t - \boldsymbol{q}_F$, the UAV's velocity $\boldsymbol{V}_t$, the SNRs at the UAV γ_{U}^t and the Rx node γ_{Rx}^t,[3] and the distance between the Tx-UAV link $d_{\mathrm{Tx,U}}^t$ at the t-th time step, i.e., $s_t = \{\boldsymbol{q}_t - \boldsymbol{q}_F, \boldsymbol{V}_t, \gamma_{\mathrm{U}}^t, \gamma_{\mathrm{Rx}}^t, d_{\mathrm{Tx,U}}^t\}$.

Action Space $\mathcal{A}$: The action space collects all possible actions, i.e., $a_t \in \mathcal{A}$. At the t-th time step, the action, a_t, consists of two parts: The reflecting coefficients of the RIS, i.e., $\{\theta_n^t\}_{n\in\mathcal{N}}$, and the acceleration of the UAV, i.e., $\boldsymbol{A}_t := [A_{xt}, A_{yt}, A_{zt}]^T$. The action a_t has $(N+3)$ dimensions. The UAV's acceleration is constrained by (9.30d)–(9.30f). Its waypoints $\boldsymbol{q}_t$ and velocities $\boldsymbol{V}_t$ depend on the accelerations, i.e., (9.30a)–(9.30c).

We exploit the normalized representation for both the reflecting coefficients of the RIS and the acceleration of the UAV, i.e., $\boldsymbol{A}_t = A_{\max}\boldsymbol{\lambda}_A$ and $\theta_n^t = \pi\lambda_\theta(n)$, where $\boldsymbol{\lambda}_A = [\lambda_A(0), \lambda_A(1), \lambda_A(2)]$ and $n \in \mathcal{N}$. Here, $-1 \leq \lambda_A(0), \lambda_A(1), \lambda_A(2), \lambda_\theta(n) \leq 1$.

Reward $\boldsymbol{r_t}$: The reward function is designed based on the objective and constraints in (9.39) to evaluate the action a_t at each time slot. The reward function has three parts:

1. *Eavesdropping Rate Maximization.* The reward for maximizing the amount of eavesdropped data while disguising the UAV's intention of eavesdropping by keeping a predefined distance, i.e., D_{cov}, from the Tx node at any time slot t, is defined as

$$r_t^{\{er\}} = \begin{cases} \zeta_{er} R_{\mathrm{U}}^t, & \text{if } \hat{d}_{\mathrm{Tx,U}}^t \geq D_{\mathrm{cov}}; \\ 0, & \text{otherwise,} \end{cases} \tag{9.16}$$

where ζ_{er} is a positive constant used to tune the relative importance of this part in the reward function.

2. *Destination Approaching.* The reward for destination approaching is designed as

$$r_t^{\{dr\}} = \zeta_{dr}\left(d_F^{t-1} - d_F^t\right), \tag{9.17}$$

where $\left(d_F^{t-1} - d_F^t\right)$ is the reduced distance from the UAV to the final location after time slot t. $d_F^{t-1} = \|\boldsymbol{q}_{t-1} - \boldsymbol{q}_F\|$ and $d_F^t = \|\boldsymbol{q}_t - \boldsymbol{q}_F\|$ are the distances from the UAV to the final location at slots $(t-1)$ and t, respectively. ζ_{dr} is a positive constant to adjust the relative importance of this part in the reward function.

With the reward $\zeta_{dr}\left(d_F^{t-1} - d_F^t\right)$, the UAV is incentivized to get closer to the final location at each time step. Moreover, $\zeta_{dr}\left(d_F^{t-1} - d_F^t\right)$ provides a positive reward if the distance from the UAV to the final location is shorter at the t-th time step than it was at slot $(t-1)$. Otherwise, a penalty occurs.

3 The UAV estimates the rate of the Tx, $\log_2(1+\gamma_{\mathrm{Rx}}^t)$, and recover γ_{Rx}^t.

When $t = T_w$, we set a reward $r_t^{\{arr\}}$ to indicate whether the UAV reaches the final location

$$r_t^{\{arr\}} = \begin{cases} 0, \; t = 1, \dots, T_w - 1; \\ \zeta_{arr}(t) r_{arr} + (1 - \zeta_{arr}(t)) \, r_{narr}, \; t = T_w, \end{cases} \tag{9.18}$$

where r_{arr} is a positive constant to motivate the arrival of the UAV within the desired region centered at the final location, and r_{narr} is a negative constant to penalize if the UAV does not arrive at the destination. $\zeta_{arr}(t) = 1$ if the UAV reaches the desired final location at the final time step (i.e., $t = T_w$). Otherwise, $\zeta_{arr}(t) = 0$.

3. *Constraints.* We also set penalties to punish infeasible actions that cannot satisfy constraints (9.30e) and (9.30f)

$$\begin{aligned} r_t^{\{v\}} &= \zeta_v(t) \cdot r_v, \quad \text{if} \quad \|\boldsymbol{V}_t\| \in [V_{\min}, V_{\max}]; \\ r_t^{\{z\}} &= \zeta_z(t) \cdot r_z, \quad \text{if} \quad z_t \in [z_{\min}, z_{\max}], \end{aligned} \tag{9.19}$$

where $\zeta_v(t)$ is the velocity constraint indicator with $\zeta_v(t) = 1$ indicating that the UAV's velocity violates constraint (9.30e) and $\zeta_v(t) = 0$ indicating otherwise; and $\zeta_z(t)$ is the altitude constraint indicator with $\zeta_z(t) = 1$ indicating that the UAV's altitude violates constraint (9.30f) and $\zeta_z(t) = 0$ indicating otherwise.

The overall reward function is

$$r_t = r_t^{\{er\}} + r_t^{\{dr\}} + r_t^{\{arr\}} + r_t^{\{v\}} + r_t^{\{z\}}. \tag{9.20}$$

Policy: A mapping from the state space $\mathcal{S}$ to the action space $\mathcal{A}$ is known as a policy, $\mu : \mathcal{S} \rightarrow \mathcal{A}$, which is a distribution $\mu(a|s) = \Pr(a_t = a | s_t = s)$ over $s \in \mathcal{S}$. A policy $a_t = \mu(s_t)$ maps state s_t to a feasible action. The agent selects the policy maximizing the accumulated reward $R_t = \sum_{n=t}^{N} \gamma^{n-t} r_t$ with $\gamma \in [0, 1]$ being a discount factor. Given state s_t, action a_t and policy $\mu(\cdot|\cdot)$, R_t is evaluated by an action-value function, i.e., the Q-function, as given by

$$Q_\mu(s_t, a_t) = \mathbb{E}_\mu[R_t | s_t, a_t], \tag{9.21}$$

which satisfies the Bellman Expectation Equation

$$Q_\mu(s_t, a_t) = \mathbb{E}_{r_t, s_{t+1} \sim \mathcal{E}} \left[r_t + \gamma \mathbb{E}_{a_{t+1} \sim \mu} \left[Q_\mu(s_{t+1}, a_{t+1}) \right] \right], \tag{9.22}$$

where $\mathcal{E}$ denotes the environment that the agent interacts with.

Experience: An experience is defined as $e_t = (s_t, a_t, r_t, s_{t+1})$ and stored in an experience replay buffer $\boldsymbol{R}$.

9.2.3.2 TD3 Model

To address the Q-value overestimation issue of the DDPG algorithm, TD3 introduces *three improvements* over DDPG, i.e., *clipped double-Q learning with two critics*, *target policy smoothing*, and *delayed policy update* [64].

- *Clipped double-Q learning with two critics:* TD3 has two critics (i.e., to produce two Q-values), and admits the smaller of the two Q-values to evaluate the target Q-values in the Bellman error loss functions.
- *Target policy smoothing:* TD3 adds noises to the target action and smooths the Q-function of the actions to make the policy unlikely to exploit the errors in the Q-function.
- *"Delayed" policy updates:* The actors are updated less frequently than the critics. It was suggested in [64] that the actors are updated after the critics are updated twice.

The TD3-based framework consists of an actor-network and a critic network. The actor-network comprises an actor and a target-actor. The critic network comprises two critics and two target-critics; see Figure 9.2. Six deep neural network (DNN) approximators are used. The actor with parameters θ_a, denoted by $\mu\,(s_t;\theta_a)$, approximates the policy function of the UAV and produces the actions. The two critics with parameters θ_1 and θ_2, denoted by $Q_1(s_t, a_t;\theta_1)$ and $Q_2(s_t, a_t;\theta_2)$, concurrently estimate two action-value functions of the actions produced by the actor [244]. The target-actor with parameter θ_a', denoted by $\mu'\,(s_t;\theta_a')$, produces the target policy. The two target-critics with parameters θ_1' and θ_2', denoted by $Q_1'(s_t, a_t;\theta_1')$ and $Q_2'(s_t, a_t;\theta_2')$, respectively, generate two Q-values. The smaller of the two Q-values is taken as the target Q-value.

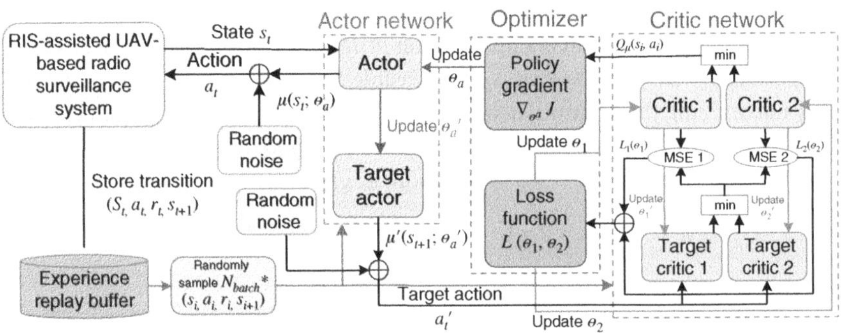

Figure 9.2 The architecture of the proposed TD3-based framework, where the actor-network comprises an actor and a target-actor, and the critic network comprises two critics and two target-critics. The experience replay buffer provides batches of samples of state transitions to train and update the networks.

The TD3 network follows the deterministic policy gradient (DPG) theorem [244] to update the parameters, θ_a, θ_1, θ_2, θ'_a, θ'_1, and θ'_2, and optimize the actions. The use of the target network (comprising the target-actor and two target-critics) helps avoid unstable learning arising from using only an actor-critic network (with one actor and one critic) [104].

Algorithm 10 TD3 model for solving (9.14)

1: **Initialization:**
2: Randomly initialize the actor μ and the two critics Q_1 and Q_2 with parameters θ_a, θ_1, and θ_2, the target-actor μ' and two target-critics Q'_1 and Q'_2 with parameters $\theta'_a \leftarrow \theta_a$, $\theta'_1 \leftarrow \theta_1$, and $\theta'_2 \leftarrow \theta_2$, the experience replay buffer $\boldsymbol{R}$, and the channel allocation and modulation-coding selection η_0.
3: Initialize the relative locations and velocities from the UAV to the final location, and estimate the eavesdropping rate of the UAV as the initial state s_0.
4: **for** $t = 1, \dots, T_w$ **do**
5: Pick action $a_t = \text{clip}\left(\mu\,(s_t; \theta_a) + \epsilon, a_{\min}, a_{\max}\right)$.
6: Execute action a_t, receive the reward r_t, perceive a new state s_{t+1}, and store transition (s_t, a_t, r_t, s_{t+1}) in $\boldsymbol{R}$.
7: Randomly sample N_{batch} historical transitions (s_i, a_i, r_i, s_{i+1}) from $\boldsymbol{R}$.
8: Update the target action after target policy smoothing based on the sampled transitions: $a'_t = \text{clip}\left(\mu'\,(s_{t+1}; \theta'_a) + \text{clip}\,(\epsilon', -\sigma_m^2, \sigma_m^2)\,, a_{\min}, a_{\max}\right)$.
9: Update the target Q-value: $y_i = r_i + \gamma \min\left(Q'_1(s_{i+1}, a'_i; \theta'_1), Q'_2(s_{i+1}, a'_i; \theta'_2)\right)$.
10: Calculate the loss function based on (9.27), and update the two critics by (9.28).
11: **if** mod $(t, 2) = 0$ **then**
12: Update the actor based on (9.25), and the target-actor and the two target-critics by (9.29).
13: **end if**
14: **end for**

9.2.3.3 Training Algorithm

Algorithm 10 summarizes the proposed TD3-based algorithm. Based on the UAV's current state, an action a_t is produced by the actor following the DPG theorem [244]. We use the tanh$(\cdot)$ activation function to output λ_A and λ_θ. A random exploration noise is added to the action to balance the exploration of new actions and the exploitation of known actions. The output action is

$$a_t = \text{clip}\,(\mu(s_t; \theta_a) + \epsilon, a_{\min}, a_{\max}), \tag{9.23}$$

where the noise ϵ is randomly sampled from a zero-mean Gaussian distribution with variance σ_e^2, i.e., $\epsilon \sim \mathcal{N}(0, \sigma_e^2)$; clip$(\cdot)$ is a clipping function to limit the actions within the upper and lower bounds of the actions, denoted by $a_{\max}$ and $a_{\min}$, respectively. The UAV executes a_t, refreshes its state, collects its reward r_t, and transits to the next state s_{t+1}. The transition (s_t, a_t, r_t, s_{t+1}) is recorded in the experience replay buffer $\boldsymbol{R}$.

Due to the overfitting issue of the deterministic policies [244], the target Q-value produced by the deterministic policy is susceptible to the inaccuracies induced by the Q-function estimation errors. *Target policy smoothing*, a regularization strategy for Q-function value learning [81] is used to reduce the inaccuracies. With N_{batch} transitions randomly sampled from $\boldsymbol{R}$, the target action after the target policy smoothing is

$$a_t' = \text{clip}(\mu'(s_{t+1}; \theta_a') + \text{clip}(\epsilon', -\sigma_m^2, \sigma_m^2), a_{\min}, a_{\max}), \tag{9.24}$$

where the noise ϵ' is randomly sampled from a zero-mean Gaussian distribution with variance σ_a^2, i.e., $\epsilon' \sim \mathcal{N}(0, \sigma_a^2)$. ϵ' is clipped to be within $[-\sigma_m^2, \sigma_m^2]$ with σ_m^2 pre-specified.

Given the probability distribution of the parameter θ_a, i.e., $J(\theta_a)$, the actor-network is updated in the direction that the estimation improves the strategy fastest. With the N_{batch} transitions from $\boldsymbol{R}$, $\nabla_{\theta_a} J(\theta_a)$ is approximated by

$$\nabla_{\theta_a} J(\theta_a) \approx \frac{1}{N_{\text{batch}}} \sum_{i=1}^{N_{\text{batch}}} \left[\nabla_{\theta_a} \mu(s_i) \nabla_a Q_\mu(s_i, a; \theta_c)|_{a=\mu(s_i)}\right]. \tag{9.25}$$

The model parameter of the actor, i.e., θ_a, is updated by using the gradient descent method [259]

$$\begin{aligned} \theta_a &\leftarrow \theta_a + \eta_a \nabla_{\theta_a} J(\theta_a) \\ &\approx \theta_a + \frac{\eta_a}{N_{\text{batch}}} \sum_{i=1}^{N_{\text{batch}}} \left[\nabla_{\theta_a} \mu(s_i) \nabla_a Q_\mu(s_i, a; \theta_c)|_{a=\mu(s_i)}\right], \end{aligned} \tag{9.26}$$

where η_a is the learning rate of the actor-network $\mu(s_t; \theta_a)$ [162]; $\nabla_{\theta_a} \mu(s)$ is the gradient of the actor $\mu(s)$ with respect to the parameter θ_a; and $\nabla_a Q_\mu(s_t, a; \theta_a)$ is the gradient of $Q_\mu(s_t, a; \theta_a)$ with respect to the action a.

The mean square error (MSE)-based loss function, $L_k(\theta_k)$, of the two critics is approximately evaluated as

$$L_k(\theta_k) \approx \frac{1}{N_{\text{batch}}} \sum_{i=1}^{N_{\text{batch}}} \left[(Q_k(s_i, a_i) - y_i)^2\right], \qquad k = 1, 2 \tag{9.27}$$

where $y_i = r_i + \gamma \min\left(Q_1'(s_{i+1}, a_i'; \theta_1'), Q_2'(s_{i+1}, a_i'; \theta_2')\right)$ is the approximate target Q-value produced by the target network based on N_{batch} randomly sampled transitions; γ is a discount factor; θ_1' and θ_2' are decayed copies of θ_1 and θ_2, respectively.

The gradient of $L_k(\theta_k)$ is given by

$$\nabla_{\theta_k} L_k(\theta_k) \approx \frac{1}{N_{\text{batch}}} \sum_{i=1}^{N_{\text{batch}}} \left[(Q_\mu\left(s_i, \mu(s_i; \theta_a); \theta_k\right) - y_i)\nabla_{\theta_k} Q_\mu(s_i, \mu(s_i; \theta_a); \theta_k)\right], \quad k = 1, 2 \tag{9.28}$$

The parameters of the two critics, i.e., θ_1 and θ_2, are updated by utilizing the stochastic gradient descent method [259].

According to the *"delayed" policy updates*, the target-actor and the two target-critics evolving from the actor and critics are updated every two iterations by running the Polyak Averaging [64], as given by

$$\theta_a' \leftarrow \rho_\tau \theta_a + (1 - \rho_\tau)\theta_a'; \theta_k' \leftarrow \rho_\tau \theta_k + (1 - \rho_\tau)\theta_k', \forall k, \tag{9.29}$$

where ρ_τ is the decaying rate of both actor and critic networks.

The model is trained iteratively until the convergence of the cumulative reward.

9.2.3.4 Complexity and Convergence

To evaluate the computational complexity, we consider the actor and critic networks separately. For the actor-network with L_a layers, the number of neurons in the m-th layer is J_m. The complexity of the m-th layer is $\mathcal{O}(J_{m-1}J_m + J_m J_{m+1})$ [82]. Hence, the complexity of the actor-network is $\mathcal{O}\left(\sum_{m=2}^{L_a-1}(J_{m-1}J_m + J_m J_{m+1})\right)$. For the critic network with L_c layers, the number of neurons is G_n in the n-th layer. The computational complexity of the n-th layer is $\mathcal{O}(G_{n-1}G_n + G_n G_{n+1})$ [82]. The complexity of the critic network is $\mathcal{O}\left(\sum_{n=2}^{L_c-1}(G_{n-1}G_n + G_n G_{n+1})\right)$. As a result, the overall complexity of Algorithm 10 is [82]

$$\mathcal{O}\left(\sum_{m=2}^{L_a-1}(J_{m-1}J_m + J_m J_{m+1}) + \sum_{n=2}^{L_c-1}(G_{n-1}G_n + G_n G_{n+1})\right).$$

In terms of convergence, the TD3-based algorithm satisfies the following conditions: (i) the network parameters θ and θ' (with the subscripts omitted for brevity) are upper bounded since they are sequentially compact following the Arzela-Ascoli theorem [35]; (ii) the state and action spaces are compact, as the sampled states and actions are bounded by the UAV's maximum speed and acceleration; (iii) the reward function (9.16) is continuous; and (iv) the training networks are feedforward fully connected neural networks (FCNNs) with twice continuously differentiable activation functions. According to [220, Lemma 2], the algorithm can asymptotically converge as the number of iterations tends to infinity since we adopt a sequence of square summable learning rates, i.e., $\sum_t \eta_a(t) = \infty$ and $\sum_t \eta_a^2(t) < \infty$. Here, t indicates a time step. $\eta_a(t)$ is the learning rate of the actor-network at the t-th time step.

9.3 RIS-assisted Jamming Rejection for UAV-borne IoT

9.3.1 System Model

As depicted in Figure 9.3, a ground BS offers wireless communication service for a fixed-wing UAV in the presence of a terrestrial jammer. Consider a 3D Cartesian coordinate system. The BS and the jammer are placed at $\boldsymbol{q}_{\mathrm{B}} = [0, 0, 0]^T$ and $\boldsymbol{q}_{\mathrm{J}} = [x_{\mathrm{J}}, y_{\mathrm{J}}, 0]^T$. An RIS is deployed to facilitate the UAV communication and suppress the jamming signals from the jammer.

We suppose that the BS, UAV, and jammer all have a single antenna for description convenience (but the proposed DRL architecture can be extended to support multi-antenna BSs and UAVs, as part of our future work). Only based on its received data rate, the UAV determines its flight path and the RIS configuration dynamically. The RIS is installed with a smart controller to receive commands from the UAV (via the cellular system) for RIS configuration [195].

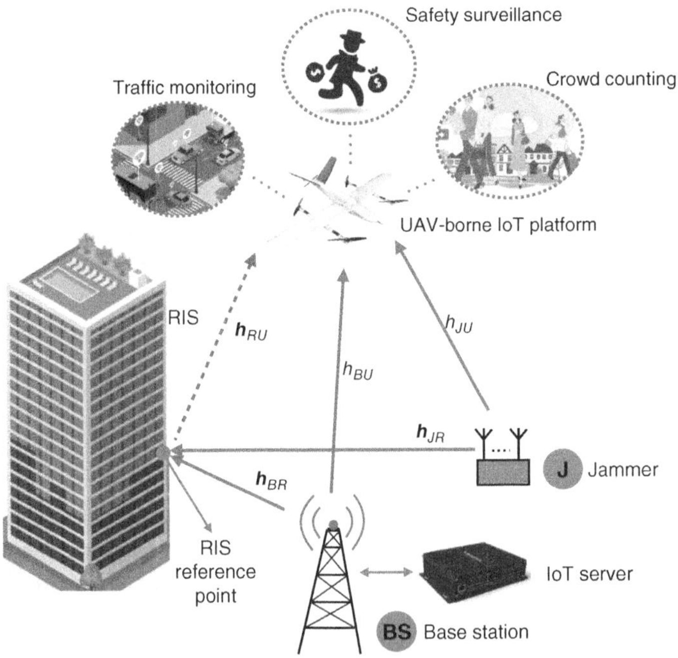

Figure 9.3 An RIS-assisted anti-jamming UAV-borne IoT platform, where an RIS is adaptively configured along with the UAV trajectory to enhance the desired signals and reject the jamming signals.

9.3.1.1 UAV Mobility Model

The UAV functions for a finite scheduling horizon of T seconds, which is split into T_w time slots indexed by t, and $t = 1, \dots, T_w$. A slot lasts $\delta = T/T_w$, which is short enough that the UAV can be viewed as stationary per slot. The UAV is traveling from a predetermined starting position $\boldsymbol{q}_0 = [x_0, y_0, z_0]^T$ to a predetermined ending position $\boldsymbol{q}_F = [x_F, y_F, z_F]^T$. The UAV's 3D coordinates are $\boldsymbol{q}_t = [x_t, y_t, z_t]^T, \forall t$. Let $\boldsymbol{V}_t := [V_{xt}, V_{yt}, V_{zt}]^T$ and $\boldsymbol{A}_t := [A_{xt}, A_{yt}, A_{zt}]^T$ collect the velocity and acceleration of the UAV per slot t, respectively. The UAV obeys some mobility constraints [251]

$$\boldsymbol{q}_{t+1} = \boldsymbol{q}_t + \boldsymbol{V}_t\delta + \frac{1}{2}\boldsymbol{A}_t\delta^2, \ \forall t, \tag{9.30a}$$

$$\boldsymbol{q}_F = \boldsymbol{q}_{T_w}, \tag{9.30b}$$

$$\boldsymbol{V}_{t+1} = \boldsymbol{V}_t + \boldsymbol{A}_t\delta, \ \forall t, \tag{9.30c}$$

$$|A_{i,t}| \le A_{\max}, i \in \{x, y, z\}, \ \forall t, \tag{9.30d}$$

$$V_{\min} \le \|\boldsymbol{V}_t\| \le V_{\max}, \ \forall t, \tag{9.30e}$$

$$V_{zt}/\|\boldsymbol{V}_t\| \le \sin\vartheta, \ \forall t, \tag{9.30f}$$

where $A_{\max}$ is the UAV's largest acceleration; $V_{\max}$ and $V_{\min}$ are the UAV's maximum and minimum speeds; and ϑ is the largest UAV pitch angle when ascending or descending.

9.3.1.2 RIS Configuration

The RIS is adhered to the outer surface of a building, which is on the (x, z)-plane and aligns with the x-axis. The RIS has a URA of $N = N_xN_y$ reflecting elements (units), and a controller to dynamically control the phase shift of each unit. Let $\boldsymbol{\Theta}_t := \text{diag}(e^{j\theta_1^t}, \dots, e^{j\theta_N^t})$ be the phase shift matrix for the RIS at time slot t, where $\theta_n^t = \theta_{(n_x-1)N_y+n_y}^t \in [-\pi, \pi), n = 1, \dots, N$, is the phase shift of the n-th reflecting unit which is located at the n_y-th row and the n_x-th column of the RIS, and $j = \sqrt{-1}$. The first element of the RIS is at the bottom right corner of the RIS, and its coordinates are $\boldsymbol{q}_\text{R} = [x_\text{R}, y_\text{R}, z_\text{R}]^T$.

9.3.1.3 Channel Model

The distances of the BS-UAV link d_BU^t, the BS-RIS link d_BR, the Jammer-RIS link d_JR, the Jammer-UAV link d_JU^t, and RIS-UAV link d_RU^t are given by

$$d_\text{BU}^t = \|\boldsymbol{q}_t\|, \ \forall t, \ d_\text{BR} = \|\boldsymbol{q}_\text{R}\|, \ d_\text{JR} = \|\boldsymbol{q}_\text{J} - \boldsymbol{q}_\text{R}\|, \tag{9.31a}$$

$$d_\text{JU}^t = \|\boldsymbol{q}_t - \boldsymbol{q}_\text{J}\|, \ d_\text{RU}^t = \|\boldsymbol{q}_t - \boldsymbol{q}_\text{R}\|, \ \forall t. \tag{9.31b}$$

Consider an LoS channel for the RIS-UAV link (i.e., the R-U link), and Rician fading channels between the BS/jammer and the UAV (i.e., the B-U and J-U links), and between the BS/jammer and the RIS (i.e., the B-R and J-R links). The channel gains of the B-U and J-U links are: $\forall t$,

$$h_{\mathrm{BU}}^t = \sqrt{\rho(d_{\mathrm{BU}}^t)^{-\kappa_1}} \left(\sqrt{\frac{\beta_t}{1+\beta_t}} g_{\mathrm{BU}}^t + \sqrt{\frac{1}{1+\beta_t}} \tilde{g}_{\mathrm{BU}}^t \right), \tag{9.32a}$$

$$h_{\mathrm{JU}}^t = \sqrt{\rho(d_{\mathrm{JU}}^t)^{-\kappa_1}} \left(\sqrt{\frac{\beta_t}{1+\beta_t}} g_{\mathrm{JU}}^t + \sqrt{\frac{1}{1+\beta_t}} \tilde{g}_{\mathrm{JU}}^t \right), \tag{9.32b}$$

where ρ stands for the path loss at the reference distance $d_0 = 1$ m with the path loss exponent $\kappa_1 > 2$; β_t is the Rician factor of the B-U and J-U links; g_{BU}^t and g_{JU}^t are the deterministic LoS components with $|g_{\mathrm{BU}}^t| = 1$ and $|g_{\mathrm{JU}}^t| = 1$; $\tilde{g}_{\mathrm{BU}}^t$ and $\tilde{g}_{\mathrm{JU}}^t$ stand for stochastic dispersion captured by a zero-mean, unit-variance CSCG random variable. The elevation-angle-reliant Rician factor β_t is captured by the following exponential function [317]

$$\beta_t = \xi_1 \exp\left(\xi_2 \arcsin(z_t/d_{\mathrm{BU}}^t)\right), \ \forall t, \tag{9.33}$$

where ξ_1 and ξ_2 are two constant coefficients dependent on the environment.

The channel gains of the B-R, J-R, and R-U links, denoted by $\boldsymbol{h}_{\mathrm{BR}} \in \mathbb{C}^{N\times 1}$, $\boldsymbol{h}_{\mathrm{JR}} \in \mathbb{C}^{N\times 1}$, and $\boldsymbol{h}_{\mathrm{RU}}^t \in \mathbb{C}^{N\times 1}$, are

$$\boldsymbol{h}_{\mathrm{BR}} = \underbrace{\sqrt{\rho d_{\mathrm{BR}}^{-\kappa_2}}}_{\text{path loss}} \underbrace{\left(\sqrt{\frac{\beta}{1+\beta}} \boldsymbol{h}_{\mathrm{BR}}^{los} + \sqrt{\frac{1}{1+\beta}} \boldsymbol{h}_{\mathrm{BR}}^{nlos} \right)}_{\text{array response \& small-scale fading}}; \tag{9.34a}$$

$$\boldsymbol{h}_{\mathrm{JR}} = \underbrace{\sqrt{\rho d_{\mathrm{JR}}^{-\kappa_2}}}_{\text{path loss}} \underbrace{\left(\sqrt{\frac{\beta}{1+\beta}} \boldsymbol{h}_{\mathrm{JR}}^{los} + \sqrt{\frac{1}{1+\beta}} \boldsymbol{h}_{\mathrm{JR}}^{nlos} \right)}_{\text{array response \& small-scale fading}}; \tag{9.34b}$$

$$\boldsymbol{h}_{\mathrm{RU}}^t = \sqrt{\rho (d_{\mathrm{RU}}^t)^{-2}} \boldsymbol{g}_{\mathrm{RU}}^t, \ \forall t. \tag{9.34c}$$

Here, β is the Rician factor of the B-R and J-R links (c.f. β_t); and $\boldsymbol{h}_{\mathrm{BR}}^{los}$ and $\boldsymbol{h}_{\mathrm{JR}}^{los}$ are the LoS components, as given by

$$\begin{aligned}\boldsymbol{h}_{\mathrm{BR}}^{los} = &\left[1, \dots, e^{-j\frac{2\pi d_x}{\lambda}(N_x-1)\phi_{\mathrm{BR}}^x}\right]^T \otimes \\ &\left[1, \dots, e^{-j\frac{2\pi d_y}{\lambda}(N_y-1)\phi_{\mathrm{BR}}^y}\right]^T,\end{aligned} \tag{9.35a}$$

$$\boldsymbol{h}_{\mathrm{JR}}^{los} = \left[1, \dots, e^{-j\frac{2\pi d_x}{\lambda}(N_x-1)\phi_{\mathrm{JR}}^{x}}\right]^T \otimes \left[1, \dots, e^{-j\frac{2\pi d_y}{\lambda}(N_y-1)\phi_{\mathrm{JR}}^{y}}\right]^T, \tag{9.35b}$$

where d_x and d_y are the antenna spacings in the directions of the x- and y-axes, respectively; $\phi_{\mathrm{BR}}^{x} = x_{\mathrm{R}}/d_{\mathrm{BR}}$ and $\phi_{\mathrm{BR}}^{y} = y_{\mathrm{R}}/d_{\mathrm{BR}}$ are the spatial frequencies corresponding to the AoAs from BS to RIS, and $\phi_{\mathrm{JR}}^{x} = (x_{\mathrm{J}} - x_{\mathrm{R}})/d_{\mathrm{JR}}$ and $\phi_{\mathrm{JR}}^{y} = (y_{\mathrm{J}} - y_{\mathrm{R}})/d_{\mathrm{JR}}$ are the spatial frequencies corresponding to the AoAs from the jammer to the RIS along the x- and y-axes, respectively [178].

In (9.34a) and (9.34b), $\boldsymbol{h}_{\mathrm{BR}}^{nlos} \in \mathbb{C}^{N\times 1}$ and $\boldsymbol{h}_{\mathrm{JR}}^{nlos} \in \mathbb{C}^{N\times 1}$ are the NLoS components with the variables independently drawn from the zero-mean, unit-variance CSCG distribution. In (9.34c), $\boldsymbol{g}_{\mathrm{RU}}^{t}$ is the array response, as given by

$$\boldsymbol{g}_{\mathrm{RU}}^{t} = \left[1, \dots, e^{-j\frac{2\pi d_x}{\lambda}(N_x-1)\phi_{\mathrm{RU},x}^{t}}\right]^T \otimes \left[1, \dots, e^{-j\frac{2\pi d_y}{\lambda}(N_y-1)\phi_{\mathrm{RU},y}^{t}}\right]^T, \quad \forall t, \tag{9.36}$$

where $\phi_{\mathrm{RU},x}^{t} = (x_t - x_{\mathrm{R}})/d_{\mathrm{RU}}^{t}$ and $\phi_{\mathrm{RU},y}^{t} = (y_t - y_{\mathrm{R}})/d_{\mathrm{RU}}^{t}$ are the spatial frequencies corresponding to the AoDs from RIS to UAV along the x- and y-axes, respectively.

It is noteworthy that the UAV does not ask for the CSI involving the RIS and the jammer to produce its flight path and configure the RIS in this chapter. Instead, the UAV only measures its own received data rate to learn the control policy of its flight path and RIS configuration. This consideration is of practical value, since the RIS-reflected channels are difficult and slow to estimate.

9.3.2 Problem Formulation

Let $P_t, \forall t$ denote the transmit power of the BS and P_{J} denote the transmit power of the jammer. The SINR at the UAV at time slot t, denoted by γ_{U}^{t}, is

$$\gamma_{\mathrm{U}}^{t} = \frac{P_t |h_{\mathrm{BU}}^{t} + (\boldsymbol{h}_{\mathrm{RU}}^{t})^H \boldsymbol{\Theta}_t \boldsymbol{h}_{\mathrm{BR}}|^2}{P_{\mathrm{J}} |h_{\mathrm{JU}}^{t} + (\boldsymbol{h}_{\mathrm{RU}}^{t})^H \boldsymbol{\Theta}_t \boldsymbol{h}_{\mathrm{JR}}|^2 + \sigma_{\mathrm{U}}^2}, \tag{9.37}$$

where σ_{U}^2 is the variance of the AWGN at the UAV. The received data rate of the UAV at time slot t is given by

$$R_{\mathrm{U}}^{t} = \log_2(1 + \gamma_{\mathrm{U}}^{t}). \tag{9.38}$$

We aim to maximize the total received data rate of the UAV from the BS over the mission duration of T_w slots. The problem considered is stated as follows.

$$\max_{\{\boldsymbol{q}_t,\boldsymbol{\Theta}_t,\forall t\}} \sum_{t=1}^{T_w} R_{\mathrm{U}}^t \tag{9.39a}$$

$$\text{s.t.} \ -\pi \le \theta_n^t < \pi, \ \forall n, t, \tag{9.39b}$$

(9.30a)–(9.30f).

Problem (9.39) is challenging for traditional convex solvers due to several reasons: first of all, the received data rate is a non-convex function of the UAV's flight path $\boldsymbol{q}_t, \forall t$ and the RIS phase shifts $\boldsymbol{\Theta}_t, \forall t$. Second, the UAV flight path waypoints are embedded in the exponents of the R-U link in (9.34c) and (9.36), making the trajectory optimization intractable for existing convex tools, such as SCA. Another reason is that the large number of RIS reflecting units can cause prohibitive overhead and complexity for radio channel estimation, acquisition, and reconfiguration. To overcome these limitations, the next section proposes using DRL to solve (9.39).

9.3.3 DRL for Anti-jamming Communication of UAV-borne IoT Platforms

The proposed method in this section aims to solve problem (9.39) by utilizing the DDPG and TD3 models. Our approach involves learning to adjust the RIS and control the UAV's trajectory, including heading and acceleration, based on changes in the received data rate of the UAV. Importantly, our method eliminates the need for precise CSI or knowledge of the RIS reflecting channels. DDPG and its variations, such as TD3, have been demonstrated to be effective in addressing problems with continuous action spaces [112, 162, 244]. In contrast, traditional DRL methods, such as deep Q-learning, can struggle and even diverge when faced with continuous action spaces.

9.3.3.1 State, Action, and Reward

Since the current UAV location only depends on its previous location and speed, the UAV trajectory (i.e., waypoints) is a MDP. The RIS configuration depends solely on the instantaneous position of the UAV. Therefore, we interpret problem (9.39) as an MDP with its state, action, and reward defined below.

- ***State Space*** $\mathcal{S}$: At time slot t, the system state $s_t \in \mathcal{S}$ is made of the relative position of the UAV with regards to its final location, $\boldsymbol{q}_t - \boldsymbol{q}_F$, the velocity of the UAV, $\boldsymbol{V}_t$, and the SINR at the UAV, γ_{U}^t, $s_t = \{\boldsymbol{q}_t - \boldsymbol{q}_F, V_t, \gamma_{\mathrm{U}}^t\}$.
- ***Action Space*** $\mathcal{A}$: It gathers all possible actions, i.e., $a_t \in \mathcal{A}$. During the t-th time step, the action a_t consists of the reflecting coefficients $\{\theta_n^t\}_{n\in\mathcal{N}}$ and the acceleration of the UAV, $\boldsymbol{A}_t := [A_{xt}, A_{yt}, A_{zt}]^T$, i.e.,

$a_t = \{\theta_n^t \in [-\pi, \pi), \forall n, A_{it} \in [-A_{\max}, A_{\max}],\ i \in \{x, y, z\}\}$. The UAV acceleration is constrained by (9.30d)–(9.30f). Given the initial location and velocity of the UAV, its future waypoints $\boldsymbol{q}_t$ and velocities $\boldsymbol{V}_t$ are decided by the accelerations, i.e., by (9.30a)–(9.30c).

- ***Reward*** r_t**:** The reward function gives positive returns per time step for implementing action a_t

$$r_t = \underbrace{R_{\mathrm{U}}^t}_{\text{communication}} + \underbrace{\zeta\left(d_F^{t-1} - d_F^t\right)}_{\text{distance to the final location}}, \tag{9.40}$$

where R_U^t is the UAV's received data rate, $d_F^{t-1} = \|\boldsymbol{q}_{t-1} - \boldsymbol{q}_F\|$ and $d_F^t = \|\boldsymbol{q}_t - \boldsymbol{q}_F\|$ are the distances from the UAV to the final location at the $(t-1)$-th and t-th time steps, respectively; and ζ is a tunable parameter during the learning process. The second element on the right-hand side of (9.40) encourages the UAV to fly toward the final location.

- ***Policy:*** A projection from the state space, $\mathcal{S}$, to the action space $\mathcal{A}$ is referred to as a policy, $\mu : \mathcal{S} \to \mathcal{A}$, a distribution $\mu(a|s) = \Pr(a_t = a|s_t = s)$ over state $s \in \mathcal{S}$.
- ***Experience:*** The experience, defined as $e_t = (s_t, a_t, r_t, s_{t+1})$, is stored in an experience replay memory $\boldsymbol{R}$.

The UAV experiences state s_t, performs action a_t, receives reward r_t, and turns to state s_{t+1}. A policy $a_t = \mu(s_t)$ maps state s_t to a possible action. The UAV chooses the policy that maximizes the cumulative reward $R_t = \sum_{n=t}^{N} \gamma^{n-t} r_t$. Here, $\gamma \in (0, 1)$ gives the discount factor. Given s_t, a_t, and μ, the Q-function evaluates R_t by

$$Q_\mu(s_t, a_t) = \mathbb{E}_\mu[R_t | s_t, a_t]. \tag{9.41}$$

The action-value function, $Q_\mu(s_t, a_t)$, follows the Bellman Expectation Equation

$$Q_\mu(s_t, a_t) = \mathbb{E}_{r_t, s_{t+1} \sim \mathcal{E}}\left[r_t + \gamma \mathbb{E}_{a_{t+1} \sim \mu}\left[Q_\mu(s_{t+1}, a_{t+1})\right]\right]. \tag{9.42}$$

Here, $\mathcal{E}$ stands for the environment the UAV experiences.

It is generally challenging to directly use an RL algorithm to solve the continuous-space, finite-horizon MDP and determine the Q-value, $Q(s_t, a_t)$, due to the continuous state and action spaces. This chapter puts forth a new DDPG-based algorithm to control the UAV's trajectory and configure the RIS, as delineated in the following section.

9.3.3.2 Actor-critic Framework-based DDPG

The DDPG-based network uses four DNN approximators, including training-actor and training-critic networks, and target-actor and target-critic networks, as shown in Figure 9.4. The training-actor network with parameters θ_a, denoted as $\mu(s_t; \theta_a)$, gives an approximate policy of the UAV and produces the actions. The training-critic network with parameters θ_c, denoted as $Q_\mu(s_t, a_t; \theta_c)$, estimates the

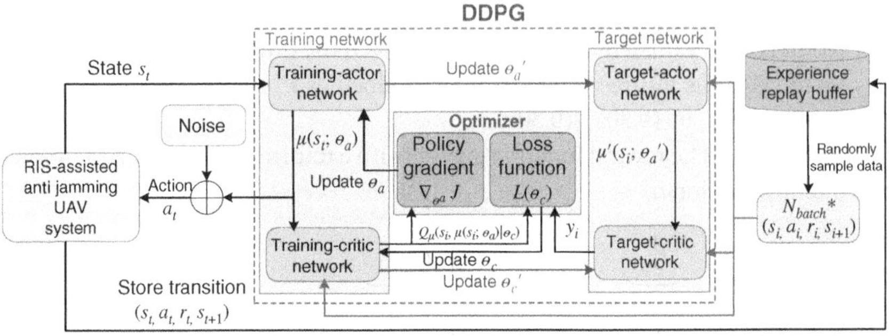

Figure 9.4 The proposed DDPG-based framework with a training network and a target network, each comprising an actor network and a critic network. The experience replay buffer gives batches of samples of state transitions for training and updating the networks.

action-value function concerning the actions created in the training-actor network [244]. The target-actor network with parameter θ_a', represented by $\mu'(s_t; \theta_a')$, and the target-critic networks with parameter θ_c', represented by $Q'_{\mu'}(s_t, a_t; \theta_c')$, generate the target Q-value for training the training-actor and training-critic networks.

The DDPG network uses the DPG theorem [244] to refresh θ_a, θ_c, θ_a', and θ_c'. It produces actions in an actor-critic setting. Additionally, the adoption of a target network (i.e., the target-actor and target-critic networks) helps prevent unstable learning, as opposed to using only a training network (with a training-actor and a training-critic network) [104].

The UAV inputs state s_t into the training-actor network. Using the DPG theorem [244], the network generates the strategy by projecting the state to an action in a deterministic fashion. This network approximates the agent's policy function and selects action a_t. A noise is added to a_t to balance between new and known actions, resulting in an output action $a_t = \mu(s_t; \theta_a) + \mathcal{N}_t$. Herein, $\mathcal{N}_t$ is a random noise process with a normal distribution. The agent is rewarded with r_t and transitions to state s_{t+1}. Then, it stores the experience $(s_t, a_t, r_t, st + 1)$ in $\boldsymbol{R}$.

The training-critic network evaluates the action-value function $Q_\mu(s_t, \mu(s_t; \theta_a); \theta_c)$ of the selected action a_t. By using a random sample from the replay memory $\boldsymbol{R}$, the network approximates the action-value function as $Q_\mu(s_i, \mu(s_i; \theta_a); \theta_c)$. We take $J(\theta_a)$ to be the probability distribution of the parameter θ_a. The training-actor network is adjusted in the direction that improves the strategy the most rapidly, i.e., in the direction of the gradient of $J(\theta_a)$ with respect to (w.r.t.) θ_a [244]

$$\nabla_{\theta_a} J(\theta_a) = \mathbb{E}_{s \sim \rho^\mu} \left[\nabla_{\theta_a} Q_\mu(s_t, \mu(s_t; \theta_a); \theta_c) \right] \tag{9.43a}$$

$$= \mathbb{E}_{s \sim \rho^\mu} \left[\nabla_{\theta_a} \mu(s_t; \theta_a) \nabla_a Q_\mu(s_t, a; \theta_c)|_{a=\mu(s_t; \theta_a)} \right], \tag{9.43b}$$

where (9.43b) uses the chain rule; ρ^{μ} provides a discounted state distribution of $\mu(s_t;\theta_a)$ [162]; $\nabla_{\theta_a}\mu(s)$ gives the gradient of the training-actor network $\mu(s)$ w.r.t. θ_a; $\nabla_a Q_\mu(s_t, a; \theta_a)$ provides the gradient of $Q_\mu(s_t, a; \theta_a)$ w.r.t. a.

By randomly drawing N_{batch} sampled historical transitions from $\boldsymbol{R}$, the gradient $\nabla_{\theta_a} J(\theta_a)$ is approximated by

$$\nabla_{\theta_a} J(\theta_a) \approx \frac{1}{N_{\text{batch}}} \sum_{i=1}^{N_{\text{batch}}} \left[\nabla_{\theta_a}\mu(s_i)\nabla_a Q_\mu(s_i, a; \theta_c)|_{a=\mu(s_i)}\right]. \tag{9.44}$$

The training-actor network parameter, i.e., θ_a, is refreshed based on the gradient ascent [259]

$$\begin{aligned} \theta_a &\leftarrow \theta_a + \eta_a \nabla_{\theta_a} J(\theta_a) \\ &\approx \theta_a + \frac{\eta_a}{N_{\text{batch}}} \sum_{i=1}^{N_{\text{batch}}} \left[\nabla_{\theta_a}\mu(s_i)\nabla_a Q_\mu(s_i, a; \theta_c)|_{a=\mu(s_i)}\right]. \end{aligned} \tag{9.45}$$

Here, η_a specifies the learning rate of the training-actor network.

The training-critic network is refreshed through minimizing the following loss function

$$L(\theta_c) = \mathbb{E}_{s_t\sim\rho^{\mu}, a_t\sim\mu(s_t;\theta_a)}\left[\left(Q_\mu\left(s_t, a_t; \theta_c\right) - y_t\right)^2\right]. \tag{9.46}$$

Here, $y_t = r_t + \gamma Q'_{\mu'}\left(s_{t+1}, \mu'\left(s_{t+1}; \theta'_a\right); \theta'_c\right)$ is the target Q-value produced by the target network under the transition (s_t, a_t, r_t, s_{t+1}). Here, the parameters of the target-actor and target-critic networks, θ'_a and θ'_c, are the respective decayed copies of θ_a and θ_c.

With N_{batch} randomly sampled transitions, the loss function, $L(\theta_c)$, is approximately evaluated by

$$L(\theta_c) \approx \frac{1}{N_{\text{batch}}} \sum_{i=1}^{N_{\text{batch}}} \left[\left(Q_\mu\left(s_i, \mu(s_i; \theta_a); \theta_c\right) - y_i\right)^2\right], \tag{9.47}$$

where $y_i = r_i + \gamma Q'_{\mu'}\left(s_{i+1}, \mu'\left(s_{i+1}; \theta'_a\right); \theta'_c\right)$ gives the approximate target Q-value that the target network generates upon N_{batch} transitions sampled at random. By differentiating $L(\theta_c)$ w.r.t. θ_c, the gradient is attained

$$\begin{aligned} \nabla_{\theta_c} L(\theta_c) \approx \frac{2}{N_{\text{batch}}} \sum_{i=1}^{N_{\text{batch}}} &\left[\left(Q_\mu\left(s_i, \mu(s_i; \theta_a); \theta_c\right) - y_i\right)\right. \\ &\left.\times \nabla_{\theta_c} Q_\mu(s_i, \mu(s_i; \theta_a); \theta_c)\right]. \end{aligned} \tag{9.48}$$

The training-critic network parameter, θ_c, is refreshed by utilizing the stochastic gradient descent method [259].

The target-actor and target-critic networks are refreshed based on the training-actor and training-critic networks

$$\begin{aligned} \theta'_a &\leftarrow \tau_a \theta_a + (1 - \tau_a)\theta'_a, \\ \theta'_c &\leftarrow \tau_c \theta_c + (1 - \tau_c)\theta'_c, \end{aligned} \tag{9.49}$$

where τ_a and τ_c are the decaying rates for the training-actor and training-critic networks, respectively.

9.3.3.3 TD3

TD3 is one of the latest extensions of DDPG and consists of a training network and a target network, where the training network is made of a training-actor and two training-critic networks, and the target network comprises a target-actor and two target-critic networks, as shown in Figure 9.5. TD3 addresses the Q-value over-estimation problem of the DDPG algorithm by incorporating *three improvements* over the classical DDPG model, namely, *clipped double-Q learning*, *target policy smoothing*, and *delayed policy update* [64, 328].

- *Clipped double-Q learning:* TD3 contains two training-critic and target-critic networks to produce two Q-values. The lesser of the two is used to evaluate the target Q-value in the Bellman error loss function. Specifically, $Q_\mu(s_t, a_t; \theta_c)$ in the DDPG is replaced by $Q_\mu(s_t, a_t) = \min\{Q_1(s_t, a_t; \theta_1), Q_2(s_t, a_t; \theta_2)\}$ in the TD3.
- *Target policy smoothing:* TD3 perturbs actions produced by the target-actor network (i.e., "target action") with noises and smooths the corresponding

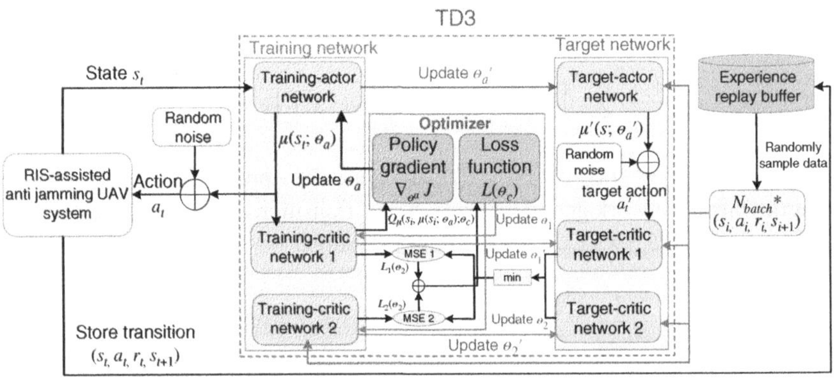

Figure 9.5 The proposed TD3-based framework with an actor network comprising an actor and a target-actor, and a critic network comprising two critics and two target-critics. The experience replay buffer gives batches of samples of state transitions for training and updating the networks.

Q-function values to enhance the resistance of the policy against erroneous Q-functions. The smoothed target action is written as

$$a'_t = \text{clip}\left(\mu'\left(s_{t+1}; \theta'_a\right) + \text{clip}\left(\epsilon', -\sigma_m^2, \sigma_m^2\right), a_{\min}, a_{\max}\right). \tag{9.50}$$

Here, the noise ϵ' is taken at random from a Gaussian distribution with zero mean and variance σ_a^2, i.e., $\epsilon' \sim \mathcal{N}(0, \sigma_a^2)$; and σ_m^2 is the maximum exploration noise supported by the environment.

- *"Delayed" policy update:* The training-actor and target-actor networks (i.e., policies) are refreshed less frequently than the training-critic and target-critic networks. For example, it was recommended in [64] that the training-actor and target-actor networks are refreshed after the training-critic and target-critic networks are refreshed twice in TD3.

In contrast, the DDPG model only contains a training-critic network and a target-critic network, and produces a Q-value. The DDPG model does not add noise to target actions. Moreover, the classical DDPG model refreshes its train-actor and target-actor networks and train-critic and target-critic networks at the same pace.

9.4 Performance Evaluation

9.4.1 RIS-assisted Radio Surveillance System

In this section, extensive experiments are conducted to evaluate the proposed approach. The distance between the Tx and Rx is $d = 100$ m. The initial and final locations of the UAV are $\boldsymbol{q}_0 = [0, -100, 30]^T$ m and $\boldsymbol{q}_F = [150, 100, 100]^T$ m, respectively. The RIS has $N = 5 \times 4 = 20$, $N = 5 \times 8 = 40$, or $N = 5 \times 12 = 60$ reflecting elements, and the reference point of the RIS is $\boldsymbol{q}_R = [50, 50, 30]^T$ m. The scheduling horizon is $T = 30$ seconds with $\delta = 0.2$ seconds per time slot. The other parameters concerning the system model are summarized in Table 9.1.

The TD3-based network is implemented by a two-layer feedforward neural network with 128 and 64 hidden nodes in the two layers. Rectified linear units (ReLUs) are used as the activation functions between the layers of the actor and critic networks. The output layers of the actor networks use the tanh(·) activation function to bound the output actions within $[-\pi, \pi)$ for the RIS configuration and $[-2, 2]$ m/s^2 for the UAV navigation. The state and action are taken as the input to the first layer of the critic networks. The learning rates of both the actor and critic networks are 10^{-4}. The exploration noises used to train the TD3 actor are sampled from the zero-mean complex Gaussian noise with a variance of 0.1.

Table 9.1 The parameters of the system model.

Parameter	Value
RIS antenna separation, d_x, d_y	$\lambda/2$
RIS amplitude coefficients, $\upsilon_{\min}$, α_0, ϕ	0.8, 1.6, 0.43π
Maximum and minimum speeds, $V_{\max}$, $V_{\min}$	60 m/s, 10 m/s
Maximum and minimum altitudes, $z_{\max}$, $z_{\min}$	120 m, 30 m
Path loss, ρ	−30 dB
Path loss exponents, κ_1, κ_2, κ_3, κ_4 , κ_5	3.5, 2.2, 2.2, 3.5, 2.2
Rician factor coefficients, ξ_1, ξ_2	1, 4.4
Noise power, σ^2_{U}, σ^2_{Rx}	−169 dBm
Covert distance between the UAV and Tx, D_{cov}	100 m

Table 9.2 The hyperparameters of the new TD3-based algorithm.

Parameter	Value
Discounting factor for future reward, γ	0.99
Learning rate for critic network, η_c	1×10^{-4}
Learning rate for actor-network, η_a	1×10^{-4}
Decaying rate for actor and critic networks, ρ_τ	5×10^{-3}
Buffer size for experience replay	1×10^5
Number of episodes, T_{ep}	3000
Number of experiences in the mini-batch, N_{batch}	128
Policy delay update frequency	2
Maximum value of the Gaussian noise, σ^2_m	0.2
Variance of the exploration noise, σ^2_e	0.1
Variance of the policy noise, σ^2_a	0.1

The policy noises that smooth the target-actor are sampled from the zero-mean complex Gaussian noise with a variance of 0.1. The maximum exploration noise variance is 0.2. The update frequency of the actor networks is 2. The TD3-based network is built on Python 3.9, PyTorch v.1.8.1, and trained on a server with an Nvidia Tesla P100 SXM2 16GB GPU. The network hyperparameters are summarized in Table 9.2.

Baseline 1: This scheme is under the assumption that the instantaneous CSI of all involved channels is perfectly known for the configuration of the RIS. Accordingly, the signals from different paths are combined coherently at the UAV to maximize the received signal power and, in turn, the achievable eavesdropping rate. The phase shifts of the RIS are configured to align the phases of the Tx-UAV link, i.e., $\arg(h^t_{\mathrm{Tx,U}})$, and the RIS-UAV link, i.e., $\arg(\boldsymbol{h}^t_{\mathrm{R,U}})$, at the UAV. To align the phase shifts, we rewrite the LoS component of the Tx-RIS link, $\boldsymbol{h}^{los}_{\mathrm{Tx,R}} \in \mathbb{C}^{N\times 1}$, as

$$\begin{aligned}\boldsymbol{h}^{los}_{\mathrm{Tx,R}} = & \Big[1, \dots, e^{-j\frac{2\pi d_y}{\lambda}(N_y-1)\phi^y_{\mathrm{Tx,R}}}, e^{-j\frac{2\pi d_x}{\lambda}\phi^x_{\mathrm{Tx,R}}}, \dots, \\ & e^{-j\frac{2\pi}{\lambda}[d_x\phi^x_{\mathrm{Tx,R}} + d_y(N_y-1)\phi^y_{\mathrm{Tx,R}}]}, \dots, \\ & e^{-j\frac{2\pi}{\lambda}[d_x(N_x-1)\phi^x_{\mathrm{Tx,R}} + d_y(N_y-1)\phi^y_{\mathrm{Tx,R}}]}\Big]^T.\end{aligned} \tag{9.51}$$

The n-th element of $\boldsymbol{h}^{los}_{\mathrm{Tx,R}}$, $n = 1, \dots, N$, is given by

$$\boldsymbol{h}^{los}_{\mathrm{Tx,R}}(n) = \exp\left\{-j\frac{2\pi}{\lambda}\left[d_x(n_x-1)\phi^x_{\mathrm{Tx,R}} + d_y(n_y-1)\phi^y_{\mathrm{Tx,R}}\right]\right\}, \tag{9.52}$$

$$n_x = \left\lfloor \frac{n-1}{N_y} \right\rfloor + 1,\ n_y = n - N_y \times \left\lfloor \frac{n-1}{N_y} \right\rfloor.$$

With the knowledge of the NLoS components $\boldsymbol{h}^{nlos}_{\mathrm{Tx,R}}$ and $\boldsymbol{h}^{nlos,t}_{\mathrm{R,U}}$ (or more explicitly, the phases of the Tx-RIS and RIS-UAV links, $\arg(\boldsymbol{h}_{\mathrm{Tx,R}})$ and $\arg(\boldsymbol{h}^t_{\mathrm{R,U}})$), the channel phases are aligned at the UAV, as given by

$$\theta^t_n - \arg(\boldsymbol{h}_{\mathrm{Tx,R}}) - \arg(\boldsymbol{h}^t_{\mathrm{R,U}}) = \arg(h^t_{\mathrm{Tx,U}}). \tag{9.53}$$

As a result, the RIS phase shifts are given by

$$\theta^t_n = \arg(h^t_{\mathrm{Tx,U}}) + \arg(\boldsymbol{h}_{\mathrm{Tx,R}}) + \arg(\boldsymbol{h}^t_{\mathrm{R,U}}). \tag{9.54}$$

Under the assumption of the perfect CSI of all channels involved, this approach is expected to provide the upper bound for the eavesdropping rate and eavesdropping success probability of the TD3-based UAV trajectory planning in the considered problem.

Baseline 2: We also consider the proposed TD3-based scheme without the assistance of the RIS (i.e., $N = 0$), referred to as "*TD3 without RIS,*" where the action space only consists of the UAV accelerations. This baseline helps to reveal the gain stemming from using the RIS.

For fair comparisons, we adopt a consistent hyperparameter setting across all three schemes; see Tables 9.1 and 9.2. In the baseline schemes, we use the TD3-based algorithm to learn the control policy of the UAV with the action $a'_t = \{A_{it} \in [-A_{\max}, A_{\max}],\ i = x, y, z\}$ at the t-th time step.

9.4.1.1 Results of Policy Learning

Figure 9.6 plots both the per-episode reward and the average reward of the proposed and baseline algorithms under $N = 20$, 40, and 60. The average reward for

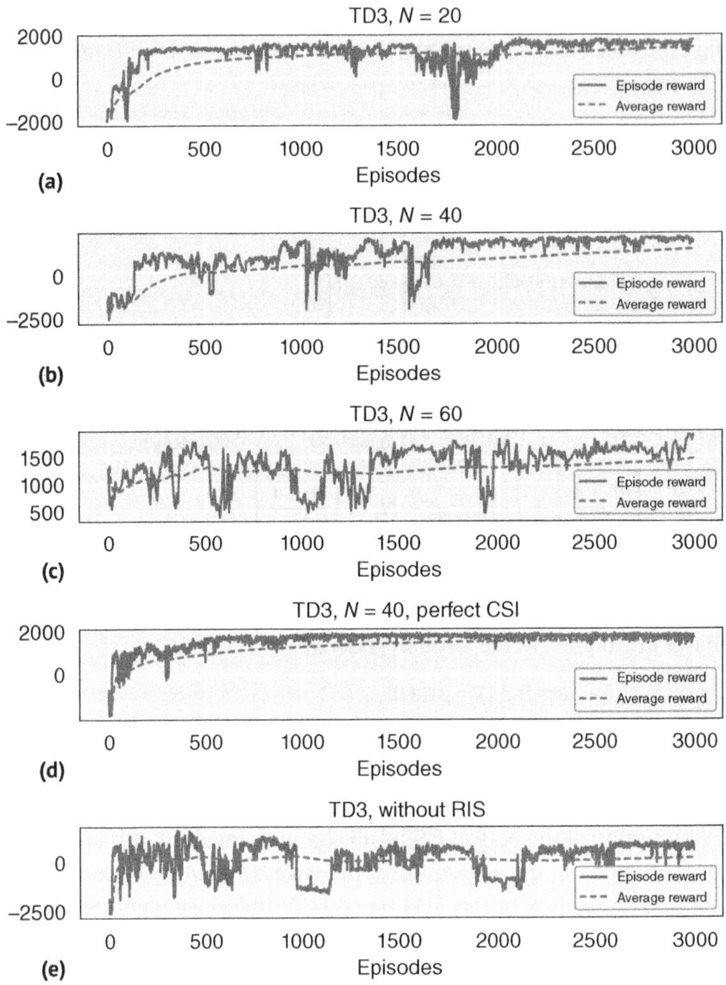

Figure 9.6 The per-episode and the average rewards of the proposed and baseline algorithms: (a) the proposed TD3 scheme under $N = 20$; (b) the TD3 scheme under $N = 40$; (c) the TD3 scheme under $N = 60$; (d) the benchmark with perfect CSI under $N = 40$; and (e) the TD3 scheme without RIS.

the training of the TD3 algorithm at each episode, denoted by $\bar{r}$, is evaluated as $\bar{r} = \frac{1}{T_w} \sum_{j=t}^{T_w} r_t^i$, where r_t^i is the step reward for each training episode; see (9.20). Figure 9.6 shows that the average reward gradually increases as the UAV control policy changes with a randomly generated target trajectory in each episode. The proposed algorithm outperforms or approaches the baselines and reaches its maximum reward at the 2505th episode when $N = 20$, at the 2635th episode when

$N = 40$, and at the 2989th episode when $N = 60$. The benchmark with perfect CSI reaches its maximum reward at the 1588th episode when $N = 40$. The fast convergence of the benchmark with perfect CSI is due to its substantially smaller action space comprising only UAV accelerations and the assumption of instantaneously available, perfect CSI of all channels. The baseline scheme without the RIS reaches its maximum reward at the 353rd episode. However, it still undergoes significant fluctuations, even after the convergence of its average reward.

9.4.1.2 Test Results of Learned Policy

According to the learning results in Section 9.4.1.1, we examine the proposed algorithm under $N = 20, 40$, and 60 at the RIS. We also test the benchmark with perfect CSI ($N = 40$), and the scheme without RIS, and examine 2000 testing episodes. Each testing episode comprises 150 steps. During the testing process, no exploration noise is added.

We test two application scenarios with different initial and final locations of the UAV: $\boldsymbol{q}_0 = [0, -100, 30]^T$ m and $\boldsymbol{q}_F = [150, 100, 100]^T$ m in *Scenario 1*, and $\boldsymbol{q}_0 = [0, -150, 30]^T$ m and $\boldsymbol{q}_F = [50, 150, 100]^T$ m in *Scenario 2*. We plot the 2D projections of the 3D UAV trajectories learned by the proposed and baseline schemes in Figure 9.7. We see that, in both Scenarios 1 and 2, the UAV can stay near the Tx while keeping the safety distance from the Tx, especially when N is large. Under both the new scheme and benchmark with perfect CSI, the UAV can reach a small region centered at the final location at the end of the surveillance mission. In contrast, the baseline scheme without the RIS cannot reach the region in the z-direction, as shown in the top three subfigures of both Figures 9.8 and 9.9.

To examine the UAV's eavesdropping performance and demonstrate the usefulness of the RIS, we plot the 3D UAV trajectory along the x-, y-, and z-axes (top three subfigures), and the cumulative distribution functions (CDFs) of the per-slot eavesdropping success probability and per-slot eavesdropping rate (bottom two subfigures) in Figures 9.8 and 9.9.

Recall that eavesdropping is successful at time slot t if the SNR received by the UAV is greater than that received by the Rx, i.e., $\gamma_{\mathrm{U}}^t \geq \gamma_{\mathrm{Rx}}^t$. At any time slot t, the per-slot eavesdropping success probability is the probability that the UAV can successfully eavesdrop on the Tx at the time slot, i.e., $\Pr\{\gamma_{\mathrm{U}}^t \geq \gamma_{\mathrm{Rx}}^t\}$ as per t. We evaluate the per-slot eavesdropping success probability of a mission by repeating the mission independently 2000 times. The channel fading is independently generated among the 2000 tests. For any time slot t, the per-slot eavesdropping success probability is evaluated as the proportion of tests with successful eavesdropping at the time slot. For each time slot, the (average) per-slot eavesdropping rate is evaluated by averaging the eavesdropping rates of the 2000 tests at the time slot.

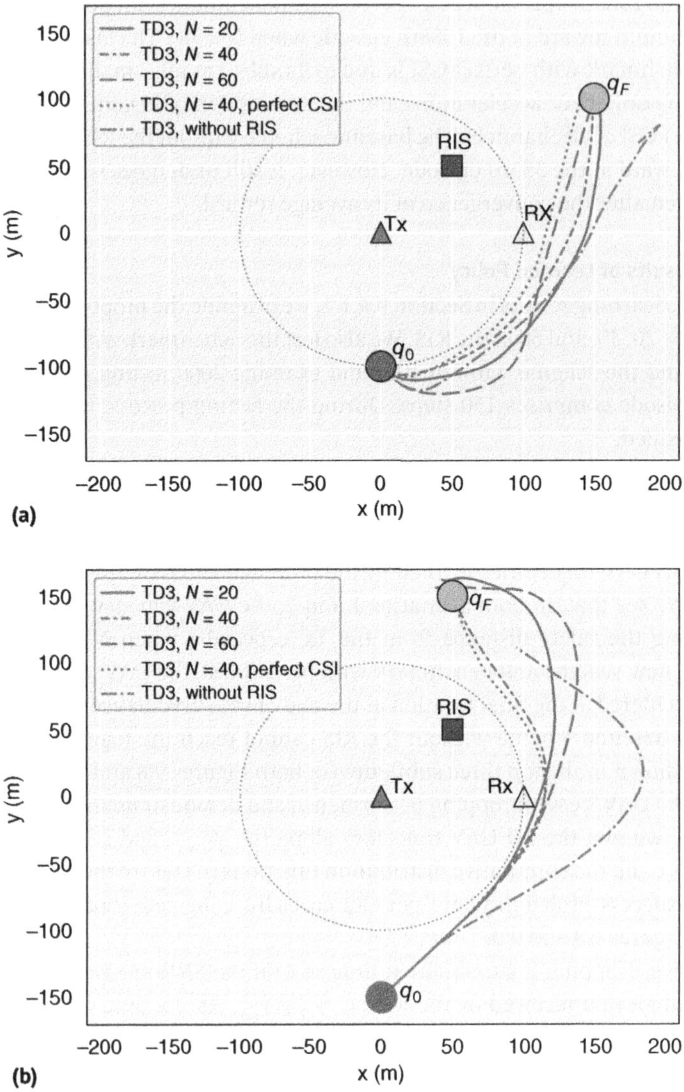

Figure 9.7 The projection of 3D UAV trajectory on the x- and y-plane, where the green and orange dots are the UAV initial and final locations, respectively; the red and yellow triangles denote the locations of the Tx and Rx, respectively; and the blue square is the RIS reference point. (a) Scenario 1: $q_0 = (0, -100, 30)$ and $q_F = (150, 100, 100)$; (b) Scenario 2: $q_0 = (0, -150, 30)$ and $q_F = (50, 150, 100)$.

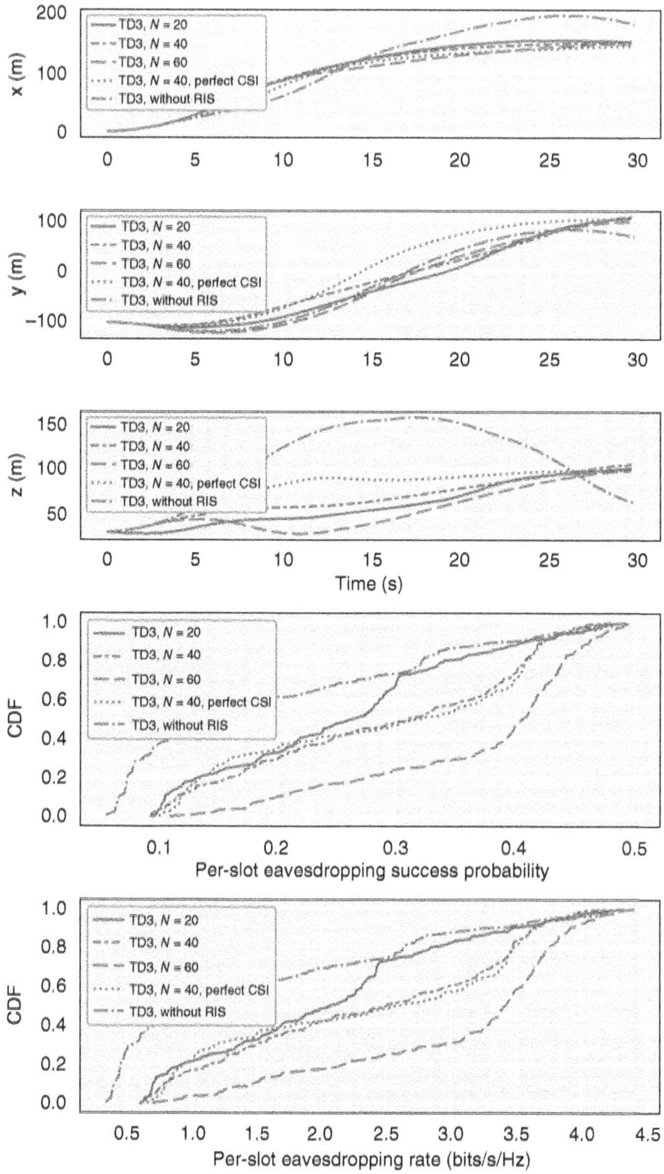

Figure 9.8 The 3D UAV trajectory along the x-, y-, and z-axes (top three figures), and the CDFs of the eavesdropping success probability and rate (bottom two figures). Scenario 1 is considered, where $\boldsymbol{q}_0 = [0, -100, 30]^T$ m and $\boldsymbol{q}_F = [150, 100, 100]^T$ m.

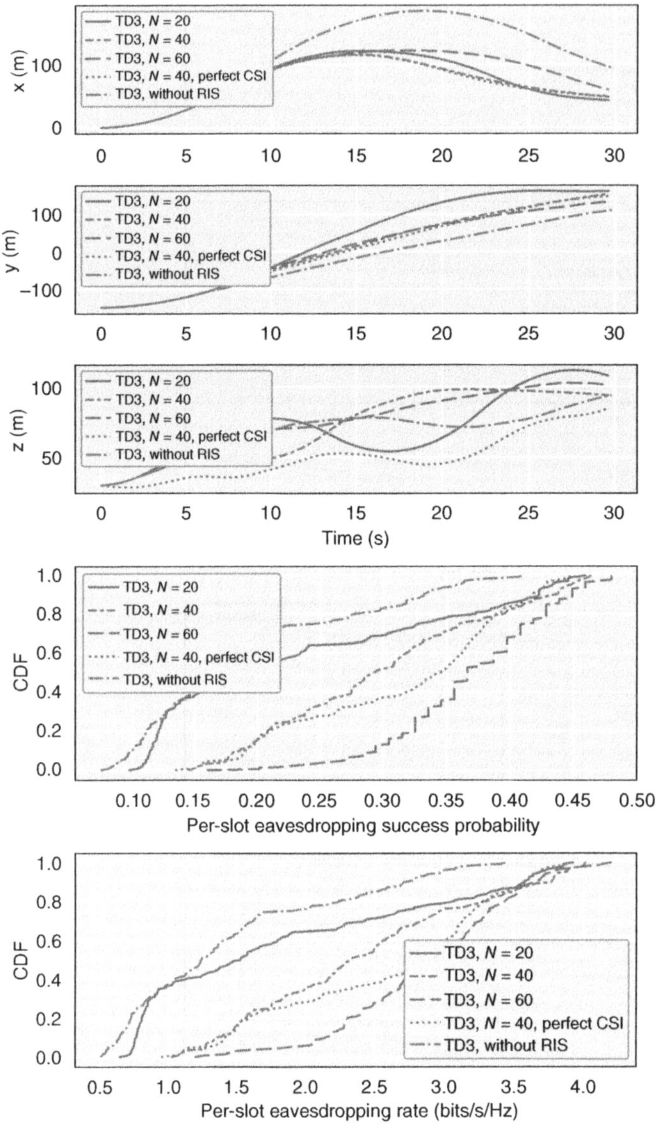

Figure 9.9 The 3D UAV trajectory along the x-, y-, and z-axes (top three figures), and the CDFs of the eavesdropping success probability and rate (bottom two figures). Scenario 2 is considered, where $\boldsymbol{q}_0 = [0, -150, 30]^T$ m and $\boldsymbol{q}_F = [50, 150, 100]^T$ m.

Figures 9.8 and 9.9 show that the proposed scheme is generally very close to the scheme under the ideal assumption of perfect CSI in the per-slot eavesdropping success probability and per-slot eavesdropping rate, showing the robustness of the proposed algorithm to channel uncertainty. The usefulness of the RIS is demonstrated in the sense that the UAV eavesdropping performance is the worst in both scenarios when there is no RIS, and is the best when the RIS has $N = 60$ reflecting units.

Last but not least, we collate the average eavesdropping success probability and the average eavesdropping rate of the proposed and baseline schemes in the two tested scenarios in Table 9.3. The average eavesdropping success probability and the average eavesdropping rate of the mission are obtained by averaging the per-slot eavesdropping success probability and the per-slot eavesdropping rate across all time slots.

Compared with the scheme without RIS, the average eavesdropping rate of the proposed scheme is improved by 37.48% when $N = 20$, by 59.20% when $N = 40$, and by 120.8% when $N = 60$. The benchmark with perfect CSI is only 3.36% better than the proposed approach only relying on the observation of the transmit rate when $N = 40$, despite the fact that the benchmark is under the ideal assumption of the instantaneous availability of the perfect CSI of all channels.

9.4.2 RIS-assisted Radio Surveillance System

We carry out extensive experiments in Python to evaluate the proposed approach. The location of the jammer is $\boldsymbol{q}_J = [-25, -25, 0]^T$ m. The UAV's initial and

Table 9.3 Comparison of eavesdropping capability between the TD3-based algorithms with and without the RIS, and under the ideal assumption of perfect CSI of all channels, where the mission duration $T = 30$ s.

	Number of RIS elements (N)	**Average eavesdropping success probability (%)**	**Average eavesdropping rate (bit/Hz/s)**
Scenario 1	$N = 20$ (proposed)	25.502	2.032
	$N = 40$ (proposed)	28.751	2.353
	$N = 60$ (proposed)	38.215	3.263
	no RIS	19.215	1.478
	$N = 40$ (perfect CSI)	28.884	2.432
Scenario 2	$N = 20$ (proposed)	23.140	1.818
	$N = 40$ (proposed)	29.842	2.417
	$N = 60$ (proposed)	33.667	2.835
	no RIS	19.106	1.441
	$N = 40$ (perfect CSI)	31.546	2.594

Table 9.4 The parameters of the system model.

Parameter	Value
RIS antenna separation, d_x, d_y	$\lambda/2$
Maximum and minimum speeds, $V_{\max}$, $V_{\min}$	40 m/s, 2 m/s
Path loss, ρ	−30 dB
Path loss exponents, κ_1, κ_2	3.5, 2.8
Rician factor coefficients, ξ_1, ξ_2	1, 4.4
Rician factor, β	3 dB
Noise power, σ_{U}^2	−169 dBm

final locations are $\boldsymbol{q}_0 = [200, -100, 5]^T$ m and $\boldsymbol{q}_F = [100, 60, 50]^T$ m. The RIS has $N = 5 \times 4 = 20$ (or $N = 5 \times 8 = 40$) reflecting elements, and the reference point is $\boldsymbol{q}_{\mathrm{R}} = [50, 50, 30]^T$ m. The scheduling horizon is $T = 30$ seconds with each time slot being $\delta = 0.1$ seconds. The other parameters concerning the system model are collated in Table 9.4.

9.4.2.1 Experiment Settings

The proposed DDPG network is composed of actor networks implemented using FCNNs with three hidden layers and learning rates of 10^{-4}. The first, second, and third layers of the actor networks have 64, 128, and 64 neurons, respectively. The output layer implements the tanh(·) activation function to bound the output actions within $[-\pi, \pi)$ for the RIS configuration and $[-2, 2]$ m/s^2 for the UAV control. Additionally, the chapter utilizes critic networks that employ FCNNs with two hidden layers and learning rates of 10^{-3}. Both hidden layers utilize the ReLU activation functions with 64 neurons in the first layer and 128 neurons in the second layer. The DDPG actor policy is trained using additive noise $\mathcal{N}$, which is sampled from a complex Gaussian noise distribution with zero mean and variance 0.2.

The proposed TD3 network is built upon the DDPG network. It includes two duplicates of the training-critic and target-critic networks; see Figure 9.5. Similar to the DDPG network, the actor in the TD3 network is trained using exploration noise that is drawn from a complex Gaussian distribution with zero mean and variance 0.2. Additionally, the target-actor in the TD3 network is smoothed using policy noise that is drawn from a complex Gaussian distribution with zero mean and variance of 0.2. The maximum exploration noise is set to 0.5, and the actor networks are refreshed every two steps. The TD3 improves the DDPG by providing faster and smoother convergence, which is especially beneficial for larger RISs.

The hyperparameters of the DDPG and TD3 networks are summarized in Table 9.5. The DDPG and TD3 networks are trained on a server equipped with a NVIDIA Tesla P100 SXM2 16GB GPU.

Table 9.5 The hyperparameters of the proposed DDPG and TD3 algorithms.

Parameter	Value
Reduction coefficient for upcoming reward, γ	0.99
Training coefficient for actor and critic networks, η_a, η_c	1×10^{-4}
Declining coefficient for actor and critic networks, τ_a, τ_c, ρ_τ	5×10^{-3}
Capacity for experience repetition	1×10^5
Quantity of episodes, T_{ep}	3000
Total steps per episodes, T_s	300
Quantity of experiences in a mini-batch, N_{batch}	128
Variance of the exploration noise, σ_e^2	0.2
Delayed policy update interval (TD3)	2
Variance of the policy noise (TD3), σ_a^2	0.2
Largest value of the Gaussian noise (TD3), σ_m^2	0.5

As discussed in Chapter 9.1, no existing works have addressed the RIS-assisted jamming rejection for a UAV-borne IoT platform. For this reason, it may not be fair or adequate to compare the proposed approach in this work with those existing works. With due diligence, we come up with two baselines to further verify the merits of the proposed approach:

Baseline 1: This baseline applies TD3 to UAV flight path planning in the absence of the RIS, referred to as "baseline without RIS" or simply "without RIS." The TD3 algorithm learns the UAV's trajectory solely based on the UAV's received data rate without CSI involving the RIS or jammer.

Baseline 2: This baseline decouples the UAV's trajectory plan from the RIS configuration by first using a TD3-based algorithm to optimize the UAV's trajectory given the RIS configuration, and then maximizing the SNR at each time slot using the Dinkelbach method under the assumption of perfect and instantaneous CSI for all involved channels. The Dinkelbach method is used to reformulate the SNR maximization problem as a fractional program defined as $F(\gamma_{\text{U}}^t) = \min_{\Theta_t} f(\Theta_t) - \gamma_{\text{U}}^t g(\Theta_t)$, s.t. $-\pi \leq \theta_n^t < \pi, \forall n, t$, where $\gamma_{\text{U}}^t = \frac{f(\Theta_t)}{g(\Theta_t)}$ and Θ_t gives the RIS configuration. Given γ_{U}^t, the fractional program can be reorganized as a quadratic program with a unit-modulus constraint and solved using MO. The value of γ_{U}^t is refreshed based on the resultant Θ_t. This process is repeated until convergence, and the convergent value of γ_{U}^t is output [66].

9.4.2.2 Results of Policy Learning

Figure 9.10 plots the per-episode and average rewards of the proposed and baseline algorithms for $N = 20$ and 40. The average reward for the i-th training episode,

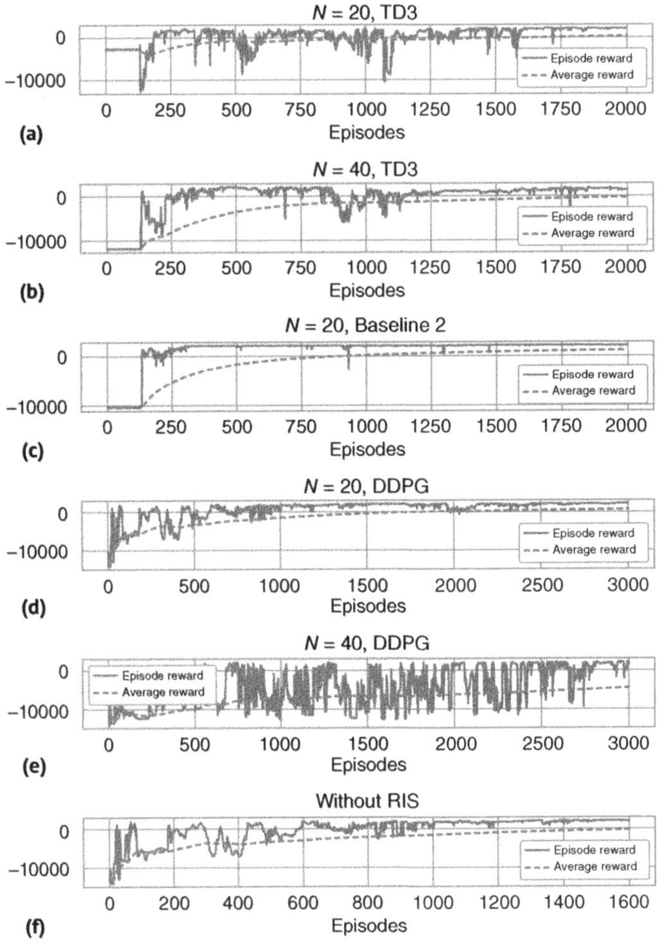

Figure 9.10 The per-episode and average rewards of the proposed DDPG and TD3 algorithms. (a) Plots the proposed TD3 scheme when $N = 20$; (b) plots the proposed TD3 scheme when $N = 40$; (c) plots Baseline 2 when $N = 20$; (d) plots the proposed DDPG scheme when $N = 20$; (e) plots the proposed DDPG scheme when $N = 40$; and (f) plots Baseline 1 "without RIS."

denoted by $\bar{r}_i$, is evaluated as $\bar{r}_i = \frac{1}{i}\sum_{j=1}^{i} r_j$, where $i = 1, \dots, T_{ep}$, and r_j is the step reward for the j-th training episode; see (9.40). The per-episode reward indicates the reward obtained by the agent in a single episode of the task, while the average reward accounts for the cumulative reward over multiple episodes. The average reward is useful when the reward function is noisy or stochastic, because the per-episode reward may fluctuate significantly over episodes, making it hard to accurately evaluate the agent's performance. By taking the average over multiple episodes, we can get a more stable estimate of the agent's performance.

Figure 9.10 shows that the average reward gradually increases, as the UAV control policy adapts to a randomly generated target trajectory in each episode. The proposed TD3 algorithm outperforms the baselines, and achieves its maximum reward at the 654th episode for $N = 20$, and at the 477th episode for $N = 40$. The proposed DDPG algorithm reaches its maximum reward at the 1607th episode when $N = 20$, and at the 2995th episode when $N = 40$. Baseline 1 without RIS reaches its maximum reward at the 280th episode. Baseline 2 reaches its maximum reward at the 1752nd episode when $N = 20$. The fast convergence of Baseline 2 is due to its substantially smaller action space of only UAV accelerations, resulting from an unrealistic assumption of perfect and instantaneous CSI of all involved channels.

In general, the TD3 converges faster and more smoothly than the DDPG. Yet, it undergoes less smooth changes in the per-episode reward when the action space is smaller, i.e., $N = 20$. This is because the TD3 has two critic networks for both the training and target networks (c.f. Figure 9.5), which could incur higher complexity and more randomness when training, especially when the action space is smaller. In contrast, DDPG is suitable for training tasks that are less complicated and have relatively smaller action spaces.

9.4.2.3 Test Results of Learned Policy

Using the learning results obtained in Section 9.4.2.2, we test the proposed DDPG-based and TD3-based algorithms for $N = 20$ and 40 at the RIS, as well as the baseline schemes for $N = 20$ at the RIS for comparison. 500 testing episodes are conducted, each consisting of 300 steps. During testing, no exploration noise is added. The proposed algorithms and baselines are evaluated in terms of the 3D UAV trajectory and received data rate. Figure 9.11 shows the 3D trajectory of the UAV, while Figure 9.12 illustrates the trajectory in the x-y plane and along the z-axis. These figures demonstrate that the UAV is able to adapt its control policy to the anti-jamming communication and successfully reach the destination.

Figure 9.13 shows the received data rate of the UAV as the mission duration increases. The results are based on an average of 500 independent testing episodes, with error bars representing the associated uncertainty. It is observed that the received data rate increases with the mission duration under all considered algorithms, and the use of an RIS improves the received data rate. The proposed TD3 and DDPG algorithms give slightly lower data rates than Baseline 2, but operate without the CSI involving the RIS or the jammer, demonstrating their ability to adapt to system changes. Additionally, the proposed TD3 algorithm achieves a substantially higher data rate than the proposed DDPG algorithm, particularly for larger numbers of RIS elements. This is attributed to the faster convergence and

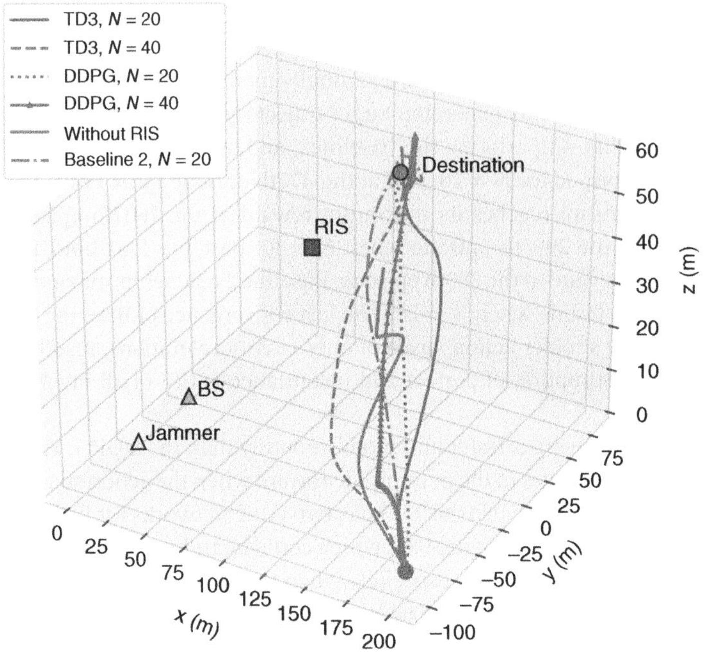

Figure 9.11 3D UAV trajectory, where the green and red dots are the UAV's initial location and its expected destination, the orange and yellow triangles denote the locations of the BS and Jammer, and the blue square is the RIS reference point.

better convergent control policy of TD3 compared to DDPG, as previously shown in Figure 9.10.

Figure 9.14 illustrates the distance between the UAV and its expected destination at the end of the mission. The results show that as the mission duration increases, the distance decreases for all algorithms. Without the use of the RIS, the UAV is unable to reach its expected final location as it must remain close to the BS to maintain a sufficient data rate. However, the proposed algorithms, such as TD3, enable the UAV to get closer or reach its final location by adjusting the RIS to enhance the desired signals, weaken the jamming signals, and extend the effective BS-UAV transmission range.

It is worth noting that as the mission duration increases, DDPG can increasingly approach TD3 in the average achievable data rate. This indicates that DDPG is suitable for a less constrained problem setting where there is sufficient time for the UAV to maneuver and explore its action space. In this case, DDPG can be a suitable solution, as it can benefit from its simpler network architecture than TD3.

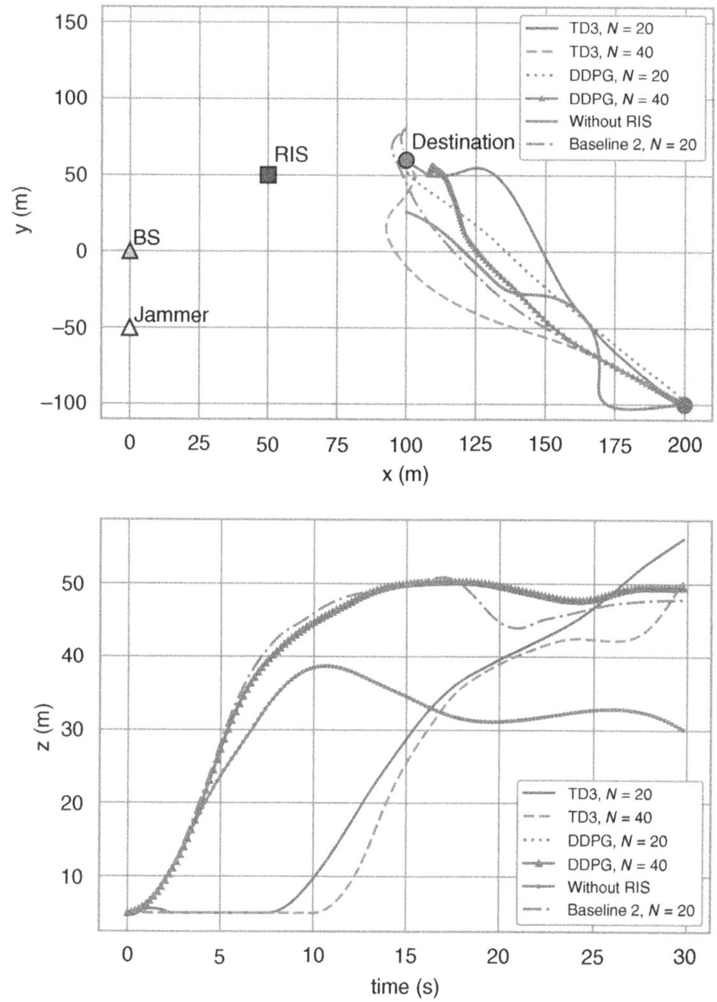

Figure 9.12 Projection of the UAV trajectory on the x-y plane, with the green and red dots being the UAV initial and final locations, the orange and yellow triangles being the locations of the BS and Jammer, and the blue square being the RIS reference point.

In contrast, TD3 demonstrates significant gains over DDPG when the time constraint is more stringent. In other words, TD3 suits better under a shorter mission duration, since it has a more complex network structure and can generate more randomness to test the action space more extensively for better solutions.

Figure 9.15 plots the CDF of the distance between the UAV and its expected destination at the end of the mission, for mission durations of $T = 30$ and 40 seconds.

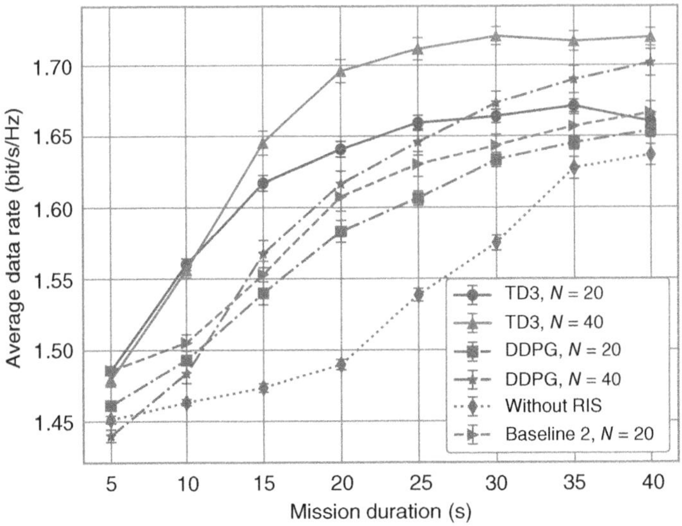

Figure 9.13 The received data rate of the UAV vs. the mission duration averaged over 500 independent testing runs.

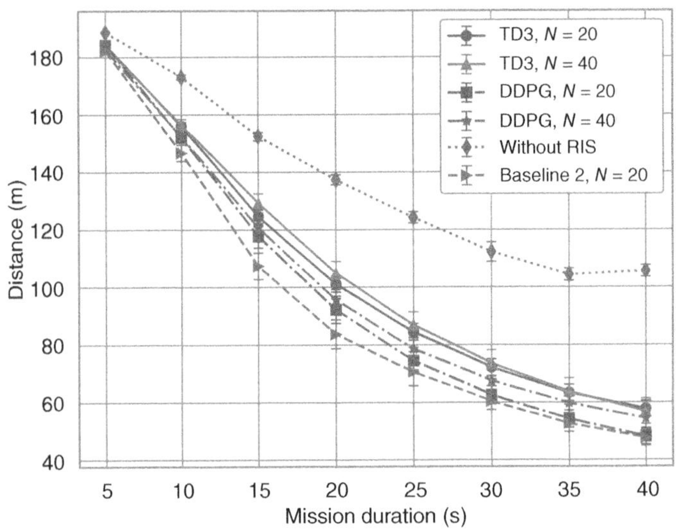

Figure 9.14 The finishing distance between the UAV and its destination at the end of the mission with the increase of the mission duration.

The results show that as the number of RIS elements (i.e., N) or the mission duration (i.e., T) increases, the proposed TD3 algorithm can get closer to or reach the destination more frequently, and performs significantly better than the case without an RIS. While perfect CSI is important for UAV trajectory planning, as seen in Baseline 2 for $N = 20$, the TD3 algorithm can produce equally effective trajectories without CSI by utilizing a larger RIS with more elements (as seen in TD3 for $N = 40$). On the other hand, the DDPG algorithm appears to suffer from overfitting, as the distance between the UAV and its expected destination has little dispersion. This indicates that DDPG is more prone to overestimating the Q-value function for a small number of possible actions, leading to a noisy gradient for policy refreshes and less effective UAV trajectories, and lower data rates.

Figure 9.16 plots the received data rate of the UAV versus the finishing distance of the UAV to the destination at the end of the mission time. The figure is produced by collating the data rates shown in Figure 9.13 and their corresponding finishing

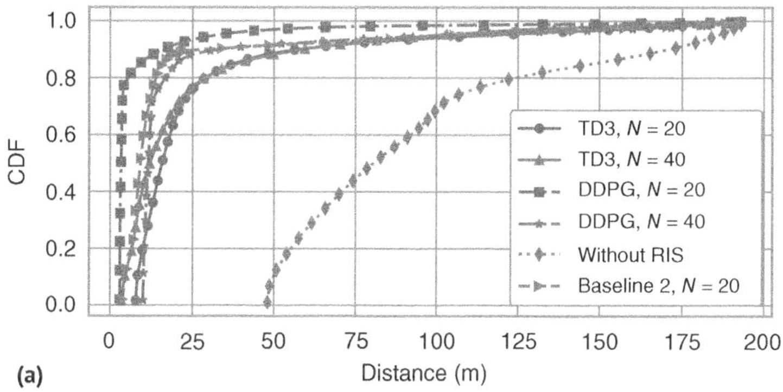

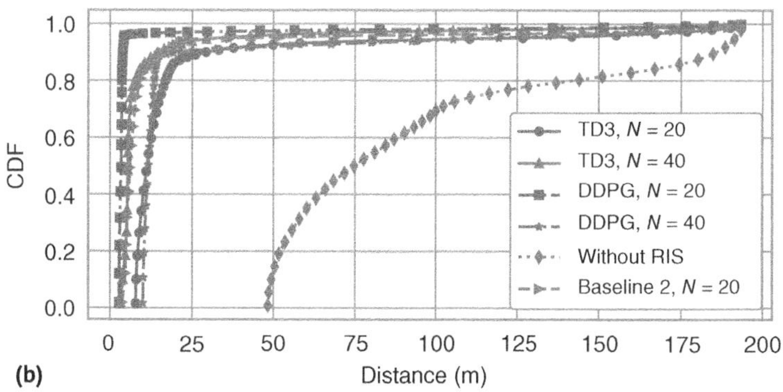

Figure 9.15 CDF of the distance from the UAV to the final location when the mission duration is $T = 30$ and 40 seconds. (a) Mission duration $T = 30$ (s); (b) Mission duration $T = 40$ (s).

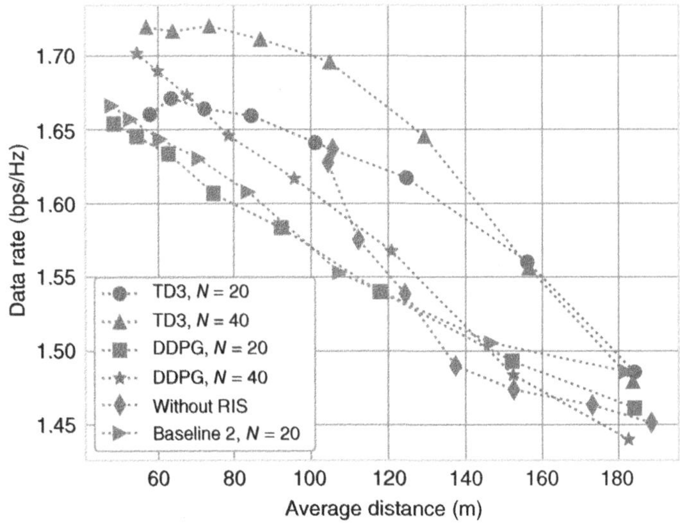

Figure 9.16 The UAV's data rate vs. its distance to the expected destination, where T ranges from 5 to 40 seconds.

distances shown in Figure 9.14. The values at the top left corner of Figure 9.16 indicate a better trade-off between the data rate and the finishing distance, where the mission duration is 25 to 40 seconds, as discussed in Figures 9.13 and 9.14. Note that the use of target networks and the noise added when exploring new actions for the DDPG and TD3 algorithms may incur more randomness and lead to an unstable performance during training, as observed in Figure 9.16.

9.5 Conclusion

This chapter presents two studies. In the first study, we developed a new DRL-driven framework for UAV-based covert radio surveillance with the assistance of a legitimate RIS, where the RIS augments the radio environment to facilitate a fixed-wing UAV's acquisition of the radio fingerprint of a suspicious Tx. A TD3 model was designed to allow the UAV to learn its trajectory and the RIS configuration only based on its observed transmit rate of the Tx. Extensive experiments demonstrated that the new scheme (Scenario 1) converges fast and smoothly. The use of the RIS allows for the improvement of over 37.48% and 59.20% in the average eavesdropping success probability and average eavesdropping rate, respectively. In the second study, we developed a new DRL-driven framework for the trajectory planning and RIS-assisted jamming rejection for a UAV-borne IoT platform. The DDPG model and its enhancement, TD3, were designed to allow the UAV to learn its trajectory and the RIS configuration only based on its received data rate, eliminating the need for CSI for learning. Extensive simulations showed that the proposed DRL algorithms offer the UAV reliable resistance against jamming. The TD3 algorithm converges faster and more smoothly than the DDPG algorithm. It also demonstrates robustness against different locations of the jammer. This is particularly important due to the difficulty in locating the jammer in practice.

10

Conclusion and Future Work

UAVs have established themselves as a cornerstone technology in modern engineering and applied sciences, with profound impacts across environmental monitoring, disaster response, logistics, infrastructure inspection, and defense. This book has presented an in-depth exploration of the foundational and emerging elements that underpin UAV communications, ranging from sensing and computing to mobility modeling and AI integration, offering a holistic understanding of how UAVs are transforming the way we collect data, connect devices, and make autonomous decisions.

Through the critical examination of 3D communication environments, this book has highlighted the unique challenges and opportunities that UAV systems face in dynamic and often unpredictable operational conditions. The transition from traditional 2D communication models to more realistic 3D network paradigms has unveiled the need for more robust and adaptive connectivity solutions that account for UAV mobility, energy constraints, and interference-limited scenarios.

AI has emerged as a transformative enabler, empowering UAVs with unprecedented autonomy and intelligence. From trajectory planning and resource management to distributed learning and swarm coordination, AI methods, particularly deep learning and reinforcement learning, have shown promise in elevating the performance and adaptability of UAV systems. Coupled with onboard edge computing and intelligent sensing technologies, AI-enhanced UAVs now support increasingly complex, multi-agent missions with minimal human intervention.

Moreover, the integration of cutting-edge technologies, such as RIS and cellular-assisted sensing, has further expanded the design space for UAV communications. These advancements offer novel solutions for improving signal propagation, enhancing sensing accuracy, and enabling more efficient task execution in urban and remote environments alike.

Artificial Intelligence for Unmanned Aerial Vehicles: Sensing, Communication, and Computing, First Edition. Shuyan Hu et al.

Despite the significant progress made, several open challenges remain, offering fertile ground for future research:

- *Scalability and Coordination in UAV Swarms:* As UAV deployments grow in scale, ensuring coordinated, interference-aware, and collision-free operation among hundreds or thousands of aerial agents remains a complex problem. New algorithms are needed to achieve reliable, distributed decision-making under uncertain environments and limited communication resources.
- *Energy-aware AI and Learning Models:* Current AI techniques used onboard UAVs are often computationally intensive and energy-demanding. Future efforts must focus on lightweight, energy-efficient learning models that maintain high performance without compromising mission duration or safety.
- *Security and Privacy in UAV Communications:* With increased autonomy and connectivity comes heightened risk [35]. Ensuring secure communications, safeguarding data privacy, and developing resilience against cyber-attacks are critical areas that require further investigation, especially in adversarial or mission-critical settings.
- *Regulation, Ethics, and Public Acceptance:* As UAVs become increasingly integrated into public and commercial spaces, harmonizing their operation with regulatory frameworks, ethical guidelines, and societal expectations will be crucial. Multi-stakeholder collaboration is essential to address legal, safety, and privacy concerns.
- *Cross-Layer Optimization and Co-Design:* The joint design of communication, control, and computation systems remains an open research frontier. Holistic frameworks that co-optimize these subsystems can yield more efficient and adaptable UAV solutions suitable for dynamic and constrained environments.
- *Digital Twins and Simulation Platforms:* The development of high-fidelity simulation platforms and digital twins for UAV communication networks will support accelerated prototyping, testing, and deployment of novel algorithms and system architectures in a risk-free, cost-effective manner.

In conclusion, UAV communications represent a rapidly evolving field at the intersection of wireless networking, AI, edge computing, and systems engineering. This book has aimed to consolidate current knowledge while illuminating the pathways for future innovation. As UAV technologies continue to mature, they are poised to play an even more pivotal role in enabling resilient, intelligent, and connected systems for both civilian and defense applications.

A

Appendix of Chapter 2

A.1 Proof of the Convexity of C_{erg1} with Respect to l

For notation simplicity, we let $\xi_1 = \frac{P}{\sigma^2}$ and $\xi_2 = \frac{2K+1}{2(1+K)^2}$, where $\xi_1 > 0$ and $\frac{21}{242} < \xi_2 < \frac{3}{8}$. C_{erg1} can be rewritten as

$$C_{\text{erg1}} = \frac{1}{\ln(2)} \left[\ln\left(1 + \xi_1 l^{-\alpha}\right) - \frac{\xi_2}{\left(1 + \frac{l^\alpha}{\xi_1}\right)^2} \right]. \tag{A.1}$$

The first derivative of C_{erg1} with respect to l can be given by

$$\frac{dC_{\text{erg1}}(l)}{dl} = \frac{1}{\ln(2)} \left[\frac{-\xi_1 \alpha l^{-\alpha-1}}{1 + \xi_1 l^{-\alpha}} - \frac{\xi_2}{\xi_1} \cdot \frac{2\alpha l^{\alpha-1}}{\left(1 + \frac{l^\alpha}{\xi_1}\right)^3} \right] \leq 0.$$

The second derivative of C_{erg1} can be given by

$$\begin{aligned}
\frac{d^2 C_{\text{erg1}}(l)}{dl^2} &= \frac{1}{\ln(2)} \left[\frac{\xi_1(\alpha^2 + \alpha) l^{-\alpha-2} + \xi_1^2 \alpha l^{-2\alpha-2}}{\left(1 + \xi_1 l^{-\alpha}\right)^2} + \frac{\xi_2}{\xi_1^2} \cdot \frac{2\alpha(2\alpha - 1) l^{2\alpha-2} - 2\xi_1 \alpha(\alpha - 1) l^{\alpha-2}}{\left(1 + \frac{l^\alpha}{\xi_1}\right)^4} \right] \\
&= \frac{1}{\ln(2)} \left\{ \underbrace{\left[\frac{\xi_1(\alpha^2 + \alpha)}{\left(\xi_1 + l^\alpha\right)^2} - \frac{2\xi_2(2\alpha^2 - \alpha)}{\xi_1\left(1 + \frac{l^\alpha}{\xi_1}\right)^4} \right]}_{a(1)} l^{\alpha-2} + \underbrace{\left[\frac{\xi_1^2 \alpha}{\left(1 + \xi_1 l^{-\alpha}\right)^2} l^{-2\alpha-2} + \frac{\xi_2 2\alpha(2\alpha - 1)}{\xi_1^2\left(1 + \frac{l^\alpha}{\xi_1}\right)^4} l^{2\alpha-2} \right]}_{a(2) \geq 0} \right\},
\end{aligned}$$

Artificial Intelligence for Unmanned Aerial Vehicles: Sensing, Communication, and Computing, First Edition. Shuyan Hu et al.

where

$$a(1) = \frac{\xi_1(\alpha^2+\alpha)}{(\xi_1+l^\alpha)^2} - \frac{2\xi_2\xi_1^3(2\alpha^2-\alpha)}{(\xi_1+l^\alpha)^4}$$

$$= \frac{\xi_1\left[(\alpha^2+\alpha)(\xi_1+l^\alpha)^2 + 2\xi_1^2\xi_2\alpha - 4\xi_1^2\xi_2\alpha^2)\right]}{(\xi_1+l^\alpha)^4}.$$

To achieve $a(1) \geq 0$, $(\alpha^2+\alpha-4\xi_2\alpha^2)\xi_1^2$ has to be larger than 0; or in other words, α must satisfy the following conditions: $\alpha \leq \frac{1}{2}$ or $\alpha \geq \frac{121}{79}$. Since α is the path loss exponent which is no less than 2 in nature, the second derivative of C_{erg1} is larger than 0, i.e., $\frac{d^2C_{\text{erg1}}(l)}{dl^2} \geq 0$. As a result, C_{erg1} is convex with respect to l.

A.2 Proof of $\mathbb{E}\left[l_{\min}\right]$

Consider that the positions of the UAVs follow a random and uniform distribution under the ST mobility models, the PDFs of the distance between a UAV and its nearest neighbor in the swarm is given by [71]

$$f_{l_{\min}}(l) = 3Nl^2\exp\left(-Nl^3\right), \tag{A.2}$$

which can be rewritten as generalized gamma distributions

$$f_{l_{\min}}(l; N^{-\frac{1}{3}}, 3, 3) = \frac{3}{N^{-\frac{1}{3}}\Gamma(1)}\left(\frac{l}{N^{-\frac{1}{3}}}\right)^2 \exp\left[-\left(\frac{l}{N^{-\frac{1}{3}}}\right)^3\right]. \tag{A.3}$$

By order statistics, the expectations of $l_{\min}^\alpha$ is given by [135]

$$\mathbb{E}\left[l_{\min}^k\right] = N^{-\frac{1}{3}k}\frac{\Gamma(1+k)}{\Gamma(1)}r = \Gamma\left(\frac{3+k}{3}\right)N^{-\frac{1}{3}k}r^k. \tag{A.4}$$

By substituting $k=1$ into (A.4), $\mathbb{E}\left[l_{\min}\right]$ is obtained in (2.3).

Collision-avoidance can be considered by enforcing a minimum safety distance between UAVs [271, Sec. III]. Existing protocols designed for a random direction (RD) mobility model, such as a sense-and-stop (S&S) protocol [173], can be applied to the ST mobility model: At any time, if the next waypoints of two UAVs are within the safety distance of each other, the UAV to the right of the other stops and hovers at its current waypoint and the other proceeds as planned until they are the safety distance away. In [173, theorem 3], the S&S protocol was proved to maintain the random uniform stationary distribution of the waypoints of a UAV in the RD mobility model. Given the identical way of the models deploying the waypoints, the waypoints are uniformly distributed in the ST mobility model. However, the joint distribution of multiple UAVs is not uniform due to the collision avoidance [173]. The average distance between a UAV and its nearest neighbor in the swarm needs to be numerically evaluated by

$$\mathbb{E}\left[l_{\min}\right] = \oiint_{\boldsymbol{u}_i, i=1,\ldots,N} \min_{i,j=1,\ldots,N, i\neq j} \{l_{ij}\} \times \underbrace{\frac{1}{\Psi} \prod_{i=1}^{N} f_{\mathrm{U}_i}(\boldsymbol{u}_i) \cdot \mathbb{1}\left(\mathbf{u}_i,\ i=1,\ldots,N\right) \prod_{i=1}^{N}}\, d\boldsymbol{u}_i, \tag{A.5}$$

where $l_{ij} = \|\boldsymbol{u}_i - \boldsymbol{u}_j\|$; $\mathbf{U}_i = (X_i, Y_i, Z_i)$ collects the coordinates of the i-th UAV; $\boldsymbol{u}_i = (x_i, y_i, z_i)$ is a realization of $\mathbf{U}_i$; $f_{\mathrm{U}_i}(\boldsymbol{u}_i) = \frac{3}{4\pi r^3}$ is the PDF of the uniform distribution for the waypoints of a UAV; and the indicator function $\mathbb{1}\left(\boldsymbol{u}_i, i=1,\ldots,N\right) = 1$ if $\min_{i,j=1,\ldots,N, i\neq j}\{l_{ij}\} > r_0$, or $\mathbb{1}\left(\boldsymbol{u}_i, i=1,\ldots,N\right) = 0$, otherwise. The part underscored is the joint PDF of the waypoints of the UAVs when collision avoidance is enforced. $\Psi = \oiint_{\boldsymbol{u}_i, i=1,\ldots,N} \left(\prod_{i=1}^{N} f_{\mathrm{U}_i}(\boldsymbol{u}_i) \cdot \mathbb{1}\left(\boldsymbol{u}_i, i=1,\ldots,N\right)\right) \prod_{i=1}^{N} d\boldsymbol{u}_i$ normalizes the joint PDF. The result of (A.5) can replace (2.1), and the rest of our analysis remains unchanged.

A.3 Proof of $\mathbb{E}[l_I]$

Let $\mathcal{P}_2$ denote the probability that the distance from the ground interferer to any UAV in the swarm is l. Then, $\mathcal{P}_2 = \left[\frac{3l^2}{2r^3} - \frac{3(2rh_0 + h_0^2)l + 3l^3}{4r^3(r+h_0)}\right] dl$. The PDF of the distance l_I is given by

$$f_{l_I}(l) = \frac{\mathcal{P}_2}{dl} = \frac{3l^2}{2r^3} - \frac{3(2rh_0 + h_0^2)l + 3l^3}{4r^3(r+h_0)}, \quad l \in [h_0, h_0 + 2r]. \tag{A.6}$$

The expectation of l_I is given by

$$\begin{aligned} \mathbb{E}(l_I) &= \int_{h_0}^{h_0+2r} l f_{l_I}(l) dl \\ &= \frac{(h_0 + 2r)^4 (3r - h_0) + h_0^4(5r - 9h_0)}{40r^3(h_0 + r)}. \end{aligned} \tag{A.7}$$

When $h_0 = 0$, $\mathbb{E}(l_I) = \frac{6}{5}r$.

B

Appendix of Chapter 3

B.1 Proof of (3.1)–(3.3)

By dividing the 3D spherical cap into two parts: R1 and R2, as shown in Figure B.1a, we calculate the PDF of l_e, denoted by $f_{l_e}(\tau)$. The volumes of R1 and R2 are $V_1 = \frac{2}{3}\pi(1+\cos\theta)r^3$ and $V_2 = \frac{1}{3}\pi r^3 \sin^2\theta \cos\theta$, respectively. Let $\mathcal{P}_1$ and $\mathcal{P}_2$ denote the probabilities that the eavesdropper is within R1 and R2, respectively. We have

$$\mathcal{P}_1 = \frac{V_1}{V_1+V_2} = \frac{2(1+\cos\theta)}{2(1+\cos\theta)+\sin^2\theta\cos\theta};$$

$$\mathcal{P}_2 = \frac{V_2}{V_1+V_2} = \frac{\sin^2\theta\cos\theta}{2(1+\cos\theta)+\sin^2\theta\cos\theta}.$$

For region R1, we define $\mathcal{P}_{\tau 1}$ as the probability that l_e is shorter than τ. $\mathcal{P}_{\tau 1}$ is equal to the probability that the eavesdropper is located within the spherical sector with radius τ, and $\mathcal{P}_{\tau 1} = \frac{\tau^3}{d^3}$. *For region R2*, we further divide the region into two parts, i.e., R2-1 and R2-2, as shown in Figure B.1(b). The volumes of R2-1 and R2-2 are $V_{21} = \frac{2}{3}\pi r^3(1-\cos\theta)\cos^3\theta$ and $V_{22} = \frac{1}{3}\pi r^3(\cos\theta - 3\cos^3\theta + 2\cos^4\theta)$, respectively. Let $\mathcal{P}_{21}$ and $\mathcal{P}_{22}$ denote the probabilities that the eavesdropper is within R2-1 and R2-2, respectively. We have

$$\mathcal{P}_{21} = \frac{V_{21}}{V_2} = \frac{2(1-\cos\theta)\cos^2\theta}{\sin^2\theta};$$

$$\mathcal{P}_{22} = \frac{V_{22}}{V_2} = \frac{1-3\cos^2\theta+2\cos^3\theta}{\sin^2\theta}.$$

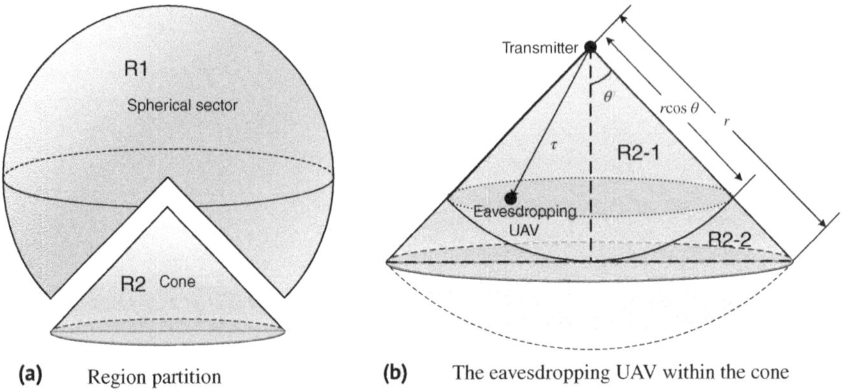

Figure B.1 Geometric manipulation for the proof of (3.1)–(3.3).

When the aerial eavesdropper is within R2-1, the probability that l_e is shorter than τ is $\mathcal{P}_{\tau 21} = \frac{\tau^3}{r^3 \cos^3\theta}$, $0 \leq \tau \leq r\cos\theta$. When the aerial eavesdropper is within R2-2, the region R2-2 plus the shaded region is used to approximate R2-2 for mathematical tractability, and the probability can be approximated by $\mathcal{P}_{\tau 22} \approx \frac{\tau^3 - r^3\cos^3\theta}{r^3 - r^3\cos^3\theta}$, $r\cos\theta < \tau \leq r$.

Now, we can obtain the probability that l_e is shorter than τ within the entire spherical cap. In the case of $0 \leq \tau \leq r\cos\theta$, the probability $\mathcal{P}_\tau$ and the PDF of l_e are given by

$$\begin{aligned}\mathcal{P}_\tau &= \mathcal{P}_1\mathcal{P}_{\tau 1} + \mathcal{P}_2\mathcal{P}_{21}\mathcal{P}_{\tau 21} \\ &= \frac{2(1+\cos\theta)\cos^2\theta + \sin^2\theta}{2(1+\cos\theta)\cos^2\theta + \sin^2\theta\cos^3\theta}\frac{\tau^3}{r^3};\end{aligned} \tag{B.1}$$

$$f_{l_e}(\tau) = \frac{d\mathcal{P}_\tau}{d\tau} = \frac{2(1+\cos\theta)\cos^2\theta + \sin^2\theta}{2(1+\cos\theta)\cos^2\theta + \sin^2\theta\cos^3\theta}\frac{3\tau^2}{r^3}. \tag{B.2}$$

In the case of $r\cos\theta \leq \tau \leq r$, $\mathcal{P}_\tau$ and $f_{l_e}(\tau)$ are given by

$$\begin{aligned}\mathcal{P}_\tau &\approx \mathcal{P}_1\mathcal{P}_{\tau 1} + \mathcal{P}_2\mathcal{P}_{22}\mathcal{P}_{\tau 22} \\ &= \frac{2(1+\cos\theta)(1-\cos^3\theta) + \sin^2\theta\cos\theta}{2(1+\cos\theta)(1-\cos^3\theta) + \sin^2\theta\cos\theta(1-\cos^3\theta)} \\ &\quad \times\frac{\tau^3 - r^3\cos^3\theta}{r^3};\end{aligned} \tag{B.3}$$

$$f_{l_e}(\tau) = \frac{d\mathcal{P}_\tau}{d\tau} \approx \frac{2(1+\cos\theta)(1-\cos^3\theta)+\sin^2\theta\cos\theta}{2(1+\cos\theta)(1-\cos^3\theta)+\sin^2\theta\cos\theta(1-\cos^3\theta)}\frac{3\tau^2}{r^3}. \tag{B.4}$$

By using (B.2) and (B.4), the expectation of l_e can be obtained. For $0<\theta<\frac{\pi}{2}$,

$$\bar{l}_e = \int_0^r \tau f_{l_e}(\tau)d\tau \approx \frac{3}{4}\left[1+\frac{\sin^2\theta\cos\theta}{\sin^2\theta\cos\theta+2(1+\cos\theta)}\right]r. \tag{B.5}$$

For $\theta = 0$ or $\theta = \frac{\pi}{2}$, $\bar{l}_e = \frac{3}{4}r$.

B.2 Proof of Lemma 3.1

Since the channel coefficient h_0 and h_e follow the Rician distribution with parameter K and K_e, respectively, the PDFs of ζ_r and ζ_e can be written as [23]

$$f_{\zeta_r}(x) = \frac{1+K}{\bar{\zeta}_r}\exp\left(-K-\frac{(1+K)x}{\bar{\zeta}_r}\right)I_0\left(2\sqrt{\frac{K(1+K)}{\bar{\zeta}_r}x}\right),$$

$$f_{\zeta_e}(y) = \frac{1+K_e}{\bar{\zeta}_e}\exp\left(-K_e-\frac{(1+K_e)y}{\bar{\zeta}_e}\right)I_0\left(2\sqrt{\frac{K_e(1+K_e)}{\bar{\zeta}_e}y}\right),$$

where $I_0(x) = \sum_{n=0}^{\infty}\frac{(x/2)^{2n}}{n!\Gamma(n+1)}$ is the zeroth order modified Bessel function of the first kind [91] and $\Gamma(z) = \int_0^\infty \frac{t^{z-1}}{e^t}dt$ is the Gamma function. $\bar{\zeta}_r$ is the expectation of ζ_r

$$\begin{aligned}
\bar{\zeta}_r &= \int_0^\infty x f_{\zeta_r}(x)dx \\
&= \frac{1+K}{\bar{\zeta}_r}\exp(-K)\sum_{n=0}^{\infty}\frac{\left(\frac{K(1+K)}{\bar{\zeta}_r}\right)^n}{(n!)^2} \\
&\qquad \int_0^\infty x^{n+1}\exp\left(-\frac{(1+K)x}{\bar{\zeta}_r}\right)dx \\
&= \frac{\zeta_r}{1+K}\exp(-K)\sum_{n=0}^{\infty}\frac{K^n}{(n!)^2}\cdot(n+1)! \\
&= \frac{\bar{\zeta}_r}{1+K}\exp(-K)\cdot(1+K)\exp(K) = \frac{PL^{-\alpha}}{\sigma_r^2}.
\end{aligned}$$

Likewise, $\bar{\zeta}_e$ is the expectation of ζ_e and $\bar{\zeta}_e = \frac{Pl_e^{-\alpha_e}}{\sigma_e^2}$ in the special case of $a_2 = 0$. As proved in Appendix B.3, $\mathbb{E}\left[\log_2\left(\frac{\mathcal{X}}{\mathcal{Y}}\right)\right] \xrightarrow{a.s.} \log_2\left(\frac{\mathbb{E}[\mathcal{X}]}{\mathbb{E}[\mathcal{Y}]}\right)$. Therefore, we can obtain

$$\begin{aligned}\mathbb{E}\left[\log_2\left(\frac{1+\zeta_r}{1+\zeta_e}\right)\right] &\xrightarrow{a.s.} \log_2\left(\frac{\mathbb{E}[1+\zeta_r]}{\mathbb{E}[1+\zeta_e]}\right)\\ &= \log_2\left(\frac{1+\bar{\zeta}_r}{1+\bar{\zeta}_e}\right)\\ &= \log_2\left(\frac{P\sigma_e^2 L^{-\alpha} + \sigma_r^2\sigma_e^2}{P\sigma_r^2 l_e^{-\alpha_e} + \sigma_r^2\sigma_e^2}\right).\end{aligned} \tag{B.6}$$

This concludes the proof of Lemma 3.1.

B.3 Proof of $\mathbb{E}\left[\log_2\left(\frac{\mathcal{X}}{\mathcal{Y}}\right)\right] \xrightarrow{a.s.} \log_2\left(\frac{\mathbb{E}[\mathcal{X}]}{\mathbb{E}[\mathcal{Y}]}\right)$

To prove this, we consider a generic scenario where the ground receiver and the aerial eavesdropper are equipped with N and N' antennas, respectively. Let x_i ($i = 1, \dots, N$) and y_j ($j = 1, \dots, N'$) be the received SNRs at the different antennas of the ground receiver and the aerial eavesdropper, respectively. They are two sequences of positive square-integrable random variables and not necessarily independent across i or j, $\frac{N'}{N} \to \varepsilon < \infty$. $\mathcal{X} = \sum_{i=1}^{N} x_i$ and $\mathcal{Y} = \sum_{j=1}^{N'} y_j$ are the SNRs of MRC at the ground receiver and the aerial eavesdropper, respectively.

Since $\mathbb{E}[x_i] < \infty$, $i = 1, \dots, N$, based on Kolmogorov's strong law of large numbers, we have [224]

$$\frac{1}{N}\sum_{i=1}^{N} x_i - \frac{1}{N}\sum_{i=1}^{N}\mathbb{E}[x_i] \xrightarrow{a.s.} 0, \text{ as } N \to \infty. \tag{B.7}$$

Since $\frac{1}{N}\sum_{i=1}^{N}\mathbb{E}[x_i] = \mathbb{E}\left[\frac{1}{N}\sum_{i=1}^{N} x_i\right] = \frac{1}{N}\mathbb{E}[\mathcal{X}]$, we have $\frac{1}{N}\mathcal{X} - \frac{1}{N}\mathbb{E}[\mathcal{X}] \xrightarrow{a.s.} 0$, as $N \to \infty$ and $\frac{N'}{N} \to \varepsilon$, i.e., $\frac{1}{N}\mathcal{X} \xrightarrow{a.s.} \frac{1}{N}\mathbb{E}[\mathcal{X}]$, as $N \to \infty$. Similarly, we also have $\frac{1}{N'}\mathcal{Y} - \frac{1}{N'}\mathbb{E}[\mathcal{Y}] \xrightarrow{a.s.} 0$, as $N' \to \infty$, i.e., $\frac{1}{N'}\mathcal{Y} \xrightarrow{a.s.} \frac{1}{N'}\mathbb{E}[\mathcal{Y}]$, as $N' \to \infty$.

By using the *Continuous mapping theorem* [36] for a few times, we have

$$\mathbb{E}\left[\log_2\left(\frac{\mathcal{X}}{\mathcal{Y}}\right)\right] \xrightarrow{a.s.} \log_2\left(\frac{\mathbb{E}[\mathcal{X}]}{\mathbb{E}[\mathcal{Y}]}\right), \quad \text{as } N \to \infty \text{ and } \frac{N'}{N} \to \varepsilon. \tag{B.8}$$

From this analysis, we can see that the *a.s.* convergence can be achieved in (B.8), in the case where N and N' are large, e.g., the ground receiver and aerial eavesdropper equipped with large number of antennas in a mmWave systems [77, 344]. In the case where a single antenna is considered at the ground receiver and the aerial eavesdropper (as discussed in this chapter), $\mathbb{E}\left[\log_2\left(\frac{\mathcal{X}}{\mathcal{Y}}\right)\right] \approx \log_2\left(\frac{\mathbb{E}[\mathcal{X}]}{\mathbb{E}[\mathcal{Y}]}\right)$, e.g., in Lemma 3.1.

B.4 Proof of Theorem 3.1

Based on Lemma 3.1, $C_{\text{erg}}(L, l_e)$ can be rewritten as

$$\mathcal{H}_1(\mu) = \log_2\left(\frac{P\sigma_e^2 L^{-\alpha} + \sigma_r^2\sigma_e^2}{P\sigma_r^2\mu + \sigma_r^2\sigma_e^2}\right);$$

$$\mathcal{H}_2(\nu) = \log_2\left[\frac{\left(P\sigma_e^2 L^{-\alpha} + \sigma_r^2\sigma_e^2\right)\nu}{P\sigma_r^2 + \sigma_r^2\sigma_e^2\nu}\right],$$

which are achieved by defining $\mu = l_e^{-\alpha_e}$ and $\nu = l_e^{\alpha_e}$. We can obtain

$$\frac{\partial\mathcal{H}_1(\mu)}{\partial\mu} = -\frac{1}{\ln 2}\cdot\frac{P\sigma_r^2}{\left(P\sigma_r^2\mu + \sigma_e^2\sigma_r^2\right)} \leq 0;$$

$$\frac{\partial^2\mathcal{H}_1(\mu)}{\partial\mu^2} = \frac{1}{\ln 2}\cdot\frac{P^2\sigma_r^4}{\left(P\sigma_r^2\mu + \sigma_e^2\sigma_r^2\right)^2} \geq 0.$$

Thus, $\mathcal{H}_1(\mu)$ is a monotonically decreasing and convex function of μ. We also find that

$$\frac{\partial\mathcal{H}_2(\nu)}{\partial\nu} = \frac{1}{\ln 2}\cdot\frac{P\sigma_r^2}{\left(P\sigma_r^2\nu + \sigma_e^2\sigma_r^2\nu^2\right)} \geq 0;$$

$$\frac{\partial^2\mathcal{H}_2(\nu)}{\partial\nu^2} = -\frac{1}{\ln 2}\cdot\frac{P^2\sigma_r^4 + 2P\sigma_r^4\sigma_e^2\nu}{\left(P\sigma_r^2\nu + \sigma_e^2\sigma_r^2\nu^2\right)^2} \leq 0.$$

Thus, $\mathcal{H}_2(\nu)$ is a monotonically increasing and concave function of μ.

Based on the Jensen's inequality (i.e., $\mathbb{E}[f(x)] \geq f(\mathbb{E}[x])$ if $f(x)$ is convex, otherwise $\mathbb{E}[f(x)] \leq f(\mathbb{E}[x])$), the convexity of $\mathcal{H}_1(\mu)$ with respect to μ, and the concavity of $\mathcal{H}_2(\nu)$ with respect to ν, we have

$$\mathbb{E}[\mathcal{H}_1(\mu)] \geq \mathcal{H}_1(\mathbb{E}[\mu]);\ \mathbb{E}[\mathcal{H}_2(\nu)] \leq \mathcal{H}_2(\mathbb{E}[\nu]). \tag{B.9}$$

Since $\mathcal{H}_1(\mu) = \mathcal{H}_2(\nu)$, we have

$$\begin{aligned} \mathcal{H}_1(\mathbb{E}[\mu]) &\le \mathbb{E}[\mathcal{H}_1(\mu)] = \mathbb{E}_{l_e}\left[\mathcal{C}_{\rm erg}(L, l_e)\right] \\ &= \mathbb{E}[\mathcal{H}_2(\nu)] \le \mathcal{H}_2(\mathbb{E}[\nu]). \end{aligned} \tag{B.10}$$

Since $\mu = l_e^{-\alpha_e}$ and $\nu = l_e^{\alpha_e}$ are both convex with regard to l_e, by using Jensen's inequality, we have

$$\mathbb{E}[\mu] \ge (\bar{l}_e)^{-\alpha_e}; \quad \mathbb{E}[\nu] \ge (\bar{l}_e)^{\alpha_e}. \tag{B.11}$$

As $\mathcal{H}_1(\mu)$ and $\mathcal{H}_2(\nu)$ are monotonically decreasing and increasing functions of μ and ν, respectively, we have

$$\begin{aligned} \mathcal{H}_1(\mathbb{E}[\mu]) &\le \mathcal{H}_1\left((\bar{l}_e)^{-\alpha_e}\right) = \mathcal{C}_{\rm erg}(L, \bar{l}_e) \\ &= \mathcal{H}_2\left((\bar{l}_e)^{\alpha_e}\right) \le \mathcal{H}_2(\mathbb{E}[\nu]). \end{aligned} \tag{B.12}$$

Comparing (B.10) and (B.12), we find that $\mathcal{C}_{\rm erg}(L, \bar{l}_e)$ lies between the upper and lower bounds of $\mathbb{E}_{l_e}\left[\mathcal{C}_{\rm erg}(L, l_e)\right]$. We can approximate

$$\begin{aligned} \mathbb{E}_{l_e}\left[\mathcal{C}_{\rm erg}(L, l_e)\right] &\approx \mathcal{C}_{\rm erg}(L, \bar{l}_e) \\ &= \log_2\left(\frac{P\sigma_e^2 L^{-\alpha} + \sigma_r^2\sigma_e^2}{P\sigma_r^2(\bar{l}_e)^{-\alpha_e} + \sigma_r^2\sigma_e^2}\right). \end{aligned} \tag{B.13}$$

By substituting the expectation of the transmitter-eavesdropper distance, $\bar{l}_e$, (B.13) can be rewritten as (3.11).

We proceed to evaluate the approximation accuracy of (B.13). By recalling (3.1) and (3.2) and referring to (B.5), the expectation of $l_e^{\alpha_e}$ and $l_e^{-\alpha_e}$ can be given by

$$\mathbb{E}[l_e^{\alpha_e}] = h_1(r) = \int_0^r \tau^{\alpha_e} f_{l_e}(\tau)d\tau \approx \frac{3\mathcal{D}_1(\theta)}{3+\alpha_e} r^{\alpha_e}; \tag{B.14}$$

$$\mathbb{E}[l_e^{-\alpha_e}] = h_2(r) = \int_0^r \tau^{-\alpha_e} f_{l_e}(\tau)d\tau \approx \frac{3\mathcal{D}_2(\theta)}{3-\alpha_e} r^{-\alpha_e}, \ \alpha_e \ne 3, \tag{B.15}$$

where $\mathcal{D}_1(\theta) = (1 - (\cos\theta)^{1+\alpha_e})\frac{2(1+\cos\theta)(1-\cos^3\theta) + \sin^2\theta\cos\theta}{[2(1+\cos\theta) + \sin^2\theta\cos\theta](1-\cos^3\theta)} + (\cos\theta)^{1+\alpha_e}\frac{2(1+\cos\theta)\cos^2\theta + \sin^2\theta}{2(1+\cos\theta) + \sin^2\theta\cos\theta}$ and $\mathcal{D}_2(\theta) = (\cos\theta)^{1-\alpha_e}\frac{2(1+\cos\theta)\cos^2\theta + \sin^2\theta}{2(1+\cos\theta) + \sin^2\theta\cos\theta} + \frac{2(1+\cos\theta)(1-\cos^3\theta) + \sin^2\theta\cos\theta}{[2(1+\cos\theta) + \sin^2\theta\cos\theta](1-\cos^3\theta)}(1 - (\cos\theta)^{3-\alpha_e})$. Then, we have

$$\begin{aligned}\mathcal{H}_1\left(\mathbb{E}\left[\mu\right]\right) &= \log_2\left(\frac{P\sigma_e^2 L^{-\alpha} + \sigma_r^2\sigma_e^2}{P\sigma_r^2 h_2(r) + \sigma_r^2\sigma_e^2}\right) \\ &\approx \log_2\left(\frac{P\sigma_e^2 L^{-\alpha} + \sigma_r^2\sigma_e^2}{\frac{3\mathcal{D}_2(\theta)}{3-\alpha_e} P\sigma_r^2 r^{-\alpha_e} + \sigma_r^2\sigma_e^2}\right);\end{aligned} \tag{B.16}$$

$$\begin{aligned}\mathcal{H}_2\left(\mathbb{E}\left[\nu\right]\right) &= \log_2\left(\frac{P\sigma_e^2 L^{-\alpha} + \sigma_r^2\sigma_e^2}{P\sigma_r^2 h_1^{-1}(r) + \sigma_r^2\sigma_e^2}\right) \\ &\approx \log_2\left(\frac{P\sigma_e^2 L^{-\alpha} + \sigma_r^2\sigma_e^2}{\frac{3+\alpha_e}{3\mathcal{D}_1(\theta)} P\sigma_r^2 r^{-\alpha_e} + \sigma_r^2\sigma_e^2}\right).\end{aligned} \tag{B.17}$$

Here, $\frac{3+\alpha_e}{3\mathcal{D}_1(\theta)}$ and $\frac{3\mathcal{D}_2(\theta)}{3-\alpha_e}$ are constants given θ and α_e. (B.14)–(B.17) hold strict equality, when $\theta = \frac{\pi}{2}$ (or in other words, the transmitter antenna height is comparatively negligible to the eavesdropper's flight radius r); see (3.3).

The approximation error of (B.13) is no greater than half of the difference between $\mathcal{H}_1\left(\mathbb{E}\left[\mu\right]\right)$ and $\mathcal{H}_2\left(\mathbb{E}\left[\nu\right]\right)$. The error is increasingly negligible with the growth of r, since both $h_1^{-1}(r)$ and $h_2(r)$ are strictly monotonically decreasing functions of r. $\mathcal{H}_1\left(\mathbb{E}\left[\mu\right]\right) \to \mathcal{H}_2\left(\mathbb{E}\left[\nu\right]\right)$ when $r \gg \left(\frac{3\mathcal{D}_2(\theta)}{3-\alpha_e}\frac{P}{\sigma_e^2}\right)^{1/\alpha_e}$ and $r \gg \left(\frac{3+\alpha_e}{3\mathcal{D}_1(\theta)}\frac{P}{\sigma_e^2}\right)^{1/\alpha_e}$ in (B.16) and (B.17).

B.5 Proof of Lemma 3.2

When h_0 follows the Rayleigh distribution ($K = 0$), i.e., $|h_0|^2 \sim \exp(1)$, and h_e follows the Rician distribution with parameter K_e, the PDF of Ψ is given by

$$f_\Psi(\psi) = \int_0^\infty f_{\mathcal{X}}(\psi + y) f_{\mathcal{Y}}(y) dy \tag{B.18a}$$

$$\begin{aligned}&= \frac{B\exp\left(-K_e + \frac{B}{2} - \frac{\psi}{\bar{\mathcal{X}}}\right)}{2K_e\bar{\mathcal{X}}} \\ &\quad \times \int_0^\infty \frac{1}{2}\exp\left(-\frac{y_1 + B}{2}\right) I_0\left(\sqrt{By_1}\right) dy_1\end{aligned} \tag{B.18b}$$

$$= \frac{B \exp\left(-K_e + \frac{B}{2} - \frac{\psi}{\bar{\mathcal{X}}}\right)}{2K_e\bar{\mathcal{X}}} Q\left(\sqrt{B}, 0\right) \tag{B.18c}$$

$$= \frac{(1+K_e)\exp\left(-\frac{K_e\bar{\mathcal{Y}}}{\bar{\mathcal{Y}}+(1+K_e)\bar{\mathcal{X}}}\right)}{\bar{\mathcal{Y}}+(1+K_e)\bar{\mathcal{X}}} \exp\left(-\frac{\psi}{\bar{\mathcal{X}}}\right), \tag{B.18d}$$

where $\bar{\mathcal{X}} = \bar{\zeta}_r$ and $\bar{\mathcal{Y}} = \bar{\zeta}_e 2^{C^S_{\text{out}}}$; (B.18b) is obtained by setting $y_1 = Ay = \frac{2[(1+K_e)\bar{\mathcal{X}}+\bar{\mathcal{Y}}]}{\bar{\mathcal{X}}\bar{\mathcal{Y}}}y$ and $B = \frac{2K_e(1+K_e)\bar{\mathcal{X}}}{\bar{\mathcal{Y}}+(1+K_e)\bar{\mathcal{X}}}$; (B.18c) is based on the definition of the first-order Marcum Q-function [23, equation 8], i.e., $Q(\sqrt{a}, \sqrt{b}) = \int_b^\infty \frac{1}{2}\exp(-\frac{x+a}{2})I_0(\sqrt{ax})dx$; (B.18d) is due to $Q\left(\sqrt{a}, 0\right) = 1$.

The ϵ-outage secrecy probability can be evaluated based on [206, equation 37], as given by

$$\begin{aligned}
\epsilon &= \Pr\left(\Psi \le 2^{C^S_{\text{out}}} - 1\right) \\
&= 1 - \Pr\left(\Psi \ge 2^{C^S_{\text{out}}} - 1\right) \\
&= 1 - \int_{2^{C^S_{\text{out}}}-1}^{\infty} \frac{(1+K_e)\exp\left(-\frac{K_e\bar{\mathcal{Y}}}{\bar{\mathcal{Y}}+(1+K_e)\bar{\mathcal{X}}}\right)}{\bar{\mathcal{Y}}+(1+K_e)\bar{\mathcal{X}}} \exp\left(-\frac{\psi}{\bar{\mathcal{X}}}\right) d\psi,
\end{aligned} \tag{B.19}$$

which, by taking the integration, leads to

$$\epsilon = 1 - \mathcal{M}\exp\left(-\frac{2^{C^S_{\text{out}}}-1}{\bar{\mathcal{X}}}\right), \tag{B.20}$$

where, for notational simplicity, $\bar{\mathcal{X}} = \frac{PL^{-\alpha}}{\sigma_r^2}$, $\bar{\mathcal{Y}} = \frac{Pl_e^{-\alpha_e}2^{C^S_{\text{out}}}}{\sigma_e^2}$, and $\mathcal{M}$ is defined as

$$\mathcal{M} = \frac{(1+K_e)\,\bar{\mathcal{X}}\exp\left(-\frac{K_e\bar{\mathcal{Y}}}{\bar{\mathcal{Y}}+(1+K_e)\bar{\mathcal{X}}}\right)}{\bar{\mathcal{Y}}+(1+K_e)\bar{\mathcal{X}}}. \tag{B.21}$$

Based on the secrecy outage probability and the definition of ϵ-outage secrecy rate, we have

$$1 - \epsilon = \mathcal{M}\exp\left(-\frac{2^{C^S_{\text{out}}}-1}{\bar{\mathcal{X}}}\right). \tag{B.22}$$

By exploiting $\exp(-x) \approx 1 - x$ when x is small, (B.20) can be approximated by

$$(1-\epsilon)\frac{1}{\mathcal{M}} \approx 1 - \frac{2^{C^S_{\text{out}}}-1}{\bar{\mathcal{X}}}. \tag{B.23}$$

Since $\bar{\mathcal{X}} = \bar{\zeta}_r$ and $\bar{\mathcal{Y}} = 2^{C^{\mathrm{S}}_{\mathrm{out}}}\bar{\zeta}_e$, (B.23) can be rewritten as

$$\frac{(1-\epsilon)\left[\bar{\zeta}_e 2^{C^{\mathrm{S}}_{\mathrm{out}}} + (1+K_e)\bar{\zeta}_r\right]}{(1+K_e)\bar{\zeta}_r}\exp\left[\frac{K_e\bar{\zeta}_e 2^{C^{\mathrm{S}}_{\mathrm{out}}}}{\bar{\zeta}_e 2^{C^{\mathrm{S}}_{\mathrm{out}}} + (1+K_e)\bar{\zeta}_r}\right] \approx 1 - \frac{2^{C^{\mathrm{S}}_{\mathrm{out}}}-1}{\bar{\zeta}_r}. \tag{B.24}$$

In the case of $K_e = 0$, (B.24) can be rewritten as

$$(1-\epsilon)\left(\bar{\zeta}_r + \bar{\zeta}_e 2^{C^{\mathrm{S}}_{\mathrm{out}}}\right) \approx \bar{\zeta}_r - \left(2^{C^{\mathrm{S}}_{\mathrm{out}}} - 1\right). \tag{B.25}$$

The ϵ-outage secrecy rate can be approximated by

$$C^{\mathrm{S}}_{\mathrm{out}} \approx \log_2\left[\frac{1+\epsilon\bar{\zeta}_r}{1+(1-\epsilon)\bar{\zeta}_e}\right]. \tag{B.26}$$

In the case of $K_e > 0$, let $\mathcal{Z} = \frac{\bar{\zeta}_e 2^{C^{\mathrm{S}}_{\mathrm{out}}} + (1+K_e)\bar{\zeta}_r}{K_e(1+K_e)\bar{\zeta}_r}$, and (B.24) can be rewritten as

$$\mathcal{Z}\exp\left(-\frac{1}{\mathcal{Z}}\right) \approx \frac{\left(1+\bar{\zeta}_r - 2^{C^{\mathrm{S}}_{\mathrm{out}}}\right)\exp(-K_e)}{K_e(1-\epsilon)\bar{\zeta}_r} \tag{B.27a}$$

$$\approx \frac{(1+\bar{\zeta}_r)\exp(-K_e)}{K_e(1-\epsilon)\bar{\zeta}_r}, \tag{B.27b}$$

where (B.27b) is due to the fact that $1+\bar{\zeta}_r \gg 2^{C^{\mathrm{S}}_{\mathrm{out}}}$.[1] By applying the Lambert W function to (B.27b), we can obtain

$$\mathcal{Z}^{-1} \approx \mathcal{W}_0\left(\frac{K_e\exp(K_e)(1-\epsilon)\bar{\zeta}_r}{1+\bar{\zeta}_r}\right) \triangleq W_0. \tag{B.28}$$

Therefore, the ϵ-secrecy outage rate is given by

$$C^{\mathrm{S}}_{\mathrm{out}} \approx \log_2\left[\frac{(1+K_e)(K_e - W_0)\bar{\zeta}_r}{W_0\bar{\zeta}_e}\right] \triangleq \mathcal{C}_{\mathrm{out}}(L, l_e). \tag{B.29}$$

We proceed to prove that the approximations in (B.26) and (B.29) are asymptotically accurate in high SNR regimes, where $P \gg \sigma_r^2$ and $P \gg \sigma_e^2$ (i.e., $\bar{\zeta}_r \gg 1$ and $\bar{\zeta}_e \gg 1$). Specifically,

$$W_0 = \mathcal{W}_0\left(\frac{K_e\exp(K_e)(1-\epsilon)PL^{-\alpha}}{\sigma_r^2 + PL^{-\alpha}}\right)$$
$$\approx \mathcal{W}_0(K_e\exp(K_e)(1-\epsilon)) \leq \mathcal{W}_0(K_e\exp(K_e)) = K_e.$$

1 Based on the definition of the ϵ-secrecy outage rate, we have $\log_2(1+\bar{\zeta}_r) - \log_2(1+\bar{\zeta}_e) > C^{\mathrm{S}}_{\mathrm{out}}$, i.e., $1+\bar{\zeta}_r > 2^{C^{\mathrm{S}}_{\mathrm{out}}}(1+\bar{\zeta}_e)$. In the high SNR regimes, $P \gg \sigma_r^2$ and $P \gg \sigma_e^2$, thus $1+\bar{\zeta}_e \gg 1$ and $1+\bar{\zeta}_r \gg 2^{C^{\mathrm{S}}_{\mathrm{out}}}$.

Considering the typical range of the Rician factor $0 < K_e \leq 10$, we have $\frac{(1+K_e)(K_e - W_0)}{W_0} = (1+K_e)\left(\frac{K_e}{W_0} - 1\right) = \Theta(1+K_e) = \Theta(1)$.[2] Since $\bar{\zeta}_e \gg 1$, we have $\frac{(1+K_e)(K_e - W_0)}{W_0\bar{\zeta}_e} = \frac{\Theta(1)}{\bar{\zeta}_e} \ll 1$. Therefore, $\frac{(1+K_e)(K_e - W_0)\bar{\zeta}_r}{W_0\bar{\zeta}_e} \ll \bar{\zeta}_r$ and $1 + \bar{\zeta}_r \gg 2^{C_{\text{out}}^{\text{S}}}$. This validates the approximation accuracy of (B.23) and, in turn, the approximation accuracy of (B.24), (B.26), and (B.27a). By substituting (B.29) into $1 + \bar{\zeta}_r - 2^{C_{\text{out}}^{\text{S}}}$, we have $1 + \bar{\zeta}_r - \frac{(1+K_e)(K_e - W_0)\bar{\zeta}_r}{W_0\bar{\zeta}_e} = 1 + \bar{\zeta}_r - \frac{\Theta(1)}{\bar{\zeta}_e}\bar{\zeta}_r \approx 1 + \bar{\zeta}_r$. The accuracy of the approximation from (B.27a) to (B.27b) is verified, and so is the approximation accuracy of (B.29).

Finally, (3.14) can be obtained by substituting $\bar{\zeta}_r = \frac{PL^{-\alpha}}{\sigma_r^2}$ and $\bar{\zeta}_e = \frac{Pl_e^{-\alpha_e}}{\sigma_e^2}$ into (B.26) and (B.29).

B.6 Proof of Theorem 3.2

We first consider the case of $K_e > 0$. Based on Lemma 3.2, (3.14) can be rewritten as

$$\mathcal{F}_1(\mu) = \log_2\left[\frac{(1+K_e)(K_e - W_0)\sigma_e^2 L^{-\alpha}}{W_0\sigma_r^2\mu}\right];$$

$$\mathcal{F}_2(\nu) = \log_2\left[\frac{(1+K_e)(K_e - W_0)\sigma_e^2 L^{-\alpha}\nu}{W_0\sigma_r^2}\right],$$

which are obtained by substituting $\mu = l_e^{-\alpha_e}$ and $\nu = l_e^{\alpha_e}$. Hence, we have $\mathcal{F}_1(\mu) = \mathcal{F}_2(\nu)$.

For $\mathcal{F}_1(\mu)$, we can find that

$$\frac{\partial\mathcal{F}_1(\mu)}{\partial\mu} = -\frac{1}{\mu\ln 2} \leq 0; \quad \frac{\partial^2\mathcal{F}_1(\mu)}{\partial\mu^2} = \frac{1}{\mu^2\ln 2} \geq 0.$$

$\mathcal{F}_1(\mu)$ is a monotonically decreasing and convex function of μ. Likewise, we can obtain

$$\frac{\partial\mathcal{F}_2(\nu)}{\partial\nu} = \frac{1}{\nu\ln 2} \geq 0; \quad \frac{\partial^2\mathcal{F}_2(\nu)}{\partial\nu^2} = -\frac{1}{\nu^2\ln 2} \leq 0.$$

---

2 We denote $g(n) = O(f(n))$ if there is a positive constant c_1 such that $\lim_{n\to\infty}\Pr\left(\frac{g(n)}{f(n)} \leq c_1\right) = 1$, and $g(n) = \Omega(f(n))$ if $f(n) = O(g(n))$. We also denote $g(n) = \Theta(f(n))$ if both $g(n) = O(f(n))$ and $g(n) = \Omega(f(n))$ hold.

$\mathcal{F}_2(\nu)$ is a monotonically increasing and concave function of ν. By referring to (B.9)–(B.12) in the proof of Theorem 3.1, we have

$$\mathbb{E}_{l_e}\left[\mathcal{C}_{\text{out}}(L, l_e)\right] \approx \mathcal{C}_{\text{out}}\left(L, \bar{l}_e\right), \text{ if } K_e > 0. \tag{B.30}$$

In the case of $K_e = 0$, both the ϵ-outage secrecy rate, i.e., $\mathcal{C}_{\text{out}}(L, l_e) = \log_2\left[\frac{\sigma_r^2\sigma_e^2 + \epsilon P\sigma_e^2 L^{-\alpha}}{\sigma_r^2\sigma_e^2 + (1-\epsilon)P\sigma_r^2 l_e^{-\alpha_e}}\right]$, and the ergodic secrecy rate, i.e., $\mathcal{C}_{\text{erg}}(L, l_e)$, exhibit same monotonicity and concavity with regards to l_e. Based on the proof of Theorem 3.1, we have

$$\mathbb{E}_{l_e}\left[\mathcal{C}_{\text{out}}(L, l_e)\right] \approx \mathcal{C}_{\text{out}}\left(L, \bar{l}_e\right), \text{ if } K_e = 0. \tag{B.31}$$

By substituting $\bar{l}_e$ into (B.30) and (B.31), we can obtain (3.15). Also, the approximation error of (3.15) can asymptotically diminish with the increase of r and the number of antennas at the legitimate receiver and the eavesdropper, especially in the high SNR regime, as can be proved in the same way in Theorem 3.1. This concludes the proof of Theorem 3.2.

B.7 Proof of (3.21)

The expectation of the distance between the aerial eavesdropper to one of the ground transceivers can be evaluated by dividing the flight region into two subregions, i.e., Subregions R_1' and R_2', as shown in Figure B.2. Each of the subregions is half of a spherical cap centered at one of the transceivers. The distances from the eavesdropper inside the subregions to a ground transceiver can be separately evaluated. Without loss of generality, we evaluate the distances from the eavesdropper to Transceiver 1, as follows.

When the eavesdropper is inside Subregion R_2', the volume of this half-spherical cap can be evaluated to be $V_1 = \frac{\pi}{6}\left(2r^3 - \frac{3}{2}Lr^2 + \frac{L^3}{8}\right)$. The probability that the eavesdropper-transceiver distance l_e (denoted by l_2 in Subregion R_2') is shorter than τ is denoted by $\mathcal{P}'$ and equal to the probability that the eavesdropper is located within the spherical cap with a radius of τ. $\mathcal{P}' = \frac{16\tau^3 - 12L\tau^2 + L^3}{16r^3 - 12Lr^2 + L^3}$, $\frac{L}{2} \leq \tau \leq r$. The PDF of the eavesdropper-transceiver distance is given by

$$f_{l_2}(\tau) = \frac{48\tau^2 - 24L\tau}{16r^3 - 12Lr^2 + L^3}, \quad \frac{L}{2} \leq \tau \leq r. \tag{B.32}$$

When the eavesdropper is inside Subregion R_1', we first plot the symmetric point of the eavesdropper with respect to point O, which is the halfway between the transceivers. The symmetric point, referred to as "virtual eavesdropper," is inside the virtual region symmetrical to Subregion R_1' with respect to the ground plane. Given the uniform stationary distribution of the eavesdropper in Subregion R_2',

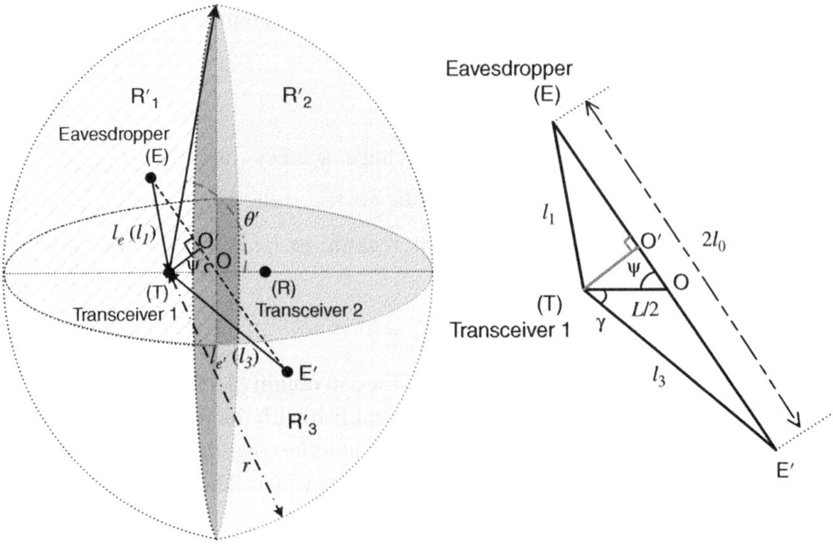

Figure B.2 Geometric interpretation for the evaluation of l_e.

the virtual eavesdropper also has a uniform stationary distribution in the virtual subregion R_3'. By using the cosine rule, we have

$$\cos\psi = \frac{(L/2)^2 + l_0^2 - l_1^2}{Ll_0}; \tag{B.33}$$

$$\cos(\pi - \psi) = \frac{(L/2)^2 + l_0^2 - l_3^{\,2}}{Ll_0}, \tag{B.34}$$

where l_1 is the distance between the eavesdropper inside Subregion R_1' and Transceiver 1; and l_3 is the distance between the virtual eavesdropper in the virtual subregion R_3' and the transceiver. l_3 has the same PDF as l_2, due to the symmetry, i.e., $f_{l_3}(\tau) = \frac{48\tau^2 - 24L\tau}{16r^3 - 12Lr^2 + L^3}, \frac{L}{2} \leq \tau \leq r$.

Since $\cos\psi = -\cos(\pi - \psi)$, by separately adding and subtracting (B.33) and (B.34), we can obtain

$$l_0 L \cos\psi = \frac{l_3^{\,2} - l_1^{\,2}}{2}; \tag{B.35}$$

$$l_3^{\,2} + l_1^{\,2} = 2l_0^2 + L^2/2. \tag{B.36}$$

According to the sine rule, we have

$$\frac{l_0}{\sin\gamma} = \frac{l_3}{\sin(\pi - \psi)}, \tag{B.37}$$

where $\gamma \in \left[0, \cos^{-1}\left(\frac{L}{2l_3}\right)\right]$ is the elevation angle between the virtual eavesdropper and the ground transmitter. γ is uniformly distributed within $\left[0, \cos^{-1}\left(\frac{L}{2l_3}\right)\right]$ given the uniform distribution of the eavesdropper in the 3D region. The PDF of γ conditioned on l_3 is $f_{\gamma|l_3}(x|\tau) = \frac{1}{\cos^{-1}\left(\frac{L}{2\tau}\right)}$. Therefore, the joint PDF of l_3 and γ can be given by

$$\begin{aligned} f_{l_3,\gamma}(\tau, x) &= f_{\gamma|l_3}(x|\tau) \cdot f_{l_3}(\tau) \\ &= \frac{1}{\cos^{-1}\left(\frac{L}{2\tau}\right)} \frac{48\tau^2 - 24L\tau}{16r^3 - 12Lr^2 + L^3}, \quad \frac{L}{2} \le \tau \le r. \end{aligned} \tag{B.38}$$

Substitute (B.35) and (B.37) into $\sin^2\psi + \cos^2\psi = 1$,

$$l_0^2 = l_3^2 \sin^2\gamma + \frac{\left(l_3^2 - l_1^2\right)^2}{4L^2}. \tag{B.39}$$

By substituting (B.39) into (B.36), we have

$$l_1^4 - 2\left(L^2 + l_3^2\right) l_1^2 + l_3^4 - 2L^2 l_3^2 (1 - 2\sin^2\gamma) = 0. \tag{B.40}$$

By using the quadratic formula, we have $l_1^2 = L^2 + {l_3}^2 - 2Ll_3 \cos\gamma$, since $l_1^2 = L^2 + {l_3}^2 + 2Ll_3 \cos\gamma > l_3^2$ does not satisfy that $l_1 < l_3$ for $\psi \in \left[0, \frac{\pi}{2}\right]$. Therefore, we have

$$l_1 = \sqrt{L^2 + {l_3}^2 - 2Ll_3 \cos\gamma}, \quad l_3 \in \left[\frac{L}{2}, r\right]. \tag{B.41}$$

By using (B.32), (B.38), and (B.41), we can obtain (3.21).

C

Appendix of Chapter 4

C.1 Proofs of Lemmas 4.1 and 4.2

Consider any UAV, say UAV i, located on the hemispherical surface with the center at the ground transmitter and radius $\tau \leq r$. $f_{l_i}(\tau)$ and $F_{l_i}(\tau)$ denote the PDF and cumulative distribution function (CDF) of l_i, $i = 1, \dots, N$, respectively. $\mathcal{P}$ denotes the probability that the distance between the ground transmitter and one of the aerial eavesdroppers is smaller than τ. We have $\mathcal{P} = F_{l_i}(\tau) = \frac{\tau^3}{r^3}$, $i = 1, \dots, N$. Therefore, $f_{l_i}(\tau)$ can be obtained, as given by

$$f_{l_i}(\tau) = \frac{d\mathcal{P}}{d\tau} = \frac{3\tau^2}{r^3},\ 0 \leq \tau \leq r;\ i = 1, \dots, N. \tag{C.1}$$

The expectation of l_i, $i = 1, \dots, N$ is obtained as

$$\mathbb{E}[l_i] = \int_0^r \tau f_{l_i}(\tau)\, d\tau = \int_0^r \frac{3\tau^3}{r^3} d\tau = \frac{3}{4} r;\ i = 1, \dots, N, \tag{C.2}$$

which proves Lemma 4.2.

By exploiting order statistics, the CDF of $\mathcal{L}$ is given by

$$F_{\mathcal{L}}(\tau) = 1 - \left[1 - F_{l_i}(\tau)\right]^N = 1 - \left(1 - \frac{\tau^3}{r^3}\right)^N. \tag{C.3}$$

By taking the derivative of (C.3), the PDF of $\mathcal{L}$ is given by

$$f_{\mathcal{L}}(\tau) = N\left[1 - F_{l_i}(\tau)\right]^{N-1} f_{l_i}(\tau) = N\frac{3\tau^2}{r^3}\left(1 - \frac{\tau^3}{r^3}\right)^{N-1}. \tag{C.4}$$

Artificial Intelligence for Unmanned Aerial Vehicles: Sensing, Communication, and Computing, First Edition. Shuyan Hu et al.

Finally, the expectation of $\mathcal{L}$ can be given by

$$\begin{aligned}\mathbb{E}[\mathcal{L}] &= \int_0^r \tau f_{\mathcal{L}}(\tau) d\tau \\ &= \int_0^r \left(1-\frac{\tau^3}{r^3}\right)^N d\tau \qquad \text{(C.5a)} \\ &= \frac{\Gamma\left(\frac{4}{3}\right)\Gamma(N+1)}{\Gamma\left(N+\frac{4}{3}\right)} r \xrightarrow{N\to\infty} \Gamma\left(\frac{4}{3}\right) N^{-\frac{1}{3}} r, \qquad \text{(C.5b)}\end{aligned}$$

where (C.5b) is due to $\lim_{N\to\infty} \frac{\Gamma(N+a)}{\Gamma(N)N^a} = 1$. Lemma 4.1 is proved.

C.2 Proof of Theorem 4.1

The channel coefficients h_0 and h_e follow the Rician distribution with parameters K and K_e, respectively. The PDFs of ζ_r and ζ_c, which are the output SNR of SC diversity for G2U links, can be written as [23]

$$f_{\zeta_k}(x) = \frac{1+K}{\bar{\zeta}_k} \exp\left(-K - \frac{(1+K)x}{\bar{\zeta}_k}\right) I_0\left(2\sqrt{\frac{K(1+K)}{\bar{\zeta}_k}x}\right), \qquad \text{(C.6)}$$

where $\bar{\zeta}_k$ is the average SNR, $k \in \{r, c\}$; and $I_0(x) = \sum_{n=0}^{\infty} \frac{(x/2)^{2n}}{n!\Gamma(n+1)}$ is the zeroth-order modified Bessel function of the first kind [91]. The expectation of ζ_k $(k \in \{r, c\})$ is

$$\begin{aligned}\mathbb{E}[\zeta_k] &= \int_0^\infty x f_{\zeta_k}(x) dx \\ &= \frac{\bar{\zeta}_k}{1+K} \exp(-K) \sum_{n=0}^{\infty} \frac{K^n}{(n!)^2} \cdot (n+1)! \qquad \text{(C.7)} \\ &= \bar{\zeta}_k = \frac{Pl^{-\alpha}}{\sigma_k^2}.\end{aligned}$$

Further, we have $\mathbb{E}[1+\zeta_r] = 1 + \bar{\zeta}_r = \frac{Pl^{-\alpha}}{\sigma_r^2} + 1$. Similarly, the expectation of $(1+\zeta_c)$ can be obtained as $\mathbb{E}[1+\zeta_c] = 1 + \bar{\zeta}_c$.

As proved in Appendix C.3, $\mathbb{E}\left[\log_2\left(\frac{\mathcal{X}}{\mathcal{Y}}\right)\right] \xrightarrow{a.s.} \log_2\left(\frac{\mathbb{E}[\mathcal{X}]}{\mathbb{E}[\mathcal{Y}]}\right)$. Therefore, we can obtain

$$\mathbb{E}\left[\log_2\left(\frac{1+\zeta_r}{1+\zeta_c}\right)\right] \xrightarrow{a.s.} \log_2\left(\frac{\mathbb{E}[1+\zeta_r]}{\mathbb{E}[1+\zeta_c]}\right) = \log_2\left(\frac{1+\bar{\zeta}_r}{1+\bar{\zeta}_c}\right). \qquad \text{(C.8)}$$

Therefore, the ergodic secrecy capacity can be given by

$$C_{\text{erg}}^{\text{sc}}(l,\mathcal{L}) \xrightarrow{a.s.} \log_2\left(\frac{P\sigma_e^2 l^{-\alpha} + \sigma_r^2\sigma_e^2}{P\sigma_r^2\mathcal{L}^{-\alpha_e} + \sigma_r^2\sigma_e^2}\right).$$

By defining $\mu = \mathcal{L}^{-\alpha_e}$ and $\nu = \mathcal{L}^{\alpha_e}$, $C_{\text{erg}}^{\text{sc}}(l,\mathcal{L})$ can be rewritten as

$$\mathcal{H}_1(\mu) = \log_2\left(\frac{P\sigma_e^2 l^{-\alpha} + \sigma_r^2\sigma_e^2}{P\sigma_r^2\mu + \sigma_r^2\sigma_e^2}\right),$$

$$\mathcal{H}_2(\nu) = \log_2\left[\frac{\left(P\sigma_e^2 l^{-\alpha} + \sigma_r^2\sigma_e^2\right)\nu}{P\sigma_r^2 + \sigma_r^2\sigma_e^2\nu}\right].$$

We can obtain that, given $\mu > 0$,

$$\frac{\partial \mathcal{H}_1(\mu)}{\partial \mu} = -\frac{1}{\ln 2}\cdot\frac{P\sigma_r^2}{\left(P\sigma_r^2\mu + \sigma_e^2\sigma_r^2\right)} \le 0;$$

$$\frac{\partial^2 \mathcal{H}_1(\mu)}{\partial \mu^2} = \frac{1}{\ln 2}\cdot\frac{P^2\sigma_r^4}{\left(P\sigma_r^2\mu + \sigma_e^2\sigma_r^2\right)^2} \ge 0.$$

Thus, $\mathcal{H}_1(\mu)$ is a monotonically decreasing and convex function of μ. Likewise, $\mathcal{H}_2(\nu)$ is a monotonically increasing and concave function of μ. Based on the Jensen's inequality (i.e., $\mathbb{E}[f(x)] \ge f(\mathbb{E}[x])$ if $f(x)$ is convex, otherwise $\mathbb{E}[f(x)] \le f(\mathbb{E}[x])$ [137]), the convexity of $\mathcal{H}_1(\mu)$ with respect to μ, and the concavity of $\mathcal{H}_2(\nu)$ with respect to ν, we have

$$\mathbb{E}[\mathcal{H}_1(\mu)] \ge \mathcal{H}_1(\mathbb{E}[\mu]);\ \mathbb{E}[\mathcal{H}_2(\nu)] \le \mathcal{H}_2(\mathbb{E}[\nu]).$$

Since $\mathcal{H}_1(\mu) = \mathcal{H}_2(\nu)$, we have

$$\mathcal{H}_1(\mathbb{E}[\mu]) \le \mathbb{E}[\mathcal{H}_1(\mu)] = \mathbb{E}[\mathcal{H}_2(\nu)] \le \mathcal{H}_2(\mathbb{E}[\nu]). \tag{C.9}$$

Since $\mu = \mathcal{L}^{-\alpha_e}$ and $\nu = \mathcal{L}^{\alpha_e}$, we have

$$\mathbb{E}[\mu] \ge (\mathbb{E}[\mathcal{L}])^{-\alpha_e};\ \mathbb{E}[\nu] \ge (\mathbb{E}[\mathcal{L}])^{\alpha_e}.$$

As $\mathcal{H}_1(\mu)$ and $\mathcal{H}_2(\nu)$ are monotonically decreasing and increasing functions of μ and ν, respectively, we have

$$\mathcal{H}_1(\mathbb{E}[\mu]) \le \mathcal{H}_1\left((\mathbb{E}[\mathcal{L}])^{-\alpha_e}\right) = C_{\text{erg}}^{\text{sc}}(l,\mathbb{E}[\mathcal{L}]); \tag{C.10}$$

$$\mathcal{H}_2(\mathbb{E}[\nu]) \ge \mathcal{H}_2\left((\mathbb{E}[\mathcal{L}])^{\alpha_e}\right) = C_{\text{erg}}^{\text{sc}}(l,\mathbb{E}[\mathcal{L}]). \tag{C.11}$$

Comparing (C.9), (C.10), and (C.11), we find that $C_{\text{erg}}^{\text{sc}}(l, \mathbb{E}[\mathcal{L}])$ lies between the lower and upper bounds of $\mathbb{E}_{l_e}\left[C_{\text{erg}}(l, \mathcal{L})\right]$. They are approximated by

$$\mathbb{E}_{\mathcal{L}}\left[C_{\text{erg}}^{\text{sc}}(l, \mathcal{L})\right] \approx C_{\text{erg}}^{\text{sc}}(l, \mathbb{E}[\mathcal{L}]). \tag{C.12}$$

We proceed to show the bounds converge and, in other words, the approximation is asymptotically accurate. Given $f_{\mathcal{L}}(\tau)$ in (C.4), the expectation of $\mathcal{L}^{\alpha_e}$ can be evaluated as

$$\begin{aligned}
\mathbb{E}[\mathcal{L}^{\alpha_e}] &= h_1(r) \\
&= \int_0^r \tau^{\alpha_e} f_{\mathcal{L}}(\tau) d\tau \\
&= \int_0^r N \frac{3\tau^{\alpha_e+2}}{r^3}\left(1 - \frac{\tau^3}{r^3}\right)^{N-1} d\tau \\
&= \frac{\Gamma\left(1+\frac{\alpha_e}{3}\right)\Gamma(1+N)}{\Gamma\left(1+\frac{\alpha_e}{3}+N\right)} r^{\alpha_e} \xrightarrow{N\to\infty} \Gamma\left(1+\frac{\alpha_e}{3}\right) N^{-\frac{\alpha_e}{3}} r^{\alpha_e}.
\end{aligned} \tag{C.13}$$

Likewise,

$$\begin{aligned}
\mathbb{E}[\mathcal{L}^{-\alpha_e}] &= h_2(r) \\
&= \frac{\Gamma\left(1-\frac{\alpha_e}{3}\right)\Gamma(1+N)}{\Gamma\left(1-\frac{\alpha_e}{3}+N\right)} r^{-\alpha_e} \\
&\xrightarrow{N\to\infty} \Gamma\left(1-\frac{\alpha_e}{3}\right) N^{\frac{\alpha_e}{3}} r^{-\alpha_e}, \quad 0 < \alpha_e < 3.
\end{aligned} \tag{C.14}$$

Substitute (C.13) and (C.14) into $\mathcal{H}_1(\mu)$ and $\mathcal{H}_2(\nu)$ separately,

$$\begin{aligned}
\mathcal{H}_1(\mathbb{E}[\mu]) &= \mathcal{H}_1(\mathbb{E}[\mathcal{L}^{-\alpha_e}]) \\
&= \log_2\left(\frac{P\sigma_e^2 l^{-\alpha} + \sigma_r^2\sigma_e^2}{P\sigma_r^2 h_2(r) + \sigma_r^2\sigma_e^2}\right) \\
&= \log_2\left(\frac{P\sigma_e^2 l^{-\alpha} + \sigma_r^2\sigma_e^2}{\frac{\Gamma\left(1-\frac{\alpha_e}{3}\right)\Gamma(1+N)}{\Gamma\left(1-\frac{\alpha_e}{3}+N\right)} P\sigma_r^2 r^{-\alpha_e} + \sigma_r^2\sigma_e^2}\right),
\end{aligned} \tag{C.15}$$

$$
\begin{aligned}
\mathcal{H}_2\left(\mathbb{E}[\nu]\right) &= \mathcal{H}_2\left(\mathbb{E}\left[\mathcal{L}^{\alpha_e}\right]\right) \\
&= \log_2\left(\frac{P\sigma_e^2 l^{-\alpha} + \sigma_r^2\sigma_e^2}{P\sigma_r^2 h_1^{-1}(r) + \sigma_r^2\sigma_e^2}\right) \\
&= \log_2\left(\frac{P\sigma_e^2 l^{-\alpha} + \sigma_r^2\sigma_e^2}{\frac{\Gamma\left(1+\frac{\alpha_e}{3}+N\right)}{\Gamma\left(1+\frac{\alpha_e}{3}\right)\Gamma(1+N)} P\sigma_r^2 r^{-\alpha_e} + \sigma_r^2\sigma_e^2}\right).
\end{aligned}
\tag{C.16}
$$

Here, $\dfrac{\Gamma\left(1-\frac{\alpha_e}{3}\right)\Gamma(1+N)}{\Gamma\left(1-\frac{\alpha_e}{3}+N\right)}$ and $\dfrac{\Gamma\left(1+\frac{\alpha_e}{3}\right)\Gamma(1+N)}{\Gamma\left(1+\frac{\alpha_e}{3}+N\right)}$ are constants given N and α_e. The approximation error of (C.12) is no greater than half of the difference between $\mathcal{H}_1\left(\mathbb{E}[\mu]\right)$ and $\mathcal{H}_2\left(\mathbb{E}[\nu]\right)$. The error is increasingly negligible with the growth of r, since both $h_1^{-1}(r)$ and $h_2(r)$ are strictly monotonically decreasing functions of r. $\mathcal{H}_1\left(\mathbb{E}[\mu]\right) \to \mathcal{H}_2\left(\mathbb{E}[\nu]\right)$, when $\sigma_r^2 = \sigma_e^2 = \sigma^2$, $r \gg \left(\dfrac{\Gamma\left(1-\frac{\alpha_e}{3}\right)\Gamma(1+N)}{\Gamma\left(1-\frac{\alpha_e}{3}+N\right)}\dfrac{P}{\sigma_e^2}\right)^{1/\alpha_e}$ and $r \gg \left(\dfrac{\Gamma\left(1+\frac{\alpha_e}{3}+N\right)}{\Gamma\left(1+\frac{\alpha_e}{3}\right)\Gamma(1+N)}\dfrac{P}{\sigma^2}\right)^{1/\alpha}$ in (C.15) and (C.16).

C.3 Proof of $\mathbb{E}\left[\log_2\left(\frac{\mathcal{X}}{\mathcal{Y}}\right)\right] \xrightarrow{a.s.} \log_2\left(\frac{\mathbb{E}[\mathcal{X}]}{\mathbb{E}[\mathcal{Y}]}\right)$

We consider a ground receiver equipped with M and N antennas, and N aerial eavesdroppers each equipped with a single antenna. Let x_i $(i = 1, \ldots, M)$ denote the received signals at the different antennas of the ground receiver, and y_j $(j = 1, \ldots, N)$ denote the received signals at the different aerial eavesdroppers. They are two sequences of positive square-integrable random variables and not necessarily independent across i or j, $\frac{M}{N} < \infty$. $\mathcal{X} = \sum_{i=1}^{M} x_i$ and $\mathcal{Y} = \sum_{j=1}^{N} y_j$ are the SNR of MRC at the ground receiver and the aerial eavesdroppers, respectively.

Since $\mathbb{E}[x_i] < \infty$, $i = 1, \ldots, N$, based on Kolmogorov's strong law of large numbers, we have [224]

$$
\frac{1}{M}\sum_{i=1}^{M} x_i - \frac{1}{M}\sum_{i=1}^{M} \mathbb{E}[x_i] \xrightarrow{a.s.} 0, \text{ as } M \to \infty. \tag{C.17}
$$

By taking the expectation on both sides of (C.17), we have

$$\begin{aligned}&\mathbb{E}\left[\frac{1}{M}\sum_{i=1}^{M}x_i-\frac{1}{M}\sum_{i=1}^{M}\mathbb{E}\left[x_i\right]\right]\\&=\mathbb{E}\left[\frac{1}{M}\sum_{i=1}^{M}x_i\right]-\frac{1}{M}\sum_{i=1}^{M}\mathbb{E}\left[x_i\right]\xrightarrow{a.s.}0,\text{ as }M\to\infty.\end{aligned}\tag{C.18}$$

Subtracting (C.18) from (C.17), we can obtain

$$\frac{1}{M}\sum_{i=1}^{M}x_i-\mathbb{E}\left[\frac{1}{M}\sum_{i=1}^{M}x_i\right]\xrightarrow{a.s.}0,\text{ as }M\to\infty.\tag{C.19}$$

In other words, $\frac{1}{M}\mathcal{X}\xrightarrow{a.s.}\frac{1}{M}\mathbb{E}\left[\mathcal{X}\right]$, as $M\to\infty$. Likewise, we have $\frac{1}{N}\mathcal{Y}-\frac{1}{N}\mathbb{E}\left[\mathcal{Y}\right]\xrightarrow{a.s.}0$, as $N\to\infty$, i.e., $\frac{1}{N}\mathcal{Y}\xrightarrow{a.s.}\frac{1}{N}\mathbb{E}\left[\mathcal{Y}\right]$, as $N\to\infty$. Given the fact that the limit of reciprocal is the reciprocal of the limit [229] and both of $\frac{1}{N}\mathcal{Y}$ and $\frac{1}{N}\mathbb{E}\left[\mathcal{Y}\right]$ are positive, we have $\frac{1}{\frac{1}{N}\mathcal{Y}}\xrightarrow{a.s.}\frac{1}{\frac{1}{N}\mathbb{E}\left[\mathcal{Y}\right]}$ as $N\to\infty$.

Based on the multiplication properties for the *a.s.* convergence [31, theorem 3.8], we have

$$\frac{\frac{1}{M}\mathcal{X}}{\frac{1}{N}\mathcal{Y}}\xrightarrow{a.s.}\frac{\frac{1}{M}\mathbb{E}\left[\mathcal{X}\right]}{\frac{1}{N}\mathbb{E}\left[\mathcal{Y}\right]},\text{ as }M\to\infty\text{ and }0<\frac{N}{M}<\infty,\tag{C.20}$$

which can be further rewritten as

$$\frac{\mathcal{X}}{\mathcal{Y}}\xrightarrow{a.s.}\frac{\mathbb{E}\left[\mathcal{X}\right]}{\mathbb{E}\left[\mathcal{Y}\right]},\text{ as }M\to\infty\text{ and }0<\frac{N}{M}<\infty.\tag{C.21}$$

According to the *Continuous mapping theorem* [36], if $X_n\xrightarrow{a.s.}X$, then $h(X_n)\xrightarrow{a.s.}h(X)$, where X_n, $n=1,\dots,N$ is a sequence of random variables and $h(\cdot)$ is a continuous function. We have

$$\log_2\left(\frac{\mathcal{X}}{\mathcal{Y}}\right)\xrightarrow{a.s.}\log_2\left(\frac{\mathbb{E}\left[\mathcal{X}\right]}{\mathbb{E}\left[\mathcal{Y}\right]}\right),\text{ as }M\to\infty\text{ and }0<\frac{N}{M}<\infty.\tag{C.22}$$

By taking the expectation on both sides of (C.22), we have

$$\mathbb{E}\left[\log_2\left(\frac{\mathcal{X}}{\mathcal{Y}}\right)\right]\xrightarrow{a.s.}\log_2\left(\frac{\mathbb{E}\left[\mathcal{X}\right]}{\mathbb{E}\left[\mathcal{Y}\right]}\right).\tag{C.23}$$

As above, the *a.s.* convergence can be achieved in the case where M and N are large. $\mathbb{E}\left[\log_2\left(\frac{\mathcal{X}}{\mathcal{Y}}\right)\right]\approx\log_2\left(\frac{\mathbb{E}\left[\mathcal{X}\right]}{\mathbb{E}\left[\mathcal{Y}\right]}\right)$ if there is a single antenna at the ground receiver and the number of eavesdroppers is not too large.

C.4 Proof of Theorem 4.2

By definition, the secrecy outage probability is

$$\mathcal{P}_{\text{out}}^{\text{s}} = \Pr\left\{\log_2(1+\zeta_r) - \log_2(1+\zeta_c) < C_{\text{out}}^{\text{s}}\right\} \tag{C.24a}$$

$$= \Pr\left\{\Psi < 2^{C_{\text{out}}^{\text{S}}} - 1\right\} = \epsilon, \tag{C.24b}$$

where (C.24b) is obtained by setting $\Psi = \zeta_r - 2^{C_{\text{out}}^{S}}\zeta_c \triangleq \mathcal{X} - \mathcal{Y}$.

The ground transmitter and receiver are likely to experience NLoS. The Rayleigh fading is typically assumed in-between, i.e., $K = 0$ [248]. The PDF of Ψ can be given by

$$f_\Psi(\psi) = \int_0^\infty f_\mathcal{X}(\psi + y) f_\mathcal{Y}(y) dy \tag{C.25a}$$

$$= \frac{B \exp\left(-K_e + \frac{B}{2} - \frac{\psi}{\bar{\mathcal{X}}}\right)}{2K_e\bar{\mathcal{X}}} \int_0^\infty \frac{1}{2} \exp\left(-\frac{y_1 + B}{2}\right) I_0\left(\sqrt{By_1}\right) dy_1 \tag{C.25b}$$

$$= \frac{B \exp\left(-K_e + \frac{B}{2} - \frac{\psi}{\bar{\mathcal{X}}}\right)}{2K_e\bar{\mathcal{X}}} Q\left(\sqrt{B}, 0\right) \tag{C.25c}$$

$$= \frac{(1+K_e)\exp\left(-\frac{K_e\bar{\mathcal{Y}}}{\bar{\mathcal{Y}}+(1+K_e)\bar{\mathcal{X}}}\right)}{\bar{\mathcal{Y}} + (1+K_e)\bar{\mathcal{X}}} \exp\left(-\frac{\psi}{\bar{\mathcal{X}}}\right), \tag{C.25d}$$

where $f_\mathcal{Z}(z)$ denotes the PDF of random variable $\mathcal{Z}$, $\bar{\mathcal{X}} = \bar{\zeta}_r$, and $\bar{\mathcal{Y}} = 2^{C_{\text{out}}^{S}}\bar{\zeta}_c$. (C.25b) is obtained by setting $y_1 = \frac{2\left[(1+K_e)\bar{\mathcal{X}} + \bar{\mathcal{Y}}\right]}{\bar{\mathcal{X}}\bar{\mathcal{Y}}} y$ and $B = \frac{2K_e(1+K_e)\bar{\mathcal{X}}}{\bar{\mathcal{Y}} + (1+K_e)\bar{\mathcal{X}}}$; (C.25c) is based on the definition of the first-order Marcum Q-function [23, equation (8)], i.e., $Q(\sqrt{a}, \sqrt{b}) = \int_b^\infty \frac{1}{2}\exp(-\frac{x+a}{2}) I_0(\sqrt{ax}) dx$; (C.25d) is obtained from $Q\left(\sqrt{a}, 0\right) = 1$. The secrecy outage probability can be evaluated based on [206, equation 37]

$$\mathcal{P}_{\text{out}}^{\text{s}} = \Pr\left(\Psi \le 2^{C_{\text{out}}^{S}} - 1\right) = 1 - \Pr\left(\Psi \ge 2^{C_{\text{out}}^{S}} - 1\right)$$

$$= 1 - \frac{(1+K_e)\bar{\mathcal{X}}\exp\left(-\frac{K_e\bar{\mathcal{Y}}}{\bar{\mathcal{Y}}+(1+K_e)\bar{\mathcal{X}}}\right)}{\bar{\mathcal{Y}} + (1+K_e)\bar{\mathcal{X}}} \exp\left(-\frac{2^{C_{\text{out}}^{S}} - 1}{\bar{\mathcal{X}}}\right). \tag{C.26}$$

By defining $\mathcal{M} = \dfrac{(1+K_e)\bar{\mathcal{X}}\exp\left(-\frac{K_e\bar{\mathcal{Y}}}{\bar{\mathcal{Y}}+(1+K_e)\bar{\mathcal{X}}}\right)}{\bar{\mathcal{Y}}+(1+K_e)\bar{\mathcal{X}}}$, (C.26) can be further rewritten as

$$1-\mathcal{P}_{\text{out}}^{\text{s}} = \mathcal{M}\exp\left(-\frac{2^{C_{\text{out}}^{\text{s}}}-1}{\bar{\mathcal{X}}}\right) = 1-\epsilon. \tag{C.27}$$

Because $\exp(-x) \approx 1-x$ when $x \to 0^+$, we have $(1-\epsilon)\frac{1}{\mathcal{M}} \approx 1 - \frac{2^{C_{\text{out}}^{\text{s}}}-1}{\bar{\mathcal{X}}}$. Moreover, $\bar{\mathcal{X}} = \bar{\zeta}_r$ and $\bar{\mathcal{Y}} = 2^{C_{\text{out}}^{\text{s}}}\bar{\zeta}_c$; see under (C.24). Therefore, we have

$$\frac{(1-\epsilon)\left[\bar{\zeta}_c 2^{C_{\text{out}}^{\text{s}}} + (1+K_e)\bar{\zeta}_r\right]}{(1+K_e)\bar{\zeta}_r}\exp\left[\frac{K_e\bar{\zeta}_c 2^{C_{\text{out}}^{\text{s}}}}{\bar{\zeta}_c 2^{C_{\text{out}}^{\text{s}}} + (1+K_e)\bar{\zeta}_r}\right] \approx 1 - \frac{2^{C_{\text{out}}^{\text{s}}}-1}{\bar{\zeta}_r}. \tag{C.28}$$

We can solve (C.28) to obtain $C_{\text{out}}^{\text{s}}$. There are two cases, as follows. In the case of $K_e > 0$, we set $\mathcal{Z} = \dfrac{\bar{\zeta}_c 2^{C_{\text{out}}^{\text{s}}} + (1+K_e)\bar{\zeta}_r}{K_e(1+K_e)\bar{\zeta}_r}$, and (C.28) can be rewritten as

$$\mathcal{Z}\exp\left(-\frac{1}{\mathcal{Z}}\right) \approx \frac{\left(1+\bar{\zeta}_r - 2^{C_{\text{out}}^{\text{s}}}\right)\exp(-K_e)}{K_e(1-\epsilon)\bar{\zeta}_r} \tag{C.29a}$$

$$\approx \frac{(1+\bar{\zeta}_r)\exp(-K_e)}{K_e(1-\epsilon)\bar{\zeta}_r}, \tag{C.29b}$$

where (C.29b) is under the assumption that $1+\bar{\zeta}_r \gg 2^{C_{\text{out}}^{\text{s}}}$.[1] Then, we can obtain the solution for $\mathcal{Z}^{-1}$ as

$$\mathcal{Z}^{-1} \approx \mathcal{W}_0\left(\frac{K_e\exp(K_e)(1-\epsilon)\bar{\zeta}_r}{1+\bar{\zeta}_r}\right) \triangleq W_0. \tag{C.30}$$

By solving $Z^{-1} = \dfrac{K_e(1+K_e)\bar{\zeta}_r}{\bar{\zeta}_c 2^{C_{\text{out}}^{\text{s}}} + (1+K_e)\bar{\zeta}_r} = W_0$, the ϵ-secrecy outage capacity can be obtained as

$$C_{\text{out}}^{\text{s}} \approx \log_2\left[\frac{(1+K_e)(K_e - W_0)\bar{\zeta}_r}{W_0\bar{\zeta}_c}\right]. \tag{C.31}$$

Note that the assumption $1+\bar{\zeta}_r \gg 2^{C_{\text{out}}^{\text{s}}}$ is taken in (C.29b) to achieve (C.31). We can verify the accuracy of the assumption in high SNR regimes as follows. First, we have

---

1 By the definition of the ϵ-secrecy outage capacity, $\log_2(1+\bar{\zeta}_r) - \log_2(1+\bar{\zeta}_c) > C_{\text{out}}^{\text{s}}$, i.e., $1+\bar{\zeta}_r > 2^{C_{\text{out}}^{\text{s}}}(1+\bar{\zeta}_e)$. In high SNR regimes, $Pl^{-\alpha} \gg \sigma_r^2$ and $Pl_e^{-\alpha_c} \gg \sigma_e^2$, thus $1+\bar{\zeta}_c \gg 1$ and $1+\bar{\zeta}_r \gg 2^{C_{\text{out}}^{\text{s}}}$.

$$W_0 = \mathcal{W}_0\left(\frac{K_e \exp(K_e)(1-\epsilon)Pl^{-\alpha}}{\sigma_r^2 + Pl^{-\alpha}}\right)$$
$$\approx \mathcal{W}_0(K_e \exp(K_e)(1-\epsilon)) \leq \mathcal{W}_0(K_e \exp(K_e)) = K_e.$$

By substituting (C.31), we have $2^{C_{\text{out}}^s} = \frac{(1+K_e)(K_e - W_0)\bar{\xi}_r}{W_0\bar{\xi}_c}$. Given the typical range of K_e, $0 < K_e \leq 10$, we have $\frac{(1+K_e)(K_e - W_0)}{W_0} = (1+K_e)\left(\frac{K_e}{W_0} - 1\right) = \Theta(1+K_e) = \Theta(1)$.[2] By following the proof of Theorem 4.1, we have $\bar{\xi}_c \approx \frac{P\mathbb{E}[\mathcal{L}]^{-\alpha_e}}{\sigma_e^2}$. When $N \to \infty$, $\bar{\xi}_c = \Theta\left(N^{\frac{\alpha_e}{3}}\right)$, and $1 + \bar{\xi}_r - \frac{(1+K_e)(K_e - W_0)\bar{\xi}_r}{W_0\bar{\xi}_c} = 1 + \bar{\xi}_r - \frac{\Theta(1)}{\bar{\xi}_c}\bar{\xi}_r \approx 1 + \bar{\xi}_r - \frac{\Theta(1)}{\Theta\left(N^{\frac{\alpha_e}{3}}\right)}\bar{\xi}_r \approx 1 + \bar{\xi}_r$. Hence, the approximation of (C.29a) to (C.29b) is accurate in the high SNR regimes.

In the case of $K_e = 0$, (C.28) can be rewritten as

$$(1-\epsilon)\left(\bar{\xi}_r + \bar{\xi}_c 2^{C_{\text{out}}^s}\right) \approx \bar{\xi}_r - \left(2^{C_{\text{out}}^s} - 1\right). \tag{C.32}$$

The ϵ-outage secrecy capacity can be approximated by

$$C_{\text{out}}^s \approx \log_2\left[\frac{1+\epsilon\bar{\xi}_r}{1+(1-\epsilon)\bar{\xi}_c}\right]. \tag{C.33}$$

By substituting $\bar{\xi}_r = \frac{Pl^{-\alpha}}{\sigma_r^2}$ and $\bar{\xi}_c = \frac{P\mathcal{L}^{-\alpha_e}}{\sigma_e^2}$ into (C.31) and (C.32), the ϵ-outage secrecy capacity under N autonomous eavesdroppers conducting SC can be approximately given by

$$C_{\text{out}}^{\text{sc}}(l, \mathcal{L}) \approx \begin{cases} \log_2\left[\frac{(1+K_e)(K_e - W_0)\sigma_e^2 l^{-\alpha}}{W_0\sigma_r^2\mathcal{L}^{-\alpha_e}}\right], & \text{if} \quad K_e > 0; \\ \log_2\left[\frac{\sigma_r^2\sigma_e^2 + \epsilon P\sigma_e^2 l^{-\alpha}}{\sigma_r^2\sigma_e^2 + (1-\epsilon)P\sigma_r^2\mathcal{L}^{-\alpha_e}}\right], & \text{if} \quad K_e = 0; \end{cases} \tag{C.34}$$

We proceed to evaluate the convexity/concavity of $C_{\text{out}}^{\text{sc}}(l, \mathcal{L})$ in the two cases of $K_e > 0$ and $K_e = 0$. In the case of $K_e > 0$, by substituting $\mu = \mathcal{L}^{-\alpha_e}$ and $\nu = \mathcal{L}^{\alpha_e}$, (C.31) can be rewritten as

$$\mathcal{F}_1(\mu) \triangleq \log_2\left[\frac{(1+K_e)(K_e - W_0)\sigma_e^2 l^{-\alpha}}{W_0\sigma_r^2\mu}\right],$$

---

2 We denote $g(n) = O(f(n))$ if there is a positive constant c_1 such that $\lim_{n\to\infty} \Pr\left(\frac{g(n)}{f(n)} \leq c_1\right) = 1$, and $g(n) = \Omega(f(n))$ if $f(n) = O(g(n))$. $g(n) = \Theta(f(n))$ if $g(n) = O(f(n))$ and $g(n) = \Omega(f(n))$.

$$\mathcal{F}_2(\nu) \triangleq \log_2\left[\frac{(1+K_e)(K_e-W_0)\,\sigma_e^2 l^{-\alpha}\nu}{W_0\sigma_r^2}\right].$$

Clearly, $\mathcal{F}_1(\mu)=\mathcal{F}_2(\nu)$. For $\mathcal{F}_1(\mu)$, we have

$$\frac{\partial \mathcal{F}_1(\mu)}{\partial \mu} = -\frac{1}{\mu \ln 2} \leq 0;\ \frac{\partial^2 \mathcal{F}_1(\mu)}{\partial \mu^2} = \frac{1}{\mu^2 \ln 2} \geq 0.$$

Thus, $\mathcal{F}_1(\mu)$ is a monotonically decreasing and convex function of μ. Likewise, we can show $\mathcal{F}_2(\nu)$ is a monotonically increasing and concave function of ν.

By following the proof of Theorem 4.1, we have

$$\mathbb{E}_{\mathcal{L}}\left[C_{\text{out}}^{\text{sc}}(l,\mathcal{L})\,|_{K_e>0}\right] \approx C_{\text{out}}^{\text{sc}}(l,\mathbb{E}[\mathcal{L}])\,|_{K_e>0}. \tag{C.35}$$

In the case of $K_e=0$, both $C_{\text{out}}^{\text{sc}}(l,\mathcal{L})$ and $C_{\text{erg}}^{\text{sc}}(l,\mathcal{L})$ exhibit the same monotonicity and concavity with regards to $\mathcal{L}$. Based on the proof of Theorem 4.1, we have

$$\mathbb{E}_{\mathcal{L}}\left[C_{\text{out}}^{\text{sc}}(l,\mathcal{L})\,|_{K_e=0}\right] \approx C_{\text{out}}^{\text{sc}}(l,\mathbb{E}[\mathcal{L}])\,|_{K_e=0}. \tag{C.36}$$

By substituting $\mathbb{E}[\mathcal{L}]$ into (C.35) and (C.36), we obtain (4.4). Therefore, Theorem 4.2 is proved.

C.5 Proof of Theorem 4.3

Based on the PDF of ζ_c^{mrc}, the expectation of ζ_c^{mrc} can be given by

$$\begin{aligned}
\mathbb{E}[\zeta_c^{\text{mrc}}] &= \int_0^\infty x f_{\zeta_c^{\text{mrc}}}(x)dx \\
&= \exp(-K_e)\sum_{k=0}^{\infty}\frac{K_e^k\left(\frac{K_e+N}{N\Omega_c}\right)^{N+k}}{k!(N+k-1)!} \\
&\quad \left[\int_0^\infty x^{N+k}\exp\left(-\frac{K_e+N}{N\Omega_c}x\right)dx\right] \\
&= \frac{N\Omega_c}{K_e+N}\exp(-K_e)\sum_{k=0}^{\infty}\left[\frac{K_e^k}{(k-1)!}+\frac{NK_e^k}{k!}\right] \\
&= \frac{N\Omega_c}{K_e+N}\exp(-K_e)\times(K_e+N)\exp(K_e) \\
&= N\Omega_c.
\end{aligned} \tag{C.37}$$

Then, we have $\mathbb{E}[1+\zeta_c^{\text{mrc}}]=1+N\Omega_c=1+\sum_{i=1}^{N}\bar{\zeta}_{e_i}=1+\sum_{i=1}^{N}\frac{Pl_i^{-\alpha_e}}{\sigma_e^2}$. Substituting $\mathbb{E}[1+\zeta_c^{\text{mrc}}]$ into (C.8), we can obtain the ergodic secrecy capacity under N autonomous eavesdroppers conducting MRC, as given by

$$\mathcal{C}_{\mathrm{erg}}^{\mathrm{s(mrc)}} \approx \log_2\left(\frac{\frac{Pl^{-\alpha}}{\sigma_r^2}+1}{\frac{P}{\sigma_e^2}\sum_{i=1}^{N} l_i^{-\alpha_e}+1}\right) \triangleq \mathcal{C}_{\mathrm{erg}}^{\mathrm{mrc}}(l, l_i). \tag{C.38}$$

Following the proof of Theorem 4.1, we obtain $\mathbb{E}_{l_i}\left[\mathcal{C}_{\mathrm{erg}}^{\mathrm{mrc}}(l, l_i)\right] \xrightarrow{a.s.} \mathcal{C}_{\mathrm{erg}}^{\mathrm{mrc}}(l, \mathbb{E}[l_i])$. Since the N aerial eavesdroppers are independently and uniformly distributed in the 3D hemisphere, $\mathbb{E}[l_i] = \frac{3}{4}r, i = 1, \dots, N$, based on Lemma 4.2. Finally, (4.6) can be obtained by substituting $\sum_{i=1}^{N}(\mathbb{E}[l_i])^{-\alpha_e} = N(\mathbb{E}[l_i])^{-\alpha_e}$ into $\mathcal{C}_{\mathrm{erg}}^{\mathrm{mrc}}(l, \mathbb{E}[l_i])$.

C.6 Proof of Theorem 4.4

By following the proof of Theorem 4.2, we let $\Phi = \zeta_r - 2^{C_{\mathrm{out}}^{\mathrm{mrc}}}\zeta_c^{\mathrm{mrc}} \triangleq \mathcal{X} - \mathcal{Y}_c$. The PDF of Φ is given by

$$f_\Phi(\phi) = \int_0^\infty f_{\mathcal{X}}(\phi + y) f_{\mathcal{Y}_c}(y)dy \tag{C.39a}$$

$$= \frac{1}{2A_c\bar{\mathcal{X}}}\left(\frac{N+K_e}{N\bar{\mathcal{Y}}_c}\right)^{\frac{N+1}{2}}\left(\frac{B_c}{A_cK_e}\right)^{\frac{N-1}{2}}\exp\left(-K_e + \frac{B_c}{2} - \frac{\phi}{\bar{\mathcal{X}}}\right)$$
$$\times \int_0^\infty \frac{1}{2}\left(\frac{y_2}{B_c}\right)^{\frac{N-1}{2}}\exp\left(-\frac{y_2+B_c}{2}\right) I_{N-1}\left(\sqrt{B_c y_2}\right)dy_1 \tag{C.39b}$$

$$= \frac{1}{2A_c\bar{\mathcal{X}}}\left(\frac{N+K_e}{N\bar{\mathcal{Y}}_c}\right)^{\frac{N+1}{2}}\left(\frac{B_c}{A_cK_e}\right)^{\frac{N-1}{2}}\exp\left(-K_e + \frac{B_c}{2} - \frac{\phi}{\bar{\mathcal{X}}}\right) Q_{N-1}\left(\sqrt{B_c}, 0\right) \tag{C.39c}$$

$$= \frac{(N+K_e)^N \bar{\mathcal{X}}^{N-1}}{\left[N\bar{\mathcal{Y}}_c + (N+K_e)\bar{\mathcal{X}}\right]^N}\exp\left(-\frac{NK_e\bar{\mathcal{Y}}_c}{N\bar{\mathcal{Y}} + (N+K_e)\bar{\mathcal{X}}}\right)\exp\left(-\frac{\phi}{\bar{\mathcal{X}}}\right), \tag{C.39d}$$

where $\bar{\mathcal{X}} = \bar{\zeta}_r$ and $\bar{\mathcal{Y}}_c = \Omega_c 2^{C_{\mathrm{out}}^{\mathrm{mrc}}}$. (C.39b) is obtained by setting $A_c = \frac{2\left[(N+K_e)\bar{\mathcal{X}} + N\bar{\mathcal{Y}}\right]}{N\bar{\mathcal{X}}\bar{\mathcal{Y}}}$, $y_2 = A_c y$, and $B_c = \frac{2K_e(N+K_e)\bar{\mathcal{X}}}{N\bar{\mathcal{Y}} + (N+K_e)\bar{\mathcal{X}}}$; (C.39c) is based on the definition of the L-th order Marcum Q-function [23, eq. 8], i.e., $Q_L(\sqrt{a}, \sqrt{b}) = \int_b^\infty \frac{1}{2}\left(\frac{x}{a}\right)^{\frac{L}{2}}\exp(-\frac{x+a}{2})I_L(\sqrt{ax})dx$; (C.39d) is because $Q_L\left(\sqrt{a}, 0\right) = 1$.

The secrecy outage probability can be evaluated based on [206, equation 37], as given by

$$\mathcal{P}_{\text{out}}^{\text{mrc}} = \Pr\left(\Phi \le 2^{C_{\text{out}}^{\text{mrc}}} - 1\right) = 1 - \Pr\left(\Phi \ge 2^{C_{\text{out}}^{\text{mrc}}} - 1\right)$$
$$= 1 - \frac{(N+K_e)^N \bar{\mathcal{X}}^N}{\left[N\bar{\mathcal{Y}}_c + (N+K_e)\bar{\mathcal{X}}\right]^N} \times \exp\left(-\frac{NK_e\bar{\mathcal{Y}}_c}{N\bar{\mathcal{Y}}_c + (N+K_e)\bar{\mathcal{X}}}\right)\exp\left(-\frac{2^{C_{\text{out}}^{\text{mrc}}}-1}{\bar{\mathcal{X}}}\right). \quad \text{(C.40)}$$

By the definition of the ϵ-outage secrecy capacity, we have

$$1 - \mathcal{P}_{\text{out}}^{\text{mrc}} = \frac{(N+K_e)^N \bar{\mathcal{X}}^N}{\left[N\bar{\mathcal{Y}}_c + (N+K_e)\bar{\mathcal{X}}\right]^N} \times \exp\left(-\frac{NK_e\bar{\mathcal{Y}}_c}{N\bar{\mathcal{Y}}_c + (N+K_e)\bar{\mathcal{X}}} - \frac{2^{C_{\text{out}}^{\text{mrc}}}-1}{\bar{\mathcal{X}}}\right) = 1 - \epsilon.$$

Since $\bar{\mathcal{X}} = \bar{\zeta}_r$ and $\bar{\mathcal{Y}}_c = 2^{C_{\text{out}}^{\text{mrc}}}\Omega_c$, we have

$$(1-\epsilon)\frac{\left[N\Omega_c 2^{C_{\text{out}}^{\text{mrc}}} + (N+K_e)\bar{\zeta}_r\right]^N}{(N+K_e)^N \bar{\zeta}_r^N} \times \exp\left(\frac{NK_e\Omega_c 2^{C_{\text{out}}^{\text{mrc}}}}{N\Omega_c 2^{C_{\text{out}}^{\text{mrc}}} + (N+K_e)\bar{\zeta}_r}\right) = \exp\left(-\frac{2^{C_{\text{out}}^{\text{mrc}}}-1}{\bar{\zeta}_r}\right). \quad \text{(C.41)}$$

We proceed to solve (C.41) for $C_{\text{out}}^{\text{mrc}}$. There are two cases, as follows. In the case of $K_e > 0$, since $\exp(-x) \approx 1 - x$ when $x \to 0^+$, the right-hand side of (C.41) can be approximated by

$$(1-\epsilon)\frac{\left[N\Omega_c 2^{C_{\text{out}}^{\text{mrc}}} + (N+K_e)\bar{\zeta}_r\right]^N}{(N+K_e)^N \bar{\zeta}_r^N} \times \exp\left(\frac{NK_e\Omega_c 2^{C_{\text{out}}^{\text{mrc}}}}{N\Omega_c 2^{C_{\text{out}}^{\text{mrc}}} + (N+K_e)\bar{\zeta}_r}\right) \approx 1 - \frac{2^{C_{\text{out}}^{\text{mrc}}}-1}{\bar{\zeta}_r}. \quad \text{(C.42)}$$

Let $\mathcal{Z}_c = \frac{N\Omega_c 2^{C_{\text{out}}^{\text{mrc}}} + (N+K_e)\bar{\zeta}_r}{K_e\,(N+K_e)\,\bar{\zeta}_r}$, (C.42) can be rewritten as

$$\mathcal{Z}_c^N \exp\left(-\frac{1}{\mathcal{Z}_c}\right) \approx \frac{\left(1+\bar{\zeta}_r - 2^{C_{\text{out}}^{\text{mrc}}}\right)\exp(-K_e)}{K_e^N\,(1-\epsilon)\,\bar{\zeta}_r} \quad \text{(C.43a)}$$
$$\approx \frac{\left(1+\bar{\zeta}_r\right)\exp(-K_e)}{K_e^N\,(1-\epsilon)\,\bar{\zeta}_r}, \quad \text{(C.43b)}$$

where (C.43b) is under the assumption that $1 + \bar{\zeta}_r \gg 2^{C_{\text{out}}^{\text{mrc}}}$. Exploiting the Lambert W function, we can obtain the solution for $\mathcal{Z}_c^{-1}$ as

$$\mathcal{Z}_c^{-1} \approx N\mathcal{W}_0\left[\frac{K_e}{N}\left(\frac{\exp(K_e)(1-\epsilon)\bar{\zeta}_r}{1+\bar{\zeta}_r}\right)^{\frac{1}{N}}\right] \triangleq W_c.$$

By solving $\mathcal{Z}_c^{-1} = \dfrac{K_e(N+K_e)\bar{\zeta}_r}{N\Omega_c 2^{C_{\text{out}}^{\text{mrc}}} + (N+K_e)\bar{\zeta}_r}$, the ϵ-secrecy outage capacity can be obtained as

$$C_{\text{out}}^{\text{mrc}} \approx \log_2\left[\frac{(N+K_e)(K_e-W_c)\bar{\zeta}_r}{N\Omega_c W_c}\right]. \tag{C.44}$$

Note that the assumption $1 + \bar{\zeta}_r \gg 2^{C_{\text{out}}^{\text{mrc}}}$ is taken in (C.43b) to achieve (C.44). We can verify the accuracy of the assumption in high SNR regimes. We first have

$$\begin{aligned} W_c &= N\mathcal{W}_0\left[\frac{K_e}{N}\left(\frac{\exp(K_e)(1-\epsilon)Pl^{-\alpha}}{\sigma_r^2 + Pl^{-\alpha}}\right)^{\frac{1}{N}}\right] \\ &\approx N\mathcal{W}_0\left[\frac{K_e}{N}\exp\left(\frac{K_e}{N}\right)(1-\epsilon)^{\frac{1}{N}}\right] \le N\mathcal{W}_0\left[\frac{K_e}{N}\exp\left(\frac{K_e}{N}\right)\right] = K_e. \end{aligned}$$

By substituting (C.44), we have $2^{C_{\text{out}}^{\text{mrc}}} = \dfrac{(N+K_e)(K_e-W_c)\bar{\zeta}_r}{NW_c\Omega_c}$. Given the typical range of K_e, i.e., $0 < K_e \le 10$, we have $\dfrac{(N+K_e)(K_e-W_0)}{NW_c} = \left(1+\dfrac{K_e}{N}\right)\left(\dfrac{K_e}{W_c}-1\right) = \Theta\left(1+\dfrac{K_e}{N}\right) = \Theta(1)$. Then, $1+\bar{\zeta}_r - \dfrac{(N+K_e)(K_e-W_c)\bar{\zeta}_r}{NW_c\Omega_c} = 1+\bar{\zeta}_r - \dfrac{\Theta(1)}{\Omega_c}\bar{\zeta}_r = 1 + \left(\dfrac{\Omega_c - \Theta(1)}{\Omega_c}\right)\bar{\zeta}_r \approx 1+\bar{\zeta}_r$ since $\Omega_c = \frac{1}{N}\sum_{i=1}^{N}\zeta_{e_i} \gg \Theta(1)$ in the high SNR regimes. Hence, the approximation of (C.43a) to (C.43b) is accurate in the high SNR regimes.

In the case of $K_e = 0$, (C.41) can be rewritten as

$$(1-\epsilon)\left(\frac{\Omega_c}{\bar{\zeta}_r}2^{C_{\text{out}}^{\text{mrc}}}+1\right)^N = \exp\left(-\frac{2^{C_{\text{out}}^{\text{mrc}}}-1}{\bar{\zeta}_r}\right). \tag{C.45}$$

which, by taking the logarithm on both sides, can be written as

$$\ln(1-\epsilon) + N\ln\left(\frac{\Omega_c}{\bar{\zeta}_r}2^{C_{\text{out}}^{\text{mrc}}}+1\right) = \frac{1-2^{C_{\text{out}}^{\text{mrc}}}}{\bar{\zeta}_r}.$$

Since $\ln(1+x) \approx x$ when $x \to 0^+$, we have $\bar{\zeta}_r \ln(1-\epsilon) + N\Omega_c 2^{C_{\text{out}}^{\text{mrc}}} \approx 1 - 2^{C_{\text{out}}^{\text{mrc}}}$.

The ϵ-outage secrecy capacity can be approximated by

$$C_{\text{out}}^{\text{mrc}} \approx \log_2 \left[\frac{1 - \bar{\zeta}_r \ln(1-\epsilon)}{1 + N\Omega_c} \right].$$

By substituting $\bar{\zeta}_r = \frac{Pl^{-\alpha}}{\sigma_r^2}$ and $\Omega_c = \frac{1}{N}\sum_{i=1}^{N} \bar{\zeta}_{e_i}$ into (C.44) and (C.45), the ϵ-outage secrecy capacity can be given by

$$C_{\text{out}}^{\text{mrc}}(l, l_i) \approx \begin{cases} \log_2 \left[\dfrac{(N+K_e)(K_e - W_c)\,\sigma_e^2 l^{-\alpha}}{W_c \sigma_r^2 \sum_{i=1}^{N} l_i^{-\alpha_e}} \right], & \text{if} \quad K_e > 0; \\ \log_2 \left[\dfrac{\sigma_r^2\sigma_e^2 - P\sigma_e^2 \ln(1-\epsilon)}{\sigma_r^2\sigma_e^2 + P\sigma_r^2 \sum_{i=1}^{N} l_i^{-\alpha_e}} \right], & \text{if} \quad K_e = 0; \end{cases} \tag{C.46}$$

where $W_c = N\mathcal{W}_0 \left[\frac{K_e}{N} \left(\frac{\exp(K_e)(1-\epsilon)Pl^{-\alpha}}{\sigma_r^2 + Pl^{-\alpha}} \right)^{\frac{1}{N}} \right]$, and $\mathcal{W}_0(\cdot)$ is the principal branch of the Lambert W function.

By following the same methods of establishing $\mathbb{E}_{\mathcal{L}}\left[C_{\text{out}}^{\text{sc}}(l, \mathcal{L})\right] \xrightarrow{a.s.} C_{\text{out}}^{\text{sc}}(l, \mathbb{E}[\mathcal{L}])$ in (4.4), we can attain $\mathbb{E}_{l_i}\left[C_{\text{out}}^{\text{mrc}}(l, l_i)\right] \xrightarrow{a.s.} C_{\text{out}}^{\text{mrc}}(l, \mathbb{E}[l_i])$. Finally, (4.7) can be obtained by substituting $\mathbb{E}[l_i] = \frac{3}{4}r$, $i = 1, \dots, N$.

C.7 Proof of Lemma 4.3

In Figure 3.1, the volume of the spherical flight region is $V_1 = \frac{2}{3}\pi r^3(1-\cos\theta) - \frac{1}{3}\pi r^3 \sin^2\theta\cos\theta = \frac{\pi}{3}\left(2r^3 - 3H_0r^2 + H_0^3\right)$, where $\theta \in [0, \frac{\pi}{2}]$ is the polar angle and $\cos\theta = \frac{H_0}{r}$. The probability that l_i is shorter than τ, denoted by $\mathcal{P}$, is equal to the probability that eavesdropper i is located within the spherical cap with radius τ. $\mathcal{P} = \frac{\frac{2}{3}\pi\tau^3(1-\cos\phi) - \frac{1}{3}\pi\tau^3\sin^2\phi\cos\phi}{V_1} \overset{(a)}{=} \frac{2\tau^3 - 3H_0\tau^2 + H_0^3}{2r^3 - 3H_0r^2 + H_0^3}$, $H_0 \le \tau \le r$, where (a) is due to $\cos\phi = \frac{H_0}{\tau}$.

As a result, the PDF of l_i, denoted by $f_{l_i}(\tau)$, is given by

$$f_{l_i}(\tau) = \frac{6\tau^2 - 6H_0\tau}{2r^3 - 3H_0r^2 + H_0^3}, \quad \text{for } H_0 \le \tau \le r.$$

The expectation of l_i is obtained as

$$\mathbb{E}[l_i] = \int_{H_0}^{r} \tau f_{l_i}(\tau)\, d\tau = \int_{H_0}^{r} \frac{6\tau^3 - 6H_0\tau^2}{2r^3 - 3H_0r^2 + H_0^3} d\tau$$
$$= \frac{3r^4 - 4H_0r^3 + H_0^4}{2\left(2r^3 - 3H_0r^2 + H_0^3\right)}.$$

By order statistics, the PDF of $\mathcal{L}$ can be given by

$$f_{\mathcal{L}}(\tau) = N(1-\mathcal{P})^{N-1} f_{l_i}(\tau)\, d\tau$$
$$= N\left(1 - \frac{2\tau^3 - 3H_0\tau^2 + H_0^3}{V_1'}\right)^{N-1} \frac{6\tau^2 - 6H_0\tau}{V_1'}, \text{ for } H_0 \leq \tau \leq r. \tag{C.47}$$

The expectation of $\mathcal{L}$ can be obtained as

$$\mathbb{E}[\mathcal{L}] = \int_{H_0}^{r} \tau N(1-\mathcal{P})^{N-1} f_{l_i}(\tau)\, d\tau$$
$$= -\tau(1-\mathcal{P})^N \Big|_{H_0}^{r} + \int_{H_0}^{r} (1-\mathcal{P})^N d\tau. \tag{C.48}$$

By substituting (C.47) into (C.48) and rewriting $1 - \frac{2\tau^3 - 3H_0\tau^2 + H_0^3}{V_1'} = \left(\frac{V_1' - H_0^3}{V_1'}\right)\left(1 + \frac{3H_0}{V_1' - H_0^3}\tau^2 - \frac{2}{V_1' - H_0^3}\tau^3\right)$ and $\left(1 - \frac{2\tau^3 - 3H_0\tau^2 + H_0^3}{V_1'}\right)^N = \left(\frac{V_1' - H_0^3}{V_1'}\right)^N \sum_{k=0}^{N} \binom{N}{k} \left(1 + \frac{3H_0}{V_1' - H_0^3}\tau^2 - \frac{2}{V_1' - H_0^3}\tau^3\right)^k$ based on the binomial theorem, (C.48) can be rewritten as

$$\mathbb{E}[\mathcal{L}] = H_0 + \left(\frac{V_1' - H_0^3}{V_1'}\right)^N \sum_{k=0}^{N} \binom{N}{k} \times \left[\int_{H_0}^{r} \left(\frac{3H_0}{V_1' - H_0^3}\tau^2 - \frac{2}{V_1' - H_0^3}\tau^3\right)^k d\tau\right], \tag{C.49}$$

which leads to (4.8).

Bibliography

1 https://www.lightreading.com/mobile/5g/china-mobile-huawei-claim-worlds-first-26ghz-4g-and-5g-integrated-network-demo-area/d/d-id/750747, May 2022.

2 3GPP TR 21.915 V15.0.0, Tech. Spec. Group Services and System Aspects. 5.5.4 radio physical layer aspects, Release 15, Sep. 2019.

3 Samith Abeywickrama, Rui Zhang, Qingqing Wu, and Chau Yuen. Intelligent reflecting surface: Practical phase shift model and beamforming optimization. *IEEE Trans. Commun.*, 68(9):5849–5863, Sep. 2020. https://doi.org/10.1109/TCOMM.2020.3001125.

4 Alemayehu Solomon Abrar, Neal Patwari, and Sneha Kumar Kasera. Quantifying interference-assisted signal strength surveillance of sound vibrations. *IEEE Trans. Inf. Forensics Secur.* 16:2018–2030, 2021. https://doi.org/10.1109/TIFS.2020.3045316.

5 I. Y. Abualhaol and M. M. Matalgah. Outage probability analysis in a cooperative uavs network over Nakagami-*m* fading channels. In *Proc. IEEE VTC*, pages 1–4, Sept. 2006. https://doi.org/10.1109/VTCF.2006.564.

6 I. Y. Abualhaol and M. M. Matalgah. Performance analysis of multi-carrier relay-based UAV network over fading channels. In *Proc. IEEE Globecom Workshops (GC Wkshps)*, pages 1811–1815, Dec. 2010. https://doi.org/10.1109/GLOCOMW.2010.5700253.

7 G. S. Aglietti, S. Redi, A. R. Tatnall, and T. Markvart. Harnessing high-altitude solar power. *IEEE Trans. Energy Conver.*, 24(2):442–451, Jun. 2009.

8 Neelima Agrawal, Ankur Bansal, Keshav Singh, and Chih-Peng Li. Performance evaluation of RIS-assisted UAV-enabled vehicular communication system with multiple non-identical interferers. *IEEE Trans. Intell. Transport. Syst.*, 23(7): 9883–9894, Jul. 2022. https://doi.org/10.1109/TITS.2021.3123072.

Artificial Intelligence for Unmanned Aerial Vehicles: Sensing, Communication, and Computing, First Edition. Shuyan Hu et al.

9 Y. Ai, M. Cheffena, A. Mathur, and H. Lei. On physical layer security of double rayleigh fading channels for vehicular communications. *IEEE Wirel. Commun. Lett.*, 7(6):1038–1041, Dec. 2018. ISSN 2162–2337. https://doi.org/10.1109/LWC.2018.2852765.

10 A. Al-Hourani, S. Kandeepan, and S. Lardner. Optimal LAP altitude for maximum coverage. *IEEE Wirel. Commun. Lett.*, 3(6):569–572, Dec. 2014. ISSN 2162–2337. https://doi.org/10.1109/LWC.2014.2342736.

11 Fadi Al-Turjman, editor. *Unmanned Aerial Vehicles in Smart Cities*. Springer, Cham, Switzerland, May 2020. https://doi.org/10.1007/978-3-030-38732-6.

12 A. Alipour-Fanid, M. Dabaghchian, N. Wang, P. Wang, L. Zhao, and K. Zeng. Machine learning-based delay-aware UAV detection and operation mode identification over encrypted Wi-Fi traffic. *IEEE Trans. Inf. Forensics Secur.* 15:2346–2360, 2020.

13 Ricardo S. Alonso, Inés Sittón-Candanedo, Óscar García, Javier Prieto, and Sara Rodríguez-González. An intelligent edge-IoT platform for monitoring livestock and crops in a dairy farming scenario. *Ad Hoc Netw.*, 98:102047, 2020.

14 E. T. Alotaibi, S. S. Alqefari, and A. Koubaa. LSAR: Multi-UAV collaboration for search and rescue missions. *IEEE Access*, 7:55817–55832, 2019.

15 M. S. Alouini and A. J. Goldsmith. Capacity of Rayleigh fading channels under different adaptive transmission and diversity-combining techniques. *IEEE Trans. Veh. Technol.*, 48(4):1165–1181, Jul. 1999. ISSN 0018–9545. https://doi.org/10.1109/25.775366.

16 Mohamed Alzenad, Amr El-Keyi, Faraj Lagum, and Halim Yanikomeroglu. 3D placement of an unmanned aerial vehicle base station (UAV-BS) for energy-efficient maximal coverage. *IEEE Wirel.Commun. Lett.*, 6(4):434–437, Aug. 2017.

17 K. An, M. Lin, and T. Liang. On the performance of multiuser hybrid satellite-terrestrial relay networks with opportunistic scheduling. *IEEE Commun. Lett.*, 19(10):1722–1725, Oct. 2015. ISSN 1089–7798. https://doi.org/10.1109/LCOMM.2015.2466535.

18 T. Andre. K. A. Hummel, A. P. Schoellig, E. Yanmaz, M. Asadpour, C. Bettstetter, P. Grippa, H. Hellwagner, S. Sand, and S. Zhang. Application-driven design of aerial communication networks. *IEEE Commun. Mag.*, 52(5):129–137, May 2014. ISSN 0163–6804. https://doi.org/10.1109/MCOM.2014.6815903.

19 Constantinos Antoniou, Moshe Ben-Akiva, and Haris N. Koutsopoulos. Nonlinear Kalman filtering algorithms for on-line calibration of dynamic traffic assignment models. *IEEE Trans. Intell. Transport. Syst.*, 8(4):661–670, Dec. 2007. https://doi.org/10.1109/TITS.2007.908569.

20 Danilo Avola, Marco Cascio, Luigi Cinque, Alessio Fagioli, and Chiara Petrioli. Person re-identification through Wi-Fi extracted radio biometric signatures.

IEEE Trans. Inf. Forensics Secur., 17:1145–1158, 2022. https://doi.org/10.1109/TIFS.2022.3158058.

21 Mohammad Mahdi Azari, Fernando Rosas, Alessandro Chiumento, and Sofie Pollin. Coexistence of terrestrial and aerial users in cellular networks. In *Proc. IEEE Globecom Workshops (GC Wkshps)*, Dec. 2017.

22 Mohammad Mahdi Azari, Fernando Rosas, K. C. Chen, and Sofie Pollin. Ultra reliable UAV communication using altitude and cooperation diversity. *IEEE Trans. Commun.*, 66(1):330–344, Jan. 2018. ISSN 0090-6778. https://doi.org/10.1109/TCOMM.2017.2746105.

23 M. M. Azari, H. Sallouha, A. Chiumento, S. Rajendran, E. Vinogradov, and S. Pollin. Key technologies and system trade-offs for detection and localization of amateur drones. *IEEE Commun. Mag.*, 56(1):51–57, Jan. 2018. ISSN 0163-6804. https://doi.org/10.1109/MCOM.2017.1700442.

24 E. Baccarelli and A. Fasano. Some simple bounds on the symmetric capacity and outage probability for QAM wireless channels with rice and Nakagami fadings. *IEEE J. Sel. Areas Commun.*, 18(3):361–368, Mar. 2000. ISSN 0733-8716. https://doi.org/10.1109/49.840195.

25 Enzo Baccarelli and Mauro Biagi. Error resistant space-time coding for emerging 4G-WLANs. In *Proc. IEEE WCNC*, volume 1, pages 72–77, Mar. 2003. https://doi.org/10.1109/WCNC.2003.1200322.

26 Enzo Baccarelli and Mauro Biagi. Performance and optimized design of space-time codes for MIMO wireless systems with imperfect channel estimates. *IEEE Trans. Signal Process.*, 52(10):2911–2923, 2004.

27 Stanley S. Baek, Hyukseong Kwon, Josiah A. Yoder, and Daniel Pack. Optimal path planning of a target-following fixed-wing UAV using sequential decision processes. In *IEEE/RSJ Int. Conf. Intell. Robots Syst.*, pages 2955–2962, Nov. 2013. https://doi.org/10.1109/IROS.2013.6696775.

28 V. N. Q. Bao, N. Linh-Trung, and M. Debbah. Relay selection schemes for dual-hop networks under security constraints with multiple eavesdroppers. *IEEE Trans. Wirel. Commun.*, 12(12):6076–6085, December 2013. ISSN 1536-1276. https://doi.org/10.1109/TWC.2013.110813.121671.

29 Michael Bartusch, Christian Bruens, Adriana Elizabeth Nuncio Quiroz, and Samuel Stettner. HRWS: The upcoming german X-band spaceborne SAR mission. In *Proc. 13th European Conference on Synthetic Aperture Radar (EUSAR)*, pages 1–4, Mar. 2021.

30 Arman Beitollahi and Parvin Azhdari. Convergence in probability and almost surely convergence in probabilistic normed spaces. *Math. Sci.*, 6(1):4, 2012.

31 F. Benkhelifa, Z. Rezki, and M. Alouini. On the low SNR capacity of maximum ratio combining over Rician fading channels with full channel state information. *IEEE Wirel. Commun. Lett.*, 2(2):191–194, Apr. 2013. ISSN 2162-2337. https://doi.org/10.1109/WCL.2013.011713.120821.

32 Basudeb Bera, Ashok Kumar Das, Sahil Garg, Md. Jalil Piran, and M. Shamim Hossain. Access control protocol for battlefield surveillance in drone-assisted IoT environment. *IEEE Internet Things J.*, 9(4):2708–2721, Feb. 2022. https://doi.org/10.1109/JIOT.2020.3049003.

33 J. Bhandari and D. Russo. Global optimality guarantees for policy gradient methods. arXiv: 1906.01786v3, Jun. 2022.

34 Siguo Bi, Kai Li, Shuyan Hu, Wei Ni, Cong Wang, and Xin Wang. Detection and mitigation of position spoofing attacks on cooperative UAV swarm formations. *IEEE Trans. Inform. Forensics Secur.*, 19:1883–1895, 2024. https://doi.org/10.1109/TIFS.2023.3341398.

35 Patrick Billingsley. *Convergence of Probability Measures*. John Wiley & Sons, 2013.

36 M. Bloch, J. Barros, M. R. D. Rodrigues, and S. W. McLaughlin. Wireless information-theoretic security. *IEEE Trans. Inform. Theory*, 54(6):2515–2534, Jun. 2008. ISSN 0018-9448. https://doi.org/10.1109/TIT.2008.921908.

37 Mohammud Z. Bocus, Carl P. Dettmann, and Justin P. Coon. An approximation of the first order Marcum q-function with application to network connectivity analysis. *IEEE Commun. Lett.*, 17(3):499–502, 2013.

38 S. Boyd and L. Vandenberghe. *Convex Optimization*. Cambridge University Press, 2004.

39 D. G. Brennan. Linear diversity combining techniques. *Proc. IEEE*, 91(2): 331–356, Feb. 2003. ISSN 0018-9219.

40 Hui Cai, Qi Zhang, Quanzhong Li, and Jiayin Qin. Proactive monitoring via jamming for rate maximization over MIMO Rayleigh fading channels. *IEEE Commun. Lett.*, 21(9):2021–2024, Sep. 2017.

41 Y. Cai, F. Cui, Q. Shi, M. Zhao, and G. Y. Li. Dual-UAV-enabled secure communications: Joint trajectory design and user scheduling. *IEEE J. Sel. Areas Commun.*, 36(9):1972–1985, Sep. 2018. ISSN 0733-8716. https://doi.org/10.1109/JSAC.2018.2864424.

42 G. C. Calafiore and L. Fagiano. Robust model predictive control via scenario optimization. *IEEE Trans. Autom. Control*, 58(1):219–224, Jan. 2013.

43 Xiaowen Cao, Jie Xu, and Rui Zhang. Mobile edge computing for cellular-connected UAV: Computation offloading and trajectory optimization. In *IEEE International Workshop on Signal Processing Advances in Wireless Communications (SPAWC)*, pages 1–5. IEEE, 2018.

44 Yashuai Cao, Tiejun Lv, Zhipeng Lin, and Wei Ni. Delay-constrained joint power control, user detection and passive beamforming in intelligent reflecting surface-assisted uplink mmWave system. *IEEE Trans. Cognit. Commun. Netw.*, 7(2):482–495, Jun. 2021.

45 L. R. G. Carrillo and K. G. Vamvoudakis. Deep-learning tracking for autonomous flying systems under adversarial inputs. *IEEE Trans. Aerosp. Electron. Syst.*, 56(2):1444–1459, Apr. 2020.

46 G. R. Chandra Mouli, P. Bauer, and M. Zeman. System design for a solar powered electric vehicle charging station for workplaces. *Appl. Energy*, 168: 434–443, 2016.

47 S. Chandrasekharan, K. Gomez, A. Al-Hourani, S. Kandeepan, T. Rasheed, L. Goratti, L. Reynaud, D. Grace, I. Bucaille, T. Wirth, and S. Allsopp. Designing and implementing future aerial communication networks. *IEEE Commun. Mag.*, 54(5):26–34, May 2016. ISSN 0163-6804. https://doi.org/10.1109/MCOM.2016.7470932.

48 N. Chatzipanagiotis and M. M. Zavlanos. Distributed scheduling of network connectivity using mobile access point robots. *IEEE Trans. Robot.*, 32(6): 1333–1346, Dec. 2016. ISSN 1552-3098. https://doi.org/10.1109/TRO.2016.2593041.

49 Chien-Yu Chen, Jung-Hao Chang, Kai-Ming Chiang, H.-L. Lin, Sheng-Yi Hsiao, and Hao-Wu Lin. Perovskite photovoltaics for dim-light applications. *Adv. Funct. Mater.*, 25(45):7064–7070, 2015.

50 M. Chen, M. Mozaffari, W. Saad, C. Yin, M. Debbah, and C. S. Hong Caching in the sky: Proactive deployment of cache-enabled unmanned aerial vehicles for optimized quality-of-experience. *IEEE J. Sel. Areas Commun.*, 35(5):1046–1061, May 2017. ISSN 0733-8716. https://doi.org/10.1109/JSAC.2017.2680898.

51 Shaxun Chen, Amit Pande, Kai Zeng, and Prasant Mohapatra. Live video forensics: Source identification in lossy wireless networks. *IEEE Trans. Inf. Forensics Secur.*, 10(1):28–39, 2015. https://doi.org/10.1109/TIFS.2014.2362848.

52 Xu Chen, Xinxin He, Zhiyong Feng, Zhiqing Wei, Qixun Zhang, Xin Yuan, and Ping Zhang. Joint localization and communication enhancement in uplink integrated sensing and communications system with clock asynchronism. *IIEEE J. Sel. Areas Commun.*, 42(10):2659–2673, 2024. https://doi.org/10.1109/JSAC.2024.3414625.

53 Yu-Jia Chen, Deng-Kai Chang, and Cheng Zhang. Autonomous tracking using a swarm of UAVs: A constrained multi-agent reinforcement learning approach. *IEEE Trans. Veh. Tech.*, 69(11):13702–13717, Nov. 2020.

54 V. V. Chetlur and H. S. Dhillon. Downlink coverage analysis for a finite 3-D wireless network of unmanned aerial vehicles. *IEEE Trans. Commun.*, 65(10): 4543–4558, Oct. 2017. ISSN 0090-6778.

55 M. M. U. Chowdhury, F. Erden, and I. Guvenc. RSS-based Q-learning for indoor UAV navigation. In *Proc. IEEE Conf. Mil. Commun.*, pages 121–126, 2019.

56 Sae-Young Chung, G. David Forney, Thomas J. Richardson, and Rüdiger Urbanke. On the design of low-density parity-check codes within 0.0045 dB of the shannon limit. *IEEE Commun. Lett.*, 5(2):58–60, 2001.

57 R. E. Collin. *Antennas and Radiowave Propagation*. McGraw-Hill, 1985.

58 Jason Cong, Bin Liu, Stephen Neuendorffer, Juanjo Noguera, Kees Vissers, and Zhiru Zhang. High-level synthesis for FPGAs: From prototyping to deployment. *IEEE Trans. Comput.-Aided Des. Integr. Circuits Syst.*, 30(4):473–491, Apr. 2011. https://doi.org/10.1109/TCAD.2011.2110592.

59 Peter Corcoran and Soumya Kanti Datta. Mobile-edge computing and the Internet of Things for consumers: Extending cloud computing and services to the edge of the network. *IEEE Consum. Electron. Mag.*, 5(4):73–74, 2016.

60 M. Crofton. Probability. *In Encyclopaedia Britannica*. Britannica Inc., 9th edition, 1885.

61 Mark Cudak, Amitabha Ghosh, Arunabha Ghosh, and Jeffrey Andrews. Integrated access and backhaul: A key enabler for 5G millimeter-wave deployments. *IEEE Commun. Mag.*, 59(4):88–94, Apr. 2021. https://doi.org/10.1109/MCOM.001.2000690.

62 M. Cui, G. Zhang, Q. Wu, and D. W. K. Ng. Robust trajectory and transmit power design for secure UAV communications. *IEEE Trans. Veh. Technol.*, 67(9): 9042–9046, Sept. 2018. ISSN 0018–9545. https://doi.org/10.1109/TVT.2018.2849644.

63 Daniel Benevides Da Costa and Sonia Aïssa. Capacity analysis of cooperative systems with relay selection in Nakagami-*m* fading. *IEEE Commun. Lett.*, 13(9):637–639, 2009.

64 Stephen Dankwa and Wenfeng Zheng. Twin-delayed DDPG: A deep reinforcement learning technique to model a continuous movement of an intelligent robot agent. In *Proc. Int. Conf. Vision, Image, Signal Process.*, pages 1–5, 2019.

65 Xianbang Diao, Jianchao Zheng, Yueming Cai, Yuan Wu, and Alagan Anpalagan. Fair data allocation and trajectory optimization for UAV-assisted mobile edge computing. *IEEE Commun. Lett.*, 23(12):2357–2361, 2019.

66 Werner Dinkelbach. On nonlinear fractional programming. *Manag. Sci.*, 13(7): 492–498, 1967.

67 Xingping Dong, Jianbing Shen, Dajiang Yu, Wenguan Wang, Jianhong Liu, and Hua Huang. Occlusion-aware real-time object tracking. *IEEE Trans. Multimed.*, 19(4):763–771, Apr. 2017. https://doi.org/10.1109/TMM.2016.2631884.

68 A. H. Abd El-Malek, A. M. Salhab, S. A. Zummo, and M. Alouini. Security-reliability trade-off analysis for multiuser SIMO mixed RF/FSO relay networks with opportunistic user scheduling. *IEEE Trans. Wirel. Commun.*, 15(9):5904–5918, Sept. 2016. ISSN 1536–1276. https://doi.org/10.1109/TWC.2016.2572681.

69 Yousef Emami, Bo Wei, Kai Li, Wei Ni, and Eduardo Tovar. Joint communication scheduling and velocity control in multi-UAV-assisted sensor networks: A deep reinforcement learning approach. *IEEE Trans. Veh. Technol.*, 70(10):10986–10998, 2021.

70 Granzotti Cristiano Roberto Fabri and Souto Martinez Alexandre. Distance statistics in random media. *Eur. Phys. J. B*, 87(4):87, Apr. 2014.

71 L. Fan, N. Yang, T. Q. Duong, M. Elkashlan, and G. K. Karagiannidis. Exploiting direct links for physical layer security in multiuser multirelay networks. *IEEE Trans. Wirel. Commun.*, 15(6):3856–3867, Jun. 2016. ISSN 1536-1276. https://doi.org/10.1109/TWC.2016.2530068.

72 Sisai Fang, Gaojie Chen, and Yonghui Li. Joint optimization for secure intelligent reflecting surface assisted UAV networks. *IEEE Wirel. Commun. Lett.*, 10(2):276–280, Feb. 2021.

73 N. Farmani, L. Sun, and D. J. Pack. A scalable multitarget tracking system for cooperative unmanned aerial vehicles. *IEEE Trans. Aerosp. Electron. Syst.*, 53(4): 1947–1961, Aug. 2017.

74 Nikos Fasarakis-Hilliard, Panos N. Alevizos, and Aggelos Bletsas. Coherent detection and channel coding for bistatic scatter radio sensor networking. *IEEE Trans. Commun.*, 63(5):1798–1810, May 2015. https://doi.org/10.1109/TCOMM.2015.2412546.

75 Jie Feng, Liqiang Zhao, Jianbo Du, Xiaoli Chu, and F. Richard Yu. Computation offloading and resource allocation in D2D-enabled mobile edge computing. In *IEEE Int. Conf. Commun. (ICC)*, pages 1–6. IEEE, 2018.

76 Z. Feng, L. Ji, Q. Zhang, and W. Li. Spectrum management for mmwave enabled UAV swarm networks: Challenges and opportunities. *IEEE Commun. Mag.*, 57(1):146–153, Jan. 2019. ISSN 0163-6804. https://doi.org/10.1109/MCOM.2018.1800087.

77 A. Filippone. *Flight Performance of Fixed and Rotary Wing Aircraft*. AIAA, Washington, DC, USA, 2006.

78 J. Fink, A. Ribeiro, and V. Kumar. Robust control for mobility and wireless communication in cyber-physical systems with application to robot teams. *Proc. IEEE*, 100(1):164–178, Jan. 2012. ISSN 0018-9219. https://doi.org/10.1109/JPROC.2011.2161427.

79 A. Fotouhi, H. Qiang, M. Ding, M. Hassan, L. Garcia-Giordano, A. G. Rodriguez, and J. Yuan. Survey on UAV cellular communications: Practical aspects, standardization advancements, regulation, and security challenges. *IEEE Commun. Surv. Tut.*, 21(4):3417–3442, 4th Quart., 2019.

80 Eric W. Frew and Timothy X. Brown. Airborne communication networks for small unmanned aircraft systems. *Proc. IEEE*, 96(12):2008–2027, 2008.

81 Scott Fujimoto, Herke Hoof, and David Meger. Addressing function approximation error in actor-critic methods. In *Int. Conf. Mach. Learn.*, pages 1587–1596. PMLR, 2018.

82 G. Gao, J. Li, and Y. Wen. DeepComfort: Energy-efficient thermal comfort control in buildings via reinforcement learning. *IEEE Internet Things J.*, 7(9): 8472–8484, Sep. 2020.

83 Amitabha Ghosh, Andreas Maeder, Matthew Baker, and Devaki Chandramouli. 5G evolution: A view on 5G cellular technology beyond 3GPP Release 15. *IEEE Access*, 7:127639–127651, Sep. 2019. https://doi.org/10.1109/ACCESS.2019.2939938.

84 Wesley M. Gifford, Moe Z. Win, and Marco Chiani. Diversity with practical channel estimation. *IEEE Trans. Wirel. Commun.*, 4(4):1935–1947, 2005.

85 Marco Giordani, Michele Polese, Arnab Roy, Douglas Castor, and Michele Zorzi. A tutorial on beam management for 3GPP NR at mmWave frequencies. *IEEE Commun. Surv. Tuts.*, 21(1):173–196, 1st Quart., 2019. https://doi.org/10.1109/COMST.2018.2869411.

86 N. Goddemeier, K. Daniel, and C. Wietfeld. Role-based connectivity management with realistic air-to-ground channels for cooperative UAVs. *IEEE J. Sel. Areas Commun.*, 30(5):951–963, June 2012. ISSN 0733–8716. https://doi.org/10.1109/JSAC.2012.120610.

87 S. Goel, V. Aggarwal, A. Yener, and A. R. Calderbank. Modeling location uncertainty for eavesdroppers: A secrecy graph approach. In *2010 IEEE Int. Symp. Inform. Theory*, pages 2627–2631, Jun. 2010.

88 Zhenzhen Gong, Omar Hashash, Yingze Wang, Qimei Cui, Wei Ni, Walid Saad, and Kei Sakaguchi. UAV-aided lifelong learning for AoI and energy optimization in nonstationary IoT networks. *IEEE Internet Things J.*, 11(24): 39206–39224, 2024. https://doi.org/10.1109/JIOT.2024.3406220.

89 Jun-Ya Gotoh, Akiko Takeda, and Katsuya Tono. DC formulations and algorithms for sparse optimization problems. *Math. Program.*, 169(1):141–176, May 2018.

90 I. S. Gradshteyn and I. M. Ryzhik. *Table of Integrals, Series, and Products*. Academic Press, 7th edition, 2000.

91 Yue Guan, Sai Zou, Haixia Peng, Wei Ni, Yanglong Sun, and Hongfeng Gao. Cooperative UAV trajectory design for disaster area emergency communications: A multiagent PPO method. *IEEE Internet Things J.*, 11(5): 8848–8859, 2024. https://doi.org/10.1109/JIOT.2023.3320796.

92 L. Gupta, R. Jain, and G. Vaszkun. Survey of important issues in UAV communication networks. *IEEE Commun. Surv. Tuts.*, 18(2):1123–1152, 2nd Quart., 2016. ISSN 1553–877X. https://doi.org/10.1109/COMST.2015.2495297.

93 I. Guvenc, F. Koohifar, S. Singh, M. L. Sichitiu, and D. Matolak. Detection, tracking, and interdiction for amateur drones. *IEEE Commun. Mag.*, 56(4): 75–81, Apr. 2018. ISSN 0163–6804.

94 M. Haenggi. *Stochastic Geometry for Wireless Networks*. Cambridge University Press, 2012.

95 Brett Hagelstein, Mehran Abolhasan, Daniel Franklin, Farzad Safaei, and Wei Ni. Analytic performance model for state-based MAC layer cooperative retransmission protocols. *IEEE Trans. Mobile Comput.*, 15(1):32–44, 2016.

96 Huang Hailong, Mohsen Eskandari, Andrey V. Savkin, and Wei Ni. Energy-efficient joint UAV secure communication and 3D trajectory optimization assisted by reconfigurable intelligent surfaces in the presence of eavesdroppers. *Def. Technol.*, 31:537–543, 2024. ISSN 2214–9147. https://doi.org/10.1016/j.dt.2022.12.010. URL https://www.sciencedirect.com/science/article/pii/S2214914722002756.

97 I. Hajnsek and T. E. Busche. TanDEM-X science plan. DLR Public Doc. TD-PD-PL-0069, no. 1.0, Jun. 2010.

98 J. Han and M. Kamber. *Data Mining: Concepts and Techniques.* Morgan Kuafmann Publisher, 2006.

99 B. He, X. Zhou, and A. L. Swindlehurst. On secrecy metrics for physical layer security over quasi-static fading channels. *IEEE Trans. Wirel. Commun.*, 15(10):6913–6924, Oct. 2016. ISSN 1536–1276. https://doi.org/10.1109/TWC.2016.2593445.

100 Yejun He, Mengna Yang, Zhou He, and Mohsen Guizani. Resource allocation based on digital twin-enabled federated learning framework in heterogeneous cellular network. *IEEE Trans. Veh. Technol.*, 72(1):1149–1158, 2023. https://doi.org/10.1109/TVT.2022.3205778.

101 Yejun He, Mengna Yang, Zhou He, and Mohsen Guizani. Computation offloading and resource allocation based on DT-MEC-assisted federated learning framework. *IEEE Trans. Cogn. Commun. Netw.*, 9(6):1707–1720, 2023.

102 Yejun He, Xiaoxu Zhong, Youhui Gan, Haixia Cui, and Mohsen Guizani. A DDPG hybrid of graph attention network and action branching for multi-scale end-edge-cloud vehicular orchestrated task offloading. *IEEE Wirel. Commun.*, 30(4):147–153, 2023. https://doi.org/10.1109/MWC.019.2100718.

103 A. N. Hong, M. Rath, E. Leitinger, S. Hinteregger, K. N. Van, and K. Witrisal Channel capacity analysis of indoor environments for location-aware communications. In *Proc. IEEE Globecom Workshops (GC Wkshps)*, pages 1–6, Dec. 2016. https://doi.org/10.1109/GLOCOMW.2016.7848856.

104 Yuenan Hou, Lifeng Liu, Qing Wei, Xudong Xu, and Chunlin Chen. A novel DDPG method with prioritized experience replay. In *Proc. IEEE Int. Conf. Systems, Man, Cybernet. (SMC)*, pages 316–321, 2017. https://doi.org/10.1109/SMC.2017.8122622.

105 Chen-Chien James Hsu, Ming-Chih Lu, and Yin-Yu Lu. Distance and angle measurement of objects on an oblique plane based on pixel number variation of CCD images. *IEEE Trans. Instrum. Meas.*, 60(5):1779–1794, 2011. https://doi.org/10.1109/TIM.2011.2108076.

106 Boyang Hu, Hui Tian, Wei Ni, Shaoshuai Fan, and Ekram Hossain. User-centric multi-static sensing for joint user and target tracking in mobile wireless systems. *IEEE Trans. Wirel. Commun.*, pages 1–1, 2025. https://doi.org/10.1109/TWC.2025.3560901.

107 Dingkun Hu, Zhang Qi, Yang Ping, and Jiayin Qin. Proactive monitoring via jamming in amplify-and-forward relay networks. *IEEE Signal Process. Lett.*, 24(11):1714–1718, Nov. 2017.

108 Qiyu Hu, Yunlong Cai, Guanding Yu, Zhijin Qin, Minjian Zhao, and Geoffrey Ye Li. Joint offloading and trajectory design for UAV-enabled mobile edge computing systems. *IEEE Internet Things J.*, 6(2):1879–1892, 2018.

109 S. Hu, X. Chen, W. Ni, X. Wang, and E. Hossain. Modeling and analysis of energy harvesting and smart grid-powered wireless communication networks: A contemporary survey. *IEEE Trans. Green Commun. Netw.*, 4(2):461–496, 2020.

110 S. Hu, Q. Wu, and X. Wang. Energy management and trajectory optimization for UAV-enabled legitimate monitoring systems. *IEEE Trans. Wirel. Commun.*, 20(1):142–155, 2021.

111 S. Hu, W. Ni, X. Wang, A. Jamalipour, and D. Ta. Joint optimization of trajectory, propulsion, and thrust powers for covert UAV-on-UAV video tracking and surveillance. *IEEE Trans. Inf. Forensics Secur.*, 16:1959–1972, Jan. 2021.

112 Shuyan Hu, Xin Yuan, Wei Ni, Xin Wang, Ekram Hossain, and H. Vincent Poor. Differentially private wireless federated learning with integrated sensing and communication. *IEEE Trans. Wirel. Commun.*, pages 1–1, 2025. https://doi.org/10.1109/TWC.2025.3555212.

113 Shuyan Hu, Xiaojing Chen, Wei Ni, Ekram Hossain, and Xin Wang. Distributed machine learning for wireless communication networks: Techniques, architectures, and applications. *IEEE Commun. Surv. Tuts.*, 23(3):1458–1493, 3rd Quart., 2021.

114 Shuyan Hu, Wei Ni, Xin Wang, and Abbas Jamalipour. Disguised tailing and video surveillance with solar-powered fixed-wing unmanned aerial vehicle. *IEEE Trans. Veh. Tech.*, 71(5):5507–5518, May 2022. https://doi.org/10.1109/TVT.2022.3157705.

115 Shuyan Hu, Xin Yuan, Wei Ni, and Xin Wang. Trajectory planning of cellular-connected UAV for communication-assisted radar sensing. *IEEE Trans. Commun.*, 70(9):6385–6396, Sep. 2022. https://doi.org/10.1109/TCOMM.2022.3195868.

116 Meng Hua, Yi Wang, Chunguo Li, Yongming Huang, and Luxi Yang. UAV-aided mobile edge computing systems with one by one access scheme. *IEEE Trans. Green Commun. Netw.*, 3(3):664–678, 2019.

117 Meng Hua, Yi Wang, Qingqing Wu, Haibo Dai, Yongming Huang, and Luxi Yang. Energy-efficient cooperative secure transmission in multi-UAV-enabled wireless networks. *IEEE Trans. Veh. Technol.*, 68(8):7761–7775, 2019.

118 Chongwen Huang, Sha Hu, George C. Alexandropoulos, Alessio Zappone, Chau Yuen, Rui Zhang, Marco Di Renzo, and Merouane Debbah. Holographic MIMO surfaces for 6G wireless networks: Opportunities, challenges, and trends. *IEEE Wirel. Commun.*, 27(5):118–125, Oct. 2020.

119 H. Huang and A. Savkin. An algorithm of reactive collision free 3-D deployment of networked unmanned aerial vehicles for surveillance and monitoring. *IEEE Trans. Ind. Informat.*, 16(1):132–140, Jan. 2020.

120 H. Huang and A. Savkin. Reactive 3D deployment of a flying robotic network for surveillance of mobile targets. *Comput. Netw.*, 161:172–182, Oct. 2019.

121 H. Huang, A. Savkin, and W. Ni. A method for covert video surveillance of a car or a pedestrian by an autonomous aerial drone via trajectory planning. In *Proc. IEEE ICCAR*, pages 1–3, Singapore, Apr. 2020.

122 Hailong Huang, Andrey V. Savkin, and Wei Ni. Energy-efficient 3D navigation of a solar-powered UAV for secure communication in the presence of eavesdroppers and no-fly zones. *Energies*, 13(6):1445, 2020.

123 Shan Huang, Qi Zhang, Quanzhong Li, and Jiayin Qin. Robust proactive monitoring via jamming with deterministically bounded channel errors. *IEEE Signal Process. Lett.*, 25(5):690–694, May 2018.

124 Y. Huang, H. Wang, and P. Yao. Energy-optimal path planning for solar-powered UAV with tracking moving ground target. *Aerospace Sci. Tech.*, 53:241–251, Jun. 2016.

125 Todd E. Humphreys, Matthew J. Murrian, and Lakshay Narula. Deep-urban unaided precise global navigation satellite system vehicle positioning. *IEEE Intell. Transport. Syst. Mag.*, 12(3):109–122, Fall 2020. https://doi.org/10.1109/MITS.2020.2994121.

126 S. Jaeckel, L. Raschkowski, S. Wu, L. Thiele, and W. Keusgen. An explicit ground reflection model for mm-Wave channels. In *Proc. Wirel. Commun. Netw. Conf. Wrkshps.*, pages 1–5, Mar. 2017.

127 F. Jameel, S. Wyne, G. Kaddoum, and T. Q. Duong. A comprehensive survey on cooperative relaying and jamming strategies for physical layer security. *IEEE Commun. Surv. Tuts.*, pages 1–1, 2018. ISSN 1553-877X. https://doi.org/10.1109/COMST.2018.2865607.

128 Rie Johnson and Tong Zhang. Accelerating stochastic gradient descent using predictive variance reduction. *Adv. Neural Inf. Process Syst.*, 26:315–323, 2013.

129 D. S. Karas, A. A. Boulogeorgos, and G. K. Karagiannidis. Physical layer security with uncertainty on the location of the eavesdropper. *IEEE Wirel. Commun. Lett.*, 5(5):540–543, Oct. 2016. ISSN 2162-2337. https://doi.org/10.1109/LWC.2016.2600323.

130 Zaheer Khan, Hamed Ahmadi, Ekram Hossain, Marceau Coupechoux, Luiz A Dasilva, and Janne J. Lehtomäki. Carrier aggregation/channel bonding in next generation cellular networks: Methods and challenges. *IEEE Netw.*, 28(6):34–40, Nov.–Dec. 2014. https://doi.org/10.1109/MNET.2014.6963802.

131 A. Kokhanovsky. Optical properties of terrestrial clouds. *Earth-Sci. Rev.*, 64(3): 189–241, Feb. 2004.

132 L. Kong and G. Kaddoum. On physical layer security over the Fisher-Snedecor $\mathcal{F}$ wiretap fading channels. *IEEE Access*, 6:39466–39472, 2018. ISSN 2169–3536. https://doi.org/10.1109/ACCESS.2018.2853700.

133 L. Kong, G. Kaddoum, and D. B. da Costa. Cascaded α-μ fading channels: Reliability and security analysis. *IEEE Access*, pages 1–1, 2018. https://doi.org/10.1109/ACCESS.2018.2833423.

134 Samuel Kotz, Narayanaswamy Balakrishnan, and Norman L. Johnson. *Continuous Multivariate Distributions, Volume 1: Models and Applications*. John Wiley & Sons, 2004.

135 Anis Koubaa and Ahmad Taher Azar, editors. *Unmanned Aerial Systems: Theoretical Foundation and Applications*. Elsevier, Amsterdam, Netherlands, January 2021.

136 Marek Kuczma. *An Introduction to the Theory of Functional Equations and Inequalities: Cauchy's Equation and Jensen's Inequality*. Springer Science & Business Media, 2009.

137 E. Kuiper and S. Nadjm-Tehrani. Geographical routing with location service in intermittently connected MANETs. *IEEE Tran. Veh. Technol.*, 60(2):592–604, Feb 2011. ISSN 0018–9545.

138 Preeti Kumari, Junil Choi, Nuria Gonzaìlez-Prelcic, and Robert W. Heath. IEEE 802.11ad-based radar: An approach to joint vehicular communication-radar system. *IEEE Trans. Veh. Tech.*, 67(4):3012–3027, Apr. 2018. https://doi.org/10.1109/TVT.2017.2774762.

139 Harrison Kurunathan, Kai Li, Wei Ni, Eduardo Tovar, and Falko Dressler. Deep reinforcement learning for persistent cruise control in UAV-aided data collection. In *IEEE Conf. Local Comp. Netw. (LCN)*, pages 347–350. IEEE, 2021.

140 H. Lee, S. Eom, J. Park, and I. Lee. UAV-aided secure communications with cooperative jamming. *IEEE Trans. Veh. Technol.*, 67(10):9385–9392, Oct. 2018. ISSN 0018–9545. https://doi.org/10.1109/TVT.2018.2853723.

141 J. Lee, H. Lou, D. Toumpakaris, and J. M. Cioffi. SNR analysis of OFDM systems in the presence of carrier frequency offset for fading channels. *IEEE Trans. Wirel. Commun.*, 5(12):3360–3364, Dec. 2006. ISSN 1536–1276. https://doi.org/10.1109/TWC.2006.256956.

142 Joo Seok Lee and Kee-Ho Yu. Optimal path planning of solar-powered UAV using gravitational potential energy. *IEEE Trans. Aerosp. Electron. Syst.*, 53(3): 1442–1451, Jun. 2017.

143 A. Li, Q. Wu, and R. Zhang. UAV-enabled cooperative jamming for improving secrecy of ground wiretap channel. *IEEE Wirel. Commun. Lett.*, 8(1):181–184, Feb. 2019. ISSN 2162–2337. https://doi.org/10.1109/LWC.2018.2865774.

144 Chen Li, Xuelei Qi, Bao Chen, Shoudong Huang, Jaime Valls Miro, Hailong Huang, Wei Ni, and Hongjun Ma. Marden-based homotopic enclosed safe motion corridor generation for UAV navigation in complex environments. *IEEE Trans. Autom. Sci. Eng.*, pages 1–15, 2024. https://doi.org/10.1109/TASE.2024.3488692.

145 Guyue Li, Chen Sun, Wei Xu, Marco Di Renzo, and Aiqun Hu. On maximizing the sum secret key rate for reconfigurable intelligent surface-assisted multiuser systems. *IEEE Trans. Inf. Forensics Secur.*, 17:211–225, 2022. https://doi.org/10.1109/TIFS.2021.3138612.

146 Kai Li, Wei Ni, Mehran Abolhasan, and Eduardo Tovar. Reinforcement learning for scheduling wireless powered sensor communications. *IEEE Trans. Green Commun. Netw.*, 3(2):264–274, 2018.

147 Kai Li, Chau Yuen, Branislav Kusy, Raja Jurdak, Aleksandar Ignjatovic, Salil S. Kanhere, and Sanjay Jha. Fair scheduling for data collection in mobile sensor networks with energy harvesting. *IEEE Trans. Mob. Comput.*, 18(6):1274–1287, 2018.

148 Kai Li, Wei Ni, Eduardo Tovar, and Abbas Jamalipour. On-board deep q-network for UAV-assisted online power transfer and data collection. *IEEE Trans. Veh. Technol.*, 68(12):12215–12226, 2019.

149 Kai Li, Razvan Christian Voicu, Salil S. Kanhere, Wei Ni, and Eduardo Tovar. Energy efficient legitimate wireless surveillance of UAV communications. *IEEE Trans. Veh. Technol.*, 68(3):2283–2293, 2019.

150 Kai Li, Yousef Emami, Wei Ni, Eduardo Tovar, and Zhu Han. Onboard deep deterministic policy gradients for online flight resource allocation of UAVs. *IEEE Netw. Lett.*, 2(3):106–110, 2020.

151 Kai Li, Wei Ni, Eduardo Tovar, and Abbas Jamalipour. Deep Q-learning based resource management in UAV-assisted wireless powered IoT networks. In *IEEE Int. Conf. Commun. (ICC)*, pages 1–6. IEEE, 2020.

152 Kai Li, Wei Ni, Eduardo Tovar, and Abbas Jamalipour. Online velocity control and data capture of drones for the Internet of Things: An onboard deep reinforcement learning approach. *IEEE Veh. Technol. Mag.*, 16(1):49–56, 2020.

153 Kai Li, Wei Ni, and Falko Dressler. Continuous maneuver control and data capture scheduling of autonomous drone in wireless sensor networks. *IEEE Trans. Mob. Comput.*, 21(8):2732–2744, 2021.

154 Kai Li, Wei Ni, and Falko Dressler. LSTM-characterized deep reinforcement learning for continuous flight control and resource allocation in UAV-assisted sensor network. *IEEE Internet Things J.*, 9(6):4179–4189, 2021.

155 Kai Li, Yilei Liang, Xin Yuan, Wei Ni, Jon Crowcroft, Chau Yuen, and Ozgur B. Akan. A novel framework of horizontal-vertical hybrid federated learning for EdgeIoT. In *IEEE Netw. Lett.*, pages 1–1, 2025. https://doi.org/10.1109/LNET.2025.3540268.

156 K. Li, W. Ni, X. Wang, R. P. Liu, S. S. Kanhere, and S. Jha. Energy-efficient cooperative relaying for unmanned aerial vehicles. *IEEE Trans. Mobile Comput.*, 15(6):1377–1386, 2016.

157 Mushu Li, Nan Cheng, Jie Gao, Yinlu Wang, Lian Zhao, and Xuemin Shen. Energy-efficient UAV-assisted mobile edge computing: Resource allocation and trajectory optimization. *IEEE Trans. Veh. Technol.*, 69(3):3424–3438, 2020.

158 R. Li, Z. Wei, L. Yang, D. W. K. Ng, J. Yuan, and J. An. Resource allocation for secure multi-UAV communication systems with multi-eavesdropper. *IEEE Trans. Commun.*, 68(7):4490–4506, Jul. 2020.

159 S. Li, B. Duo, X. Yuan, Y. Liang, and M. Di Renzo. Reconfigurable intelligent surface assisted UAV communication: Joint trajectory design and passive beamforming. *IEEE Wirel. Commun. Lett.*, 9(5):716–720, May 2020.

160 Sixian Li, Bin Duo, Marco Di Renzo, Meixia Tao, and Xiaojun Yuan. Robust secure UAV communications with the aid of reconfigurable intelligent surfaces. *IEEE Trans. Wirel. Commun.*, 20(10):6402–6417, Oct. 2021.

161 X. Li, X. Chen, J. Zhang, Y. Liang, and Y. Liu. Capacity analysis of α-η-κ-μ fading channels. *IEEE Commun. Lett.*, PP(99):1–1, Jun. 2017. ISSN 1089-7798. https://doi.org/10.1109/LCOMM.2017.2672960.

162 Timothy P. Lillicrap, Jonathan J. Hunt, Alexander Pritzel, Nicolas Heess, Tom Erez, Yuval Tassa, David Silver, and Daan Wierstra. Continuous control with deep reinforcement learning. In *ICLR (Poster)*, 2016. URL http://arxiv.org/abs/1509.02971.

163 X. Lin, V. Yajnanarayana, S. D. Muruganathan, S. Gao, H. Asplund, H.-L. Maattanen, M. Bergstrom, S. Euler, and Y.-P. E. Wang. The sky is not the limit: LTE for unmanned aerial vehicles. *IEEE Commun. Mag.*, 56(4):204–210, Apr. 2018.

164 Bin Liu, Wei Ni, Ren Ping Liu, Y. Jay Guo, and Hongbo Zhu. Privacy-preserving routing and charging scheduling for cellular-connected unmanned aerial vehicles. *IEEE Trans. Syst. Man Cybern.: Syst.*, 54(8):4929–4941, 2024. https://doi.org/10.1109/TSMC.2024.3392011.

165 Bin Liu, Wei Ni, Ren Ping Liu, Y. Jay Guo, and Hongbo Zhu. Delay-sensitive goods delivery and in-situ sensing using a multi-task drone. *IEEE Trans. Mob. Comput.*, pages 1–14, 2025. https://doi.org/10.1109/TMC.2025.3570437.

166 Dingtao Liu, Yuan Gao, Shuyan Hu, Wei Ni, and Xin Wang. Trajectory design for integrated sensing and communication enabled by cellular-connected UAV. *IEEE Wirel. Commun. Lett.*, 13(7):1973–1977, 2024. doi: 10.1109/LWC.2024.3399268.

167 Fan Liu, Christos Masouros, Ang Li, Huafei Sun, and Lajos Hanzo. MU-MIMO communications with MIMO radar: From co-existence to joint transmission. *IEEE Trans. Wirel. Commun.*, 17(4):2755–2770, Apr. 2018. https://doi.org/10.1109/TWC.2018.2803045.

168 Fan Liu, Longfei Zhou, Christos Masouros, Ang Li, Wu Luo, and Athina Petropulu. Toward dual-functional radar-communication systems: Optimal waveform design. *IEEE Trans. Signal Process.*, 66(16):4264–4279, Aug. 2018. https://doi.org/10.1109/TSP.2018.2847648.

169 H. Liu, S. Chen, and N. Kubota. Intelligent video systems and analytics: A survey. *IEEE Trans. Ind. Informat.*, 9(3):1222–1233, Aug. 2013.

170 Lingling Liu, Aimin Wang, Geng Sun, and Jiahui Li. Multiobjective optimization for improving throughput and energy efficiency in UAV-enabled IoT. *IEEE Internet Things J.*, 9(20):20763–20777, Oct. 2022. https://doi.org/10.1109/JIOT.2022.3175712.

171 M. Liu, Y. Wan, and F. L. Lewis. Analysis of the random direction mobility model with a sense-and-avoid protocol. In *2017 IEEE Globecom Workshops (GC Wkshps)*, pages 1–6, Dec. 2017. https://doi.org/10.1109/GLOCOMW.2017.8269071.

172 Qian Liu, Long Shi, Linlin Sun, Jun Li, Ming Ding, and Feng Shu. Path planning for UAV-mounted mobile edge computing with deep reinforcement learning. *IEEE Trans. Veh. Technol.*, 69(5):5723–5728, 2020.

173 Wei Liu, Changbin Yu, Xiangke Wang, YongWei Zhang, and YangGuang Yu. The altitude hold algorithm of UAV based on millimeter wave radar sensors. In *IEEE Int. Conf. Intell. Human-Mach. Syst. Cybernet. (IHMSC)*, volume 1, pages 436–439. IEEE, 2017.

174 X. Liu, Y. Liu, and Y. Chen. Reinforcement learning in multiple-UAV networks: Deployment and movement design. *IEEE Trans. Veh. Technol.*, 68(8):8036–8049, Aug. 2019. ISSN 1939-9359.

175 Xiao Liu, Yuanwei Liu, and Yue Chen. Machine learning empowered trajectory and passive beamforming design in UAV-RIS wireless networks. *IEEE J. Sel. Areas Commun.*, 39(7):2042–2055, Jul. 2021.

176 Yuanwei Liu, Xiao Liu, Xidong Mu, Tianwei Hou, Jiaqi Xu, Marco Di Renzo, and Naofal Al-Dhahir. Reconfigurable intelligent surfaces: Principles and opportunities. *IEEE Commun. Surv. Tut.*, 23(3):1546–1577, 3rd Quart., 2021.

177 Ze Liu, Yingfeng Cai, Hai Wang, Long Chen, Hongbo Gao, Yunyi Jia, and Yicheng Li. Robust target recognition and tracking of self-driving cars with radar and camera information fusion under severe weather conditions. *IEEE Trans. Intell. Transport. Syst.*, 23(7):6640–6653, Jul. 2022. https://doi.org/10.1109/TITS.2021.3059674.

178 Haiquan Lu, Haiyang Zhang, Haibo Dai, Wei Wu, and Baoyun Wang. Proactive eavesdropping in UAV-aided suspicious communication systems. *IEEE Trans. Veh. Tech.*, 68(2):1993–1997, Feb. 2019.

179 Haiquan Lu, Yong Zeng, Shi Jin, and Rui Zhang. Aerial intelligent reflecting surface: Joint placement and passive beamforming design with 3D beam flattening. *IEEE Trans. Wirel. Commun.*, 20(7):4128–4143, Jul. 2021.

180 Yuyue Luo, J. Andrew Zhang, Xiaojing Huang, Wei Ni, and Jin Pan. Optimization and quantization of multibeam beamforming vector for joint communication and radio sensing. *IEEE Trans. Commun.*, 67(9):6468–6482, Sep. 2019. https://doi.org/10.1109/TCOMM.2019.2923627.

181 Jiangbin Lyu, Yong Zeng, Rui Zhang, and Teng Joon Lim. Placement optimization of UAV-mounted mobile base stations. *IEEE Commun. Lett.*, 21(3): 604–607, Mar. 2017.

182 Chao Ma, Jia-Bin Huang, Xiaokang Yang, and Ming-Hsuan Yang. Robust visual tracking via hierarchical convolutional features. *IEEE Trans. Pattern Anal. Mach. Intell.*, 41(11):2709–2723, Nov. 2019. https://doi.org/10.1109/TPAMI.2018.2865311.

183 Seyed Mojtaba Marvasti-Zadeh, Li Cheng, Hossein Ghanei-Yakhdan, and Shohreh Kasaei. Deep learning for visual tracking: A comprehensive survey. *IEEE Trans. Intell. Transport. Syst.*, 23(5):3943–3968, May 2022. https://doi.org/10.1109/TITS.2020.3046478.

184 Michael Mayer and Ilya Molchanov. Limit theorems for the diameter of a random sample in the unit ball. *Extremes*, 10(3):129–150, 2007.

185 Arvind Merwaday, Adem Tuncer, Abhaykumar Kumbhar, and Ismail Guvenc. Improved throughput coverage in natural disasters: Unmanned aerial base stations for public-safety communications. *IEEE Veh. Technol. Mag.*, 11(4): 53–60, 2016.

186 Volodymyr Mnih, Adria Puigdomenech Badia, Mehdi Mirza, Alex Graves, Timothy Lillicrap, Tim Harley, David Silver, and Koray Kavukcuoglu. Asynchronous methods for deep reinforcement learning. In *Int. Conf. Mach. Learn.*, pages 1928–1937. PMLR, 2016.

187 Antonio Moccia and Alfredo Renga. Spatial resolution of bistatic synthetic aperture radar: Impact of acquisition geometry on imaging performance. *IEEE Trans. Geosci. Remote Sens.*, 49(10):3487–3503, Oct. 2011.

188 Jihwan Moon, Sang Hyun Lee, Hoon Lee, Seunghwan Baek, and Inkyu Lee. Deep learning-based proactive eavesdropping for wireless surveillance. In *IEEE Intl. Conf. Commun. (ICC)*, pages 1–5, 2019.

189 Philippe Morere, Roman Marchant, and Fabio Ramos. Sequential bayesian optimization as a POMDP for environment monitoring with UAVs. In *IEEE Int. Conf. Robot. Autom. (ICRA)*, pages 6381–6388, 2017. https://doi.org/10.1109/ICRA.2017.7989754.

190 Naser Hossein Motlagh, Miloud Bagaa, and Tarik Taleb. UAV-based IoT platform: A crowd surveillance use case. *IEEE Commun. Mag.*, 55(2):128–134, Feb. 2017.

191 M. Mozaffari, W. Saad, M. Bennis, and M. Debbah. Unmanned aerial vehicle with underlaid device-to-device communications: Performance and tradeoffs. *IEEE Trans. Wirel. Commun.*, 15(6):3949–3963, Jun. 2016.

192 M. Mozaffari, W. Saad, M. Bennis, Y.-H. Nam, and M. Debbah. A tutorial on UAVs for wireless networks: Applications, challenges, and open problems. *IEEE Commun. Surv. Tut.*, 21(3):2334–2360, 3rd Quart., 2019.

193 Mohammad Mozaffari, Walid Saad, Mehdi Bennis, and Mérouane Debbah. Efficient deployment of multiple unmanned aerial vehicles for optimal wireless coverage. *IEEE Commun. Lett.*, 20(8):1647–1650, 2016.

194 Mohammad Mozaffari, Walid Saad, Mehdi Bennis, and Merouane Debbah. Mobile unmanned aerial vehicles (UAVs) for energy-efficient internet of things communications. *IEEE Trans. Wirel. Commun.*, 16(11):7574–7589, Nov. 2017.

195 Xidong Mu, Yuanwei Liu, Li Guo, Jiaru Lin, and H. Vincent Poor. Intelligent reflecting surface enhanced multi-UAV NOMA networks. *IEEE J. Sel. Areas Commun.*, 39(10):3051–3066, Oct. 2021.

196 Ofir Nachum, Mohammad Norouzi, Kelvin Xu, and Dale Schuurmans. Bridging the gap between value and policy based reinforcement learning. *Adv. Neural Inf. Process Syst.*, 30:2772–2782, 2017.

197 Kamesh Namuduri, Serge Chaumette, Jae H. Kim, and James P. G. Sterbenz, editors. *UAV Networks and Communications*. Cambridge University Press, Cambridge, UK, 2017.

198 Stephen G. Nash and Ariela Sofer. On the complexity of a practical interior-point method. *SIAM J. Optim.*, 8(3):833–849, Aug. 1998. https://doi.org/10.1137/S1052623496306620.

199 A. S. Nemirovskii. Interior point polynomial time methods in convex programming, Lecture notes, Faculty of Industrial Engineering and Management, Technion, The Israel Institute of Technology, 1994.

200 Y. Nesterov. Gradient methods for minimizing composite functions. *Math. Program.*, 140(1):125–161, Aug. 2013.

201 Yurii Nesterov. Dual extrapolation and its applications to solving variational inequalities and related problems. *Math. Program.*, 109(2):319–344, Mar. 2007.

202 Wei Ni, Mehran Abolhasan, Brett Hagelstein, Ren Ping Liu, and Xin Wang. A new trellis model for MAC layer cooperative retransmission protocols. *IEEE Trans. Veh. Tech.*, 66(4):3448–3461, 2017.

203 Zhitong Ni, J. Andrew Zhang, Kai Yang, Xiaojing Huang, and Theodoros A. Tsiftsis. Multi-metric waveform optimization for multiple-input single-output joint communication and radar sensing. *IEEE Trans. Commun.*, 70(2): 1276–1289, Feb. 2022. https://doi.org/10.1109/TCOMM.2021.3132368.

204 A Nuttall. Some integrals involving the Q_M function (corresp.). *IEEE Trans. Inform. Theory*, 21(1):95–96, 1975.

205 Hyondong Oh, Hyo-Sang Shin, Seungkeun Kim, and Wen-Hua Chen. Communication-aware trajectory planning for unmanned aerial vehicles in urban environments. *J. Guid. Control. Dyn.*, 41(10):2271–2282, 2018.

206 Mariya Ouaissa, Inam Ullah Khan, Mariyam Ouaissa, Zakaria Boulouard, and Syed Bilal Hussain Shah, editors. *Computational Intelligence for Unmanned Aerial Vehicles Communication Networks*. Springer, Cham, Switzerland, March 2022. https://doi.org/10.1007/978-3-030-94349-2.

207 Jie Ouyang, Yueling Che, Jie Xu, and Kaishun Wu. Throughput maximization for laser-powered UAV wireless communication systems. In *IEEE Intl. Conf. Commun. (ICC)*, pages 1–5, 2018.

208 Xinhong Pan, Zhihai Zhang, Heyun Lin, and Jiaying Zhang. RIS relaying UAV-aided WPCN for throughput maximization. In *2023 6th World Con. Comput. Commun. Technol. (WCCCT)*, pages 100–105, Jan. 2023. https://doi.org/10.1109/WCCCT56755.2023.10052414.

209 Savvas Papaioannou, Panayiotis Kolios, Theocharis Theocharides, Christos G. Panayiotou, and Marios M. Polycarpou. Integrated guidance and gimbal control for coverage planning with visibility constraints. *IEEE Trans. Aerosp. Electron. Syst.*, 59(2):1276–1291, Apr. 2023. https://doi.org/10.1109/TAES.2022.3199196.

210 Z. Peng, Z. Zhang, L. Kong, C. Pan, L. Li, and J. Wang Deep reinforcement learning for RIS-aided multiuser full-duplex secure communications with hardware impairments. *IEEE Internet Things J.*, 9(21):21121–21135, Nov. 2022. https://doi.org/10.1109/JIOT.2022.3177705.

211 A. Pitarokoilis, S. K. Mohammed, and E. G. Larsson. Uplink performance of time-reversal MRC in massive MIMO systems subject to phase noise. *IEEE Trans. Wirel. Commun.*, 14(2):711–723, Feb. 2015. ISSN 1536–1276. https://doi.org/10.1109/TWC.2014.2359018.

212 Gopika Premsankar, Mario Di Francesco, and Tarik Taleb. Edge computing for the Internet of Things: A case study. *IEEE Internet Things J.*, 5(2):1275–1284, 2018.

213 Robert Price. Some non-central F-distributions expressed in closed form. *Biometrika*, 51(1/2):107–122, 1964.

214 M. L. Puterman. *Markov Decision Processes: Discrete Stochastic Dynamic Programming*. John Wiley & Sons, New York, NY, USA, 2014.

215 S. A. P. Quintero and J. P. Hespanha. Vision-based target tracking with a small UAV: Optimization-based control strategies. *Control Eng. Pract.*, 32:28–42, 2014.

216 N. Ben Rached, A. Kammoun, M. S. Alouini, and R. Tempone. Unified importance sampling schemes for efficient simulation of outage capacity over generalized fading channels. *IEEE J. Sel. Topics Signal Process.*, 10(2):376–388, Mar. 2016. ISSN 1932–4553. https://doi.org/10.1109/JSTSP.2015.2500201.

217 A. Rahmati, X. He, I. Guvenc, and H. Dai. Dynamic mobility-aware interference avoidance for aerial base stations in cognitive radio networks. In *IEEE INFOCOM 2019*, pages 595–603, April 2019. https://doi.org/10.1109/INFOCOM.2019.8737472.

218 A. Rahmati, S. Hosseinalipour, Y. Yapici, X. He, I. Guvenc, H. Dai, and A. Bhuyan. Interference avoidance in UAV-assisted networks: Joint 3D trajectory design and power allocation. In *2019 IEEE Globecom*, Dec. 2019.

219 Ali Ranjha and Georges Kaddoum. URLLC facilitated by mobile UAV relay and RIS: A joint design of passive beamforming, blocklength, and UAV positioning. *IEEE Internet Things J.*, 8(6):4618–4627, Mar. 2021. https://doi.org/10.1109/JIOT.2020.3027149.

220 A. Redder, A. Ramaswamy, and H. Karl. Asymptotic convergence of deep multi-agent actor-critic algorithms. arXiv: 2201.00570, Jan. 2022.

221 Jinke Ren, Yinghui He, Guan Huang, Guanding Yu, Yunlong Cai, and Zhaoyang Zhang. An edge-computing based architecture for mobile augmented reality. *IEEE Netw.*, 33(4):162–169, 2019.

222 Pál Révész. *The Laws of Large Numbers*, volume 4. Academic Press, 2014.

223 A. M. Rezende, V. M. Goncalves, and L. C. Pimenta. Constructive time-varying vector fields for robot navigation. *IEEE Trans. Robot.*, 38(2):852–867, Apr. 2022.

224 Z. Rezki and M. S. Alouini. On the capacity of Nakagami-*m* fading channels with full channel state information at low SNR. *IEEE Wirel. Commun. Lett.*, 1(3):253–256, Jun. 2012. ISSN 2162–2337. https://doi.org/10.1109/WCL.2012.041012.120127.

225 Marc Rodriguez-Cassola, Stefan V. Baumgartner, Gerhard Krieger, and Alberto Moreira. Bistatic TerraSAR-X/F-SAR spaceborne–airborne SAR experiment: Description, data processing, and results. *IEEE Trans. Geosci. Remote Sens.*, 48(2):781–794, Feb. 2010.

226 Taniza Roy, D. Meena, and L. G. M. Prakasam. FPGA based digital beam forming for radars. In *IEEE Radar Conf.*, pages 1–5, 2009. https://doi.org/10.1109/RADAR.2009.4977109.

227 Halsey Lawrence Royden and Patrick Fitzpatrick. *Real Analysis*, volume 32. Macmillan, New York, NY, 1988.

228 E. M. Royer, P. M. Melliar-Smith, and L. E. Moser. An analysis of the optimum node density for ad hoc mobile networks. In *IEEE ICC*, volume 3, pages 857–861, Jun. 2001. https://doi.org/10.1109/ICC.2001.937360.

229 Walid Saad, Zhu Han, Tamer Basar, Mérouane Debbah, and Are Hjorungnes. A selfish approach to coalition formation among unmanned air vehicles in wireless networks. In *Proc. Int. Conf. Game Theory Netw. (GameNets)*, pages 259–267, May 2009.

230 H. Sallouha, M. M. Azari, and S. Pollin. Energy-constrained UAV trajectory design for ground node localization. In *IEEE Global Commun. Conf. (GLOBECOM)*, pages 1–7, 2018.

231 Andrey V. Savkin, Wei Ni, and Mohsen Eskandari. Effective UAV navigation for cellular-assisted radio sensing, imaging, and tracking. *IEEE Trans. Veh. Technol.*, 72(10):13729–13733, 2023. https://doi.org/10.1109/TVT.2023.3277426.

232 Schewe and F. Phillip. Newton's Second Law of Motion. *Physics Today*, 60(6): 28–28, 2007.

233 S. Schopferer, M. Brizon, C. Liersch, and S. Froese. Evaluating the energy balance of high altitude platforms at early design stages. In *Proc. Int. Conf. Unmanned Aircraft Syst.*, Arlington, VA, USA, Jun. 2016.

234 S. Schramm, J. Rangel, D. A. Salazar, R. Schmoll, and A. Kroll. Target analysis for the multispectral geometric calibration of cameras in visual and infrared spectral range. *IEEE Sens. J.*, 21(2):2159–2168, 2021.

235 J. M. Seddon and S. Newman. *Basic Helicopter Aerodynamics*. Wiley, Hoboken, NJ, USA, 3rd 2011.

236 Gursel Serpen and Zhenning Gao. Complexity analysis of multilayer perceptron neural network embedded into a wireless sensor network. *Procedia Comput. Sci.*, 36:192–197, 2014.

237 V. Shaferman and T. Shima. Unmanned aerial vehicles cooperative tracking of moving ground target in urban environments. *J. Guid. Control Dyn.*, 31(5): 1360–1371, 2008.

238 Joshua A. Shaffer, Estefany Carrillo, and Huan Xu. Hierarchal application of receding horizon synthesis and dynamic allocation for UAVs fighting fires. *IEEE Access*, 6:78868–78880, 2018. https://doi.org/10.1109/ACCESS.2018.2885455.

239 P. K. Sharma and D. I. Kim. Coverage probability of 3D UAV networks with RWP Mobility-based altitude control. In *IEEE ICC Workshops*, pages 1–6, May 2018.

240 Vishal Sharma, Mehdi Bennis, and Rajesh Kumar. UAV-assisted heterogeneous networks for capacity enhancement. *IEEE Commun. Lett.*, 20(6):1207–1210, 2016.

241 X. Shi, C. Yang, W. Xie, C. Liang, Z. Shi, and J. Chen. Anti-drone system with multiple surveillance technologies: Architecture, implementation, and challenges. *IEEE Commun. Mag.*, 56(4):68–74, Apr. 2018. ISSN 0163-6804.

242 Jaw-Kuen Shiau, Der-Ming Ma, Pin-Ying Yang, Geng-Feng Wang, and Jhij Hua Gong. Design of a solar power management system for an experimental UAV. *IEEE Trans. Aerospace Electron. Syst.*, 45(4):1350–1360, Oct. 2009. https://doi.org/10.1109/TAES.2009.5310303.

243 Kyusung Shim, Nhu Tri Do, and Beongku An. Performance analysis of physical layer security of opportunistic scheduling in multiuser multirelay cooperative networks. *Sensors*, 17(2):377, 2017.

244 David Silver, Guy Lever, Nicolas Heess, Thomas Degris, and Martin Riedmiller. Deterministic policy gradient algorithms. In *Proc. 31st Int. Conf. Mach. Learn.*, volume 32, pages 387–395, 2014.

245 Sandeep Kumar Singh, Kamal Agrawal, Keshav Singh, Chih-Peng Li, and Zhiguo Ding. NOMA enhanced hybrid RIS-UAV-assisted full-duplex communication system with imperfect SIC and CSI. *IEEE Trans. Commun.*, 70(11):7609–7627, Nov. 2022. https://doi.org/10.1109/TCOMM.2022.3212729.

246 B. Sklar. Rayleigh fading channels in mobile digital communication systems .I. characterization. *IEEE Commun. Mag.*, 35(7):90–100, Jul. 1997. ISSN 0163-6804.

247 T. Sosnowski, G. Bieszczad, H. Madura, and M. Kastek. Thermovision system for flying objects detection. In *Baltic URSI Symposium (URSI)*, pages 141–144, May 2018.

248 Larry B. Stotts, Sherman Karp, and Joseph M. Aein. The origins of miniature global positioning system-based navigation systems [SP history]. *IEEE Signal Process. Mag.*, 31(6):114–117, Nov. 2014. https://doi.org/10.1109/MSP.2014.2343985.

249 E. Stump, A. Jadbabaie, and V. Kumar. Connectivity management in mobile robot teams. In *2008 IEEE Int. Conf. Roboti. Automati.*, pages 1525–1530, May 2008. https://doi.org/10.1109/ROBOT.2008.4543418.

250 Christian Sturm and Werner Wiesbeck. Waveform design and signal processing aspects for fusion of wireless communications and radar sensing. *Proc. IEEE*, 99(7):1236–1259, Jul. 2011. https://doi.org/10.1109/JPROC.2011.2131110.

251 C. Sun, W. Ni, and X. Wang. Joint computation offloading and trajectory planning for UAV-assisted edge computing. *IEEE Trans. Wirel. Commun.*, 20(8):5343–5358, Aug. 2021.

252 Chao Sun, Wei Ni, Zhiyong Bu, and Xin Wang. Energy minimization for intelligent reflecting surface-assisted mobile edge computing. *IEEE Trans. Wirel. Commun.*, 21(8):6329–6344, 2022. https://doi.org/10.1109/TWC.2022.3148296.

253 Li Sun, Yong Zhang, and A. Lee Swindlehurst. Alternate-jamming-aided wireless physical-layer surveillance: Protocol design and performance analysis. *IEEE Trans. Inf. Forensics Secur.*, 16:1989–2003, 2021. https://doi.org/10.1109/TIFS.2020.3046880.

254 Peng Sun and Azzedine Boukerche. Performance modeling and analysis of a UAV path planning and target detection in a UAV-based wireless sensor network. *Comput. Netw.*, 146:217–231, 2018.

255 Y. Sun, D. Xu, D. W. K. Ng, L. Dai, and R. Schober. Optimal 3D-trajectory design and resource allocation for solar-powered UAV communication systems. *IEEE Trans. Commun.*, 67(6):4281–4298, June 2019. ISSN 1558-0857. https://doi.org/10.1109/TCOMM.2019.2900630.

256 Yin Sun, Árpád Baricz, and Shidong Zhou. On the monotonicity, log-concavity, and tight bounds of the generalized marcum and nuttall Q-functions. *IEEE Trans. Inf. Theory*, 56(3):1166–1186, 2010.

257 Zhichao Sun, Junjie Wu, Zhongyu Li, Hongyang An, and Xun He. Geosynchronous spaceborne–airborne bistatic SAR data focusing using a novel range model based on one-stationary equivalence. *IEEE Trans. Geosci. Remote Sens.*, 59(2):1214–1230, Feb. 2021. https://doi.org/10.1109/TGRS.2020.3002900.

258 Zhichao Sun, Junjie Wu, Jianyu Yang, Yulin Huang, Caipin Li, and Dongtao Li. Path planning for GEO-UAV bistatic SAR using constrained adaptive multiobjective differential evolution. *IEEE Trans. Geosci. Remote Sens.*, 54(11): 6444–6457, Nov. 2016.

259 Richard S. Sutton, David A. McAllester, Satinder P. Singh, Yishay Mansour. Policy gradient methods for reinforcement learning with function approximation. *NIPS*, 99:1057–1063, 1999.

260 Jinchuan Tang, Gaojie Chen, and Justin P. Coon. Secrecy performance analysis of wireless communications in the presence of UAV jammer and randomly located UAV eavesdroppers. *IEEE Trans. Inf. Forensics Secur.*, 14(11):3026–3041, Nov. 2019.

261 Wankai Tang, Ming Zheng Chen, Xiangyu Chen, Jun Yan Dai, Yu Han, Marco Di Renzo, Yong Zeng, Shi Jin, Qiang Cheng, and Tie Jun Cui. Wireless communications with reconfigurable intelligent surface: Path loss modeling and experimental measurement. *IEEE Trans. Wirel. Commun.*, 20(1):421–439, Jan. 2021. https://doi.org/10.1109/TWC.2020.3024887.

262 Abhishek Tiwari, Anurag Ganguli, Ashwin Sampath, D. Scott Anderson, and David Krzysiak. Mobility aware routing for the airborne network backbone. In *Proc. Mil. Commun. Conf.*, pages 1–7, Nov. 2008. https://doi.org/10.1109/MILCOM.2008.4753313.

263 Yan Kyaw Tun, Yu Min Park, Nguyen H. Tran, Walid Saad, Shashi Raj Pandey, and Choong Seon Hong. Energy-efficient resource management in UAV-assisted mobile edge computing. *IEEE Commun. Lett.*, 25(1):249–253, 2020.

264 Dimitrios Tyrovolas, Prodromos-Vasileios Mekikis, Sotiris A. Tegos, Panagiotis D. Diamantoulakis, Christos K. Liaskos, and George K. Karagiannidis. Energy-aware design of UAV-mounted RIS networks for IoT data collection. *IEEE Trans. Commun.*, 71(2):1168–1178, 2023. https://doi.org/10.1109/TCOMM.2022.3229672.

265 G. E. Uhlenbeck and L. S. Ornstein. On the theory of the brownian motion. *Phys. Rev.*, 36: 823–841, Sep. 1930. URL https://link.aps.org/doi/10.1103/PhysRev.36.823.

266 F. T. Ulaby, R. K. Moore, and A. K. Fung. *Microwave Remote Sensing Active and Passive*, volume II, pages 569–572. Addison-Wesley, 1982.

267 S. Vuppala and G. Kaddoum. Secrecy capacity analysis in D2D underlay cellular networks: Colluding eavesdroppers. In *IEEE PIMRC*, pages 1–7, Oct. 2017. https://doi.org/10.1109/PIMRC.2017.8292188.

268 Y. Wan, K. Namuduri, Y. Zhou, and S. Fu. A smooth-turn mobility model for airborne networks. *IEEE Trans. Veh. Technol.*, 62(7):3359–3370, Sept. 2013. ISSN 0018-9545. https://doi.org/10.1109/TVT.2013.2251686.

269 D. Wang, P. Hu, J. Du, P. Zhou, T. Deng, and M. Hu. Routing and scheduling for hybrid truck-drone collaborative parcel delivery with independent and truck-carried drones. *IEEE Internet Things J.*, 6(6):10483–10495, Dec. 2019.

270 Jingrong Wang, Kaiyang Liu, and Jianping Pan. Online UAV-mounted edge server dispatching for mobile-to-mobile edge computing. *IEEE Internet Things J.*, 7(2):1375–1386, 2019.

271 L. Wang, N. Yang, M. Elkashlan, P. L. Yeoh, and J. Yuan. Physical layer security of maximal ratio combining in two-wave with diffuse power fading channels. *IEEE Trans. Inf. Forensics Secur.*, 9(2):247–258, Feb. 2014. ISSN 1556–6013. https://doi.org/10.1109/TIFS.2013.2296991.

272 Lin Wang, Lei Jiao, Ting He, Jun Li, and Max Mühlhäuser. Service entity placement for social virtual reality applications in edge computing. In *IEEE INFOCOM*, pages 468–476. IEEE, 2018.

273 Q. Wang, Z. Chen, W. Mei, and J. Fang. Improving physical layer security using UAV-enabled mobile relaying. *IEEE Wirel. Commun. Lett.*, 6(3):310–313, Jun. 2017. ISSN 2162–2337. https://doi.org/10.1109/LWC.2017.2680449.

274 Shuaijun Wang, Fan Jiang, Bin Zhang, Rui Ma, and Qi Hao. Development of UAV-based target tracking and recognition systems. *IEEE Trans. Intell. Transp. Syst.*, 21(8):3409–3422, Aug. 2020.

275 Wei Wang, Xiaohong Guan, Beizhan Wang, and Yaping Wang. A novel mobility model based on semi-random circular movement in mobile ad hoc networks. *Inform. Sci.*, 180(3):399–413, 2010.

276 Wen-Qin Wang and Jingye Cai. Antenna directing synchronization for bistatic synthetic aperture radar systems. *IEEE Antennas Wirel. Propag. Lett.*, 9:307–310, Apr. 2010. https://doi.org/10.1109/LAWP.2010.2047490.

277 Y. Wang, M. C. Ertürk, J. Liu, I. -h. Ra, R. Sankar, and S. Morgera. Throughput and delay of single-hop and two-hop aeronautical communication networks. *J. Commun. Netw.*, 17(1):58–66, Feb. 2015. ISSN 1229–2370. https://doi.org/10.1109/JCN.2015.000010.

278 Yu Wang, Chong Tang, Shuo Wang, Long Cheng, Rui Wang, Min Tan, and Zengguang Hou. Target tracking control of a biomimetic underwater vehicle through deep reinforcement learning. *IEEE Trans. Neural Netw. Learn. Syst.*, 33(8):3741–3752, Aug. 2022. https://doi.org/10.1109/TNNLS.2021.3054402.

279 Zhanze Wang, Feifeng Liu, Tao Zeng, and Chenghao Wang. A novel motion compensation algorithm based on motion sensitivity analysis for mini-UAV-based BiSAR system. *IEEE Trans. Geosci. Remote Sens.*, 60:1–13, Jan. 2022. https://doi.org/10.1109/TGRS.2021.3071460.

280 Z. Wei, H. Wu, X. Yuan, S. Huang, and Z. Feng. Achievable capacity scaling laws of three-dimensional wireless social networks. *IEEE Trans. Veh. Technol.*, 67(3): 2671–2685, Mar. 2018. ISSN 0018–9545.

281 Z. Wei, Z. Feng, X. Yuan, X. Feng, Q. Zhang, and X. Wang. The achievable capacity scaling laws of 3D cognitive radio networks. In *IEEE ICC*, pages 1–6, May 2016.

282 Zhiqing Wei, Hanyang Qu, Yuan Wang, Xin Yuan, Huici Wu, Ying Du, Kaifeng Han, Ning Zhang, and Zhiyong Feng. Integrated sensing and communication signals toward 5G-A and 6G: A survey. *IEEE Internet Things J.*, 10(13): 11068–11092, 2023. https://doi.org/10.1109/JIOT.2023.3235618.

283 Zhiqing Wei, Rubing Yao, Xin Yuan, Huici Wu, Qixun Zhang, and Zhiyong Feng. Precoding optimization for MIMO-OFDM integrated sensing and communication systems. *IEEE Trans. Cogn. Commun. Netw.*, 11(1):288–29, 2025. https://doi.org/10.1109/TCCN.2024.3445376.

284 Bo Wen, Xiaojun Chen, and Ting Kei Pong. A proximal difference-of-convex algorithm with extrapolation. *Comput. Optim. Appl.*, 69(2):297–324, Mar. 2018.

285 Bo-Rong Wu and Li-Chun Wang. Learning from UAV experimental results for performance modeling of reconfigurable intelligent surface flying platform. In *30th Wirel. Opt. Commun. Conf. (WOCC)*, pages 245–250, 2021. https://doi.org/10.1109/WOCC53213.2021.9603190.

286 Huici Wu, Yi Fang, Na Li, Xin Yuan, Zhiqing Wei, Guoshun Nan, and Xiaofeng Tao. Secret key generation with untrusted internal eavesdropper: Token-based anti-eavesdropping. *IEEE Trans. Inf. Forensics Secur.*, 20:2523–2537, 2025. https://doi.org/10.1109/TIFS.2025.3542959.

287 Q. Wu and R. Zhang. Common throughput maximization in UAV-enabled OFDMA systems with delay consideration. *IEEE Trans. Commun.*, 66(12): 6614–6627, Dec. 2018.

288 Q. Wu and R. Zhang. Beamforming optimization for wireless network aided by intelligent reflecting surface with discrete phase shifts. *IEEE Trans. Commun.*, 68(3):1838–1851, Mar. 2020.

289 Q. Wu and R. Zhang. Intelligent reflecting surface enhanced wireless network via joint active and passive beamforming. *IEEE Trans. Wirel. Commun.*, 18(11):5394–5409, Nov. 2019.

290 Q. Wu, W. Mei, and R. Zhang. Safeguarding wireless networks with UAV: A physical layer security perspective. *IEEE Wirel. Commun.*, 26(5):12–18, Oct. 2019.

291 Q. Wu, Y. Zeng, and R. Zhang. Joint trajectory and communication design for multi-UAV enabled wireless networks. *IEEE Trans. Wirel. Commun.*, 17(3): 2109–2121, Mar. 2018.

292 Qingqing Wu, Jie Xu, and Rui Zhang. Capacity characterization of UAV-enabled two-user broadcast channel. *IEEE J. Sel. Areas Commun.*, 36(9):1955–1971, Sep. 2018.

293 Qingqing Wu, Liang Liu, and Rui Zhang. Fundamental tradeoffs in communication and trajectory design for UAV-enabled wireless network. *IEEE Wirel. Commun.*, 26(1):36–44, Feb. 2019.

294 Qingqing Wu, Shuowen Zhang, Beixiong Zheng, Changsheng You, and Rui Zhang. Intelligent reflecting surface-aided wireless communications: A tutorial. *IEEE Trans. Commun.*, 69(5):3313–3351, May 2021.

295 Zonghan Wu, Shirui Pan, Fengwen Chen, Guodong Long, Chengqi Zhang, and S. Yu Philip. A comprehensive survey on graph neural networks. *IEEE Trans. Neural Netw. Learn. Syst.*, 32(1):4–24, 2020.

296 J. Xie, Y. Wan, J. H. Kim, S. Fu, and K. Namuduri. A survey and analysis of mobility models for airborne networks. *IEEE Commun. Surv. Tuts.*, 16(3): 1221–1238, 3rd Quart., 2014. ISSN 1553-877X. https://doi.org/10.1109/SURV.2013.111313.00138.

297 J. Xie, Y. Wan, B. Wang, S. Fu, K. Lu, and J. H. Kim. A comprehensive 3-Dimensional random mobility modeling framework for airborne networks. *IEEE Access*, 6:22849–22862, 2018. ISSN 2169-3536. https://doi.org/10.1109/ACCESS.2018.2819600.

298 Xinxuan Xiong, Chao Sun, Wei Ni, and Xin Wang. Three-dimensional trajectory design for unmanned aerial vehicle-based secure and energy-efficient data collection. *IEEE Trans. Veh. Technol.*, 72(1):664–678, Jan. 2023. https://doi.org/10.1109/TVT.2022.3203714.

299 J. Xu, Lingjie Duan, and R. Zhang. Transmit optimization for symbol-level spoofing. *IEEE Trans. Wirel. Commun.*, 17(1):41–55, Jan. 2018.

300 Jie Xu, Lingjie Duan, and Rui Zhang. Proactive eavesdropping via cognitive jamming in fading channels. *IEEE Trans. Wirel. Commun.*, 16(5):2790–2806, May 2017.

301 Jie Xu, Lingjie Duan, and Rui Zhang. Surveillance and intervention of infrastructure-free mobile communications: A new wireless security paradigm. *IEEE Wirel. Commun.*, 24(4):152–159, Aug. 2017.

302 Jie Xu, Yong Zeng, and Rui Zhang. UAV-enabled wireless power transfer: Trajectory design and energy optimization. *IEEE Trans. Wirel. Commun.*, 17(8):5092–5106, Aug. 2018.

303 Sai Xu, Jiajia Liu, and Yurui Cao. Intelligent reflecting surface empowered physical-layer security: Signal cancellation or jamming? *IEEE Internet Things J.*, 9(2):1265–1275, Jan. 2022. https://doi.org/10.1109/JIOT.2021.3079325.

304 Xiaoli Xu, Yong Zeng, Yongliang Guan, and Rui Zhang. Overcoming endurance issue: UAV-enabled communications with proactive caching. *IEEE J. Sel. Areas Commun.*, 36(6):1231–1244, Jun. 2018.

305 Y. Xu, T. Zhang, Y. Liu, D. Yang, L. Xiao, and M. Tao. Computation capacity enhancement by joint UAV and RIS design in IoT. *IEEE Internet Things J.*, 9(20):20590–20603, Oct. 2022. https://doi.org/10.1109/JIOT.2022.3178983.

306 R. Irem Bor Yaliniz, Amr El-Keyi, and Halim Yanikomeroglu. Efficient 3-D placement of an aerial base station in next generation cellular networks. In *IEEE Intl. Conf. Commun. (ICC)*, pages 1–6, 2016.

307 J. S. Yan, C. Q. Hua, C. L. Chen, and X. P. Guan. The capacity of aeronautical ad-hoc networks. *Wirel. Netw.*, 20(7):2123–2130, 2014.

308 Helin Yang, Zehui Xiong, Jun Zhao, Dusit Niyato, Qingqing Wu, H. Vincent Poor, and Massimo Tornatore. Intelligent reflecting surface assisted anti-jamming communications: A fast reinforcement learning approach. *IEEE Trans. Wirel. Commun.*, 20(3):1963–1974, Mar. 2021.

309 N. Yang, H. A. Suraweera, I. B. Collings, and C. Yuen. Physical layer security of TAS/MRC with antenna correlation. *IEEE Trans. Inf. Forensics Secur.*, 8(1):254–259, Jan. 2013. ISSN 1556-6013. https://doi.org/10.1109/TIFS.2012.2223681.

310 P. Yang, Y. Wu, and H. Yang. Capacity of Nakagami-*m* fading channel with BPSK/QPSK modulations. *IEEE Commun. Lett.*, 21(3):564–567, Mar. 2017. ISSN 1089-7798. https://doi.org/10.1109/LCOMM.2016.2636826.

311 Peng Yang, Xianbin Cao, Chao Yin, Zhenyu Xiao, Xing Xi, and Dapeng Wu. Proactive drone-cell deployment: Overload relief for a cellular network under flash crowd traffic. *IEEE Trans. Intell. Transp. Syst.*, 18(10):2877–2892, Oct. 2017.

312 Y. Yang, M. Pesavento, S. Chatzinotas, and B. Ottersten. Successive convex approximation algorithms for sparse signal estimation with nonconvex regularizations. *IEEE J. Sel. Topics Signal Process.*, 12(6):1286–1302, Dec. 2018.

313 Zhaohui Yang, Cunhua Pan, Kezhi Wang, and Mohammad Shikh-Bahaei. Energy efficient resource allocation in UAV-enabled mobile edge computing networks. *IEEE Trans. Wirel. Commun.*, 18(9):4576–4589, Sep. 2019.

314 P. Yao, Z. Xie, and P. Ren. Optimal UAV route planning for coverage search of stationary target in river. *IEEE Trans. Control Syst. Tech.*, 27(2):822–829, Mar. 2019.

315 J. Ye, C. Zhang, H. Lei, G. Pan, and Z. Ding. Secure UAV-to-UAV systems with spatially random UAVs. *IEEE Wirel. Commun. Lett.*, pages 1–1, 2018. ISSN 2162-2337. https://doi.org/10.1109/LWC.2018.2879842.

316 K. Yoshikawa, H. Kawasaki, W. Yoshida, T. Irie, K. Konishi, K. Nakano, T. Uto, D. Adachi, M. Kanematsu, and H. Uzu. Silicon heterojunction solar cell with interdigitated back contacts for a photoconversion efficiency over 26%. *Nat. Energy*, 2:17032, 2017.

317 Changsheng You and Rui Zhang. 3D trajectory optimization in Rician fading for UAV-enabled data harvesting. *IEEE Trans. Wirel. Commun.*, 18(6):3192–3207, Jun. 2019.

318 Feiran You, Xin Yuan, Wei Ni, and Abbas Jamalipour. A novel privacy-preserving incentive mechanism for multi-access edge computing. *IEEE Trans. Cogn. Commun. Netw.,*, 10(5):1928–1943, 2024. https://doi.org/10.1109/TCCN.2024.3391303.

319 Feiran You, Xin Yuan, Wei Ni, and Abbas Jamalipour. Privacy-preserving multi-agent deep reinforcement learning for effective resource auction in multi-access edge computing. In *IEEE Trans. Cogn. Commun. Netw.*, pages 1–1, 2024. https://doi.org/10.1109/TCCN.2024.3499342.

320 H. Yu, K. Meier, M. Argyle, and R. W. Beard. Cooperative path planning for target tracking in urban environments using unmanned air and ground vehicles. *IEEE/ASME Trans. Mechatronics*, 20(2):541–552, Apr. 2015.

321 X. Yu, J. Zhang, M. Haenggi, and K. B. Letaief. Coverage analysis for millimeter wave networks: The impact of directional antenna arrays. *IEEE J. Sel. Areas in Commun.*, 35(7):1498–1512, July 2017. ISSN 0733-8716. https://doi.org/10.1109/JSAC.2017.2699098.

322 Zhe Yu, Yanmin Gong, Shimin Gong, and Yuanxiong Guo. Joint task offloading and resource allocation in UAV-enabled mobile edge computing. *IEEE Internet Things J.*, 7(4):3147–3159, 2020.

323 Chi Yuan, Zhixiang Liu, and Youmin Zhang. Learning-based smoke detection for unmanned aerial vehicles applied to forest fire surveillance. *J. Intell. Robot. Syst.*, 93(1):337–349, Feb. 2019.

324 X. Yuan, Z. Feng, W. Ni, R. P. Liu, J. A. Zhang, and W. Xu. Secrecy performance of terrestrial radio links under collaborative aerial eavesdropping. *IEEE Trans. Inf. Forensics Secur.*, 15:604–619, 2020.

325 X. Yuan, Z. Feng, W. Xu, W. Ni, J. A. Zhang, Z. Wei, and R. P. Liu. Capacity analysis of UAV communications: Cases of random trajectories. *IEEE Trans. Veh. Technol.*, 67(8):7564–7576, Aug. 2018. ISSN 0018-9545. https://doi.org/10.1109/TVT.2018.2829726.

326 Xin Yuan, Zhiyong Feng, Wei Ni, Zhiqing Wei, Ren Ping Liu, and J. Andrew Zhang. Secrecy rate analysis against aerial eavesdropper. *IEEE Trans. Commun.*, 67(10):7027–7042, 2019. https://doi.org/10.1109/TCOMM.2019.2927449.

327 Xin Yuan, Wei Ni, Ming Ding, Kang Wei, Jun Li, and H. Vincent Poor. Amplitude-varying perturbation for balancing privacy and utility in federated learning. *IEEE Trans. Inf. Forensics Secur.*, 18:1884–1897, 2023. https://doi.org/10.1109/TIFS.2023.3258255.

328 Xin Yuan, Shuyan Hu, Wei Ni, Xin Wang, and Abbas Jamalipour. Empowering reconfigurable intelligent surfaces with artificial intelligence to secure air-to-ground Internet-of-Things. *IEEE Internet Things Mag.*, 7(2):14–21, 2024. https://doi.org/10.1109/IOTM.001.2300129.

329 Xin Yuan, Shuyan Hu, Wei Ni, Ren-Ping Liu, and Xin Wang. Joint user, channel, modulation-coding selection, and RIS configuration for jamming resistance in multiuser OFDMA systems. *IEEE Trans. Commun.*, 71(3):1631–1645, Mar. 2023.

330 Alessio Zappone, Emil Bjornson, Luca Sanguinetti, and Eduard Jorswieck. Globally optimal energy-efficient power control and receiver design in wireless networks. *IEEE Trans. Signal Process.*, 65(11):2844–2859, Jun. 2017.

331 P. J. Zarco-Tejada, R. Diaz-Varela, V. Angileri, and P. Loudjani. Tree height quantification using very high resolution imagery acquired from an unmanned aerial vehicle (UAV) and automatic 3D photo-reconstruction methods. *Eur. J. Agron.*, 55:89–99, 2014.

332 Y. Zeng and R. Zhang. Energy-efficient UAV communication with trajectory optimization. *IEEE Trans. Wirel. Commun.*, 16(6):3747–3760, Jun. 2017.

333 Y. Zeng, R. Zhang, and T. J. Lim. Wireless communications with unmanned aerial vehicles: Opportunities and challenges. *IEEE Commun. Mag.*, 54(5): 36–42, May 2016.

334 Y. Zeng, X. Xu, and R. Zhang. Trajectory optimization for completion time minimization in UAV-enabled multicasting. *IEEE Trans. Wirel. Commun.*, 17(4):2233–2246, Apr. 2018.

335 Y. Zeng, Q. Wu, and R. Zhang. Accessing from the sky: A tutorial on UAV communications for 5G and beyond. *Proc. IEEE*, 107(12):2327–2375, Dec. 2019.

336 Yong Zeng and Rui Zhang. Wireless information surveillance via proactive eavesdropping with spoofing relay. *IEEE J. Sel. Top. Signal Process.*, 10(8): 1449–1461, Dec. 2016.

337 Yong Zeng, Rui Zhang, and Teng Joon Lim. Throughput maximization for UAV-enabled mobile relaying systems. *IEEE Trans. Commun.*, 64(12): 4983–4996, Dec. 2016.

338 Yong Zeng, Jie Xu, and Rui Zhang. Energy minimization for wireless communication with rotary-wing UAV. *IEEE Trans. Wirel. Commun.*, 18(4): 2329–2345, Apr. 2019.

339 Yong Zeng, Xiaoli Xu, Shi Jin, and Rui Zhang. Simultaneous navigation and radio mapping for cellular-connected UAV with deep reinforcement learning. *IEEE Trans. Wirel. Commun.*, 20(7):4205–4220, Jul. 2021. https://doi.org/10.1109/TWC.2021.3056573.

340 Cheng Zhan, Han Hu, Xiufeng Sui, Zhi Liu, and Dusit Niyato. Completion time and energy optimization in the UAV-enabled mobile-edge computing system. *IEEE Internet Things J.*, 7(8):7808–7822, 2020.

341 C. Zhang, W. Zhang, W. Wang, L. Yang, and W. Zhang. Research challenges and opportunities of UAV millimeter-wave communications. *IEEE Wirel. Commun.*, 26(1):58–62, Feb. 2019. ISSN 1536-1284. https://doi.org/10.1109/MWC.2018.1800214.

342 G. Zhang, Q. Wu, M. Cui, and R. Zhang. Securing UAV communications via joint trajectory and power control. *IEEE Trans. Wirel. Commun.*, 18(2):1376–1389, Feb. 2019. ISSN 1536-1276. https://doi.org/10.1109/TWC.2019.2892461.

343 J. Andrew Zhang, Md Lushanur Rahman, Kai Wu, Xiaojing Huang, Y. Jay Guo, Shanzhi Chen, and Jinhong Yuan. Enabling joint communication and radar sensing in mobile networks: A survey. *IEEE Commun. Surveys Tuts.*, 24(1): 306–345, 1st Quart., 2022. https://doi.org/10.1109/COMST.2021.3122519.

344 J. Andrew Zhang, Xiaojing Huang, Y. Jay Guo, Jinhong Yuan, and Robert W. Heath. Multibeam for joint communication and radar sensing using steerable analog antenna arrays. *IEEE Trans. Veh. Tech.*, 68(1):671–685, Jan. 2019. https://doi.org/10.1109/TVT.2018.2883796.

345 Jingjun Zhang and V. Aalo. Effect of macrodiversity on average-error probabilities in a Rician fading channel with correlated lognormal shadowing. *IEEE Trans. Commun.*, 49(1):14–18, Jan. 2001. ISSN 0090–6778. https://doi.org/10.1109/26.898244.

346 Kaihua Zhang, Qingshan Liu, Yi Wu, and Ming-Hsuan Yang. Robust visual tracking via convolutional networks without training. *IEEE Trans. Image Process.*, 25(4):1779–1792, Apr. 2016. https://doi.org/10.1109/TIP.2016.2531283.

347 L. Zhang, F. Deng, J. Chen, Y. Bi, S. K. Phang, X. Chen, and B. M. Chen. Vision-based target three-dimensional geolocation using unmanned aerial vehicles. *IEEE Trans. Ind. Electron.*, 65(10):8052–8061, Oct. 2018.

348 Lu Zhang, Zi-Yan Zhang, Luo Min, Chao Tang, Hong-Ying Zhang, Ya-Hong Wang, and Peng Cai. Task offloading and trajectory control for UAV-assisted mobile edge computing using deep reinforcement learning. *IEEE Access*, 9:53708–53719, 2021.

349 Qixun Zhang, Jingran Chen, Lei Ji, Zhiyong Feng, Zhu Han, and Zhiyong Chen. Response delay optimization in mobile edge computing enabled UAV swarm. *IEEE Trans. Veh. Technol.*, 69(3):3280–3295, 2020.

350 S. Zhang, Q. Wu, S. Xu, and G. Y. Li. Fundamental green tradeoffs: Progresses, challenges, and impacts on 5G networks. *IEEE Commun. Surv. Tut.*, 19(1): 33–56, 1st Quart., 2017.

351 Shuangxi Zhang, Yuxin Gao, Mengdao Xing, Rui Guo, Junli Chen, and Yanyang Liu. Ground moving target indication for the geosynchronous-low earth orbit bistatic multichannel SAR system. *IEEE J. Sel. Topics Appl. Earth Observ. Remote Sens.*, 14:5072–5090, May 2021. https://doi.org/10.1109/JSTARS.2021.3077003.

352 Tiankui Zhang, Yu Xu, Jonathan Loo, Dingcheng Yang, and Lin Xiao. Joint computation and communication design for UAV-assisted mobile edge computing in IoT. *IEEE Trans. Ind. Inform.*, 16(8):5505–5516, 2019.

353 Wei Zhang, Ke Song, Xuewen Rong, and Yibin Li. Coarse-to-fine UAV target tracking with deep reinforcement learning. *IEEE Trans. Autom. Sci. Eng.*, 16(4): 1522–1530, Oct. 2019.

354 X. Zhang, Y. Fang, X. Zhang, J. Jiang, and X. Chen. A novel geometric hierarchical approach for dynamic visual servoing of quadrotors. *IEEE Trans. Ind. Electron.*, 67(5):3840–3849, May 2020.

355 Yun Zhang, Huilin Mu, Yicheng Jiang, Chang Ding, and Yong Wang. Moving target tracking based on improved GMPHD filter in circular SAR system. *IEEE Geosci. Remote Sens. Lett.*, 16(4):559–563, Apr. 2019. https://doi.org/10.1109/LGRS.2018.2878467.

356 Ke Zhao, Haoran Mei, Shuai Lyu, and Limei Peng. Joint optimization of multiple UAV-mounted RISs deployment and RIS elements allocation. In *13th Int. Conf. Inform. Commun. Tech. Converg. (ICTC)*, pages 1193–1197, Oct. 2022. https://doi.org/10.1109/ICTC55196.2022.9952436.

357 Zhedong Zheng, Tao Ruan, Yunchao Wei, Yi Yang, and Tao Mei. VehicleNet: Learning robust visual representation for vehicle re-identification. *IEEE Trans. Multimed.*, 23:2683–2693, 2021. https://doi.org/10.1109/TMM.2020.3014488.

358 Jie Zhou, Ganqu Cui, Shengding Hu, Zhengyan Zhang, Cheng Yang, Zhiyuan Liu, Lifeng Wang, Changcheng Li, and Maosong Sun. Graph neural networks: A review of methods and applications. *AI Open*, 1:57–81, 2020.

359 Z. Zhou, C. Zhang, C. Xu, F. Xiong, Y. Zhang, and T. Umer. Energy-efficient industrial Internet of UAVs for power line inspection in smart grid. *IEEE Trans. Ind. Informat.*, 14(6):2705–2714, Jun. 2018.

360 Zhenyu Zhou, Junhao Feng, Lu Tan, Yejun He, and Jie Gong. An air-ground integration approach for mobile edge computing in IoT. *IEEE Commun. Mag.* 56(8):40–47, 2018.

361 Y. Zhu, G. Zheng, and M. Fitch. Secrecy rate analysis of UAV-enabled mmWave networks using matérn hardcore point processes. *IEEE J. Sel. Areas Commun.*, 36(7):1397–1409, July 2018. ISSN 0733-8716. https://doi.org/10.1109/JSAC.2018.2825158.

362 M. Zinaddinov, S. Mil'shtein, and D. Kazmer. Design of light-weight solar panels. In *IEEE 46th Photovoltaic Special. Conf. (PVSC)*, pages 0582–0587, 2019. https://doi.org/10.1109/PVSC40753.2019.8980588.

363 K. Li, W. Ni, X. Yuan, A. Noor and A. Jamalipour. Deep-graph-based reinforcement learning for joint cruise control and task offloading for aerial Edge Internet of Things (EdgeIoT). *IEEE Internet Things J.*, 9(21):21676–21686, 2022. https://doi.org/10.1109/JIOT.2022.3182119.

Index

Note: Page numbers in *italics* and **bold** refers to figures and tables respectively.

f

t

Printed and bound by CPI Group (UK) Ltd, Croydon, CR0 4YY
17/09/2026
14964235-0005